S0-AVW-365

OXFORD SERIES ON SYNCHROTRON RADIATION

Series Editors

J. Chikawa J. R. Helliwell
S. W. Lovesey

OXFORD SERIES ON SYNCHROTRON RADIATION

1. S. W. Lovesey, S. P. Collins: *X-ray scattering and absorption by magnetic materials*
2. J. R. Helliwell, P. M. Rentzepis: *Time-resolved diffraction*
3. P. J. Duke: *Synchrontron radiation: production and properties*
4. J. A. Clarke: *The science and technology of undulators and wigglers*
5. M. J. Cooper, P. Mijnarends, N. Shiotani, N. Sakai, A. Bansil: *X-ray Compton scattering*
6. D. M. Paganin: *Coherent X-ray optics*

Coherent X-Ray Optics

David M. Paganin
School of Physics
Monash University
Australia

OXFORD
UNIVERSITY PRESS

Great Clarendon Street, Oxford OX2 6DP

Oxford University Press is a department of the University of Oxford.
It furthers the University's objective of excellence in research, scholarship,
and education by publishing worldwide in

Oxford New York

Auckland Cape Town Dar es Salaam Hong Kong Karachi
Kuala Lumpur Madrid Melbourne Mexico City Nairobi
New Delhi Shanghai Taipei Toronto

With offices in

Argentina Austria Brazil Chile Czech Republic France Greece
Guatemala Hungary Italy Japan Poland Portugal Singapore
South Korea Switzerland Thailand Turkey Ukraine Vietnam

Oxford is a registered trade mark of Oxford University Press
in the UK and in certain other countries

Published in the United States
by Oxford University Press Inc., New York

First published 2006

British Library Cataloguing in Publication Data
Data available

Library of Congress Cataloging in Publication Data
Data available

Printed in Great Britain
on acid-free paper by
Biddles Ltd, King's Lynn

ISBN 0–19–856728–6 978–0–19–856728–8

1 3 5 7 9 10 8 6 4 2

Preface

X-ray optics has undergone something of a renaissance in recent years, largely spurred by the coming on-line of third-generation synchrotron radiation sources. This has fostered a plethora of applications, spanning an impressive variety of fields ranging from structural biology through to fundamental atomic physics and femtosecond chemistry, which continues to proliferate with time. Indeed, it is scarcely hyperbole to compare the advances resulting from the stunning increase in X-ray source coherence, over the past quarter of a century, to those which resulted from the earlier revolutions in visible-light optics that were enabled by the advent of the optical laser.

It is timely that a self-contained text be written, dealing specifically with those advances that have been made possible by the advent of X-ray sources with a hitherto unprecedented degree of coherence. That is the goal of this book, which strives to present a unified treatment of certain core features of the science underlying coherent X-ray optics. The discussion is self-contained insofar as the great majority of its results are derived using a chain of logic which ultimately stems from the Maxwell equations of classical electrodynamics. It is intended to be of utility to both professionals in the field and to beginning graduate students and advanced undergraduates, together with scientists from other fields who wish to learn more about the foundations of coherent X-ray optics. While the great majority of results are developed from first principles, some prior acquaintance with modern physical optics (e.g. at the level of the texts by Hecht or Lipson and Lipson) will be advantageous.

The opening chapter deals with the free-space diffraction theory of X-ray wave-fields. Taking the Maxwell equations as a starting point, the vacuum wave equations for electromagnetic fields are derived. We then briefly discuss the transition from a vector theory of the electromagnetic field, which describes this disturbance by the electric and magnetic field vectors at each point in space and at each instant of time, to a scalar theory which describes the disturbance by a single complex function of position and time. The associated spectral decomposition and analytic signal are then introduced. We are then ready for the so-called angular spectrum representation of X-ray wave-fields, which decomposes such fields into a superposition of plane waves travelling in different directions. Based on this representation, the Fresnel and Fraunhofer approximations are explored. We then give a description of the Kirchhoff and Rayleigh–Sommerfeld theories of diffraction, emphasizing their connection with the angular spectrum. We move onto the foundations of the theory of partially coherent fields, including the topics of spatial coherence, temporal coherence, mutual coherence, mutual intensity, and cross spectral density. We also examine the propagation of various two-point measures of coherence, including a derivation of the van Cittert–Zernike theorem. Higher-order measures of coherence are briefly discussed.

The second chapter deals with the interactions of X-rays with matter. Again taking the Maxwell equations as a starting point, both vectorial and scalar forms of the wave equations in the presence of scatterers are obtained. The notion of refractive index is seen to emerge in a natural manner. We then outline the projection approximation for the scattering of X-rays from inhomogeneous media. A Green-function analysis is given, of X-ray interactions with matter, outlining the distinction between the kinematical and dynamical theories of X-ray scattering. Connections are established, between this Green function analysis, and both the angular spectrum and the Ewald sphere. The Ewald sphere is used to study the physics of kinematical X-ray scattering from both crystalline and non-crystalline objects. The multislice and eikonal approximations are introduced, outlining how the latter methodology allows one to make a smooth conceptual transition from wave optics to geometric optics. The connection between scattering, refractive index, and electron density is then explored, including a treatment of how Thomson scattering leads to an X-ray refractive index which is less than unity. Inelastic scattering and absorption are then considered, before giving a brief treatment of the information content of scattered X-ray fields.

The third chapter deals with three topics: X-ray sources, optical elements, and detectors. (i) The notions of brightness and emittance are introduced, as criteria for the coherence of the radiation produced by X-ray sources. Laboratory sources, such as fixed-anode and rotating-anode devices, are introduced. We then pass onto synchrotron sources, the mainstay of coherent X-ray optics. Free-electron lasers, energy-recovering linear accelerators, and soft X-ray lasers are also discussed. (ii) Having covered coherent X-ray sources, we introduce the elements of X-ray optics and detectors. Typically, such elements are qualitatively rather different from their visible-light counterparts, notwithstanding the fact that both X-rays and visible light are forms of electromagnetic radiation. We discuss the major classes of diffractive, reflective, and refractive X-ray optics: diffraction gratings, Fresnel zone plates, analyser crystals, crystal monochromators, crystal beam-splitters, the Bonse–Hart interferometer, Bragg–Fresnel optics, free space, capillary optics, square-channel arrays, X-ray mirrors, X-ray prisms, compound refractive lenses, and virtual optical elements. (iii) We then outline the major types of X-ray detector which are in current use. Broadly speaking, these may be separated into the two classes of 'counting detectors' and 'integrating detectors'. We outline the critical parameters for both classes of detector, describing several of the more important examples of each: film, image plates, CCD cameras, gas detectors, scintillators, and pixel detectors. A comparison is then made between counting and integrating detectors, before examining the relationship between detectors and coherence.

Chapter 4 is devoted to coherent X-ray imaging. The operator theory of imaging, discussed in detail, provides the formalism underlying the remainder of the chapter. We then consider Montgomery self-imaging, the Talbot effect, inline holography, off-axis holography, Fourier holography, Zernike phase contrast, differential interference contrast, analyser-based phase contrast, propagation-based phase contrast, and hybrid phase contrast. We then move onto the question of X-ray phase retrieval, including the Gerchberg–Saxton algorithm and its later variants, the transport-of-intensity equation and one-dimensional phase retrieval using the logarithmic Hilbert transform. Interferometry in its various guises is then considered, including Bonse–Hart interferometric imaging, coherence measurement using the Young interferometer, and coherence measurement using the intensity interferometer (Hanbury Brown–Twiss effect). The use of virtual optical elements for coherent X-ray imaging is also examined.

The fascinating albeit immature subject of singular X-ray optics is treated in the final chapter. Here, we examine the singularities of the wave and ray theories for coherent X-ray optics. Certain singularities of the wave theory, threads of zero intensity known as nodal lines, are examined first. We see that such nodal lines are associated with vortices in the X-ray wave-field. The dynamics of vortices are examined, together with means for producing them in X-ray optical systems: interference of three plane waves, synthetic holograms, spiral phase plates, and spontaneous vortex formation via free-space propagation. Non-vortical singularities of the wave theory are also briefly examined, together with the singularities of the wave theory known as caustics. The relation, between the singularities of the ray and wave theories, is then discussed in the context of the so-called singularity hierarchy.

In addition to serving as a medium to communicate my joy in a beautiful field of study, this text strives to emphasize the fundamental inter-connectness of the many seemingly disparate threads which comprise the rich and evolving tapestry of coherent X-ray optics. It is my earnest hope that the material given here may serve to inspire workers to both study and add to this rich fabric of knowledge.

David Paganin
Monash University
June 5, 2005

Acknowledgements

I have been fortunate indeed to have so many world-class X-ray scientists in my immediate geographical vicinity. I acknowledge the great deal that I have learned from all of them over the years, together with the inspiration they have provided. In particular I must single out Tim Gureyev (CSIRO), for providing a seemingly inexhaustible supply of deep, living knowledge on all aspects of diffraction physics and X-ray optics. I am also particularly grateful to the leaders of the three Australian X-ray groups with which I have worked: Steve Wilkins (CSIRO), Keith Nugent (University of Melbourne), and Rob Lewis (Monash University). Each have contributed immensely to my development as an optical physicist.

Colleagues who provided useful comments on parts of the manuscript are gratefully acknowledged: Tim Gureyev, Sally Irvine, Rob Lewis, Rosemary Mardling, Michael Morgan, Anthony Norton, Keith Nugent, Konstantin Pavlov, Andrew Peele, Andrew Pogany, and Steve Wilkins. I am particularly grateful to Michael Morgan, for a very detailed reading of the penultimate draft.

Thankyou, also, to those scientists who granted permission for figures from their papers to be reproduced in the text: Peter Cloetens, Stefan Eisebitt, Enzo Di Fabrizio, Marcus Kitchen, Wolfram Leitenberger, Phil McMahon, John Miao, Atsushi Momose, Ulrich Neuhäusler, Andrew Peele, Anatoly Snigirev, Steve Wilkins, and Naoto Yagi. A computer programme, written by Andrew Stevenson, was used to produce the data displayed in Fig. 2.14. Figure 5.3(a) was drawn for the author by Rotha Yu.

I am grateful to Barry Muddle, former head of the former School of Physics and Materials Engineering, for both the wonderful opportunity of being hired as a lecturer to Monash University, and for giving me the freedom to pursue my researches in such a dynamic and stimulating environment. I am also grateful to George Simon, head of the former School of Physics and Materials Engineering, together with Michael Morgan, head of the present School of Physics, for their continued support and encouragement.

Sonke Adlung, Lynsey Livingston, and Anita Petrie, from Oxford University Press, have been wonderful to work with. Their contributions, and those of the other staff involved with the production of this text, are gratefully acknowledged.

Thankyou to my parents, Maurice and Clara Paganin, for their support during the many years of my education. Thanks and apologies to Jane Micallef, for her patience during the many evenings and weekends which this endeavour took away from us.

Contents

1

X-ray wave-fields in free space

The free-space diffraction theory of classical electromagnetic waves is a rich and beautiful subject, elements of which will be given in the present chapter. This serves as a pre-requisite for our later discussions on coherent X-ray optics, by first considering the behaviour of such fields in the absence of optical elements.

The derivation of wave equations, governing the evolution of electromagnetic fields in free space, is the first task of the present chapter. Taking the Maxwell equations as a starting point, we derive the vacuum wave equations governing the spatial and temporal evolution of electromagnetic fields in free space. Having done so, we make a transition from a vector theory to a scalar theory of optics: in the former, the free-space electromagnetic disturbance is described by its electric and magnetic field vectors at each point in space–time, whereas the latter describes the disturbance by a single real number at each space–time point.

Having obtained the time-dependent vacuum wave equation governing the evolution of a scalar electromagnetic field, we consider the question of how to appropriately represent such a field as a superposition of monochromatic fields of all possible frequencies. Such a representation is spoken of, rather naturally, as a spectral decomposition. There is some freedom in how this may be done, on account of the fact that electromagnetic fields are intrinsically real, with a complex representation being introduced for the sake of mathematical convenience. One can make use of this freedom, whose form will be more precisely outlined, to demand that there be no negative temporal frequencies present in the spectral decomposition of a given complex scalar representation of an electromagnetic field. This results in the so-called analytic signal describing such a field, a notion due to Gabor. Each monochromatic component, of the spectral decomposition of the analytic signal for the scalar electromagnetic field, will be shown to obey the Helmholtz equation, this being a time-independent form of the time-dependent vacuum wave equation.

The subject of free-space diffraction, of a given monochromatic component in the spectral decomposition of the analytic signal, forms the focus of several subsequent sections. Since such monochromatic fields may be superposed to produce a polychromatic field, or a particular instance of a partially coherent field, it is profitable to study the free-space diffraction of strictly monochromatic fields. We consider this problem from a multiplicity of viewpoints, emphasizing the fundamental unity of these treatments. Formalisms discussed include the operator and convolution forms for free-space diffraction of waves obeying the Helmholtz equation, together with the three important special cases of Fresnel, Fraunhofer, and Kirchhoff diffraction. Central to all of these discussions is the angular-spectrum formulation, which represents free-space monochromatic scalar electromagnetic fields as a superposition of elementary plane waves, of a specified frequency and a range of different propagation directions.

We then turn to the subject of partially coherent fields, a notion linked with the question of the extent to which different parts of the field are able to interfere with one another. Partially coherent fields are those which lie between the limiting extremes of incoherence and coherence: regarding the former limit, no interference fringes are observed when incoherent fields are superposed, whereas in the latter limit of perfect coherence, superposition of fields results in maximally strong interference fringes. While it is artificial to do so, we introduce our discussions on partial coherence with separate treatments of the notions of spatial and temporal coherence. Loosely, the spatial coherence of a field is a measure of the ability of the electromagnetic disturbances from two space–time points to interfere with one another, with the time coordinate of each space–time point being taken as equal. If, in the previous sentence, one takes the spatial rather than the temporal coordinates of these two space–time points as equal, this gives a measure of temporal coherence. Having separately treated the questions of spatial and temporal coherence, we then launch into a treatment of partially coherent fields from the point of view of correlation functions. This discussion revolves around the two-point correlation functions of second-order coherence theory, which obey a pair of coupled wave equations. Based on these equations, one can formulate a diffraction theory for the two-point correlation functions, with two such formalisms being outlined. As an important limiting case, we obtain the van Cittert–Zernike theorem for the propagation of certain two-point correlations from an extended incoherent planar source. Note, in this context, that partially coherent radiation may be produced by an incoherent source, by the act of free-space propagation. We close our discussions on partial coherence with a brief introduction to higher-order degrees of coherence.

1.1 Vacuum wave equations for electromagnetic fields

Here we obtain the vacuum wave equations that govern the spatial and temporal evolution of the electromagnetic field in free space. Our starting point is the free-space Maxwell equations, which in Système-Internationale (SI) units[1] are given by (see, for example, Jackson (1999)):

$$\nabla \cdot \mathbf{E}(x, y, z, t) = 0, \tag{1.1}$$

$$\nabla \cdot \mathbf{B}(x, y, z, t) = 0, \tag{1.2}$$

$$\nabla \times \mathbf{E}(x, y, z, t) + \frac{\partial}{\partial t}\mathbf{B}(x, y, z, t) = \mathbf{0}, \tag{1.3}$$

$$\nabla \times \mathbf{B}(x, y, z, t) - \varepsilon_0 \mu_0 \frac{\partial}{\partial t}\mathbf{E}(x, y, z, t) = \mathbf{0}. \tag{1.4}$$

[1]Note that SI units are used throughout the text.

Here $\mathbf{B}$ is the magnetic induction, $\mathbf{E}$ is the electric field, ε_0 and μ_0 are respectively equal to the electrical permittivity and magnetic permeability of free space, ∇ and $\nabla\times$ respectively denote the three-dimensional gradient and curl operators, (x, y, z) are Cartesian coordinates in three-dimensional space, t is time, and $\mathbf{0}$ is a zero-length vector. Note the convention, which will be used throughout the book, of vector quantities being indicated using boldface type. Equation (1.1) is the free-space form of Gauss' Law, which asserts the zero electric flux through all closed surfaces that do not contain any enclosed electric charge. Equation (1.2) is the magnetic equivalent of Gauss' Law, with the exception that the right side is zero even in the presence of matter, amounting to an assertion of the nonexistence of magnetic monopoles. Equation (1.3) is Faraday's Law of induction, while eqn (1.4) is the free-space form of Ampère's Law as modified by Maxwell.[2]

With view to obtaining the free-space wave equation for the electric field, take the curl of eqn (1.3). Next, make use of the following vector identity:

$$\nabla \times [\nabla \times \mathbf{g}(x, y, z)] = \nabla [\nabla \cdot \mathbf{g}(x, y, z)] - \nabla^2 \mathbf{g}(x, y, z), \tag{1.5}$$

for a well-behaved vector field $\mathbf{g}(x, y, z)$, to arrive at:

$$\nabla [\nabla \cdot \mathbf{E}(x, y, z, t)] - \nabla^2 \mathbf{E}(x, y, z, t) + \nabla \times \frac{\partial}{\partial t} \mathbf{B}(x, y, z, t) = \mathbf{0}. \tag{1.6}$$

The first term of this equation vanishes, due to the free-space form (1.1) of Gauss' Law. In the third term, interchange the order of the operators $\nabla\times$ and $\partial/\partial t$, and then make use of the free-space Ampère Law (1.4). This results in the vacuum field equation for the electric field, namely the d'Alembert wave equation:

$$\left(\varepsilon_0 \mu_0 \frac{\partial^2}{\partial t^2} - \nabla^2\right) \mathbf{E}(x, y, z, t) = \mathbf{0}. \tag{1.7}$$

One can obtain the vacuum field equation for the magnetic induction, using a similar line of reasoning to that which gave eqn (1.7). Specifically, take the curl of the free-space Ampère Law (1.4), make use of the vector identity (1.5), use the non-existence of magnetic monopoles (1.2) to eliminate one of the resulting terms, and then invoke Faraday's Law of induction (1.3). This chain of logic yields the vacuum field equation:

$$\left(\varepsilon_0 \mu_0 \frac{\partial^2}{\partial t^2} - \nabla^2\right) \mathbf{B}(x, y, z, t) = \mathbf{0}. \tag{1.8}$$

Based on the d'Alembert wave equations (1.7) and (1.8) one can determine the speed of propagation of electric and magnetic disturbances in free space. To this end, consider any propagating monochromatic plane-wave solution to

[2]These elements of classical electrodynamics are treated in the introductory accounts of Ohanian (1988) and Good (1999). More advanced treatments are given by Panofsky and Phillips (1962) and Jackson (1999).

eqns (1.7) or (1.8). Aligning the z-axis with the direction of propagation of this plane wave, such a solution has the form $\mathbf{C}\cos(kz - \omega t + \phi)$, where $\mathbf{C}$ is a real non-zero constant vector, ϕ is a real number, $k = 2\pi/\lambda$ is the radiation wave-number corresponding to a wavelength of λ, $\omega = 2\pi\nu$ is the angular frequency of the radiation corresponding to frequency ν, and t is time. By substituting this plane-wave solution into the d'Alembert equation, one finds that:

$$\varepsilon_0\mu_0\omega^2 - k^2 = 0, \tag{1.9}$$

which is equivalent to the statement that $\nu\lambda = 1/(\sqrt{\varepsilon_0\mu_0})$. Since $c = \nu\lambda$, where c is the speed at which the electric and magnetic disturbances propagate in vacuum, we have:

$$c = \frac{1}{\sqrt{\mu_0\varepsilon_0}}. \tag{1.10}$$

Using experimentally measured values for the permittivity and permeability of free space, it was found that this speed corresponds with that of light in vacuum. In turn, this led to the momentous discovery that light was an electromagnetic disturbance.

Given formula (1.10), the vacuum field equations become:

$$\left(\frac{1}{c^2}\frac{\partial^2}{\partial t^2} - \nabla^2\right)\mathbf{E}(x, y, z, t) = \mathbf{0}, \tag{1.11}$$

$$\left(\frac{1}{c^2}\frac{\partial^2}{\partial t^2} - \nabla^2\right)\mathbf{B}(x, y, z, t) = \mathbf{0}. \tag{1.12}$$

These equations state that each of the three components of the free-space electric field, and each of the three components of the free-space magnetic induction, are uncoupled from one another. Each of these six components obey a scalar form of the d'Alembert wave equation, corresponding to a disturbance which propagates at the speed of light in vacuum.

We close this section with a very brief discussion of the transition from a vector theory of the electromagnetic field in free space, to a scalar theory. In the former theory the electromagnetic disturbance is specified by two vector fields, namely the electric field and the magnetic induction, at each point in space–time. In the latter theory one considers the disturbance to be characterized by a single scalar field $\Psi(x, y, z, t)$, which is a function of both position and time. In free space, this function is taken to be a solution of the d'Alembert equation, so that:

$$\left(\frac{1}{c^2}\frac{\partial^2}{\partial t^2} - \nabla^2\right)\Psi(x, y, z, t) = 0. \tag{1.13}$$

Given the preceding arguments, it is evident that the wave equation above is obeyed by each component of the electromagnetic disturbance in free space.

Notwithstanding this, it is not immediately apparent why one can use a single scalar field to describe the electromagnetic disturbance in free space, the squared modulus of which corresponds to the optical intensity. Indeed the question of how one makes a transition, from a vector to a scalar theory of electromagnetic optics, is rather involved and will not be treated in detail here. Entry points to the literature on this subject include papers by Green and Wolf (1953), Wolf (1959), and Marathay and Parrent (1970), together with the texts of Nieto-Vesperinas (1991) and Born and Wolf (1999). The upshot of these investigations is that, in free space, one can indeed work with a single *complex* scalar function $\Psi(x, y, z, t)$, which obeys the d'Alembert equation (1.13). When sources are present, one can also work with the d'Alembert equation, provided that an appropriate source term is included.

1.2 Spectral decomposition and the analytic signal

Consider a particular instance of a light field in a given volume of free space, with the field being described by the complex scalar function $\Psi(x, y, z, t)$, as introduced at the end of the previous section. We spectrally decompose this wave-function,[3] as a superposition of monochromatic fields, using the Fourier integral:

$$\Psi(x, y, z, t) = \frac{1}{\sqrt{2\pi}} \int_0^\infty \psi_\omega(x, y, z) \exp(-i\omega t) d\omega. \tag{1.14}$$

In this decomposition, each monochromatic component $\psi_\omega(x, y, z) \exp(-i\omega t)$ of the field has been written as a product of a spatial wave-function $\psi_\omega(x, y, z)$ with the usual harmonic time factor $\exp(-i\omega t)$. The use of an ω subscript, on the spatial wave-function, indicates functional dependence on this quantity. Note, further, that it is only the real part of the above disturbance which is of physical significance.

The reader will note that the range of integration, in eqn (1.14), includes no negative angular frequencies ω. This choice results in certain useful analyticity[4] properties if one chooses to formally extend $\Psi(x, y, z, t)$ to be a function of a complex variable, by considering time to be a complex number. Accordingly, the spectral decomposition (1.14) is known as the 'analytic signal' corresponding to the scalar electromagnetic disturbance, a notion due to Gabor (1946). For a fuller account of the analytic signal, see, for example, Mandel and Wolf (1995).

Our next task is to determine the time-independent differential equation, which governs the evolution of the spatial wave-function $\psi_\omega(x, y, z)$ associated with a given monochromatic component of the spectral decomposition (1.14).

[3]We will often refer to $\Psi(x, y, z, t)$ as a wave-function, because it is a function describing a wave-field, but this usage should not be confused with the appearance of the same term in the context of quantum-mechanical fields.

[4]Here, the term 'analytic' is used in the sense of complex analysis, this being the calculus of functions of a complex variable. Elementary accounts of this theory, which are sufficient for the use of complex analysis made in this text, include Spiegel (1964) and Kreysig (1983).

Accordingly, substitute eqn (1.14) into the d'Alembert wave equation (1.13), interchanging the order of differentiation and integration, and then performing the differentiation with respect to time. This results in:

$$\int_0^\infty \left[\left(\nabla^2 + \frac{\omega^2}{c^2}\right)\psi_\omega(x,y,z)\right]\exp(-i\omega t)d\omega = 0. \tag{1.15}$$

The quantity in square brackets must vanish. We therefore arrive at the desired time-independent equation for the spatial component $\psi_\omega(x,y,z)$ of a monochromatic field:

$$\left(\nabla^2 + k^2\right)\psi_\omega(x,y,z) = 0, \quad k = \omega/c. \tag{1.16}$$

This is known as the 'Helmholtz equation'.[5] It is a central equation of the scalar diffraction theory, which concerns itself with constructing solutions to the Helmholtz equation in a given volume of space, subject to specified boundary values and boundary conditions for the field $\psi_\omega(x,y,z)$. In the next four sections we will outline various exact and approximate methods for solving diffraction problems based on the Helmholtz equation.

We close this section by noting that, for scalar disturbances which are not monochromatic, one can solve the diffraction problem for each monochromatic component $\psi_\omega(x,y,z)$ of the field, substituting the resulting spatial wave-functions into the spectral decomposition (1.14) in order to obtain the time-dependent complex scalar disturbance $\Psi(x,y,z,t)$. Typically, both the modulus and phase of this field will exhibit an extremely complex dependence on space and time, on account of the time-dependent 'beating' between monochromatic components of different frequencies.

1.3 Angular spectrum of plane waves

Consider the scenario shown in Fig. 1.1. Here, sources in the 'half-space' $z < 0$ radiate a scalar electromagnetic field, indicated by the series of wavy lines. This field propagates into a half-space $z \geq 0$, which is assumed to be free of both sources and matter. Suppose that one is given the value of a particular monochromatic component ψ_ω, in the spectral decomposition (1.14) of this coherent scalar electromagnetic disturbance, over the plane $z = 0$. Suppose, further, that one knows the wave-field to be forward propagating, which amounts to the *a priori* knowledge that there are no points on the plane $z = 0$ at which there is a flow of optical energy from right to left. Given knowledge of the forward-propagating wave-field over the plane $z = 0$, the angular spectrum formalism will allow us to construct an operator, which may be applied to this disturbance, yielding

[5] This equation is mathematically identical in form to the time-independent free-space Schrödinger equation for spinless non-relativistic particles. Hence, if the effects of electron spin can be ignored, the diffraction operator we are about to develop may be used to describe the free-space diffraction of monoenergetic forward-propagating electron beams from plane to parallel plane.

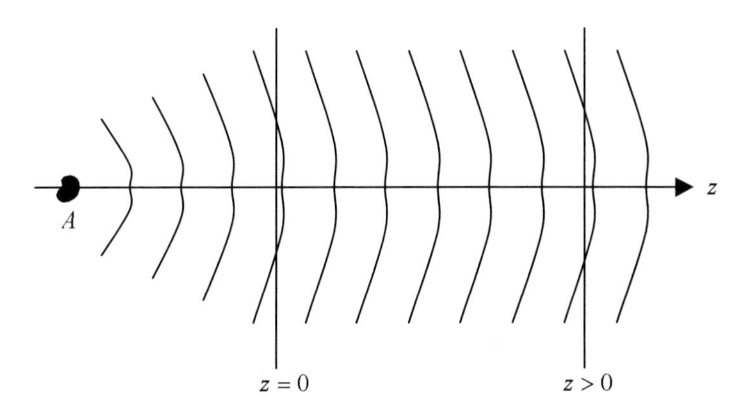

Fig. 1.1. Formulation of a diffraction problem for scalar electromagnetic wave-fields. All sources, denoted by A, are assumed to lie in the half-space $z < 0$. Here, z denotes a nominal optic axis, with the half-space $z \geq 0$ assumed to be vacuum. The wave-field generated by these sources is indicated by the wavy lines. For a given monochromatic component of this field, characterized by the spatial wave-function $\psi_\omega(x, y, z)$, a diffraction problem consists in determining the value of ψ_ω over some plane $z =$ constant > 0, given the value of ψ_ω over the plane $z = 0$.

the propagated disturbance over any parallel plane that lies downstream of it. This procedure dates back at least as far as Lord Rayleigh, with more recent treatments—together with appropriate references—in Goodman (1968), Montgomery (1968, 1969), Saleh and Teich (1991), Nieto-Vesperinas (1991), and Mandel and Wolf (1995). The particular development, given here, is due to Gureyev (1994).

Erect a Cartesian coordinate system (x, y, z), with the positive z-axis furnishing a nominal optic axis. Consider two parallel planes, $z = 0$ and $z = \Delta$, where $\Delta > 0$. Between these two planes there is only vacuum, implying that within this volume the Helmholtz equation (1.16) will be obeyed. Our present task is to make use of this equation to find an explicit form for the operator which may be applied to the forward-propagating field $\psi_\omega(x, y, z = 0)$ in order to yield the propagated wave-field $\psi_\omega(x, y, z = \Delta)$, where $\Delta > 0$.

As can be readily seen by direct substitution, the elementary plane waves:

$$\psi_\omega^{(\mathrm{PW})}(x, y, z) = \exp[i(k_x x + k_y y + k_z z)] \tag{1.17}$$

are solutions to the Helmholtz equation (1.16), provided that:

$$k_x^2 + k_y^2 + k_z^2 = k^2. \tag{1.18}$$

Here, the triple of numbers (k_x, k_y, k_z) respectively give the x, y, and z components of the wave-vector $\mathbf{k}$ of the plane wave. This wave-vector, which has a magnitude equal to 2π divided by the wavelength λ of the radiation, points in the direction of propagation of the plane wave.

Isolate k_z^2 in the previous expression, and take the positive square root, to give:

$$k_z = \sqrt{k^2 - k_x^2 - k_y^2}. \tag{1.19}$$

Taking the positive square root, in the above equation, amounts to the assumption that we are only considering plane waves which are forward propagating with respect to the optic axis z. Equivalently, we demand that the real part of k_z be non-negative.[6]

Our elementary plane-wave solutions to the Helmholtz equation can now be written in the form:

$$\psi_\omega^{(\mathrm{PW})}(x, y, z) = \exp\left[i(k_x x + k_y y)\right] \exp\left[iz\sqrt{k^2 - k_x^2 - k_y^2}\right]. \tag{1.20}$$

If we set $z = 0$ in the above equation, we obtain:

$$\psi_\omega^{(\mathrm{PW})}(x, y, z = 0) = \exp\left[i(k_x x + k_y y)\right]. \tag{1.21}$$

While it may seem trite to articulate such a simple point, we have now solved the trivial diffraction problem for a single monochromatic plane-wave solution to the Helmholtz equation. Given the value $\psi_\omega^{(\mathrm{PW})}(x, y, z = 0)$ which this plane wave takes over the plane $z = 0$, we can evaluate its propagated value over the plane $z > 0$ by multiplying the unpropagated disturbance with the propagation factor $\exp[iz(k^2 - k_x^2 - k_y^2)^{1/2}]$. This factor will henceforth be termed the 'free space propagator'.

Now that we have considered the propagation of elementary plane waves, from plane to parallel plane, we are ready to consider the problem of how to propagate ψ_ω from the plane $z = 0$ to the plane $z = \Delta$, where $\Delta \geq 0$, and there is assumed to be vacuum in the slab of free space between the planes $z = 0$ and $z = \Delta$. To this end, let us write the unpropagated field $\psi_\omega(x, y, z = 0)$ as a two-dimensional Fourier integral (see Appendix A)[7]:

[6] This statement implies that k_z may be complex, which is indeed the case. When $k_x^2 + k_y^2 \leq k^2$, k_z will be real and our plane wave $\exp[i(k_x x + k_y y + k_z z)]$ will be a propagating plane wave. When $k_x^2 + k_y^2 > k^2$, k_z will be a purely imaginary complex number, and our plane wave $\exp[i(k_x x + k_y y + k_z z)]$ will be exponentially damped in the z direction. Such plane waves are known as 'evanescent waves', and their exponential damping in the z direction corresponds with the fact that sub-wavelength information, imprinted in the field, is rapidly lost upon free-space propagation. Note also that the angular-spectrum formalism, as described here, provides a rigorous solution to the Helmholtz equation which accounts for both propagating and evanescent plane waves.

[7] The reader is assumed to be familiar with the elements of Fourier analysis outlined in Appendix A. Those unfamiliar with this material may wish to consult one of the many excellent texts which discuss Fourier analysis, in the context of optics. These include Goodman (1968), Lipson and Lipson (1981), Hecht (1987), and Reynolds *et al.* (1989).

$$\psi_\omega(x, y, z=0) = \frac{1}{2\pi} \iint \breve{\psi}_\omega(k_x, k_y, z=0) \exp\left[i(k_x x + k_y y)\right] dk_x dk_y. \quad (1.22)$$

Here, $\breve{\psi}_\omega(k_x, k_y, z = z_0)$ denotes the Fourier transform of $\psi_\omega(x, y, z = 0)$ with respect to x and y, and k_x, k_y are the usual Fourier-space variables which are conjugate to the position coordinates x, y.

From a physical point of view, eqn (1.22) decomposes the unpropagated wave-field into a linear combination of plane waves. Each such plane wave, namely $\exp[i(k_x x + k_y y)]$, may be thought of as the boundary value (over the plane $z = 0$) of a three-dimensional plane wave $\exp[i(k_x x + k_y y + k_z z)]$ which solves the Helmholtz equation. As we learned earlier, in order to propagate this section of a plane wave through the distance $\Delta \geq 0$, one need only multiply by the free-space propagator $\exp[i\Delta(k^2 - k_x^2 - k_y^2)^{1/2}]$. Therefore, to obtain the propagated disturbance $\psi_\omega(x, y, z = \Delta)$ from the unpropagated disturbance $\psi_\omega(x, y, z = 0)$, we multiply each two-dimensional plane wave $\exp[i(k_x x + k_y y)]$ in its Fourier decomposition (1.22) by the free-space propagator, leading to:

$$\begin{aligned}\psi_\omega(x, y, z=\Delta) =& \frac{1}{2\pi} \iint \breve{\psi}_\omega(k_x, k_y, z=0) \exp\left[i\Delta\sqrt{k^2 - k_x^2 - k_y^2}\right] \\ & \times \exp\left[i(k_x x + k_y y)\right] dk_x dk_y, \quad \Delta \geq 0. \end{aligned} \quad (1.23)$$

This expression is known as the angular-spectrum representation for the propagated wave-field. It solves the boundary value problem of determining the propagated field $\psi_\omega(x, y, z = \Delta)$, which results when the forward-propagating field $\psi_\omega(x, y, z = 0)$ is free-space propagated by a distance $z = \Delta \geq 0$.

Next, we seek to use the angular-spectrum representation to obtain an operator that describes the process of free-space diffraction. This diffraction operator, denoted $\mathcal{D}_\Delta$, acts upon the unpropagated wave-field in order to produce the propagated wave-field, with the propagation being through a distance $z = \Delta$ of free space. Formally, we write free-space propagation as:

$$\psi_\omega(x, y, z=\Delta) = \mathcal{D}_\Delta \psi_\omega(x, y, z=0), \quad \Delta \geq 0. \quad (1.24)$$

An explicit form, for the diffraction operator, is directly obtained from eqn (1.23). In verbal terms, this equation implies the following series of steps in order to obtain the propagated field $\psi_\omega(x, y, z = \Delta)$ given the unpropagated field $\psi_\omega(x, y, z = 0)$: (i) Take the Fourier transform of the unpropagated field $\psi_\omega(x, y, z = 0)$, with respect to x and y, to give $\breve{\psi}_\omega(k_x, k_y, z = 0)$. (ii) Multiply the result by the free-space propagator $\exp[i\Delta(k^2 - k_x^2 - k_y^2)^{1/2}]$. (iii) Take the inverse Fourier transform, with respect to k_x and k_y, of the resulting expression. Given the above verbal description, we can immediately write down the following expression for the desired diffraction operator:

$$\mathcal{D}_{\Delta} = \mathcal{F}^{-1} \exp\left[i\Delta\sqrt{k^2 - k_x^2 - k_y^2}\right] \mathcal{F}. \qquad (1.25)$$

Note that cascaded operators are taken to act from right to left.

Some remarks: (i) When the propagation distance is set to zero, the diffraction operator is equal to unity, thereby meeting a consistency requirement that propagated and unpropagated fields should be equal when the propagation distance is set to zero. (ii) If there are no evanescent plane-wave components present in the unpropagated field, the diffraction operator is unitary (see, for example, Montgomery (1981)). (iii) If there are no evanescent waves present in the propagated field, then the diffraction operator in eqn (1.25) may also be used to propagate the field through a negative distance (Wolf and Shewell 1967; Shewell and Wolf 1968; Lalor 1968; Sherman 1968).[8] (iv) The operator form (1.25) of the angular spectrum can be readily implemented numerically, by making use of the efficient numerical means of calculating Fourier transforms known as the fast Fourier transform (see, for example, Press *et al.* (1992)).

1.4 Fresnel diffraction

The angular spectrum formalism, outlined in the previous section, provides a rigorous solution to a certain boundary-value problem of the Helmholtz equation (1.16), for the free-space propagation of a coherent scalar X-ray wave-field from plane to plane. As a much-used special case of this formalism, we here give an outline of the Fresnel diffraction theory. This approximate theory, which will be obtained by using the angular spectrum as a starting point, is valid when a given free-space coherent scalar wave-field is 'paraxial'—that is, when all of the non-negligible plane-wave components of the field make a small angle with respect to a given optic axis.

Our presentation, of the Fresnel diffraction theory, is broken into two parts. The first part gives an operator formulation for Fresnel diffraction, with the second section giving an equivalent treatment based on the convolution integral. Note that this latter treatment is related to the Huygens–Fresnel notion that an unpropagated optical disturbance, over a given two-dimensional surface, may be regarded as a sum of infinitely many point sources of secondary waves, each of which propagate through free space to a given surface downstream of the unpropagated disturbance. According to the Huygens–Fresnel principle, the coherent superposition of all such secondary waves yields the resulting propagated field.

[8]Propagation through a negative distance has been termed 'inverse diffraction', in the just-cited papers. As pointed out there, if evanescent waves are not excluded from the forward-propagating field to be inverse diffracted, evanescent waves give lie to their name by being exponentially amplified upon propagation through a negative distance (rather than exponentially damped, which would be the case for diffraction through a positive distance), and the inverse wave propagator becomes singular. This singularity is removed if evanescent waves are strictly excluded from the field to be inverse diffracted.

1.4.1 *Operator formulation*

The angular spectrum formalism, outlined in the previous section, furnishes an exact solution to the boundary-value problem of free-space propagation of a monochromatic scalar wave-field from plane to parallel plane. In this section, we obtain a useful approximate form for the associated diffraction operator, which is equivalent to the famous Fresnel diffraction integral. Note that this equivalence will be demonstrated in Section 1.4.2.

Return, once more, to the scenario sketched in Fig. 1.1. We again focus attention on the spatial part $\psi_\omega(x,y,z)$, of a particular monochromatic component of the wave-field radiated by the sources, which propagates into a half-space $z \geq 0$ that is free of both sources and matter. We assume $\psi_\omega(x,y,z=0)$ to be paraxial, implying that all directions of propagation, for each of the non-negligible plane-wave components in the angular-spectrum decomposition of the field over the plane $z=0$, make a small angle with respect to the positive z-axis. Note that, if one were to adopt a ray picture, the requirement of paraxiality would simply be the statement that all rays should make a small angle with respect to the optic axis.

Since the field is paraxial, we see that, for all non-negligible plane-wave components in the half-space $z \geq 0$, $|k_x|$ and $|k_y|$ will both be much less than k_z (see eqn (1.20).[9] Accordingly, one can make the following binomial approximation:

$$\sqrt{k^2 - k_x^2 - k_y^2} \approx k - \frac{k_x^2 + k_y^2}{2k}, \tag{1.26}$$

so that the diffraction operator in (1.25) becomes:

$$\mathcal{D}_\Delta \approx \mathcal{D}_\Delta^{(\mathrm{F})} \equiv \exp(ik\Delta)\mathcal{F}^{-1} \exp\left[\frac{-i\Delta(k_x^2 + k_y^2)}{2k}\right] \mathcal{F}. \tag{1.27}$$

Here, the 'F' superscript indicates that this is the diffraction operator corresponding to Fresnel diffraction. Substituting into eqn (1.24), we arrive at an operator form for the Fresnel diffraction integral (see, for example, Nazarathy and Shamir (1980) and Saleh and Teich (1991)):

[9] As a sufficient condition for this to be so, we may demand that $k_x^2 + k_y^2 \ll k^2$, for the largest value of $k_x^2 + k_y^2$ for which $\breve{\psi}_\omega(k_x, k_y, z=0)$ is non-negligible in modulus. Denote this maximum value of $k_x^2 + k_y^2$ by $k_{\max}^2 \equiv (2\pi/a)^2$, where a represents the smallest non-negligible length scale present in the unpropagated disturbance. Taking the reciprocal of the square root of the previously mentioned sufficient condition, and making use of the definition introduced in the previous sentence, one arrives at the condition $a \gg \lambda$. This is a sufficient condition for the validity of the Fresnel approximation, which is about to be introduced in the main text. This sufficient condition amounts to the statement that the smallest characteristic length scale, over which the unpropagated disturbance varies appreciably, is much larger in size than the wavelength of the radiation.

$$\psi_\omega(x,y,z=\Delta) \approx \mathcal{D}_\Delta^{(\mathrm{F})}\psi_\omega(x,y,z=0) \tag{1.28}$$
$$= \exp(ik\Delta)\mathcal{F}^{-1}\exp\left[\frac{-i\Delta(k_x^2+k_y^2)}{2k}\right]\mathcal{F}\psi_\omega(x,y,z=0), \quad \Delta \geq 0.$$

In words, this equation indicates the following sequence of steps in order to obtain the Fresnel approximation to the propagated field $\psi_\omega(x,y,z=\Delta)$, from the unpropagated field $\psi_\omega(x,y,z=0)$: (i) Take the Fourier transform of the unpropagated field $\psi_\omega(x,y,z=0)$, with respect to x and y, to yield a function of the corresponding Fourier-space variables k_x and k_y. (ii) Multiply the resulting function by the free-space propagator $\exp[-i\Delta(k_x^2+k_y^2)/(2k)]$, which is known as the Fresnel propagator. (iii) Take the inverse Fourier transform, with respect to k_x and k_y, of the resulting expression. (iv) Multiply by the constant phase factor $\exp(ik\Delta)$. Again, if required, this sequence of steps may be readily implemented numerically, by making use of the fast Fourier transform.

1.4.2 *Convolution formulation*

Here we recast eqn (1.28) in the form of a convolution integral (see Appendix A). The resulting expression is interpreted in terms of the Huygens–Fresnel principle.

Recall that the two-dimensional convolution $f(x,y)\star g(x,y)$, of two suitably well-behaved functions $f(x,y)$ and $g(x,y)$ of Cartesian coordinates x and y, is given by:

$$f(x,y)\star g(x,y) \equiv \iint_{-\infty}^{\infty} f(x',y')g(x-x',y-y')dx'dy'. \tag{1.29}$$

The two-dimensional form of the convolution theorem is:

$$\mathcal{F}\left[f(x,y)\star g(x,y)\right] = 2\pi\left\{\mathcal{F}\left[f(x,y)\right]\right\}\times\left\{\mathcal{F}\left[g(x,y)\right]\right\}, \tag{1.30}$$

where $\mathcal{F}$ denotes Fourier transformation with respect to x and y, under the following convention (see Appendix A):

$$f(x,y) = \frac{1}{2\pi}\iint_{-\infty}^{\infty}\breve{f}(k_x,k_y)\exp[i(k_x x+k_y y)]dk_x dk_y,$$
$$\breve{f}(k_x,k_y) \equiv \mathcal{F}[f(x,y)], \tag{1.31}$$
$$\breve{f}(k_x,k_y) = \frac{1}{2\pi}\iint_{-\infty}^{\infty} f(x,y)\exp[-i(k_x x+k_y y)]dx dy,$$
$$f(x,y) \equiv \mathcal{F}^{-1}[\breve{f}(k_x,k_y)]. \tag{1.32}$$

Here, $\breve{f}(k_x,k_y)$ denotes the Fourier transform of $f(x,y)$ with respect to x and y, as a function of the coordinates k_x and k_y dual to x and y. As stated earlier, the corresponding two-dimensional Fourier transform operator is denoted by $\mathcal{F}$. The associated inverse transformation is denoted by $\mathcal{F}^{-1}$.

It will prove convenient to write down an alternative form of eqn (1.30), obtained by taking the inverse Fourier transform of both sides of this expression:

$$f(x,y) \star g(x,y) = 2\pi\mathcal{F}^{-1}\left(\{\mathcal{F}[f(x,y)]\} \times \{\mathcal{F}[g(x,y)]\}\right). \tag{1.33}$$

This may be compared to eqn (1.28) for Fresnel diffraction, which we rewrite in the following suggestive form:

$$\begin{aligned}
&\psi_\omega(x,y,z=\Delta) \qquad\qquad (1.34)\\
&= 2\pi\mathcal{F}^{-1}\left(\frac{\exp(ik\Delta)}{2\pi}\exp\left[\frac{-i\Delta(k_x^2+k_y^2)}{2k}\right] \times \{\mathcal{F}[\psi_\omega(x,y,z=0)]\}\right)\\
&= 2\pi\mathcal{F}^{-1}\left(\left\{\mathcal{F}\mathcal{F}^{-1}\frac{\exp(ik\Delta)}{2\pi}\exp\left[\frac{-i\Delta(k_x^2+k_y^2)}{2k}\right]\right\} \times \{\mathcal{F}[\psi_\omega(x,y,z=0)]\}\right).
\end{aligned}$$

Comparing eqns (1.33) and (1.34), we conclude that the Fresnel diffraction integral (1.28), for free-space propagation from plane to plane, can be written as the following convolution integral:

$$\psi_\omega(x,y,z=\Delta\geq 0) = \psi_\omega(x,y,z=0) \star P(x,y,\Delta). \tag{1.35}$$

The function $P(x,y,\Delta)$, which is known as the real-space form of the 'Fresnel propagator', is given by:

$$P(x,y,\Delta) \equiv \frac{1}{2\pi}\exp(ik\Delta)\mathcal{F}^{-1}\left\{\exp\left[\frac{-i\Delta(k_x^2+k_y^2)}{2k}\right]\right\}. \tag{1.36}$$

Prior to giving a physical interpretation of the above pair of equations, we seek an explicit form for the Fresnel propagator. Accordingly, use formula (1.31) for inverse two-dimensional Fourier transformation, to rewrite eqn (1.36) as:

$$\begin{aligned}
P(x,y,\Delta) =& \frac{\exp(ik\Delta)}{4\pi^2}\iint_{-\infty}^{\infty}\exp\left[\frac{-i\Delta(k_x^2+k_y^2)}{2k}\right]\exp[i(k_x x+k_y y)]dk_x dk_y\\
=& \frac{\exp(ik\Delta)}{4\pi^2}\left\{\int_{-\infty}^{\infty}\exp\left[i\left(k_x x-\frac{\Delta}{2k}k_x^2\right)\right]dk_x\right\}\\
&\times\left\{\int_{-\infty}^{\infty}\exp\left[i\left(k_y y-\frac{\Delta}{2k}k_y^2\right)\right]dk_y\right\}.
\end{aligned} \tag{1.37}$$

Next, we need to evaluate each of the integrals which appear in the braces of the two lowest lines of this equation. Since they are mathematically identical in form, let us concentrate on the first of these integrals. Completing the square of the exponent in the corresponding integrand, and then changing the variable of integration from k_x to $\kappa \equiv [k_x - (kx/\Delta)]\sqrt{\Delta/(\pi k)}$, we see that:

$$\begin{aligned}
&\int_{-\infty}^{\infty} \exp\left[i\left(k_x x - \frac{\Delta}{2k}k_x^2\right)\right] dk_x \\
&= \exp\left(\frac{ikx^2}{2\Delta}\right)\int_{-\infty}^{\infty} \exp\left[-\frac{i\Delta}{2k}\left(k_x - \frac{kx}{\Delta}\right)^2\right] dk_x \\
&= 2\sqrt{\frac{\pi k}{\Delta}} \exp\left(\frac{ikx^2}{2\Delta}\right)\int_{0}^{\infty} \exp\left(-i\frac{\pi}{2}\kappa^2\right) d\kappa. \qquad (1.38)
\end{aligned}$$

The integral, in the bottom line of the above series of equations, is equal to $\frac{1}{2}(1-i) = \exp(-i\pi/4)/\sqrt{2}$ (note that this result can be obtained by considering the limiting form of the so-called Fresnel integrals[10]). Thus:

$$\int_{-\infty}^{\infty} \exp\left[i\left(k_x x - \frac{\Delta}{2k}k_x^2\right)\right] dk_x = \sqrt{\frac{2\pi k}{\Delta}} \exp\left[i\left(\frac{kx^2}{2\Delta} - \frac{\pi}{4}\right)\right]. \qquad (1.39)$$

Equipped with this formula, together with the corresponding expression that is obtained when k_x is replaced by k_y and x is replaced by y, expression (1.37) can be evaluated to give the following explicit form for the Fresnel propagator $P(x, y, \Delta)$:

$$P(x, y, \Delta) = -\frac{ik\exp(ik\Delta)}{2\pi\Delta} \exp\left[\frac{ik(x^2 + y^2)}{2\Delta}\right]. \qquad (1.40)$$

Inserting this into eqn (1.35), we obtain the required convolution formulation of the Fresnel diffraction integral:

$$\begin{aligned}
\psi_\omega(x, y, z = \Delta \geq 0) &= \psi_\omega(x, y, z = 0) \star P(x, y, \Delta) \\
&= -\frac{ik\exp(ik\Delta)}{2\pi\Delta}\left\{\psi_\omega(x, y, z = 0) \star \exp\left[\frac{ik(x^2 + y^2)}{2\Delta}\right]\right\}. \qquad (1.41)
\end{aligned}$$

With a view to interpreting this expression, consider the simple case of a thin black two-dimensional screen, which is located in the plane $z = 0$ in Fig. 1.1. Suppose this screen to be pierced at a single point located at $(x, y, z) = (x_0, y_0, 0)$, where (x, y) denotes the usual Cartesian coordinates in the plane perpendicular to the optic axis z. We consider the punctured screen to be illuminated by

[10]The Fresnel integrals are respectively given by $\mathscr{C}(s) \equiv \int_0^s \cos(\pi\kappa^2/2)d\kappa$ and $\mathscr{S}(s) \equiv \int_0^s \sin(\pi\kappa^2/2)d\kappa$. Both $\mathscr{C}(s)$ and $\mathscr{S}(s)$ tend to $\frac{1}{2}$ as s tends to infinity, leading to the result given in the main text. For a thorough discussion of the Fresnel integrals, which includes the limiting forms quoted here, see, for example, Born and Wolf (1999).

a z-directed monochromatic scalar wave-field, and approximate the unpropagated disturbance, at the exit surface of the screen, as being proportional to the two-dimensional Dirac delta $\delta(x - x_0, y - y_0)$. Thus, we take the unpropagated disturbance, $\psi_\omega(x, y, z = 0)$, to be:

$$\psi_\omega(x, y, z = 0) = \psi_\omega(x_0, y_0, z = 0)\delta(x - x_0, y - y_0), \tag{1.42}$$

where $\psi_\omega(x_0, y_0, z = 0)$ is the incident disturbance at the entrance to the pinhole. Inserting the above approximate expression into eqn (1.41), and making use of the sifting property of the Dirac delta,[11] we obtain the following formula for the propagated disturbance over the plane $z = \Delta$:

$$\begin{aligned}\psi_\omega(x, y, z = \Delta) &= \psi_\omega(x_0, y_0, z = 0)P(x - x_0, y - y_0, \Delta) \\ &= -\frac{ik\psi_\omega(x_0, y_0, z = 0)\exp(ik\Delta)}{2\pi\Delta} \\ &\quad \times \exp\left\{\frac{ik[(x - x_0)^2 + (y - y_0)^2]}{2\Delta}\right\}.\end{aligned} \tag{1.43}$$

This immediately leads us to a simple interpretation for the Fresnel propagator: $P(x - x_0, y - y_0, \Delta)$ gives the amplitude of the propagated wave-field at the point $(x, y, z = \Delta)$, which is due to a unit disturbance at the point $(x_0, y_0, z = 0)$, with $\psi_\omega(x_0, y_0, z = 0)P(x - x_0, y - y_0, \Delta)$ giving the propagated wave-field at the point $(x, y, z = \Delta)$ which is due to a disturbance $\psi_\omega(x_0, y_0, z = 0)$ at the point $(x_0, y_0, z = 0)$. As the reader may readily show, this propagator—as given by the lower two lines of eqn (1.43)—is proportional to a spherical wave emanating from the point $(x_0, y_0, z = 0)$, for the case where x and y have magnitudes that are much smaller than the propagation distance Δ.

Suppose that, rather than being pierced with a small pinhole at one point, the previously mentioned thin black screen is pierced with small pinholes at two distinct points. One will then have a Young interferometer, which will be of some importance in this chapter's later discussions on the notion of partial coherence. In the present context, however, we note that the propagated disturbance will be given by a superposition of two disturbances, each of which propagate from a given pinhole. If the two pinholes are taken to be respectively located at (x_0^A, y_0^A) and (x_0^B, y_0^B), then the propagated disturbance over the plane $z = \Delta$ will evidently be given by:

$$\begin{aligned}\psi_\omega(x, y, z = \Delta) =& \psi_\omega(x_0^A, y_0^A, z = 0)P(x - x_0^A, y - y_0^A, \Delta) \\ &+ \psi_\omega(x_0^B, y_0^B, z = 0)P(x - x_0^B, y - y_0^B, \Delta).\end{aligned} \tag{1.44}$$

Thus, for the case where our thin black screen is punctured by two small pinholes, the propagated disturbance at a given point $(x, y, z = \Delta)$ is equal to the sum

[11] This sifting property may be expressed as $\iint_{-\infty}^{\infty} \delta(x - x_0, y - y_0) f(x, y) dx dy = f(x_0, y_0)$, for any well-behaved function $f(x, y)$ and any real numbers x_0 and y_0. See Appendix A.

of the disturbances which are propagated from each of the two holes at the exit surface of the screen. Each term in this sum is obtained by multiplying the disturbance incident on a given pinhole, by the appropriate value of the corresponding Fresnel propagator.

One can imagine continuing this process, of puncturing the thin black screen with an increasing number of small pinholes, until one reaches the limiting case of there being no screen at all. Over the plane $z = 0$, one will then have the unpropagated disturbance $\psi_\omega(x, y, z = 0)$ which exists due to the incident field. The resulting propagated field, over the plane $z = \Delta$, is then given by eqn (1.41). This convolution formulation of Fresnel diffraction expresses the propagated field as the sum of the propagated disturbances, which are due to each of the points on the incident wavefront over the plane $z = 0$. This represents a mathematical embodiment of the Huygens–Fresnel principle, which views the propagated disturbance as a sum of the propagated disturbances that emanate from each point on the initial wavefront.

We close this sub-section by writing down a certain explicit form of the Fresnel diffraction integral, which often provides a convenient means for numerically computing coherent diffraction patterns. Of more immediate relevance is the fact that the resulting expression will allow us to make a smooth transition to the so-called 'far-field' regime, in Section 1.5, by taking the propagation distance to be very large. With these ends in mind, take eqn (1.41) and then explicitly write its convolution integral using eqn (1.29), to give:

$$\begin{aligned}\psi_\omega(x, y, z = \Delta \geq 0) \\ = -\frac{ik\exp(ik\Delta)}{2\pi\Delta}\iint_{-\infty}^{\infty}\psi_\omega(x', y', z = 0) \\ \times\exp\left\{\frac{ik}{2\Delta}[(x - x')^2 + (y - y')^2]\right\}dx'dy'.\end{aligned} \qquad (1.45)$$

Expanding the exponent in the integrand, subsequently factoring out a term that does not depend on the variables of integration, one obtains:

$$\begin{aligned}\psi_\omega(x, y, z = \Delta \geq 0) = -\frac{ik\exp(ik\Delta)}{2\pi\Delta}\exp\left[\frac{ik}{2\Delta}(x^2 + y^2)\right] \\ \times\iint_{-\infty}^{\infty}\psi_\omega(x', y', z = 0)\exp\left[\frac{ik}{2\Delta}(x'^2 + y'^2)\right] \\ \times\exp\left[\frac{-ik}{\Delta}(xx' + yy')\right]dx'dy'.\end{aligned} \qquad (1.46)$$

1.5 Fraunhofer diffraction

Here, we consider a limiting form of the Fresnel diffraction integral, for the case of propagation distances that are very large compared to the characteristic length scale of the unpropagated wave-field. Such propagated wave-fields are said to be in the 'far field', or in the 'far zone'. As we shall see, subject to certain approximations which will be made explicit in due course, the diffracted field in the far zone takes the form of a modulated expanding spherical wave, with the modulation being a transversely scaled form of the two-dimensional Fourier transform of the unpropagated disturbance. Such diffraction patterns are known as 'Fraunhofer diffraction patterns', with the corresponding diffraction integral being known as the 'Fraunhofer diffraction integral'.

Let us return to eqn (1.46), with all symbols as previously defined. Assume the unpropagated disturbance, in the plane $z = 0$, to be non-negligible only over a region of diameter b. Introduce the dimensionless Fresnel number, N_{F}, via:

$$N_{\mathrm{F}} \equiv \frac{b^2}{\lambda\Delta} = \frac{kb^2}{2\pi\Delta}. \tag{1.47}$$

Assume the propagation distance Δ to be sufficiently large that the Fresnel number is much less than unity:

$$N_{\mathrm{F}} \ll 1. \tag{1.48}$$

This is a sufficient condition for us to ignore the first exponent which appears in the Fresnel diffraction integral (1.46). We therefore obtain the following limiting form for the propagated disturbance, which is known as the 'Fraunhofer diffraction integral'[12]:

$$\begin{aligned} \psi_\omega(x, y, z = \Delta \geq 0) \to & -\frac{ik\exp(ik\Delta)}{2\pi\Delta} \exp\left[\frac{ik}{2\Delta}(x^2 + y^2)\right] \\ \times \iint_{-\infty}^{\infty} & \psi_\omega(x', y', z = 0) \exp\left[\frac{-ik}{\Delta}(xx' + yy')\right] dx'dy', \quad N_{\mathrm{F}} \ll 1. \end{aligned} \tag{1.49}$$

To write this result in a more compact form, make use of eqn (A.5)[13] for the two-dimensional Fourier transform, leading to:

[12]Note that it is the largest length scale, in the unpropagated disturbance, which appears in the sufficiency condition (1.48) for the validity of the Fraunhofer diffraction formula. This is natural, since it this largest length scale—given by the diameter b of the unpropagated disturbance—which will be the most slowly diffracting, and therefore require the largest propagation distance in order to be placed in the far zone. This may be compared to the sufficiency condition for the validity of the Fresnel approximation, which is governed by the *smallest* length scale in the unpropagated disturbance. Note, also, that both of these sufficient conditions were assumed in obtaining eqn (1.48).

[13]Note that the lettered prefix, in this equation, indicates that it is to be found in Appendix A. This notation will be adopted throughout the remainder of the text.

$$\psi_\omega(x, y, z = \Delta \geq 0) \to -\frac{ik\exp(ik\Delta)}{\Delta}\exp\left[\frac{ik}{2\Delta}(x^2+y^2)\right]$$
$$\times\, \breve{\psi}_\omega\left(k_x = \frac{kx}{\Delta}, k_y = \frac{ky}{\Delta}, z = 0\right), \quad N_{\rm F} \ll 1. \quad (1.50)$$

Clearly, this has the form of a paraxial modulated spherical wave $\exp[ik(x^2+y^2)/(2\Delta)]$, emanating from the origin of coordinates in the plane of the unpropagated disturbance, with the modulation being proportional to a transversely scaled form of the two-dimensional Fourier transform of the unpropagated disturbance. This important and rather simple result will be used on a number of occasions, in later chapters, to calculate far-field coherent diffraction patterns for paraxial scalar X-ray fields.

1.6 Kirchhoff and Rayleigh–Sommerfeld diffraction theory

The present section opens with a treatment of the Helmholtz–Kirchhoff integral theorem for treating the diffraction of coherent scalar electromagnetic waves. The resulting diffraction integral yields an expression for the diffracted disturbance at a given point Q, given the value of both the field and its normal derivative over a smooth closed surface which encloses Q (see Fig. 1.2). Note that this surface, together with its interior, is assumed to be in vacuum. The Helmholtz–Kirchhoff integral may be modified, using certain assumptions due to Kirchhoff that we shall outline, to arrive at the Kirchhoff diffraction formula, which allows one to propagate a given coherent scalar field into a vacuum-filled half-space $z > 0$, given knowledge of both the field and its normal derivative over the plane $z = 0$.

The requirement, of needing to know both the field and its normal derivative over a plane in order to propagate it into a vacuum-filled half-space, is removed in passing from the Kirchhoff to the Rayleigh–Sommerfeld diffraction integrals. We show how, by appropriate choice of Green functions, one can obtain the Rayleigh–Sommerfeld integrals from the Kirchhoff integral. We shall also demonstrate the equivalence of one of the Rayleigh–Sommerfeld integrals, to the angular-spectrum formalism outlined in Section 1.3.

1.6.1 *Kirchhoff diffraction integral*

Consider, once again the diffraction scenario sketched in Fig. 1.1. The vacuum-filled half-space $z > 0$ supports a coherent scalar electromagnetic field $\psi_\omega(x, y, z > 0)$, at every point of which the field satisfies the Helmholtz equation (1.16). With reference to Fig. 1.2, consider a smooth closed surface $\partial\Omega$ which bounds a given volume Ω, entirely contained within the vacuum-filled half-space $z > 0$, and inside which lies a given point Q. The outward-pointing normal vector of unit length, at any point A on the surface $\partial\Omega$, is denoted by $\hat{\mathbf{n}}$. We now give a derivation of the integral theorem of Helmholtz and Kirchhoff, which expresses the value of the disturbance at Q, given the value of both the disturbance and its normal derivative at every point on $\partial\Omega$.

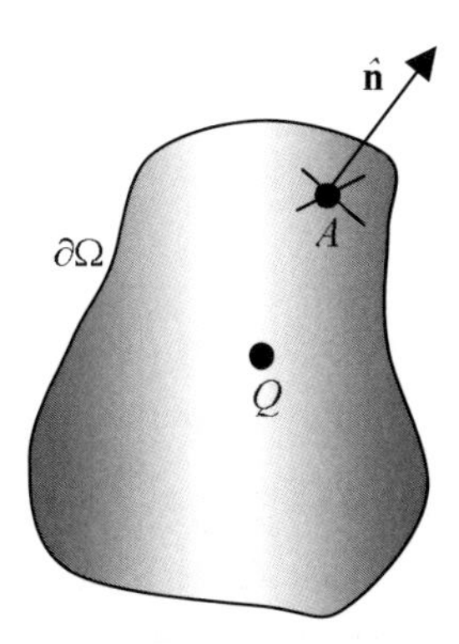

Fig. 1.2. A fictitious smooth closed surface $\partial\Omega$ encloses the volume Ω, together with the interior point Q. Every point within and on this surface is occupied by free space. At any given point A on $\partial\Omega$, $\hat{\mathbf{n}}$ denotes the outward pointing normal unit vector. The Helmholtz–Kirchhoff integral theorem allows one to determine the value of a coherent complex scalar electromagnetic disturbance at Q, given knowledge of both the field and its normal derivative over the surface $\partial\Omega$.

Consider a pair of continuous scalar functions $f(x,y,z)$ and $g(x,y,z)$ which are sufficiently well-behaved to have continuous first and second partial derivatives with respect to x, y, and z. Making use of the following readily proved identity:

$$\nabla \cdot [f(x,y,z)\nabla g(x,y,z)] = f\nabla^2 g(x,y,z) + \nabla f(x,y,z)\cdot\nabla g(x,y,z), \qquad (1.51)$$

and dropping explicit functional dependence on (x,y,z) for clarity, we see that:

$$\nabla\cdot(f\nabla g - g\nabla f) = f\nabla^2 g - g\nabla^2 f. \qquad (1.52)$$

Integrating over the volume Ω, and then applying the Gauss divergence theorem to the integral on the left side of the resulting expression, leads to Green's theorem:

$$\oiint_{\partial\Omega} \left(f\frac{\partial g}{\partial n} - g\frac{\partial f}{\partial n}\right) d\sigma = \iiint_{\Omega} (f\nabla^2 g - g\nabla^2 f)dV. \qquad (1.53)$$

Here $d\sigma$ denotes a differential surface area element on $\partial\Omega$, dV denotes a differential volume element within Ω, and $\partial/\partial n = (\hat{\mathbf{n}}\cdot\nabla)$ denotes differentiation in the direction of the outward unit normal $\hat{\mathbf{n}}$ (see Fig. 1.2).

In Green's theorem, let f be an expanding spherical wave[14] of angular frequency ω which emanates from the point Q, and let g be the complex scalar

[14]The reader may object that the expanding spherical wave does not satisfy the assumption, of being twice differentiable, which forms a sufficient condition for the validity of Green's

disturbance ψ_ω. Since it coincides with the free-space outgoing Green function for the Helmholtz equation (cf. Section 2.3), the expanding spherical wave will be denoted by G^Q_ω. With these identifications we see that Green's theorem becomes:

$$\oiint_{\partial\Omega} \left(G^Q_\omega \frac{\partial \psi_\omega}{\partial n} - \psi_\omega \frac{\partial G^Q_\omega}{\partial n} \right) d\sigma = \iiint_{\Omega} (G^Q_\omega \nabla^2 \psi_\omega - \psi_\omega \nabla^2 G^Q_\omega) dV. \quad (1.54)$$

By adding and then subtracting $k^2 G^Q_\omega \psi_\omega$ from the integrand on the right side of eqn (1.54), we see that:

$$\oiint_{\partial\Omega} \left(G^Q_\omega \frac{\partial \psi_\omega}{\partial n} - \psi_\omega \frac{\partial G^Q_\omega}{\partial n} \right) d\sigma = \iiint_{\Omega} G^Q_\omega (\nabla^2 + k^2) \psi_\omega dV - \iiint_{\Omega} \psi_\omega (\nabla^2 + k^2) G^Q_\omega dV. \quad (1.55)$$

The first integral, on the right side of this equation, vanishes on account of the fact that ψ_ω obeys the Helmholtz equation (1.16). To simplify the second integral on the right side, we make use of the fact that G^Q_ω obeys the equation:

$$(\nabla^2 + k^2) G^Q_\omega = -4\pi \delta^Q, \quad (1.56)$$

where δ^Q is a three-dimensional Dirac delta located at the point Q.[15] Bearing the above two points in mind, we see that eqn (1.55) becomes:

$$\oiint_{\partial\Omega} \left(G^Q_\omega \frac{\partial \psi_\omega}{\partial n} - \psi_\omega \frac{\partial G^Q_\omega}{\partial n} \right) d\sigma = 4\pi \iiint_{\Omega} \psi_\omega \delta^Q dV = 4\pi \psi_\omega(Q), \quad (1.57)$$

where $\psi_\omega(Q)$ denotes the value of ψ_ω at the point Q. Rearranging, we arrive at the Helmholtz–Kirchhoff integral theorem:

theorem. More precisely, this requirement is not met at the point Q from which the spherical waves emanates. Accordingly, in a more rigorous argument than that presented here, one would exclude the point Q from the volume of integration when deriving the Helmholtz–Kirchhoff integral theorem (see, for example, Baker and Copson (1950) or Born and Wolf (1999)).

[15] The above equation may be obtained in a number of ways, including that which is outlined in Section 2.3 of the next chapter. As a simpler alternative which suffices for the purposes of the present chapter, one may apply the Helmholtz operator $(\nabla^2 + k^2)$ to the explicit form $\exp(ikr_Q)/r_Q$ for G^Q_ω, where r_Q denotes radial distance from the point Q, and then consider the integral of the resulting function over a vanishingly small volume that encloses Q. The resulting integral is readily seen to be equal to -4π, no matter how small one makes the non-zero volume enclosing the point Q. This indicates that $(\nabla^2 + k^2) G^Q_\omega$ is indeed equal to -4π multiplied by a three-dimensional Dirac delta centred at Q, as stated in eqn (1.56).

$$\psi_\omega(Q) = \frac{1}{4\pi} \oiint_{\partial\Omega} \left(G_\omega^Q \frac{\partial \psi_\omega}{\partial n} - \psi_\omega \frac{\partial G_\omega^Q}{\partial n} \right) d\sigma. \tag{1.58}$$

As outlined at the beginning of this section, the Helmholtz–Kirchhoff integral theorem allows one to determine the value of a coherent scalar electromagnetic disturbance at a specified point Q, given the values of both the field and its normal derivative at every point on a smooth closed surface that completely encloses Q, provided that the surface and its interior are in vacuum.

In many instances, it is more convenient to have a diffraction integral that allows one to determine the value of a coherent scalar disturbance at a specified point Q, given knowledge of the disturbance and its normal derivative over a plane bounding a vacuum-filled half-space which contains Q. In this context, consider the construction sketched in Fig. 1.3. Here all sources of the coherent field[16] lie in the half-space $z < 0$, together with any optical elements which may be present, with the volume $z > 0$ being vacuum. The spatial coordinate of Q with respect to a given origin O is denoted by $\mathbf{x}_Q$. The closed surface $\partial\Omega$ is taken to be the hemisphere of radius R, which is indicated in the diagram; the flat part of the hemisphere makes contact with the plane $z = 0$, with the curved part lying in the space $z \geq 0$. Denote the flat part of the hemisphere by $\partial\Omega_1$, and the curved part of the hemisphere by $\partial\Omega_2$. The point Q lies within the surface $\partial\Omega = \partial\Omega_1 + \partial\Omega_2$. For this surface, the Helmholtz–Kirchhoff integral theorem becomes:

$$\begin{aligned} \psi_\omega(Q) = & \frac{1}{4\pi} \iint_{\partial\Omega_1} \left(G_\omega^Q \frac{\partial \psi_\omega}{\partial n} - \psi_\omega \frac{\partial G_\omega^Q}{\partial n} \right) d\sigma \\ & + \frac{1}{4\pi} \iint_{\partial\Omega_2} \left(G_\omega^Q \frac{\partial \psi_\omega}{\partial n} - \psi_\omega \frac{\partial G_\omega^Q}{\partial n} \right) d\sigma. \end{aligned} \tag{1.59}$$

We now assume that, in this limit as R tends to infinity, the field over the surface $\partial\Omega_2$ behaves as a modulated outgoing spherical wave emanating from the origin O. Under this assumption, one can readily show that the second integral vanishes, on the right side of the above equation. Regarding the first integral on the right side of this equation, we note that $\partial/\partial n$ can be replaced by $-\partial/\partial z$, since the outward normal vector to $\partial\Omega_1$ points in the negative z direction. Further, in the same integral, the areal element $d\sigma$ can be re-written as $dxdy$, where x and y are the usual Cartesian coordinates in the plane $z = 0$. Since we have taken R as tending to infinity, the range of integration extends over all x and y. Bearing all of the above in mind, we see that:

[16] More generally, one could consider the sources to produce a polychromatic field, in which case we restrict consideration to a single monochromatic component of the polychromatic field.

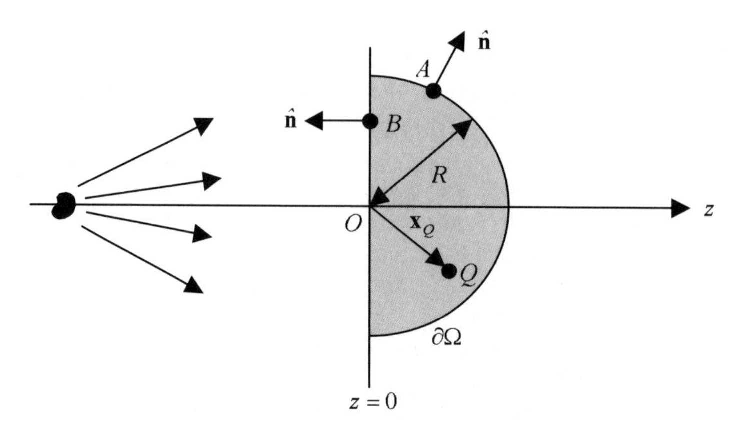

Fig. 1.3. A fictitious ball of radius R is located at the origin O. The intersection of this ball with the half-space $z \geq 0$ gives the volume Ω, the surface of which is denoted by $\partial\Omega$. Every point in and on this surface is taken to be vacuum. The outward pointing normal unit vector, at any point on the surface, is denoted by $\hat{\mathbf{n}}$; two such vectors are indicated, at the points A and B. The point Q, with coordinate vector $\mathbf{x}_Q$, is enclosed by the surface $\partial\Omega$. All sources and optical elements are taken to lie in the half space $z < 0$. The Kirchhoff integral allows one to determine the value of the propagated coherent scalar electromagnetic field, at any point Q within $\partial\Omega$, given the value of both the unpropagated field and its normal derivative over the plane $z = 0$. Note that this formula requires R to be infinitely large, with the propagated field obeying the Sommerfeld radiation condition.

$$\psi_\omega(Q) = \frac{1}{4\pi} \iint_{-\infty}^{\infty} \left(\psi_\omega \frac{\partial G_\omega^Q}{\partial z} - G_\omega^Q \frac{\partial \psi_\omega}{\partial z} \right) dx dy. \tag{1.60}$$

Notwithstanding the fact that it is a slight abuse of the conventional terminology, we will refer to the above expression as the 'Kirchhoff diffraction formula'.[17] It allows the propagated disturbance $\psi_\omega(Q)$ to be determined at a specified point Q in the half-space downstream of the plane $z = 0$, over which both the field and its z derivative are known.

[17]More properly, this term refers to a certain special case of eqn (1.60). In this special case, a thin black screen is considered to be located in the plane $z = 0$, with the two-dimensional integral then being restricted to the aperture over this screen. The so-called Kirchhoff approximation considers the disturbance to be identically zero over the opaque portion of the exit-surface of the screen, with the disturbance and its normal derivative over the remainder of the screen being equal to that which would be present in the absence of the screen. The resulting special case, of eqn (1.60), is what should be more properly referred to as the Kirchhoff diffraction formula.

1.6.2 *Rayleigh–Sommerfeld diffraction integrals*

Evidently, the Kirchhoff diffraction formula over-specifies the boundary conditions required to solve the problem of propagating a given forward-travelling complex disturbance, from a plane into a vacuum-filled region. One may conclude that this is so, based on the fact that our earlier discussions on the angular-spectrum formalism allowed us to solve the same boundary value problem given only the value of the unpropagated disturbance over the plane $z = 0$, without the need to specify its derivative normal to this plane.

With this in mind, consider eqn (1.60) once more. The Green function, appearing in this equation, obeys the defining relation (1.56). While the Green function was taken to be an outgoing spherical wave emanating from the point Q, this is not the only choice of Green function that can be made. Indeed, by choosing appropriate alternative Green functions that solve eqn (1.56), one can eliminate either the first or the second term, which appears in the integrand of the Kirchhoff diffraction formula. The resulting pair of diffraction formulae, respectively known as the Rayleigh–Sommerfeld diffraction integrals of the first and second kind, allow one to solve the diffraction problem given knowledge of either the field, or its normal derivative, over the plane $z = 0$. In this sub-section, we will show how the Kirchhoff diffraction integral can be transformed into the Rayleigh–Sommerfeld diffraction integral of the first kind, and the corresponding Rayleigh–Sommerfeld diffraction integral of the second kind.

Consider Fig. 1.4. Here we see the usual observation point Q, in the vacuum-filled space downstream of the plane $z = 0$ from which we wish to propagate a given coherent scalar field. Denote the mirror image of Q by Q^*, this being obtained by reflecting Q through the plane $z = 0$. The coordinate vectors of Q and Q^* are respectively denoted by $\mathbf{x}_Q$ and $\mathbf{x}_Q^*$, with the point A in the plane $z = 0$ specified by the vector $\mathbf{x}$.

Previously, we took the Green function G_ω^Q to be an outgoing spherical wave emanating from the point Q. In the present more explicit notation, this may be written as $G_\omega^Q(\mathbf{x}) = \exp(ik|\mathbf{x} - \mathbf{x}_Q|)/|\mathbf{x} - \mathbf{x}_Q|$, where $\mathbf{x}$ is any point in the space $z > 0$ (note that we will later restrict this point to lie in the plane $z = 0$, as indicated in Fig. 1.4.). Suppose, instead, that one considers a Green function $G_\omega^{Q,\mathrm{D}}$, formed by subtracting an outgoing spherical wave emanating from the point Q^*, from the outgoing spherical wave emanating from Q[18]:

$$G_\omega^{Q,\mathrm{D}}(\mathbf{x}) = \frac{\exp(ik|\mathbf{x} - \mathbf{x}_Q|)}{|\mathbf{x} - \mathbf{x}_Q|} - \frac{\exp(ik|\mathbf{x} - \mathbf{x}_Q^*|)}{|\mathbf{x} - \mathbf{x}_Q^*|}. \tag{1.61}$$

[18]Note that the 'D' superscript indicates that this is a so-called Dirichlet Green function, a terminology which will be justified once we see that its use serves to eliminate the dependence of the Kirchhoff diffraction integral upon the normal derivative of the unpropagated field. One will then have a solution to the so-called Dirichlet boundary-value problem, in which one solves a partial differential equation given knowledge of the value of the function of interest over a specified boundary. This may be contrasted to the Neumann boundary-value problem, in which one solves a partial differential equation given knowledge of the normal derivative of the function over a specified boundary.

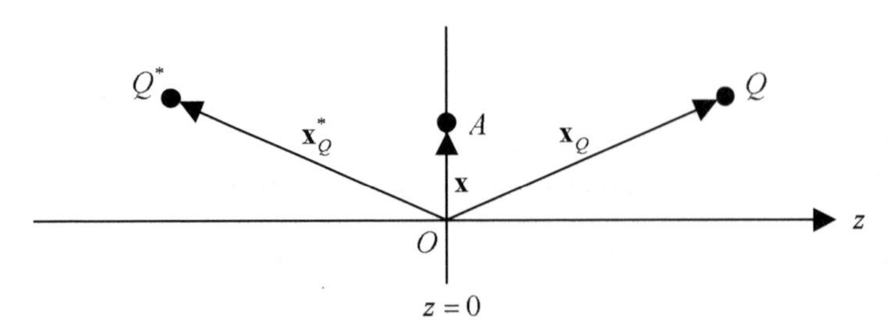

Fig. 1.4. The point A, which lies in the plane $z = 0$, has coordinate vector $\mathbf{x}$ with respect to the origin of coordinates O. The point Q, with coordinate vector $\mathbf{x}_Q$, lies in the half-space $z > 0$. Its mirror image, reflected through the plane $z = 0$, is denoted by Q^*, with corresponding coordinate vector $\mathbf{x}_Q^*$. This construction will be used in the derivation, based on the method of images, of the Dirichlet Green function which allows one to transform the Kirchhoff diffraction integral into the Rayleigh–Sommerfeld diffraction integral of the first kind.

In the half-space $z > 0$, the above Green function obeys the requisite equation:

$$(\nabla^2 + k^2) G_\omega^{Q,\mathrm{D}}(\mathbf{x}) = -4\pi\delta(\mathbf{x} - \mathbf{x}_Q). \tag{1.62}$$

Moreover, given the symmetry inherent in the method of images by which it was constructed, it is clear that $G_\omega^{Q,\mathrm{D}}(\mathbf{x})$ vanishes at any point A in the plane $z = 0$. Therefore, if one makes use of the Green function $G_\omega^{Q,\mathrm{D}}(\mathbf{x})$ in the Kirchhoff diffraction formula (1.60), its dependence on the normal derivative of the field evidently vanishes, leaving:

$$\psi_\omega(\mathbf{x}_Q) = \frac{1}{4\pi} \iint_{-\infty}^{\infty} \psi_\omega(x, y, z = 0) \times \frac{\partial}{\partial z} \left[\frac{\exp(ik|\mathbf{x} - \mathbf{x}_Q|)}{|\mathbf{x} - \mathbf{x}_Q|} - \frac{\exp(ik|\mathbf{x} - \mathbf{x}_Q^*|)}{|\mathbf{x} - \mathbf{x}_Q^*|} \right] dxdy. \tag{1.63}$$

To simplify this formula, we again appeal to the symmetry of the situation sketched in Fig. 1.4. At any point A on the mirror plane $z = 0$, it is evident that the z derivative, of a spherical wave emanating from the point Q^*, will be equal to the negative of the z derivative of a spherical wave emanating from the point Q. Therefore, in the square brackets which appear in eqn (1.63), we may delete the second term and then double the resulting expression, to arrive at the Rayleigh–Sommerfeld diffraction integral of the first kind:

$$\psi_\omega(\mathbf{x}_Q) = \frac{1}{2\pi} \iint_{-\infty}^{\infty} \psi_\omega(x, y, z = 0) \frac{\partial}{\partial z} \frac{\exp(ik|\mathbf{x} - \mathbf{x}_Q|)}{|\mathbf{x} - \mathbf{x}_Q|} dxdy. \tag{1.64}$$

This diffraction integral expresses the value of the propagated complex disturbance in the space $z > 0$, given its boundary value over the plane $z = 0$.

One can obtain the Rayleigh–Sommerfeld diffraction integral of the second kind, which allows one to determine the propagated field from knowledge of its normal derivative over the plane $z = 0$, using a variation on the chain of argument presented above. One uses the method of images, based on the construction in Fig. 1.4, to construct the Neumann Green function $G_{\omega}^{Q,\mathrm{N}}$. This is formed by adding an outgoing spherical wave emanating from the point Q^* to the outgoing spherical wave emanating from Q:

$$G_{\omega}^{Q,\mathrm{N}}(\mathbf{x}) = \frac{\exp(ik|\mathbf{x} - \mathbf{x}_Q|)}{|\mathbf{x} - \mathbf{x}_Q|} + \frac{\exp(ik|\mathbf{x} - \mathbf{x}_Q^*|)}{|\mathbf{x} - \mathbf{x}_Q^*|}. \tag{1.65}$$

Again, the symmetry of this scenario makes it evident that, over the plane $z = 0$, the normal derivative of the Neumann Green function vanishes. Making use of this Neumann Green function in the Kirchhoff diffraction formula (1.60), one obtains:

$$\psi_{\omega}(\mathbf{x}_Q) = -\frac{1}{4\pi} \iint_{\infty}^{\infty} \left[\frac{\exp(ik|\mathbf{x} - \mathbf{x}_Q|)}{|\mathbf{x} - \mathbf{x}_Q|} + \frac{\exp(ik|\mathbf{x} - \mathbf{x}_Q^*|)}{|\mathbf{x} - \mathbf{x}_Q^*|} \right] \\ \times \left[\frac{\partial \psi_{\omega}(x, y, z)}{\partial z} \right]_{z=0} dx dy. \tag{1.66}$$

Invoking mirror symmetry once more, we deduce the equality of each of the two terms that appear in the first pair of square brackets in the above expression. Thus, we can delete the spherical wave emanating from the point Q^*, and then double the resulting expression, to arrive at the Rayleigh–Sommerfeld diffraction integral of the second kind:

$$\psi_{\omega}(\mathbf{x}_Q) = -\frac{1}{2\pi} \iint_{-\infty}^{\infty} \frac{\exp(ik|\mathbf{x} - \mathbf{x}_Q|)}{|\mathbf{x} - \mathbf{x}_Q|} \left[\frac{\partial \psi_{\omega}(x, y, z)}{\partial z} \right]_{z=0} dx dy. \tag{1.67}$$

This diffraction integral expresses the value of the propagated complex disturbance in the space $z > 0$, given the z-derivative of its boundary value over the plane $z = 0$.

We close this discussion with two remarks: (i) The Kirchhoff diffraction formula may be regained once more, by taking the average of the Rayleigh–Sommerfeld diffraction integrals of the first and second kinds. (ii) By making use of the Weyl expansion for a spherical wave, which is discussed in the next chapter (see eqn (2.83)), one can readily demonstrate the equivalence of the Rayleigh–Sommerfeld diffraction integral of the first kind, and the angular-spectrum formalism outlined earlier in this chapter (Section 1.3).

1.7 Partially coherent fields

For most of this chapter, we have concentrated on the free-space propagation of strictly monochromatic scalar electromagnetic fields. While this idealization is sufficient for an analysis of many scenarios in coherent X-ray optics, as shall be evident from the numerous applications of these ideas in later chapters, there are a number of circumstances in which we shall need to invoke the more general theory of partially coherent fields. This theory explicitly takes into account the random character of the fields produced by realistic sources. Elements of this theory, for partially coherent fields, will occupy us for the remainder of this chapter.

The present section offers an introduction to some fundamental notions of partial coherence, laying the requisite conceptual groundwork prior to launching into a treatment of the more formal theory of partial coherence in subsequent sections. These introductory discussions are broken into four parts. First, we give a brief outline of some key notions regarding random processes. Second, we apply these notions to the end of providing a general overview of what a partially coherent field is, and how such fields can be considered as a more general case of both the monochromatic and polychromatic fields which have been encountered earlier in this chapter. We then give separate introductory discussions on the concepts of spatial and temporal coherence, in the respective contexts of the Young and Michelson interferometers, with the understanding that these concepts cannot, in general, be considered in isolation from one another.

1.7.1 *Random variables and random processes*

Consider a fair die, the six faces of which are labelled with the integers one through six. Suppose that one were to consider a hypothetical experiment in which the die was thrown an infinite number of times, leading to an infinite ordered sequences of integers, each one of which was randomly chosen from the set $\{1, 2, 3, 4, 5, 6\}$ with equal probability. The probability, of achieving a particular outcome when the die is thrown, is termed the probability distribution governing the random variable. Any particular infinite sequence of the resulting integers, obtained under the conditions listed above, would be spoken of as a realization of this random variable. For later convenience, we assume that the die is thrown once per second, so that the random sequence of obtained numbers may be considered as a sequence in time.

Evidently, there are infinitely many realizations of this random variable. Some of these realizations may be exceptional, such as an infinite sequence of ones or an infinite sequence of twos, which serves to make the point that a given random variable will not, in general, be completely characterized by any one of its realizations. The set of all possible realizations, some of which will be exceptional in the sense specified above, is termed the ensemble of realizations for the random variable.

Suppose, now, that one wanted to determine the average number which appeared on the die. One would not, in general, be justified in simply taking the average value of the sequence of numbers obtained in a particular realization of the random variable, as there are exceptional sequences—such as the two exceptional sequences mentioned in the previous paragraph—which will not yield the 'obvious' mean value of $\frac{1}{6}(1+2+3+4+5+6)=3\frac{1}{2}$. Rather than averaging over a particular realization of our random variable in order to determine the mean value taken by the die, one must instead average over the ensemble of all possible realizations of the random variable. To make this notion of ensemble average more precise, let the mth realization of the random variable be denoted by the infinite sequence of numbers $\{\cdots, a_{-2}^{(m)}, a_{-1}^{(m)}, a_0^{(m)}, a_1^{(m)}, a_2^{(m)}, \cdots\}$. The ensemble average, of the value appearing on the die, would then be given by $\lim_{M\to\infty}[M^{-1}\sum_{m=1}^{M} a_b^{(m)}]$, where b is any fixed integer which specifies a particular location in the random sequence.[19] Further, since the die is fair, the outcome of any one throw will be statistically independent of any other throw. Thus, there will be no correlation between a given outcome at a particular location in the random sequence, and the outcome at some other point in the same sequence. Lastly, note that the probability distribution, governing the outcome at a particular location in a particular realization of the random sequence, is independent of the location of that outcome within the sequence. Stated differently, the statistics of the outcome are independent of the time at which the die is thrown.

We now consider a simple variation on the above random variable. Rather than considering an infinite sequence of throws of a fair die, let us instead consider an infinite sequence of throws of a pair of unfair dice. These unfair dice, one of which is green in colour and the other of which is gold in colour, will each be considered to have a small permanent magnet embedded within. Our modified random variable is as follows: (i) Throw the gold die onto a table, discarding the value which is obtained. (ii) With the gold die sitting on the table, throw the green die and record the number that appears on its uppermost face. (iii) Pick up the gold die, leaving the green die on the table, subsequently throwing the gold die and recording the number which appears on its uppermost face. (iv) Pick up the green die, leaving the gold die on the table, subsequently throwing the green die and recording the number which appears on its uppermost face. (v) Go to step iii, and repeat *ad infinitum*.

As was the case previously, the mth realization of this random variable is denoted by the infinite sequence of numbers $\{\cdots, a_{-2}^{(m)}, a_{-1}^{(m)}, a_0^{(m)}, a_1^{(m)}, a_2^{(m)}, \cdots\}$. Once again, there exists a time-independent probability distribution for the variable, which gives the probability of obtaining a particular number at a particular point in the random sequence. However, unlike the case of consecutive throws of a fair die, there will now exist a correlation between the outcomes of consecutive

[19] Here, b can correspond to any point on the sequence, because we have assumed that the statistics governing the random variable do not depend on when the die is thrown. If this were not the case, then the stated ensemble average would, in general, depend on the choice for b. Note also that we have implicitly assumed all members of the ensemble to carry an equal statistical weight. This assumption is readily dropped, if necessary, although there is no need to introduce this complication in the present context.

throws. This is due to the fact that the dice interact with one another, through the presence of the magnets embedded within them—the outcomes of consecutive throws are no longer statistically independent. As a corollary to this, we see that the probability distribution alone (for a single throw) is no longer sufficient to completely characterize the random variable, as it contains no information regarding the non-zero correlations which now exist between any pair of locations within a particular realization of the random sequence. Such correlation functions, between the outcomes at any two pairs of points in the sequence, are termed two-point correlation functions. Further, since the statistics of the random variable are assumed to be independent of the origin of time, the two-point correlation functions will only depend on the difference in the times corresponding to the two points in the sequence, between which the correlations are to be determined. Also, on account of the fact that exceptional sequences exist, both the probability distribution and the two-point correlation functions may be obtained via a suitable ensemble average, but not by averaging over a particular realization of the random sequence.

The modified random variable, involving magnetically coupled dice, is not completely characterized by specifying both the probability distribution and all two-point correlation functions. Indeed, this pair of quantities gives no information regarding any three-point correlations which may exist. Pushing this argument further, we state that one needs to provide the probability distribution, together with an infinite hierarchy of n-point correlation functions ($n = 2, 3, \cdots$), in order to completely specify the random variable generated by our pair of magnetically coupled dice. This infinite hierarchy of correlation functions, together with the probability distribution (which may be regarded as a one-point correlation function), may in principle be obtained via suitably averaging over the ensemble of all possible realizations of the random variable.

We stated, earlier, that the statistics of the random variable are assumed to be independent of the origin of time. Thus, the probability distribution is independent of location in the sequence. Similarly, the two-point correlation functions will depend only on the relative position of two points in the sequence. Analogous statements hold for the higher-order correlation functions. When one says that the statistics of the random variable are independent of the origin of time, this amounts to the rather strong requirement that all correlation functions are invariant under a shift in the origin of time, this origin of time taken as corresponding to the zeroth member of the random sequence. A random variable, that meets these strong requirements, is said to be statistically stationary. As a weaker form of statistical stationarity, one refers to a variable as being statistically stationary in the wide sense, if its first- and second-order correlation functions are invariant with respect to a shift in the origin of time.

Rather than having a discrete random process that generates random sequences of numbers, which are labelled by a discrete integer index, one can have a continuous random process that generates random sequences of numbers, which are labelled by a real continuous index. Such continuous random processes, also known as stochastic processes, will occupy our attention for the remainder of this sub-section.

Many of the notions, introduced in the context of discrete random processes, may be directly carried over to the case of continuous random processes: the ensemble of all possible realizations, the existence of exceptional realizations of the process together with the associated need for ensemble averaging, statistical stationarity in the strict and wide senses, the existence of correlations within each realization of the process, and the need to specify correlation functions of all orders in order to completely characterize the random process.

To this list let us add the concept of ergodicity. A random process is said to be ergodic if all ensemble averages can be replaced by their corresponding time averages. For an ergodic field, exceptional realizations do exist, but with vanishingly small probability. All information regarding the ergodic random process can then be considered to be contained within any single realization of the process. Evidently, the set of all ergodic stochastic processes is a subset of the set of all statistically stationary processes, with the set of all statistically stationary processes being a subset of the set of all stochastic processes which are statistically stationary in the wide sense.

1.7.2 *Intermediate states of coherence*

Our motivation, for the above introduction of certain core ideas regarding the theory of stochastic processes, lies in the fact that a realistic X-ray field is an example of such a process. One speaks of such stochastic fields as being partially coherent, with the case of monochromatic or polychromatic fields forming a special case in which the ensemble of all realizations contains but a single member.

Here we restrict ourselves to a scalar theory for such realistic disturbances. Consider, then, the complex X-ray scalar disturbance $\Psi(\mathbf{x}, t)$, which is generated within a specified volume of points in free space that have coordinate vectors $\mathbf{x}$, the field being due to a given realistic X-ray source together with its interaction with any optical elements and/or samples, which lie between the source and the points of observation. A given realization of the experiment will yield a specified complex function $\Psi(\mathbf{x}, t)$, which will, in general, exhibit random fluctuations in both space and time. This randomness is due to the inherently probabilistic nature of the quantum and thermal processes that govern the emission of X-ray radiation by the sources of the field, together with the probabilistic nature of the interactions of this field with any optical elements or samples which may be present. A different instance, of the same experiment, will, in general, yield a different realization of the underlying stochastic process.

Let us assume this process to be statistically stationary, in the wide sense. It will then be partially characterized by the totality of all one-point and two-point correlation functions, with the one-point correlation functions being evaluated via appropriate ensemble averaging at all given space–time points $(\mathbf{x}, t)$, and the two-point correlation functions being evaluated between any two space–time points $(\mathbf{x}_1, t_1)$ and $(\mathbf{x}_2, t_2)$. Since the process is assumed to be stationary in the wide sense, the one-point correlation functions are independent of t, with the two-point correlation functions depending only on the time difference $t_2 - t_1 \equiv \tau$. Note that the one-point correlation functions, that is, the field averages, are often of no interest on account of the fact that the average value of the complex disturbance will be zero in most cases of practical interest. One can also form higher-order correlation functions between more than two space–time points; if the process is statistically stationary in the strict sense, then these functions will be unchanged under any shift in the origin of time. Further, if the partially coherent field is not stationary in any sense, as will be the case for pulsed X-ray sources such as free-electron lasers, then all correlation functions will, in general, depend upon the origin of time.

An important special class, of partially coherent disturbance, is given by the so-called 'quasi-monochromatic' field. The temporal Fourier transform, of any particular realization of such a field which is truncated to a sufficiently long but not infinite time, will be non-negligible only over a range of angular frequencies that is very small compared to the mean angular frequency of the disturbance. From a more intuitive point of view, a given realization of a quasi-monochromatic field is one that has a very narrow spectral range. Notwithstanding this narrow spectral range, the behaviour of quasi-monochromatic radiation is, in general, very different from that of strictly monochromatic radiation. As an important example of this, the ability of quasi-monochromatic radiation, to form interference fringes, may be very different from that of a corresponding strictly monochromatic field. This notion will be of some importance in the discussions of the following two sub-sections, where we separately treat the concepts of spatial and temporal coherence for quasi-monochromatic fields from extended sources.

1.7.3 *Spatial coherence*

The Young interferometer, sketched in Fig. 1.5, provides a powerful archetype by which much of the theory of partial coherence is ultimately inspired. In the present section, we use this interferometer to introduce the notion of the spatial coherence properties of a partially coherent field. As a fringe benefit of these discussions, we shall see some hints of a theory for the enhancement of coherence by the act of free-space propagation. A more complete formulation of this appealing idea is reserved for later in the chapter. Note also that the account of this sub-section follows a rather orthodox line of development, which is given in many texts on modern optics, such as Lipson and Lipson (1981) and Hecht (1987).

We work in two spatial dimensions for simplicity, considering an extended incoherent uniform quasi-monochromatic line source S, which is perpendicularly bisected by an optic axis z (see Fig. 1.5). The length of the line source is taken to be a, with the coordinate transverse to the optic axis labelled by x. The incoherent nature of the source amounts to the assumption that the field, at any point on the source, is completely uncorrelated with the field at any other distinct point

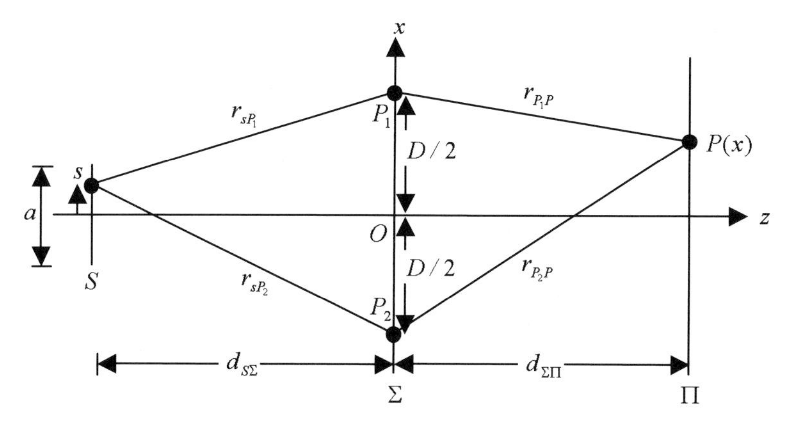

Fig. 1.5. A thin one-dimensional black screen Σ is pierced by two small pinholes P_1 and P_2, which are equidistant from the optic axis z and separated by a distance D. This pierced screen is illuminated by an extended incoherent quasi-monochromatic line source S of diameter a, with the source being perpendicularly bisected by the optic axis. The distance from the source S to the pierced screen Σ is denoted by $d_{S\Sigma}$, with $d_{\Sigma\Pi}$ denoting the distance from the pierced screen to the plane of observation Π. A point s on this source is indicated, located at coordinates $(x, z) = (x', -d_{S\Sigma})$, with the distances from this point to P_1 and P_2 being respectively denoted by r_{sP_1} and r_{sP_2}. The distances, from P_1 and P_2 to the observation point $P(x)$ at $(x, z) = (x, d_{\Sigma\Pi})$, are respectively denoted by r_{P_1P} and r_{P_2P}. We shall determine the visibility of the Young interference fringes which are formed over the observation plane Π, as a function of the size of the incoherent source.

on the source.[20] This source illuminates a thin black one-dimensional screen Σ, which is punctured with two small pinholes P_1 and P_2 that are a distance D apart, and symmetrically placed with respect to the optic axis. The resulting interference pattern is observed over the screen Π, at an observation point P which is a function of the transverse position x across the screen. The distances, from a point s on the source to the first and second pinholes, are respectively denoted by r_{sP_1} and r_{sP_2}; similarly, r_{P_1P} and r_{P_2P} respectively denote the distances from the first and second pinholes, to the observation point P. Lastly, $d_{S\Sigma}$ denotes the distance from the source to the punctured screen, with $d_{\Sigma\Pi}$ denoting the distance from the punctured screen to the screen of observation.

[20]Note that this assumption, of the absence of correlations between disturbances at two distinct space–time points—both of whose spatial coordinates correspond to distinct points on the source—does not imply the absence of correlations between disturbances at distinct space–time points, both of whose spatial coordinates do not coincide with the location of the source. Indeed, as we shall see, the presence of the latter correlations is necessary for the presence of interference fringes over the observation screen of the Young interferometer.

Consider a point $s(x')$ on the source, where x' is the x coordinate of s. Let $\eta(x', x)$ be the difference of two path lengths, with the first path length being that from $s(x')$ to the observation point $P(x)$, which is obtained by passing through the first pinhole, and the second path length being that from $s(x')$ to the observation point $P(x)$, which is obtained by passing through the second pinhole. Thus:

$$\begin{aligned}\eta(x', x) &= r_{sP_2} + r_{P_2P} - r_{sP_1} - r_{P_1P} \\ &= \sqrt{d_{s\Sigma}^2 + \left(\frac{D}{2} + x'\right)^2} - \sqrt{d_{s\Sigma}^2 + \left(\frac{D}{2} - x'\right)^2} \\ &+ \sqrt{d_{\Sigma\Pi}^2 + \left(\frac{D}{2} + x\right)^2} - \sqrt{d_{\Sigma\Pi}^2 + \left(\frac{D}{2} - x\right)^2}.\end{aligned} \tag{1.68}$$

Assume that D, x', and x are all much smaller, in magnitude, than both the distance from the source to the pinholes, and the distance from the pinholes to the observation screen. One may therefore make a first-order binomial approximation to the square roots in the above expression, yielding:

$$\eta(x', x) \approx \frac{x'D}{d_{s\Sigma}} + \frac{Dx}{d_{\Sigma\Pi}}. \tag{1.69}$$

To proceed further, we convert this expression for path difference $\eta(x', x)$ into one for the phase difference $\Delta\phi(x', x)$ between radiation which traverses the path from s to P_1 to P, and the radiation which traverses the path from s to P_2 to P. This phase difference will be equal to 2π multiplied by the number of wavelengths which are contained in the path difference $\eta(x', x)$. Denoting the mean wavelength of the quasi-monochromatic radiation by $\overline{\lambda}$, we see that:

$$\Delta\phi(x', x) = \frac{2\pi\eta(x', x)}{\overline{\lambda}} \approx \frac{2\pi D}{\overline{\lambda}}\left(\frac{x'}{d_{s\Sigma}} + \frac{x}{d_{\Sigma\Pi}}\right). \tag{1.70}$$

The intensity of the interference fringes, due to the point radiator located at $s(x')$ on the source, as a function of position x on the screen, is then given by[21]:

[21]Here, we give a brief justification for this expression. Suppose the respective complex amplitudes at a given point x on the observation screen, which are due to the disturbances emanating from each of the pinholes as a result of them being illuminated by a single point $s(x')$ on the source, to be given by $\psi(x)$ and $\psi(x)\exp[i\Delta\phi(x', x)]$. Here, we have implicitly assumed that the amplitudes of the two contributions are equal, as will be the case if the pinholes are sufficiently close to the optic axis, and are of the same size and shape. The intensity of the resulting disturbance is then equal to $|\psi(x) + \psi(x)\exp[i\Delta\phi(x', x)]|^2 = 2I_0[1 + \cos\Delta\phi(x', x)]$, where $I_0 \equiv |\psi|^2$.

$$I(x', x) = 2I_0 \left\{ 1 + \cos\left[\frac{2\pi D}{\overline{\lambda}} \left(\frac{x'}{d_{s\Sigma}} + \frac{x}{d_{\Sigma\Pi}} \right) \right] \right\}. \tag{1.71}$$

Here, I_0 is the intensity which would have been measured at x if either of the pinholes had been blocked off. Note that the quality of the fringes is maximal, in the sense that the minima of intensity correspond to complete darkness. Note also that, to within the binomial approximation upon which this expression is predicated (see eqn (1.69)), a transverse displacement of the radiator (i.e. a change in x' from one point on the source to a different point on the source) will lead to a proportional transverse displacement in the fringes over the observation screen that are produced by that radiator.

Having obtained an expression for the interference fringes produced by one of the radiators in the extended source, we are ready to calculate an approximate expression for the interference pattern which is produced by the entire extended incoherent uniform quasi-monochromatic line source S. The assumption, that the source is incoherent, amounts to the assumption that one can simply add the intensities due to each radiator in the source, in order to obtain the intensity of the interference pattern produced by the whole source. Since the source is taken to be uniform over its length, the number of radiators per unit length is independent of position on the source; each of these radiators is taken to radiate with the same power. Further, bearing in mind the last sentence of the previous paragraph, the fringes produced by different points on the source will not, in general, have coinciding peaks and troughs. One might therefore expect the quality of the fringes to be degraded as the size of the incoherent line source is increased. As we shall see, this expectation is not quite true, for reasons that will become clearer shortly. For the moment, let us integrate over each of the radiators in the source, to obtain the following expression for the interference pattern $I(x)$ due to the entire extended source:

$$\begin{aligned} I(x) &= \int_{x'=-a/2}^{x'=a/2} I(x', x) dx' \\ &= 2I_0 \int_{x'=-a/2}^{x'=a/2} \left\{ 1 + \cos\left[\frac{2\pi D}{\overline{\lambda}} \left(\frac{x'}{d_{s\Sigma}} + \frac{x}{d_{\Sigma\Pi}} \right) \right] \right\} dx'. \end{aligned} \tag{1.72}$$

Evaluating the above expression, by first making use of the expansion formula for the cosine of the sum of two quantities, and then computing the resulting integral, we arrive at:

$$I(x) = 2I_0 a + \frac{2I_0 \overline{\lambda} d_{S\Sigma}}{\pi D} \sin\left(\frac{a\pi D}{\overline{\lambda} d_{S\Sigma}} \right) \cos\left(\frac{2\pi D x}{\overline{\lambda} d_{\Sigma\Pi}} \right). \tag{1.73}$$

Unlike the case of the interference pattern produced by a single radiator in the source, the quality of the interference pattern is now no longer necessarily maximal, as the minima of the interference pattern are now no longer completely dark.

Qualitatively, this is to be expected, owing to the non-coincidence, in general, of the maxima and minima of the interference fringes produced by independent radiators in the source.

To render more precise the notion of the 'quality' of the interference fringes, we invoke Michelson's definition of fringe visibility $\mathscr{V}$:

$$\mathscr{V} \equiv \frac{I_{\mathrm{MAX}} - I_{\mathrm{MIN}}}{I_{\mathrm{MAX}} + I_{\mathrm{MIN}}}. \tag{1.74}$$

Here, I_{MAX} and I_{MIN} respectively denote the maximum and minimum intensities of the fringes in a given area of the interference pattern.[22] Evidently, when the intensity minima are equal to zero and the intensity maxima are non-zero, the visibility is maximal and equal to unity. At the opposite extreme, if the intensity maxima and minima have the same value, then there are no fringes, with a corresponding visibility of zero. The former limit corresponds to complete coherence of the radiation at the location of the two pinholes, with the latter limit corresponding to complete incoherence between the radiation at the two pinholes. Intermediate states of fringe visibility may then be taken as indicating partial coherence between the disturbances at the location of each of the two pinholes. We shall return to this important point a little later in the present sub-section.

Returning to the main thread of the argument, let us calculate the fringe visibility for the Young interferometer sketched in Fig. 1.5. From eqn (1.73), we see that I_{MAX} and I_{MIN} are respectively given by:

$$\begin{aligned} I_{\mathrm{MAX}} &= 2I_0 a + \frac{2I_0\bar{\lambda} d_{S\Sigma}}{\pi D}\left|\sin\left(\frac{a\pi D}{\bar{\lambda} d_{S\Sigma}}\right)\right|, \\ I_{\mathrm{MIN}} &= 2I_0 a - \frac{2I_0\bar{\lambda} d_{S\Sigma}}{\pi D}\left|\sin\left(\frac{a\pi D}{\bar{\lambda} d_{S\Sigma}}\right)\right|. \end{aligned} \tag{1.75}$$

Substituting into expression (1.74) for the fringe visibility $\mathscr{V}$, we see that:

$$\mathscr{V} = \frac{|\sin \varrho|}{\varrho}, \qquad \varrho \equiv \frac{a\pi D}{\bar{\lambda} d_{S\Sigma}}. \tag{1.76}$$

A plot of this expression, for the fringe visibility $\mathscr{V}$ versus the dimensionless parameter ϱ, is given in Fig. 1.6.

There is an immense amount of physics to be learned from the above result, some of which we now detail. The visibility $\mathscr{V}$ is seen to be maximal only when $\varrho = 0$. This will be the case, for example, if the size a of the source is taken to zero. Keeping fixed all variables in the expression for ϱ, with the exception of a, we see that an increase in a from zero leads to a progressive decrease in the visibility of the interference fringes, up until the point where $\varrho = \pi$. More precisely, in the

[22] Implicit, in this definition, is the assumption that the fringes vary over a length scale that is sufficiently smaller than the envelope which multiplies these fringes.

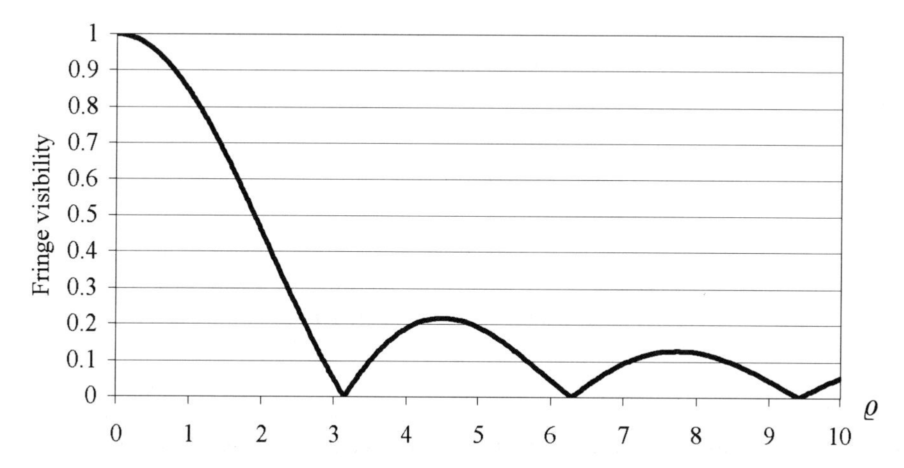

Fig. 1.6. Visibility $\mathscr{V}$ of fringes for the Young interferometer sketched in Fig. 1.5, as a function of the dimensionless parameter ϱ given in eqn (1.76).

regime of source sizes $0 \leq a \leq \overline{\lambda} d_{S\Sigma}/D$, the visibility monotonically decreases from its maximal value of unity, to its minimal value of zero. As the source size increases from $a = 0$ across this interval, the progressive degradation in visibility may be pictured as due to the blurring of the resulting interference fringes, which result from the superposition of the laterally displaced fringes that are formed by each of the independent radiators in the source. At the critical source size where $\varrho = \pi$, this cancellation is complete and all fringes are 'washed out'. However, as the source size is increased beyond this critical value, the fringes are seen to re-appear, with the previous maxima being interchanged with minima and vice-versa.[23] For $\pi \leq \varrho \leq 2\pi$, the visibility of these inverted fringes rises from zero up to a maximum value which is less than unity, and then diminishes to zero once more. This process continues, in an analogous manner, as ϱ is increased further.

This scenario, of progressively increasing the source size while observing the effects of this increase on the visibility of the fringes produced by a Young interferometer, gives a convenient entry point to the notion of spatial coherence. When the source size is taken to be vanishingly small, the resulting disturbance at the location of the two pinholes is said to be 'spatially coherent', on account of the fact that the interference of the fields from these spatially separated points is able to yield fringes with maximal visibility. As the source size is increased from zero, such that $\varrho \leq \pi$, there is a corresponding decrease in the spatial coherence of the radiation at the location of the two pinholes. When the fringe visibility

[23]This reversal is evident from the sine term in eqn (1.73). Note, also, that the disappearance of the fringes at $\varrho = \pi$ allows one to determine the angular size $\Delta\theta = a/d_{S\Sigma}$ of the source given both the mean wavelength $\overline{\lambda}$ of the radiation and the separation D between the pinholes: $\varrho = \pi$ implies that $\Delta\theta = a/d_{S\Sigma} = \overline{\lambda}/D$. This observation forms the basis of Michelson's stellar interferometer, used to determine the angular diameter of stars.

is zero, the disturbances at the two spatially separated pinholes are said to be spatially incoherent with one another. Intermediate states of visibility then correspond to intermediate ('partial') states of spatial coherence. Historically, the use of such a real quantity (visibility), as a measure of the degree of coherence of two points in a field, was introduced by Zernike (1938).

Within the regime $\varrho < \pi$, we have seen that an increase in the size of the source a leads to a corresponding decrease in the spatial coherence of the radiation over the two pinholes of the Young interferometer (see Fig. 1.5). Instead, one can consider a to be fixed, along with the mean radiation wavelength $\overline{\lambda}$ and the slit separation D, so as to investigate the effect of changing the distance $d_{S\Sigma}$ from the source to the screen containing the pinholes. In this case, and once again restricting ourselves to the regime where $\varrho < \pi$, we see that an increase in $d_{S\Sigma}$ leads to an increase in the spatial coherence of the radiation at the location of the two pinholes. Further, we see that the spatial coherence, of the radiation at the two pinholes, can be made maximally large in the limit as the source becomes infinitely distant. While this observation may initially appear mundane, there is really something rather profound going on here—the coherence properties of the radiation, at the location of the two pinholes, is altered by the process of propagation from the source to the pinholes. Stated differently, the spatial coherence properties of the radiation are altered upon free-space propagation. This point will be expanded upon in Section 1.9. For the moment, let us now shift our attention from the notion of spatial coherence, to that of temporal coherence.

1.7.4 *Temporal coherence*

In the preceding section we examined the visibility of interference fringes produced by the superposition of radiation from two spatially separated points in a given quasi-monochromatic field. This fringe visibility was taken as a measure of the degree of spatial coherence of the field at the two given points. The Young interferometer, which divides an incident wavefront into two components before recombining them to yield an interference pattern, was seen to yield data that could be used to determine the spatial coherence properties of the field at a given pair of points in the plane of the interferometer.

A parallel presentation exists, which considers temporal rather than spatial coherence. Instead of considering the ability of the field at spatially separated points to interfere, one may consider interference between the field at temporally separated points. Rather than using a division-of-wavefront interferometer such as the Young pinholes, a division-of-amplitude interferometer, such as the Michelson interferometer, can be considered. One can then examine the visibility of the interference fringes formed when combining the quasi-monochromatic radiation from a given point in space at a given time t, with the radiation from the same point in space at a later time $t+\tau$. In turn, this leads directly to a measure for the degree of temporal coherence exhibited by a given wide-sense-stationary quasi-monochromatic field, with the so-called coherence time being the time-lag τ which is required in order for the visibility of the time-averaged fringes to be significantly reduced from the maximal value which is obtained at $\tau = 0$.

Rather than further pursuing the notion of temporal coherence via an analysis of the visibility of the interference fringes, which are generated by division-of-amplitude interferometry, we pick up on a generalization that demands to be made. Specifically, the properties of both spatial and temporal coherence have been discussed in terms of the ability of two parts of a given wave-field to interfere. In the former case, we considered the visibility of the interference fringes formed when disturbances from spatially separated points of the field are superposed. In the latter case, for which only the briefest of outlines was given, temporal rather than spatial separation was considered. This can evidently be generalized, to consider the interference of a partially coherent field which exists at two points that are separated in both space and time. It is to such a generalized treatment, of the theory of partially coherent fields, that we now turn.

1.8 The mutual coherence function

We again return to the Young double slit experiment, in order to arrive at a more sophisticated theory of partial coherence which generalizes the treatment given earlier. Rather than following Zernike (1938) in considering the fringe visibility $\mathscr{V}$ as a measure of the degree of coherence, we shall instead consider a closely related complex degree of coherence known as the 'mutual coherence function'. The results of this section were reported in a classic series of papers by Wolf, which laid the theoretical foundations for the modern theory of partially coherent fields (Wolf 1954a,b, 1955).

Consider the Young double slit experiment sketched in Fig. 1.7. Upstream of a planar screen Σ, we see a series of sources radiating statistically stationary quasi-monochromatic radiation of mean wavelength $\overline{\lambda}$. The screen is pierced by two small pinholes P_1 and P_2, respectively located at the position vectors $\mathbf{x}_{P_1}$ and $\mathbf{x}_{P_2}$, with $\mathbf{D} \equiv \mathbf{x}_{P_2} - \mathbf{x}_{P_1}$ denoting the relative position vector of the second pinhole with respect to the first, and $D \equiv |\mathbf{D}|$ denoting the separation of the pinholes. The resulting interference fringes are to be sampled at an observation point P which lies in the vacuum-filled half-space downstream of the screen containing the pinholes, with r_{P_1P} and r_{P_2P} respectively denoting the distance from the first and second pinholes to the point P, which is located at the position vector $\mathbf{x}_P$.

Consider a given realization $\Psi(\mathbf{x}, t)$ of the quasi-monochromatic wave-field incident on the screen, as a function of both position vector $\mathbf{x}$ and time t. Such a field can be Fourier decomposed[24] into a superposition of monochromatic fields $\psi_\omega(\mathbf{x}) \exp(-i\omega t)$, the angular frequencies ω of which all lie within a narrow range of the mean angular frequency $\overline{\omega}$. Assuming that all non-zero monochromatic

[24]More properly, we should truncate the field to be non-zero only over a given finite time interval. This avoids mathematical difficulties due to the fact that, if the field is assumed to exist for all time, it will not be square integrable, with the corollary that its temporal Fourier transform is not well defined. Such subtleties do not affect the conclusions arrived at here.

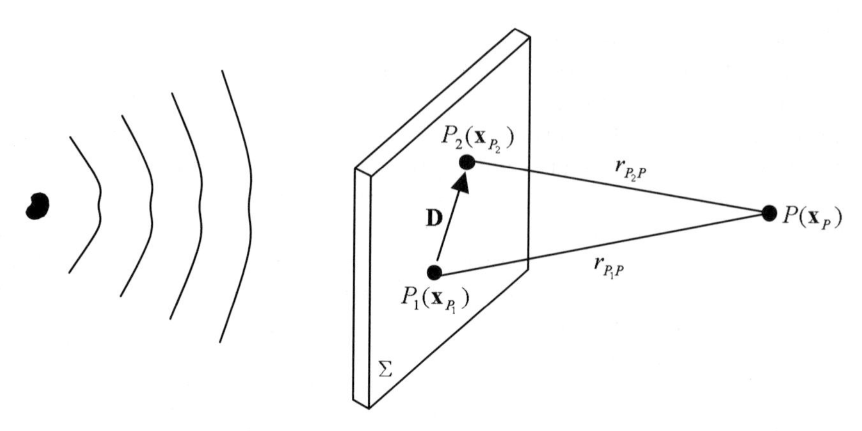

Fig. 1.7. A thin two-dimensional black screen Σ is pierced by two small pinholes P_1 and P_2, which are respectively located at position vectors $\mathbf{x}_{P_1}$ and $\mathbf{x}_{P_2}$ with respect to a given origin. The vector, pointing from P_1 to P_2, is denoted by $\mathbf{D}$. All quasi-monochromatic sources and optical elements are assumed to lie in the half-space to the left of the pierced screen. The visibility of the resulting Young's interference pattern is to be determined at the point P, with position vector $\mathbf{x}_P$, which lies in the vacuum-filled half-space to the right of the screen. The distances, from P_1 and P_2 to P, are respectively denoted by r_{P_1P} and r_{P_2P}.

components have angular frequencies ω, which are confined to the interval $\overline{\omega} - \frac{1}{2}\Delta\omega \leq \omega \leq \overline{\omega} + \frac{1}{2}\Delta\omega$, we may write down the following Fourier integral for the complex analytic signal (see Section 1.2) of the associated quasi-monochromatic disturbance:

$$\Psi(\mathbf{x}, t) = \frac{1}{\sqrt{2\pi}} \int_{\omega=\overline{\omega}-(\Delta\omega/2)}^{\omega=\overline{\omega}+(\Delta\omega/2)} \psi_\omega(\mathbf{x}) \exp(-i\omega t) d\omega. \tag{1.77}$$

If we now let $\omega = \overline{\omega} + \delta\omega$, with $\delta\omega$ indicating the difference of a given monochromatic component's angular frequency ω from the mean angular frequency $\overline{\omega}$ of the quasi-monochromatic field, the above expression becomes:

$$\begin{aligned}\Psi(\mathbf{x}, t) &= \exp(-i\overline{\omega}t)\mathscr{A}(\mathbf{x}, t),\\ \mathscr{A}(\mathbf{x}, t) &\equiv \frac{1}{\sqrt{2\pi}} \int_{\delta\omega=-\Delta\omega/2}^{\delta\omega=\Delta\omega/2} \psi_{\overline{\omega}+\delta\omega}(\mathbf{x}) \exp[-i(\delta\omega)t] d(\delta\omega).\end{aligned} \tag{1.78}$$

Since the integral, in this expression, is a superposition of monochromatic fields whose angular frequency $\delta\omega$ is small compared to the mean angular frequency $\overline{\omega}$ of the quasi-monochromatic field, $\mathscr{A}(\mathbf{x}, t)$ is evidently a complex envelope that varies slowly in comparison to the harmonic time factor $\exp(-i\overline{\omega}t)$, which it multiplies. At a given position $\mathbf{x}$ in space, this envelope will drift in both amplitude and phase, over timescales that are long in comparison to the mean period $2\pi/\overline{\omega}$ of the radiation.

We therefore expect the fields, diffracted by each of the two pinholes, to slowly vary in both relative phase and amplitude. The associated interference pattern will be time varying, with its time average corresponding to the intensity distribution which is measured at the observation point P. With a view to determining the time-averaged intensity at the observation point P, let us first write down the following expression for the time-dependent complex disturbance that exists at this point, due to the superposition of the two fields which respectively emanate from each of the pinholes:

$$\Psi(\mathbf{x}_P, t) = K_1 \Psi\left(\mathbf{x}_{P_1}, t - \frac{r_{P_1 P}}{c}\right) + K_2 \Psi\left(\mathbf{x}_{P_2}, t - \frac{r_{P_2 P}}{c}\right). \tag{1.79}$$

Here, $\Psi(\mathbf{x}_{P_1}, t) = \exp(-i\overline{\omega}t)\mathscr{A}(\mathbf{x}_{P_1}, t)$ and $\Psi(\mathbf{x}_{P_2}, t) = \exp(-i\overline{\omega}t)\mathscr{A}(\mathbf{x}_{P_2}, t)$ respectively denote the complex disturbance at the entrance surface of the first and second pinholes, as a function of time. The respective retardations, of the time arguments appearing in each of these expressions, is due to the finite speed c of light in vacuum: the disturbance at time t, which is measured at the point P due to pinhole 1, was emanated at a time $t - r_{P_1 P}/c$ from this pinhole, with an analogous time delay for the light from the second pinhole. Lastly, the complex numbers K_1 and K_2 are propagators that depend on the shape of the pinholes, and the angles between the rays incident upon and diffracted from the pinholes. We state without proof that these propagators are purely imaginary, so that we can write $K_1 = i\kappa_1$ and $K_2 = i\kappa_2$, with κ_1 and κ_2 being non-zero real numbers which have the same sign, so that $\kappa_1\kappa_2 > 0$ (Wolf 1954a).

By taking the squared modulus of eqn (1.79), and then forming the time average (as indicated by angular brackets) of the result, we arrive at the following expression for the time-averaged intensity $\overline{I}(\mathbf{x}_P)$ at the observation point P:

$$\begin{aligned} \overline{I}(\mathbf{x}_P) &= \left\langle \left| K_1 \Psi\left(\mathbf{x}_{P_1}, t - \frac{r_{P_1 P}}{c}\right) + K_2 \Psi\left(\mathbf{x}_{P_2}, t - \frac{r_{P_2 P}}{c}\right) \right|^2 \right\rangle \\ &= \left\langle \left| K_1 \Psi\left(\mathbf{x}_{P_1}, t - \frac{r_{P_1 P}}{c}\right) \right|^2 \right\rangle + \left\langle \left| K_2 \Psi\left(\mathbf{x}_{P_2}, t - \frac{r_{P_2 P}}{c}\right) \right|^2 \right\rangle \\ &\quad + 2\,\mathrm{Re} \left\langle K_1 K_2^* \Psi\left(\mathbf{x}_{P_1}, t - \frac{r_{P_1 P}}{c}\right) \Psi^*\left(\mathbf{x}_{P_2}, t - \frac{r_{P_2 P}}{c}\right) \right\rangle. \end{aligned} \tag{1.80}$$

Imagine, now, that the second pinhole were to be blocked off. The second and third terms on the right side of the above expression would then vanish, leading to the following equation for the time-averaged intensity $\overline{I}_1(\mathbf{x}_P)$ which would be measured at the observation point P, due to the field emanating from the first pinhole alone:

$$\overline{I}_1(\mathbf{x}_P) = \left\langle \left| K_1 \Psi\left(\mathbf{x}_{P_1}, t - \frac{r_{P_1 P}}{c}\right)\right|^2 \right\rangle. \tag{1.81}$$

Similarly, if the first pinhole were to be blocked off, one has the following expression for the time-averaged intensity $\overline{I}_2(\mathbf{x}_P)$, which would be measured at the observation point P, due to the field emanating from the second pinhole alone:

$$\overline{I}_2(\mathbf{x}_P) = \left\langle \left| K_2 \Psi\left(\mathbf{x}_{P_2}, t - \frac{r_{P_2 P}}{c}\right)\right|^2 \right\rangle. \tag{1.82}$$

Making use of the above two equations, together with the fact that $K_1 K_2^* = |K_1 K_2^*|$ is positive and real, we see that eqn (1.80) can be re-written as:

$$\begin{aligned} \overline{I}(\mathbf{x}_P) &= \overline{I}_1(\mathbf{x}_P) + \overline{I}_2(\mathbf{x}_P) \\ &\quad + 2|K_1 K_2^*| \mathrm{Re} \left\langle \Psi\left(\mathbf{x}_{P_1}, t - \frac{r_{P_1 P}}{c}\right) \Psi^*\left(\mathbf{x}_{P_2}, t - \frac{r_{P_2 P}}{c}\right)\right\rangle. \end{aligned} \tag{1.83}$$

To proceed further, we make use of the previously stated assumption that the field is statistically stationary, that is, that its statistical properties are independent of the origin of time. Under this assumption, the time average in eqn (1.83) is unchanged under the transformation $t \to t + c^{-1} r_{P_2 P}$, so that:

$$\begin{aligned} &\left\langle \Psi\left(\mathbf{x}_{P_1}, t - \frac{r_{P_1 P}}{c}\right) \Psi^*\left(\mathbf{x}_{P_2}, t - \frac{r_{P_2 P}}{c}\right)\right\rangle \\ &\quad = \left\langle \Psi\left(\mathbf{x}_{P_1}, t - \frac{r_{P_1 P} - r_{P_2 P}}{c}\right) \Psi^*\left(\mathbf{x}_{P_2}, t\right)\right\rangle. \end{aligned} \tag{1.84}$$

Introducing the time difference τ, between the travel time of light from the first and second pinholes to the observation point:

$$\tau = \frac{r_{P_2 P} - r_{P_1 P}}{c}, \tag{1.85}$$

we re-write eqn (1.84) as:

$$\left\langle \Psi\left(\mathbf{x}_{P_1}, t - \frac{r_{P_1 P}}{c}\right) \Psi^*\left(\mathbf{x}_{P_2}, t - \frac{r_{P_2 P}}{c}\right)\right\rangle = \left\langle \Psi\left(\mathbf{x}_{P_1}, t + \tau\right) \Psi^*\left(\mathbf{x}_{P_2}, t\right)\right\rangle. \tag{1.86}$$

Equation (1.80) therefore becomes:

$$\overline{I}(\mathbf{x}_P) = \overline{I}_1(\mathbf{x}_P) + \overline{I}_2(\mathbf{x}_P) + 2|K_1 K_2^*| \mathrm{Re}\left[\Gamma(\mathbf{x}_{P_1}, \mathbf{x}_{P_2}, \tau)\right], \tag{1.87}$$

where we have introduced the following definition for the so-called mutual coherence function of the field at the locations of the two pinholes:

$$\Gamma(\mathbf{x}_{P_1}, \mathbf{x}_{P_2}, \tau) \equiv \langle \Psi(\mathbf{x}_{P_1}, t+\tau)\, \Psi^*(\mathbf{x}_{P_2}, t) \rangle . \tag{1.88}$$

The third term, on the right-hand side of eqn (1.87), is evidently an interference term. This interference term is proportional to the real part of the mutual coherence function. Importantly, $\Gamma(\mathbf{x}_{P_1}, \mathbf{x}_{P_2}, \tau)$ is a measure of the correlation which exists between the disturbances at the exit surfaces of each of the two pinholes. If the pair of disturbances is uncorrelated, then no interference fringes will be visible in the Young interferometer. A necessary condition, for the existence of such fringes, is that there be some degree of correlation between the radiation at the exit surfaces of each of the pinholes. Perfect correlation implies maximal fringe visibility and perfect coherence; zero correlation implies zero fringe visibility and incoherence; an intermediate degree of correlation implies that the visibility of the fringes is intermediate between maximal and minimal, corresponding to a partially coherent field.

At this point the reader may be wondering as to the quantitative relation of the mutual coherence function, to the visibility of fringes which are observed in a Young double slit experiment. Let us now turn to a chain of argument that will lead to a formulation of this link. To this end, set the two spatial arguments of Γ equal to the location $\mathbf{x}_{P_1}$ of the first pinhole, and let the time lag τ be zero. With the aid of eqn (1.81), together with the fact that the field is statistically stationary, we see that:

$$\begin{aligned}\overline{I}_1(\mathbf{x}_P) &= |K_1|^2 \left\langle \Psi\left(\mathbf{x}_{P_1}, t - \frac{r_{P_1 P}}{c}\right) \Psi^*\left(\mathbf{x}_{P_1}, t - \frac{r_{P_1 P}}{c}\right)\right\rangle \\ &= |K_1|^2 \langle \Psi(\mathbf{x}_{P_1}, t)\, \Psi^*(\mathbf{x}_{P_1}, t)\rangle \\ &= |K_1|^2 \Gamma(\mathbf{x}_{P_1}, \mathbf{x}_{P_1}, \tau = 0).\end{aligned} \tag{1.89}$$

Similarly, we have:

$$\overline{I}_2(\mathbf{x}_P) = |K_2|^2 \Gamma(\mathbf{x}_{P_2}, \mathbf{x}_{P_2}, \tau = 0). \tag{1.90}$$

Equipped with the above pair of expressions, eqn (1.87) becomes:

$$\begin{aligned}\overline{I}(\mathbf{x}_P) &= \overline{I}_1(\mathbf{x}_P) + \overline{I}_2(\mathbf{x}_P) \\ &\quad + 2\sqrt{\overline{I}_1(\mathbf{x}_P)\overline{I}_2(\mathbf{x}_P)} \frac{\mathrm{Re}\left[\Gamma(\mathbf{x}_{P_1}, \mathbf{x}_{P_2}, \tau)\right]}{\sqrt{|K_1|^{-2}\overline{I}_1(\mathbf{x}_P)|K_2|^{-2}\overline{I}_2(\mathbf{x}_P)}} \\ &= \overline{I}_1(\mathbf{x}_P) + \overline{I}_2(\mathbf{x}_P) \\ &\quad + 2\sqrt{\overline{I}_1(\mathbf{x}_P)\overline{I}_2(\mathbf{x}_P)}\, \mathrm{Re}\left[\frac{\Gamma(\mathbf{x}_{P_1}, \mathbf{x}_{P_2}, \tau)}{\sqrt{\Gamma(\mathbf{x}_{P_1}, \mathbf{x}_{P_1}, \tau=0)\Gamma(\mathbf{x}_{P_2}, \mathbf{x}_{P_2}, \tau=0)}}\right].\end{aligned} \tag{1.91}$$

If we now introduce the following definition for the normalized mutual coherence function, otherwise known as the 'complex degree of coherence':

$$\gamma(\mathbf{x}_{P_1}, \mathbf{x}_{P_2}, \tau) \equiv \frac{\Gamma(\mathbf{x}_{P_1}, \mathbf{x}_{P_2}, \tau)}{\sqrt{\Gamma(\mathbf{x}_{P_1}, \mathbf{x}_{P_1}, \tau = 0)\Gamma(\mathbf{x}_{P_2}, \mathbf{x}_{P_2}, \tau = 0)}}, \tag{1.92}$$

then we see that the generalized interference law (1.91) assumes the form:

$$\overline{I}(\mathbf{x}_P) = \overline{I}_1(\mathbf{x}_P) + \overline{I}_2(\mathbf{x}_P) + 2\sqrt{\overline{I}_1(\mathbf{x}_P)\overline{I}_2(\mathbf{x}_P)}\mathrm{Re}\left[\gamma(\mathbf{x}_{P_1}, \mathbf{x}_{P_2}, \tau)\right]. \tag{1.93}$$

With a view to gaining a deeper understanding of the meaning of this expression, including why we speak of it as a generalized interference law, let us investigate the limiting case of strictly monochromatic radiation. In this limit, and in an obvious notation, one can write $\Psi(\mathbf{x}_{P_1}, t) = \sqrt{I(\mathbf{x}_{P_1})}\exp\{i[\phi(\mathbf{x}_{P_1}) - \omega t]\}$ and $\Psi(\mathbf{x}_{P_2}, t) = \sqrt{I(\mathbf{x}_{P_2})}\exp\{i[\phi(\mathbf{x}_{P_2}) - \omega t]\}$; from eqn (1.88), we see that $\Gamma(\mathbf{x}_{P_1}, \mathbf{x}_{P_2}, \tau) = \sqrt{I(\mathbf{x}_{P_1})I(\mathbf{x}_{P_2})}\exp\{i[\phi(\mathbf{x}_{P_1}) - \phi(\mathbf{x}_{P_2}) - \omega\tau]\}$, which may be substituted into eqn (1.92) to give $\gamma(\mathbf{x}_{P_1}, \mathbf{x}_{P_2}, \tau) = \exp\{i[\phi(\mathbf{x}_{P_1}) - \phi(\mathbf{x}_{P_2}) - \omega\tau]\}$. Hence, in the limit of strict monochromaticity, eqn (1.93) becomes the following familiar expression for the interference of two monochromatic fields, emanating from the two pinholes of a Young interferometer:

$$\overline{I}(\mathbf{x}_P) = \overline{I}_1(\mathbf{x}_P) + \overline{I}_2(\mathbf{x}_P) + 2\sqrt{\overline{I}_1(\mathbf{x}_P)\overline{I}_2(\mathbf{x}_P)}\cos[\phi(\mathbf{x}_{P_1}) - \phi(\mathbf{x}_{P_2}) - \omega\tau]. \tag{1.94}$$

Here, the presence of the interference term—namely, the third term on the right-hand side of eqn (1.94)—was obtained via the correlation function (complex degree of coherence) for a monochromatic field, with the perfect correlation between the fields at the two pinholes being a direct consequence of the strict harmonic time dependence of such a field. This derivation of the interference term, via a correlation function, is conceptually rather different to elementary physical-optics discussions of Young's double-slit experiment for the case of strictly monochromatic radiation, which can also be used to derive this expression without the need to invoke field correlation functions. Regarding the position of the fringes associated with this expression, we note that this quantity is related to the phase of the previously mentioned correlation function; again, note the conceptual difference between this statement and that based on the elementary physical-optics treatment which makes no mention of correlations. Regarding the visibility $\mathscr{V}$ of the fringes associated with the above expression: by making use of the fact that the cosine of any number must lie between -1 and 1, together with the assumption that the envelope of the fringes has a much slower spatial variation than the fringes themselves,[25] we see that (see eqn (1.74)):

[25] Indeed, the notions of both fringe visibility and fringe envelope break down when this assumption is not met.

$$\mathscr{V} = \frac{2\sqrt{\overline{I}_1(\mathbf{x}_P)\overline{I}_2(\mathbf{x}_P)}}{\overline{I}_1(\mathbf{x}_P) + \overline{I}_2(\mathbf{x}_P)}. \tag{1.95}$$

Again, this expression for fringe visibility has been obtained via the construction of a correlation function for the strictly monochromatic field. This visibility tends to its maximal value of unity if the intensity over each of the pinholes is equal, that is, if $\overline{I}_1(\mathbf{x}_P) = \overline{I}_2(\mathbf{x}_P)$. This maximal fringe visibility is a direct consequence of the fact that the complex degree of coherence has modulus unity, this being indicative of the previously mentioned perfect correlation that exists between any two space–time points in a strictly monochromatic complex scalar electromagnetic field.

Let us now consider how the argument, of the above paragraph, generalizes to the case of quasi-monochromatic radiation. Adopting the notation of eqn (1.78), one can write $\Psi(\mathbf{x}_{P_1}, t) = \exp(-i\overline{\omega}t)\mathscr{A}(\mathbf{x}_{P_1}, t)$ and $\Psi(\mathbf{x}_{P_2}, t) = \exp(-i\overline{\omega}t)\mathscr{A}(\mathbf{x}_{P_2}, t)$, where the amplitude and phase of the complex envelopes $\mathscr{A}(\mathbf{x}_{P_1}, t)$ and $\mathscr{A}(\mathbf{x}_{P_2}, t)$ have a temporal evolution with a characteristic timescale $2\pi/\Delta\omega$ that is much larger than the mean period $2\pi/\overline{\omega}$ of the quasi-monochromatic radiation. From eqn (1.88), we see that the mutual coherence function is given by:

$$\Gamma(\mathbf{x}_{P_1}, \mathbf{x}_{P_2}, \tau) = \exp(-i\overline{\omega}\tau) \left\langle \mathscr{A}(\mathbf{x}_{P_1}, t+\tau)\mathscr{A}^*(\mathbf{x}_{P_2}, t) \right\rangle, \tag{1.96}$$

which may be substituted into eqn (1.92) to give:

$$\gamma(\mathbf{x}_{P_1}, \mathbf{x}_{P_2}, \tau) = \frac{\exp(-i\overline{\omega}\tau) \left\langle \mathscr{A}(\mathbf{x}_{P_1}, t+\tau)\mathscr{A}^*(\mathbf{x}_{P_2}, t) \right\rangle}{\sqrt{\langle |\mathscr{A}(\mathbf{x}_{P_1}, t)|^2\rangle \langle |\mathscr{A}(\mathbf{x}_{P_2}, t)|^2\rangle}}. \tag{1.97}$$

Using the Schwarz inequality, one can readily see that the modulus, of this complex degree of coherence, is less than unity[26]:

$$|\gamma(\mathbf{x}_{P_1}, \mathbf{x}_{P_2}, \tau)| \leq 1. \tag{1.98}$$

As shall now be shown, this inequality is directly related to the range of fringe visibilities $\mathscr{V}$ which are possible, when our Young interferometer is illuminated with quasi-monochromatic radiation. With this end in mind, let us once again assume that the spatial variation of the fringes is sufficiently faster than that of their envelope for the notion of fringe visibility to be meaningful; we leave it as an exercise to the reader to show that the resulting fringe visibility is given by:

[26] For two complex functions of time $a(t)$ and $b(t)$, the Schwarz inequality is $|\langle a(t)b^*(t)\rangle| \leq [\langle |a(t)|^2\rangle\langle |b(t)|^2\rangle]^{1/2}$. Setting $a(t) = \mathscr{A}(\mathbf{x}_{P_1}, t+\tau)$ and $b(t) = \mathscr{A}(\mathbf{x}_{P_2}, t)$, this inequality becomes $|\langle \mathscr{A}(\mathbf{x}_{P_1}, t+\tau)\mathscr{A}^*(\mathbf{x}_{P_2}, t)\rangle| \leq [\langle |\mathscr{A}(\mathbf{x}_{P_1}, t+\tau)|^2\rangle\langle |\mathscr{A}(\mathbf{x}_{P_2}, t)|^2\rangle]^{1/2}$. Invoking statistical stationarity for the first time average appearing on the right side of this inequality, we see that $|\langle \mathscr{A}(\mathbf{x}_{P_1}, t+\tau)\mathscr{A}^*(\mathbf{x}_{P_2}, t)\rangle| \leq [\langle |\mathscr{A}(\mathbf{x}_{P_1}, t)|^2\rangle\langle |\mathscr{A}(\mathbf{x}_{P_2}, t)|^2\rangle]^{1/2}$, that is, $|\langle \mathscr{A}(\mathbf{x}_{P_1}, t+\tau)\mathscr{A}^*(\mathbf{x}_{P_2}, t)\rangle|/[\langle |\mathscr{A}(\mathbf{x}_{P_1}, t)|^2\rangle\langle |\mathscr{A}(\mathbf{x}_{P_2}, t)|^2\rangle]^{1/2} \leq 1$. According to the modulus of eqn (1.97), the left side of this last inequality is equal to the modulus of the complex degree of coherence, leading immediately to eqn (1.98).

$$\mathscr{V} = \frac{2\sqrt{\overline{I}_1(\mathbf{x}_P)\overline{I}_2(\mathbf{x}_P)}}{\overline{I}_1(\mathbf{x}_P) + \overline{I}_2(\mathbf{x}_P)} |\gamma(\mathbf{x}_{P_1}, \mathbf{x}_{P_2}, \tau)|. \tag{1.99}$$

Note that this differs from eqn (1.95), by the presence of a multiplicative factor equal to the modulus of the complex degree of coherence. In the case of equality between the time-averaged intensities at the location of the two pinholes, so that $\overline{I}_1(\mathbf{x}_P) = \overline{I}_2(\mathbf{x}_P)$, the above expression for the fringe visibility assumes the very simple form:

$$\mathscr{V} = |\gamma(\mathbf{x}_{P_1}, \mathbf{x}_{P_2}, \tau)|. \tag{1.100}$$

This is an important result, which makes the promised connection between fringe visibility and the complex degree of coherence: the former is equal to the modulus of the latter, when treating the fringe visibility of a Young's double-slit experiment in which the time-averaged intensity at the location of each of the two pinholes is equal, subject to assumptions which were stated in the process of arriving at this result (small near-identical pinholes, slowly varying envelope on the fringes, etc.). Since we know that the modulus of the fringe visibility is less than or equal to unity, we arrive at the admittedly obvious statement that the fringe visibility lies between zero and unity. In turn, as was the case with the previous analysis, the modulus of the complex degree of coherence may be used to classify the coherence properties of the quasi-monochromatic field under study: complete coherence corresponds to $|\gamma(\mathbf{x}_{P_1}, \mathbf{x}_{P_2}, \tau)| = 1$ for any pair of points $(\mathbf{x}_{P_1}, \mathbf{x}_{P_2})$ and any time delay τ; complete incoherence corresponds to $|\gamma(\mathbf{x}_{P_1}, \mathbf{x}_{P_2}, \tau)| = 0$ for any pair of points $(\mathbf{x}_{P_1}, \mathbf{x}_{P_2})$ and any time delay τ; otherwise, the field is partially coherent. Note that the extremes of complete coherence and complete incoherence are limiting cases which are never strictly realized in practice, with the corollary that all realistic electromagnetic fields are partially coherent.

Let us now turn to the question of the observability of both the mutual coherence function and the complex degree of coherence, in a chain of argument which will rapidly lead to the conclusion that such correlation functions can be measured with the aid of a Young interferometer. At optical and higher frequencies, the rapidity of wave-field oscillations is so great that their temporal evolution is not directly measurable with existing apparatus. The field quantities, or functions thereof to which a given experiment may be sensitive, are therefore averaged over many cycles of oscillation; the characteristic timescale for these oscillations is proportional to the reciprocal of the mean angular frequency of the radiation. This may be used as the basis for a criticism, that the fundamental quantities of the optical theory are not amenable to direct observation with existing technology. As has been emphasized by Wolf, in the previously cited series of papers which laid the foundation for the modern formulation of the theory of partial coherence for statistically stationary fields, the correlation functions

$\Gamma(\mathbf{x}_{P_1}, \mathbf{x}_{P_2}, \tau)$ and $\gamma(\mathbf{x}_{P_1}, \mathbf{x}_{P_2}, \tau)$ do not suffer from this limitation. In such correlation functions, the offending time variable is integrated out, being replaced by the time lag τ. The correlation functions themselves may be measured using the Young interferometer, according to the following argument. The modulus of the complex degree of coherence can be determined by rearranging eqn (1.99) into the form:

$$|\gamma(\mathbf{x}_{P_1}, \mathbf{x}_{P_2}, \tau)| = \frac{\mathscr{V}\left[\overline{I}_1(\mathbf{x}_P) + \overline{I}_2(\mathbf{x}_P)\right]}{2\sqrt{\overline{I}_1(\mathbf{x}_P)\overline{I}_2(\mathbf{x}_P)}}, \tag{1.101}$$

with a similar expression existing for the determination of the modulus of the mutual coherence function. All quantities, on the right-hand sides of these two expressions, are directly measurable as time-averaged intensities. The phase of $\Gamma(\mathbf{x}_{P_1}, \mathbf{x}_{P_2}, \tau)$ and $\gamma(\mathbf{x}_{P_1}, \mathbf{x}_{P_2}, \tau)$ can evidently be measured by recording the location of the maxima in the interference pattern of a Young interferometer, for a given time lag τ, which can be arranged by placing a transparent slab of suitable thickness in one of the arms of the interferometer. Since both the modulus and phase of the correlation functions are observable quantities, one thereby has an 'optics in terms of observable quantities' (Wolf 1954b).

In the analysis thus far, these observable quantities—namely, the mutual coherence function and the complex degree of coherence—have been considered in the restricted context of the double-slit experiment for quasi-monochromatic light. This restriction can be weakened: indeed, these correlation functions are meaningful for a free field in the absence of a Young interferometer. More precisely, for a statistically stationary complex scalar X-ray field which occupies a given volume Ω of free space, the mutual coherence function $\Gamma(\mathbf{x}_1, \mathbf{x}_2, \tau)$ may be defined via eqn (1.88), for any position vectors $\mathbf{x}_1, \mathbf{x}_2 \in \Omega$. This correlation function—together with its normalized form given by the complex degree of coherence—is therefore a property of free statistically stationary partially coherent complex scalar X-ray wave-fields, irrespective of whether or not one chooses to introduce a suitable Young's double-slit experiment in order to make measurements which are sensitive to these correlations. In addition to so liberating the mutual coherence function and the complex degree of coherence from the Young interferometer, this pair of correlation functions evidently remains well defined for the case of statistically stationary complex disturbances whose spectrum is not sufficiently narrow for them to be classified as quasi-monochromatic.

We close this section by introducing the useful notion of mutual intensity $J(\mathbf{x}_1, \mathbf{x}_2)$, due to Zernike (1938). Such a quantity may be obtained by setting the time-lag to zero in the mutual coherence function (Wolf 1955):

$$J(\mathbf{x}_1, \mathbf{x}_2) \equiv \Gamma(\mathbf{x}_1, \mathbf{x}_2, \tau = 0). \tag{1.102}$$

To motivate this definition for the mutual intensity, let us once again restrict ourselves to the case of quasi-monochromatic radiation. Recall eqn (1.96), which states that $\Gamma(\mathbf{x}_1, \mathbf{x}_2, \tau) = \exp(-i\overline{\omega}\tau)\langle\mathscr{A}(\mathbf{x}_1, t+\tau)\mathscr{A}^*(\mathbf{x}_2, t)\rangle$. Two very different characteristic timescales appear on the right-hand side of this expression: $\exp(-i\overline{\omega}\tau)$ varies appreciably over time-lag scales $\tau \approx 2\pi/\overline{\omega}$ on the order of the reciprocal of the mean angular frequency $\overline{\omega}$ of the radiation, whereas $\langle\mathscr{A}(\mathbf{x}_1, t+\tau)\mathscr{A}^*(\mathbf{x}_2, t)\rangle$ varies appreciably only over much longer time-lag scales $\tau \approx 2\pi/\Delta\omega$, the order of which is given by the reciprocal of the mean spread $\Delta\omega$ in angular frequencies of the quasi-monochromatic radiation. Since $\Delta\omega \ll \overline{\omega}$ for quasi-monochromatic radiation, the former time-lag scale is much smaller than the latter.

Accordingly, suppose that one performs a given optical experiment using quasi-monochromatic radiation, which only introduces path-length differences having corresponding time-lags τ that are much less than $2\pi/\Delta\omega$. In eqn (1.96), one may therefore replace $\langle\mathscr{A}(\mathbf{x}_1, t+\tau)\mathscr{A}^*(\mathbf{x}_2, t)\rangle$ by the value that this quantity takes when $\tau = 0$, so that:

$$\begin{aligned}\Gamma(\mathbf{x}_1, \mathbf{x}_2, \tau) &= \exp(-i\overline{\omega}\tau)\,\langle\mathscr{A}(\mathbf{x}_1, t+\tau)\mathscr{A}^*(\mathbf{x}_2, t)\rangle \\ &\approx \exp(-i\overline{\omega}\tau)\,\langle\mathscr{A}(\mathbf{x}_1, t)\mathscr{A}^*(\mathbf{x}_2, t)\rangle\,, \quad \tau \ll 2\pi/\Delta\omega. \end{aligned} \tag{1.103}$$

Making use of expression (1.102) for the mutual intensity, together with eqn (1.96), we see that:

$$\Gamma(\mathbf{x}_1, \mathbf{x}_2, \tau) \approx \exp(-i\overline{\omega}\tau) J(\mathbf{x}_1, \mathbf{x}_2), \quad \tau \ll 2\pi/\Delta\omega. \tag{1.104}$$

Under the stated conditions, we conclude that the mutual coherence function factorizes into a product of two functions, one of which is a harmonic function that depends only on the time-lag τ, with the other being the mutual intensity that depends only on the spatial coordinates $\mathbf{x}_1, \mathbf{x}_2$ of the two points for which the mutual coherence is to be determined. As shall be seen in the following section, this provides a useful simplification in contexts where the condition $\tau \ll 2\pi/\Delta\omega$ is satisfied.

Note, also, that the mutual intensity $J(\mathbf{x}_1, \mathbf{x}_2) \equiv \Gamma(\mathbf{x}_1, \mathbf{x}_2, \tau = 0)$ evidently embodies the spatial coherence properties of a given partially coherent field, as it governs the time-averaged fringe visibility, which results when one combines the statistically stationary optical disturbances from two points in space, with zero time-lag between the two. Similarly, the diagonal components $\Gamma(\mathbf{x}_1, \mathbf{x}_1, \tau)$ embody the temporal coherence properties of the disturbance, at a given point $\mathbf{x}_1$ in space. In general, however, the information contained in the mutual coherence function cannot be cleanly separated into properties pertaining to spatial and temporal coherence.

1.9 Propagation of two-point correlation functions

In our earlier discussions on spatial coherence in the context of a Young interferometer illuminated by an extended incoherent quasi-monochromatic source, we saw some hint of the notion that field correlations may be induced by the act of free-space propagation. Specifically, in this example, we began with a source which by assumption possessed no correlations between the fields at any two distinct points on the source. Yet, when the resulting field was propagated through free space so as to impinge on a black screen pierced by two pinholes, the possibility of non-zero correlations was seen to exist between the resulting disturbances at this pair of pinholes. In the present section, we further investigate this notion of propagation-induced correlations, by outlining a diffraction theory describing the free-space propagation of both the mutual coherence function and the mutual intensity. Note that we speak of these as 'two-point correlation functions', on account of (i) the fact that they involve the correlation between the field at two space–time points; and (ii) the fact that this term distinguishes these correlations from the higher-order correlation functions to be discussed in the final section of this chapter.

Our presentation, of the propagation of two-point correlation functions, is broken into four sub-sections. We begin by deriving the Wolf equations, these being a pair of d'Alembert wave equations obeyed by the mutual coherence function for a free field. We shall also obtain a Fourier representation of these vacuum wave equations, together with their limiting case for fields well described by the mutual intensity. Just as a diffraction theory for coherent scalar fields can be obtained by solution of an appropriate boundary-value problem for the Helmholtz equation (see eqn (1.16)), one may formulate a diffraction theory for two-point correlation functions by taking their associated vacuum equations as a starting point. The second sub-section gives such a diffraction theory, this being an appropriate generalization of the angular-spectrum formulation for coherent diffraction given in Section 1.3. As an alternative yet equivalent formulation of the same theory, the third sub-section presents Parrent's generalization of one of the Rayleigh–Sommerfeld diffraction integrals (see Section 1.6.2) to the case of the mutual coherence function. Both the second and third sub-sections treat the problem of how the mutual coherence function may be propagated from plane to parallel plane, the unpropagated mutual coherence function being a function of any two points in the initial plane, with the propagated mutual coherence function being determined for any pair of points in the final plane downstream of the initial plane. Finally, in the fourth sub-section we show how Parrent's formula reduces to the famous van Cittert–Zernike theorem for the propagation of mutual intensity, under suitable approximations which shall be outlined in due course.

1.9.1 *Vacuum wave equations for propagation of two-point correlation functions*

Here, we give a derivation of the Wolf equations (Wolf 1954b, 1955). As mentioned earlier, these are a pair of vacuum wave equations governing the free-space evolution of the mutual coherence function $\Gamma(\mathbf{x}_1, \mathbf{x}_2, \tau)$. We shall also obtain a Fourier representation of the same, leading to a pair of Helmholtz equations governing the Fourier transform of the mutual coherence function with respect to the time-lag.[27] Lastly, we will obtain the vacuum wave equations for the mutual intensity, as a special case of the Wolf equations. Each of these three sets of differential equations will be of utility in our later discussions on the propagation of two-point correlations.

1.9.1.1 *Wolf equations for mutual coherence* With ∇_1^2 denoting the three-dimensional Laplacian with respect to the spatial coordinate $\mathbf{x}_1$, we apply the d'Alembert operator $(c^{-2}\partial^2/\partial\tau^2 - \nabla_1^2)$ to both sides of definition (1.88) for the mutual coherence function. Bringing this operator inside the angular brackets on the right side of the resulting expression, and noting that this operator does not act on the variables $\mathbf{x}_2$ or t, we see that:

$$\left(\frac{1}{c^2}\frac{\partial^2}{\partial\tau^2} - \nabla_1^2\right)\Gamma(\mathbf{x}_1,\mathbf{x}_2,\tau) = \left\langle \left[\left(\frac{1}{c^2}\frac{\partial^2}{\partial\tau^2} - \nabla_1^2\right)\Psi\left(\mathbf{x}_1, t+\tau\right)\right]\Psi^*\left(\mathbf{x}_2, t\right)\right\rangle. \tag{1.105}$$

Since the d'Alembert wave equation (1.13) is invariant with respect to a shift in the origin of time, the above square-bracketed quantity must vanish. This yields the first Wolf equation:

$$\left(\frac{1}{c^2}\frac{\partial^2}{\partial\tau^2} - \nabla_1^2\right)\Gamma(\mathbf{x}_1,\mathbf{x}_2,\tau) = 0. \tag{1.106}$$

The second Wolf equation is arrived at in a similar fashion. With ∇_2^2 denoting the three-dimensional Laplacian with respect to $\mathbf{x}_2$, apply the d'Alembert operator $(c^{-2}\partial^2/\partial\tau^2 - \nabla_2^2)$ to both sides of eqn (1.88). Invoke statistical stationarity, to re-write $\langle\Psi\left(\mathbf{x}_1, t+\tau\right)\Psi^*\left(\mathbf{x}_2, t\right)\rangle$ as $\langle\Psi\left(\mathbf{x}_1, t\right)\Psi^*\left(\mathbf{x}_2, t-\tau\right)\rangle$, on the right side of the resulting expression. The d'Alembert operator can then be brought inside the angular brackets, on the right side, so that:

$$\left(\frac{1}{c^2}\frac{\partial^2}{\partial\tau^2} - \nabla_2^2\right)\Gamma(\mathbf{x}_1,\mathbf{x}_2,\tau) = \left\langle \Psi\left(\mathbf{x}_1, t\right)\left[\left(\frac{1}{c^2}\frac{\partial^2}{\partial\tau^2} - \nabla_2^2\right)\Psi^*\left(\mathbf{x}_2, t-\tau\right)\right]\right\rangle. \tag{1.107}$$

Once again, we invoke certain symmetries of the d'Alembert wave equation (1.13). Specifically, this equation is invariant with respect to: (i) complex conjugation of the field; (ii) reversal of the direction of time; (iii) a shift in the origin of time. This again implies the vanishing of the quantity in square brackets above, leading to the second Wolf equation:

$$\left(\frac{1}{c^2}\frac{\partial^2}{\partial\tau^2} - \nabla_2^2\right)\Gamma(\mathbf{x}_1,\mathbf{x}_2,\tau) = 0. \tag{1.108}$$

[27]This Fourier transform is known as the 'cross-spectral density'.

1.9.1.2 *Helmholtz equations for cross-spectral density* The mutual coherence function is an example of a correlation function existing in the space–time domain. As shall be seen in due course, it is often convenient to work in the corresponding space-frequency domain, the Wolf equations thereby being transformed into a pair of Helmholtz equations.

With this formulation in mind, let us represent the mutual coherence function as a Fourier integral with respect to the time-lag τ (Wolf 1955; Parrent 1959; Mandel and Wolf 1965):

$$\Gamma(\mathbf{x}_1, \mathbf{x}_2, \tau) = \int_0^\infty W(\mathbf{x}_1, \mathbf{x}_2, \nu) \exp(-2\pi i \nu \tau) d\nu. \tag{1.109}$$

The Fourier transform $W(\mathbf{x}_1, \mathbf{x}_2, \nu)$, of the mutual coherence function with respect to the time-lag τ, is known as the 'cross-spectral density'. Note that the above Fourier integral only extends over positive frequencies, on account of the fact that the mutual coherence function is an analytic signal in the sense described in Section 1.2. Further, note that we have departed from our usual convention of representing temporal Fourier integrals as an integral over angular frequency, opting instead for an integral over frequency, as the above expression conforms with a usage that has become standard in the literature. Lastly, for future reference, we note that eqn (1.109) has the corresponding inverse transform:

$$W(\mathbf{x}_1, \mathbf{x}_2, \nu) = \int_{-\infty}^\infty \Gamma(\mathbf{x}_1, \mathbf{x}_2, \tau) \exp(2\pi i \nu \tau) d\tau. \tag{1.110}$$

Returning to the main thread of the argument, which will lead from the Wolf equations for the mutual coherence function to the promised pair of Helmholtz equations for the cross-spectral density, substitute the Fourier integral (1.109) into eqn (1.106). Interchange the order of differentiation and integration, allow $\partial^2/\partial\tau^2$ to act on $\exp(-2\pi i \nu \tau)$, and then make use of the fact that $k = 2\pi\nu/c$ in vacuum. Thus:

$$\int_0^\infty \left[\left(\nabla_1^2 + k^2\right) W(\mathbf{x}_1, \mathbf{x}_2, \nu)\right] \exp(-2\pi i \nu \tau) d\nu = 0. \tag{1.111}$$

The quantity in square brackets must vanish. This leads to the following Helmholtz equation for the cross-spectral density, this being the space-frequency counterpart to the first Wolf equation:

$$\left(\nabla_1^2 + k^2\right) W(\mathbf{x}_1, \mathbf{x}_2, \nu) = 0. \tag{1.112}$$

Similarly, the space-frequency counterpart to the second Wolf equation is:

$$\left(\nabla_2^2 + k^2\right) W(\mathbf{x}_1, \mathbf{x}_2, \nu) = 0. \tag{1.113}$$

1.9.1.3 *Helmholtz equations for mutual intensity* As a special case of the Wolf equations, suppose a given statistically stationary complex scalar electromagnetic field to be quasi-monochromatic, with all relevant time-lags being much smaller that the reciprocal of the spread of temporal frequencies present in the field. The associated mutual coherence function will then be well approximated by eqn (1.104). Substitute this expression into the Wolf equations (1.106) and (1.108), allow the operator $\partial^2/\partial\tau^2$ to act on the harmonic factor $\exp(-i\overline{\omega}\tau)$, and then introduce the average wave-number $\overline{k}$ via $\overline{k} = \overline{\omega}/c$. Cancelling the harmonic time factor from the resulting expression, we arrive at the following pair of Helmholtz equations governing the mutual intensity:

$$\left(\nabla_a^2 + \overline{k}^2\right) J(\mathbf{x}_1, \mathbf{x}_2) = 0, \quad \tau \ll 2\pi/\Delta\omega, \quad a = 1, 2. \tag{1.114}$$

1.9.2 *Operator formulation for propagation of two-point correlation functions* We outline a rigorous diffraction theory for the propagation of two-point correlation functions, due to Marchand and Wolf (1972). Specifically, we develop plane-to-plane diffraction operators for three different two-point correlation functions, namely the cross-spectral density, mutual intensity, and mutual coherence function. Such diffraction operators form a natural counterpart to the diffraction operator, for coherent scalar electromagnetic fields, arrived at during our discussions on the angular spectrum in Section 1.3.

With reference to Fig. 1.1, let us erect the usual Cartesian coordinate system (x, y, z), with nominal optic axis z. All sources, of a given statistically stationary complex scalar electromagnetic field, are assumed to lie in the half-space $z < 0$, with the space $z \geq 0$ being filled with vacuum. Assume that one of the three previously mentioned two-point correlation functions is known, at all pairs of points which lie in the plane $z = 0$. The diffraction problem, with which we are here concerned, is to develop an expression for the same two-point correlation function as a function of any pair of points that lie in the space $z \geq 0$. This propagated two-point correlation function must obey the appropriate vacuum equations for any pair of points in $z \geq 0$, and reduce to the specified boundary value when both pairs of points lie in the plane $z = 0$.

1.9.2.1 *Operator formulation for propagation of cross-spectral density* Let $\mathbf{x}_1 \equiv (x_1, y_1, z_1)$ and $\mathbf{x}_2 \equiv (x_2, y_2, z_2)$ denote any pair of points in the space $z \geq 0$, where $z_1, z_2 \geq 0$. Consider the following six-dimensional elementary plane wave $W^{(\mathrm{PW})}(x_1, y_1, z_1, x_2, y_2, z_2, \nu)$, which is a solution to the Helmholtz equations (1.112) and (1.113) obeyed by the cross-spectral density:

$$\begin{aligned} &W^{(\mathrm{PW})}(x_1, y_1, z_1, x_2, y_2, z_2, \nu) \\ &\quad = \exp[i(k_{1x}x_1 + k_{1y}y_1 + k_{1z}z_1 + k_{2x}x_2 + k_{2y}y_2 + k_{2z}z_2)]. \end{aligned} \tag{1.115}$$

The above function is evidently a product of two three-dimensional plane waves, the first being a function of the spatial coordinates (x_1, y_1, z_1) with wave-vector

(k_{1x}, k_{1y}, k_{1z}), and the second being a function of spatial coordinates (x_2, y_2, z_2) with wave-vector (k_{2x}, k_{2y}, k_{2z}). By substituting expression (1.115) into eqns (1.112) and (1.113), we see that:

$$\begin{aligned} k_{1x}^2 + k_{1y}^2 + k_{1z}^2 &= k^2, \\ k_{2x}^2 + k_{2y}^2 + k_{2z}^2 &= k^2. \end{aligned} \tag{1.116}$$

Solve these equations for k_{1z} and k_{2z}, respectively:

$$\begin{aligned} k_{1z} &= \sqrt{k^2 - k_{1x}^2 - k_{1y}^2}, \\ k_{2z} &= -\sqrt{k^2 - k_{2x}^2 - k_{2y}^2}. \end{aligned} \tag{1.117}$$

Note the choice of signs in the above equations, which are taken to ensure the correct asymptotic behaviour for our six-dimensional plane wave: eqn (1.88) implies that it must be outgoing in (x_1, y_1, z_1) as $|(x_1, y_1, z_1)| \to \infty$ in the half-space $z_1 > 0$, and incoming in (x_2, y_2, z_2) as $|(x_2, y_2, z_2)| \to \infty$ in the half-space $z_2 > 0$. Note, also, that evanescent waves are assumed to be absent.

Using eqns (1.117), the elementary plane wave (1.115) becomes:

$$\begin{aligned} &W^{(\mathrm{PW})}(x_1, y_1, z_1, x_2, y_2, z_2, \nu) \\ &\quad = \exp[i(k_{1x}x_1 + k_{1y}y_1 + k_{2x}x_2 + k_{2y}y_2)] \\ &\qquad \times \exp\left[i\left(z_1\sqrt{k^2 - k_{1x}^2 - k_{1y}^2} - z_2\sqrt{k^2 - k_{2x}^2 - k_{2y}^2}\right)\right]. \end{aligned} \tag{1.118}$$

Next, set $z_1 = z_2 = 0$ in the above expression to give:

$$\begin{aligned} &W^{(\mathrm{PW})}(x_1, y_1, z_1 = 0, x_2, y_2, z_2 = 0, \nu) \\ &\quad = \exp[i(k_{1x}x_1 + k_{1y}y_1 + k_{2x}x_2 + k_{2y}y_2)]. \end{aligned} \tag{1.119}$$

The above pair of equations represent the solution to a certain trivial propagation problem, for six-dimensional plane waves which obey the pair of Helmholtz equations (1.112) and (1.113). If the unpropagated plane wave, over the space $z_1 = z_2 = 0$, is known to have the functional form given by (1.119), then eqn (1.118) tells us that it can be propagated to any $z_1, z_2 \geq 0$ by multiplying the stated functional form by the free-space transfer function $\exp\{i[z_1(k^2 - k_{1x}^2 - k_{1y}^2)^{1/2} - z_2(k^2 - k_{2x}^2 - k_{2y}^2)^{1/2}]\}$.

Put this result to one side for the moment, passing onto the problem of how the unpropagated cross-spectral density $W(x_1, y_1, z_1 = 0, x_2, y_2, z_2 = 0, \nu)$ can be used to determine the propagated cross-spectral density $W(x_1, y_1, z_1, x_2, y_2, z_2, \nu)$, for any $z_1, z_2 \geq 0$. To this end, write the unpropagated cross-spectral density as the following fourfold Fourier integral:

$$\begin{aligned}
&W(x_1, y_1, z_1 = 0, x_2, y_2, z_2 = 0, \nu) \\
&= \frac{1}{4\pi^2} \iiiint \breve{W}(k_{1x}, k_{1y}, z_1 = 0, k_{2x}, k_{2y}, z_2 = 0, \nu) \\
&\quad \times \exp[i(k_{1x}x_1 + k_{1y}y_1 + k_{2x}x_2 + k_{2y}y_2)] dk_{1x} dk_{1y} dk_{2x} dk_{2y}. \qquad (1.120)
\end{aligned}$$

Here, $\breve{W}(k_{1x}, k_{1y}, z_1 = 0, k_{2x}, k_{2y}, z_2 = 0, \nu)$ denotes the Fourier transform of the unpropagated cross-spectral density $W(x_1, y_1, z_1 = 0, x_2, y_2, z_2 = 0, \nu)$, with respect to x_1, y_1, x_2, and y_2.

The Fourier integral in eqn (1.120) expresses $W(x_1, y_1, z_1 = 0, x_2, y_2, z_2 = 0, \nu)$ as a linear combination of the unpropagated elementary plane waves given by eqn (1.119). Given the conclusion immediately following the latter equation, the linearity of eqns (1.112) and (1.113) implies that one may propagate $W(x_1, y_1, z_1 = 0, x_2, y_2, z_2 = 0, \nu)$ into $z_1, z_2 \geq 0$ by multiplying the harmonic factor $\exp[i(k_{1x}x_1 + k_{1y}y_1 + k_{2x}x_2 + k_{2y}y_2)]$, in the integrand of its Fourier representation (1.120), by the free-space transfer function $\exp\{i[z_1(k^2 - k_{1x}^2 - k_{1y}^2)^{1/2} - z_2(k^2 - k_{2x}^2 - k_{2y}^2)^{1/2}]\}$. We thereby arrive at the following integral which solves the desired diffraction problem for the cross-spectral density:

$$\begin{aligned}
&W(x_1, y_1, z_1 \geq 0, x_2, y_2, z_2 \geq 0, \nu) \\
&= \frac{1}{4\pi^2} \iiiint \breve{W}(k_{1x}, k_{1y}, z_1 = 0, k_{2x}, k_{2y}, z_2 = 0, \nu) \\
&\quad \times \exp\left[i\left(z_1\sqrt{k^2 - k_{1x}^2 - k_{1y}^2} - z_2\sqrt{k^2 - k_{2x}^2 - k_{2y}^2}\right)\right] \\
&\quad \times \exp[i(k_{1x}x_1 + k_{1y}y_1 + k_{2x}x_2 + k_{2y}y_2)] dk_{1x} dk_{1y} dk_{2x} dk_{2y}. \qquad (1.121)
\end{aligned}$$

In words, the above equation indicates the following series of steps, which may be used to determine the propagated cross-spectral density $W(x_1, y_1, z_1 \geq 0, x_2, y_2, z_2 \geq 0, \nu)$ from its unpropagated form $W(x_1, y_1, z_1 = 0, x_2, y_2, z_2 = 0, \nu)$: (i) take the unpropagated cross-spectral density $W(x_1, y_1, z_1 = 0, x_2, y_2, z_2 = 0, \nu)$, then Fourier transform it with respect to x_1, y_1, x_2, and y_2, so as to form $\breve{W}(k_{1x}, k_{1y}, z_1 = 0, k_{2x}, k_{2y}, z_2 = 0, \nu)$; (ii) multiply the resulting function by the free-space transfer function $\exp\{i[z_1(k^2 - k_{1x}^2 - k_{1y}^2)^{1/2} - z_2(k^2 - k_{2x}^2 - k_{2y}^2)^{1/2}]\}$; (iii) take the inverse Fourier transform of the result, with respect to k_{1x}, k_{1y}, k_{2x}, and k_{2y}.

Bearing the above verbal description in mind, we can immediately write down a diffraction operator for the cross-spectral density:

$$W(x_1, y_1, z_1 \geq 0, x_2, y_2, z_2 \geq 0, \nu) = \mathcal{D}_{z_1, z_2, \nu} W(x_1, y_1, z_1 = 0, x_2, y_2, z_2 = 0, \nu), \qquad (1.122)$$

where (cf. eqn (1.25))[28]:

[28]This diffraction operator is readily implemented numerically, by making use of the fast Fourier transform (see, for example, Press *et al.* (1992)).

$$\mathcal{D}_{z_1,z_2,\nu} = \mathcal{F}_4^{-1} \exp\left[i\left(z_1\sqrt{k^2 - k_{1x}^2 - k_{1y}^2} - z_2\sqrt{k^2 - k_{2x}^2 - k_{2y}^2}\right)\right]\mathcal{F}_4, \quad z_1, z_2 \geq 0. \quad (1.123)$$

Here, $\mathcal{F}_4$ denotes Fourier transformation with respect to x_1, y_1, x_2, and y_2, with $\mathcal{F}_4^{-1}$ denoting the corresponding inverse Fourier transformation. Note that cascaded operators are assumed to act from right to left.

1.9.2.2 *Operator formulation for propagation of mutual intensity* Both the mutual intensity and the cross-spectral density obey a pair of Helmholtz equations, as given in (1.112) and (1.113) for the cross-spectral density, and (1.114) for the mutual intensity. The former pair of equations may be mapped onto the latter, by replacing the cross-spectral density with the mutual intensity, and replacing the frequency ν with the mean frequency $\overline{\nu}$ of a quasi-monochromatic field. The diffraction operator (1.123) may therefore also be used to solve the diffraction problem for mutual intensities, for the case of quasi-monochromatic fields for which all relevant time-lags are sufficiently small for eqn (1.104) to be a good approximation, so that:

$$J(x_1, y_1, z_1 \geq 0, x_2, y_2, z_2 \geq 0, \overline{\nu}) = \mathcal{D}_{z_1,z_2,\overline{\nu}} J(x_1, y_1, z_1 = 0, x_2, y_2, z_2 = 0, \overline{\nu}), \quad z_1, z_2 \geq 0. \quad (1.124)$$

1.9.2.3 *Operator formulation for propagation of mutual coherence* Equation (1.110) may be written as:

$$W(\mathbf{x}_1, \mathbf{x}_2, \nu) = \mathcal{F}_1 \Gamma(\mathbf{x}_1, \mathbf{x}_2, \tau), \quad (1.125)$$

where $\mathcal{F}_1$ denotes Fourier transformation with respect to τ using the convention in eqn (1.110). Use the above equation to convert both the unpropagated and propagated cross-spectral densities, in eqn (1.122), into their corresponding mutual coherence functions. Apply the inverse $\mathcal{F}_1^{-1}$ of $\mathcal{F}_1$ to both sides of the resulting expression, giving:

$$\Gamma(x_1, y_1, z_1 \geq 0, x_2, y_2, z_2 \geq 0, \tau) = \mathcal{D}_{z_1,z_2} \Gamma(x_1, y_1, z_1 = 0, x_2, y_2, z_2 = 0, \tau), \quad z_1, z_2 \geq 0. \quad (1.126)$$

Here, the diffraction operator $\mathcal{D}_{z_1,z_2}$ for the mutual coherence function is:

$$\mathcal{D}_{z_1,z_2} = \mathcal{F}_1^{-1} \mathcal{D}_{z_1,z_2,\nu} \mathcal{F}_1. \quad (1.127)$$

1.9.3 *Green function formulation for propagation of two-point correlation functions*

In the previous sub-section, we used an angular-spectrum formalism to treat the diffraction problem for three different two-point correlation functions, namely

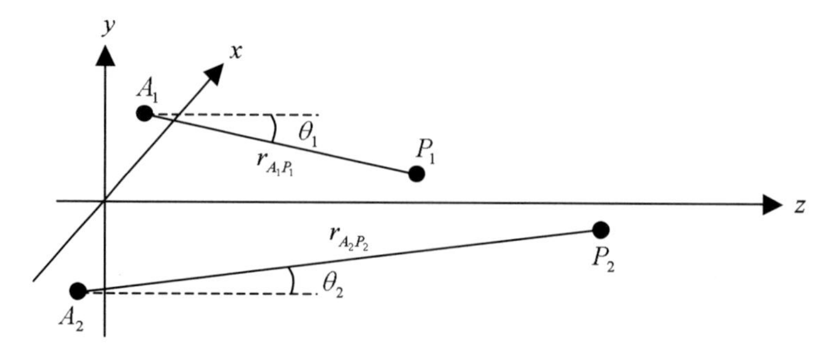

Fig. 1.8. Geometry for propagation of two-point correlation functions. Suppose that a given two-point correlation function (mutual coherence, mutual intensity, or cross-spectral density) is specified for all pairs of points A_1 and A_2 in the plane $z = 0$, where (x, y, z) denotes the usual Cartesian coordinate system with nominal optic axis z. All sources, of the partially coherent statistically stationary scalar electromagnetic field associated with the specified two-point correlation function, are assumed to lie in the half-space $z \leq 0$. The diffraction problem is to determine the corresponding two-point correlation function, for all pairs of points P_1 and P_2, in the vacuum-filled half-space $z \geq 0$. The line from A_1 to P_1, whose length is $r_{A_1P_1}$, makes an angle of θ_1 with respect to the positive z-axis. The quantities $r_{A_2P_2}$ and θ_2 are similarly defined.

the cross-spectral density, mutual coherence, and mutual intensity. Here, we give an alternative treatment of these diffraction problems, in a Green function formulation that mirrors the development of the Rayleigh–Sommerfeld diffraction integral (of the first kind) for the case of coherent scalar radiation. Again, we see some degree of parallel between various theories for the propagation of strictly monochromatic scalar electromagnetic waves, and the various two-point correlation functions associated with statistically stationary, partially coherent scalar electromagnetic waves.[29] The analysis of the present sub-section is due to Parrent (1959) (see also Beran and Parrent 1964).

1.9.3.1 *Green function formulation for propagation of cross-spectral density* Consider Fig. 1.8, with all symbols as defined in the associated caption. Suppose the cross-spectral density to be specified, for all frequencies ν, for all pairs of points A_1 and A_2 in the plane $z = 0$. The point A_1 is taken to be located

[29]We have just outlined the angular-spectrum formalism for the propagation of two-point correlation functions, this being the counterpart to the angular-spectrum formalism for the propagation of strictly monochromatic scalar radiation, which was given in Section 1.3. In the present section, we give the partially coherent counterpart to the Rayleigh–Sommerfeld diffraction integral of the first kind, which parallels the corresponding treatment for monochromatic scalar waves given in Section 1.6.2. Note also that there exists a counterpart to the Kirchhoff integral, for the case of two-point correlation functions, which will not be outlined in this book: see Wolf (1955) or Born and Wolf (1999).

at $\mathbf{x}'_1 = (x'_1, y'_1, z'_1 = 0)$, with the point A_2 located at $\mathbf{x}'_2 = (x'_2, y'_2, z'_2 = 0)$. We assume all sources of the field to lie in the half-space $z < 0$. Under this assumption, we wish to solve the diffraction problem of determining the cross-spectral density at any two points P_1 and P_2 in the vacuum-filled half-space $z \geq 0$, which lies downstream of the plane $z = 0$ over which the unpropagated cross-spectral density is specified. The point P_1 is located at $\mathbf{x}_1 = (x_1, y_1, z_1 \geq 0)$, with P_2 located at $\mathbf{x}_2 = (x_2, y_2, z_2 \geq 0)$.

Recall that the cross-spectral density $W(\mathbf{x}_1, \mathbf{x}_2, \nu)$ satisfies the pair of Helmholtz equations (1.112) and (1.113). If $\mathbf{x}_2$ is considered to be held fixed, then $W(\mathbf{x}_1, \mathbf{x}_2, \nu)$ obeys the single Helmholtz equation (1.112). Similarly, if $\mathbf{x}_1$ is considered to be held fixed, then $W(\mathbf{x}_1, \mathbf{x}_2, \nu)$ obeys the single Helmholtz equation (1.113). Having noted the above statements, we recall that a single Helmholtz equation (1.16) is obeyed by a monochromatic scalar electromagnetic field $\psi_\omega(\mathbf{x})$; this was used to construct the first Rayleigh–Sommerfeld integral (1.64), for propagation of the field $\psi_\omega(\mathbf{x})$ into a half-space bounded by a plane over which the unpropagated field is specified.

Bearing all of the above in mind, we can immediately write down a generalized form of the Rayleigh–Sommerfeld diffraction integral of the first kind, yielding a solution to the diffraction problem for the cross-spectral density:

$$W(\mathbf{x}_1, \mathbf{x}_2, \nu) = \frac{1}{(2\pi)^2} \iiiint W(\mathbf{x}'_1, \mathbf{x}'_2, \nu) \left[\frac{\partial}{\partial z_1} \frac{\exp(ikr_{A_1P_1})}{r_{A_1P_1}} \right] \times \left[\frac{\partial}{\partial z_2} \frac{\exp(-ikr_{A_2P_2})}{r_{A_2P_2}} \right] d\mathbf{x}'_1 d\mathbf{x}'_2. \tag{1.128}$$

The above integral is obtained by the following two-step process: (i) Write the first Rayleigh–Sommerfeld integral as a solution to eqn (1.112), with $\mathbf{x}_2$ considered fixed. In writing this expression, make use of the expanding/outgoing spherical wave $\exp(ik|\mathbf{x}_1|)/|\mathbf{x}_1|$ as the appropriate Green function, on account of the fact that $W(\mathbf{x}_1, \mathbf{x}_2, \nu)$ must be outgoing at infinity, in the coordinate $\mathbf{x}_1$. Note that this is the same Green function used to obtain the first Rayleigh–Sommerfeld integral, as given in eqn (1.64). (ii) Using the resulting equation, write the first Rayleigh–Sommerfeld integral as a solution to eqn (1.113), with $\mathbf{x}_1$ considered fixed. In writing this expression, make use of the collapsing/incoming spherical wave $\exp(-ik|\mathbf{x}_2|)/|\mathbf{x}_2|$ as the appropriate Green function, on account of the fact that $W(\mathbf{x}_1, \mathbf{x}_2, \nu)$ must be incoming at infinity, in the coordinate $\mathbf{x}_2$. This requirement is evident from the definition of the mutual coherence function (1.88), from which the cross-spectral density is obtained via eqn (1.110). Note that this is a different Green function to that used to obtain the first Rayleigh–Sommerfeld integral, as given in eqn (1.64).

To proceed further, let us evaluate each of the square brackets appearing in eqn (1.128). Using the quotient rule for differentiation, we have:

$$\frac{\partial}{\partial z_1}\frac{\exp(ikr_{A_1P_1})}{r_{A_1P_1}} = \frac{\exp(ikr_{A_1P_1})}{r_{A_1P_1}}\left(ik - \frac{1}{r_{A_1P_1}}\right)\frac{\partial}{\partial z_1}r_{A_1P_1}. \quad (1.129)$$

By making use of the fact that:

$$r_{A_1P_1} = \sqrt{(x_1 - x_1')^2 + (y_1 - y_1')^2 + z_1^2}, \quad (1.130)$$

one can readily see that:

$$\frac{\partial}{\partial z_1}r_{A_1P_1} = \frac{z_1}{r_{A_1P_1}} = \cos\theta_1, \quad (1.131)$$

where θ_1 is the angle which the line from A_1 to P_1 makes with the positive z-axis (see Fig. 1.8). Substitute the above result into eqn (1.129), to obtain the following expression for the first square-bracketed quantity appearing in eqn (1.128):

$$\frac{\partial}{\partial z_1}\frac{\exp(ikr_{A_1P_1})}{r_{A_1P_1}} = \frac{\exp(ikr_{A_1P_1})}{r_{A_1P_1}}\left(ik - \frac{1}{r_{A_1P_1}}\right)\cos\theta_1. \quad (1.132)$$

In a similar manner, one obtains the following for the second square-bracketed quantity in eqn (1.128):

$$\frac{\partial}{\partial z_2}\frac{\exp(-ikr_{A_2P_2})}{r_{A_2P_2}} = -\frac{\exp(-ikr_{A_2P_2})}{r_{A_2P_2}}\left(ik + \frac{1}{r_{A_2P_2}}\right)\cos\theta_2. \quad (1.133)$$

Here, θ_2 is the angle which the line from A_2 to P_2 makes with the positive z-axis (see Fig. 1.8). Inserting the above two equations into expression (1.128), we obtain the desired propagation law for the cross-spectral density:

$$W(\mathbf{x}_1, \mathbf{x}_2, \nu) = \frac{1}{(2\pi)^2}\iiiint W(\mathbf{x}_1', \mathbf{x}_2', \nu)\frac{\exp[ik(r_{A_1P_1} - r_{A_2P_2})]}{r_{A_1P_1}r_{A_2P_2}} \times \cos\theta_1\cos\theta_2\left(\frac{1}{r_{A_1P_1}} - ik\right)\left(\frac{1}{r_{A_2P_2}} + ik\right)d\mathbf{x}_1'd\mathbf{x}_2'. \quad (1.134)$$

This formula is an integral expression, which independently integrates over the unpropagated cross-spectral density as a function of all pairs of points in the plane $z = 0$, so as to yield the propagated cross-spectral density for any pair of points in the vacuum-filled half space $z \geq 0$. The cross-spectral densities, for each distinct frequency ν, are seen to propagate independently of one another.

1.9.3.2 *Green function formulation for propagation of mutual coherence* Recall eqn (1.109), which furnishes an invertible mapping from the cross-spectral density to the mutual coherence function. Bearing this mapping in mind, take the propagation law (1.134) for the cross-spectral density, multiply both sides by $\exp(-2\pi i\nu\tau)$, and then integrate over all positive frequencies ν. Make use of eqn

(1.109) on the left side of the resulting expression, and then use the fact that $k = 2\pi\nu/c$ on the right side, to give:

$$\Gamma(\mathbf{x}_1, \mathbf{x}_2, \tau) = \frac{1}{(2\pi)^2} \int_0^\infty \iiiint W(\mathbf{x}'_1, \mathbf{x}'_2, \nu) \frac{\exp[2\pi i\nu c^{-1}(r_{A_1P_1} - r_{A_2P_2})]}{r_{A_1P_1} r_{A_2P_2}} \times \left[\frac{1}{r_{A_1P_1}} + \frac{(-2\pi i\nu)}{c}\right]\left[\frac{1}{r_{A_2P_2}} - \frac{(-2\pi i\nu)}{c}\right] \times \cos\theta_1 \cos\theta_2 \exp(-2\pi i\nu\tau)\, d\mathbf{x}'_1 d\mathbf{x}'_2 d\nu. \tag{1.135}$$

In what amounts to a use of the Fourier derivative theorem in reverse, make the replacement $(-2\pi i\nu) \rightarrow \partial/\partial\tau$, in each of the square brackets on the middle line of the above equation. Then, interchange the order of the quadruple integral over space and the integral over frequency. This leaves:

$$\Gamma(\mathbf{x}_1, \mathbf{x}_2, \tau) = \frac{1}{(2\pi)^2} \iiiint \frac{\cos\theta_1 \cos\theta_2}{r_{A_1P_1} r_{A_2P_2}} \left[\frac{1}{r_{A_1P_1}} + \frac{1}{c}\frac{\partial}{\partial\tau}\right]\left[\frac{1}{r_{A_2P_2}} - \frac{1}{c}\frac{\partial}{\partial\tau}\right] \times \left\{\int_0^\infty W(\mathbf{x}'_1, \mathbf{x}'_2, \nu) \exp\left[-2\pi i\nu\left(\tau - \frac{r_{A_1P_1} - r_{A_2P_2}}{c}\right)\right] d\nu\right\} d\mathbf{x}'_1 d\mathbf{x}'_2. \tag{1.136}$$

Use eqn (1.109) to re-write the quantity in braces in terms of the mutual coherence function, thereby arriving at Parrent's formula for the propagation of mutual coherence (Parrent 1959):

$$\Gamma(\mathbf{x}_1, \mathbf{x}_2, \tau) = \frac{1}{(2\pi)^2} \iiiint \frac{\cos\theta_1 \cos\theta_2}{r_{A_1P_1} r_{A_2P_2}} \times \left[\frac{1}{r_{A_1P_1}} + \frac{1}{c}\frac{\partial}{\partial\tau}\right]\left[\frac{1}{r_{A_2P_2}} - \frac{1}{c}\frac{\partial}{\partial\tau}\right] \times \Gamma\left(\mathbf{x}'_1, \mathbf{x}'_2, \tau - \frac{r_{A_1P_1} - r_{A_2P_2}}{c}\right) d\mathbf{x}'_1 d\mathbf{x}'_2. \tag{1.137}$$

This formula is an integral expression that independently integrates over all pairs of points $\mathbf{x}'_1, \mathbf{x}'_2$ in the plane of the unpropagated mutual coherence function (together with its first and second derivative with respect to the time-lag τ), so as to yield the propagated mutual coherence function for any pair of points $\mathbf{x}_1, \mathbf{x}_2$ in the vacuum-filled half-space $z \geq 0$. The mutual coherence functions, for each distinct τ, evidently do not propagate independently of one another.

1.9.3.3 *Green function formulation for propagation of mutual intensity* Recall that the mutual intensity and the cross-spectral density obey a pair of Helmholtz equations, as given in (1.112) and (1.113) for the cross-spectral density, and

(1.114) for the mutual intensity. As has already been noted, the former pair of equations may be mapped onto the latter, by replacing the cross-spectral density with the mutual intensity, and replacing the frequency ν with the mean frequency $\bar{\nu}$ of a quasi-monochromatic field. Therefore, using an analogous argument to that presented in Section 1.9.2.2, the propagation law (1.134) for the cross-spectral density may be adapted to give the following propagation law for the mutual intensity, for the case of quasi-monochromatic fields for which all relevant time-lags are sufficiently small for eqn (1.104) to be a good approximation:

$$J(\mathbf{x}_1, \mathbf{x}_2) = \frac{1}{(2\pi)^2} \iiiint J(\mathbf{x}'_1, \mathbf{x}'_2) \frac{\exp[i\bar{k}(r_{A_1P_1} - r_{A_2P_2})]}{r_{A_1P_1} r_{A_2P_2}} \times \cos\theta_1 \cos\theta_2 \left(\frac{1}{r_{A_1P_1}} - i\bar{k}\right)\left(\frac{1}{r_{A_2P_2}} + i\bar{k}\right) d\mathbf{x}'_1 d\mathbf{x}'_2. \tag{1.138}$$

1.9.4 *Van Cittert–Zernike theorem for propagation of mutual intensity*

Here, we obtain two approximate forms for the propagation law governing mutual intensity, both of which shall be obtained from eqn (1.138). The first of these approximate forms is a formula for propagating the mutual intensity through a distance that is much greater than the mean wavelength of the radiation, subject to the paraxial approximation. The second approximate form, known as the van Cittert–Zernike theorem, specializes the first approximate formula to the case of an incoherent extended source in the plane $z = 0$.

With reference to Fig. 1.8, make the small-angle (paraxial) approximation that all relevant angles θ_1 and θ_2 have magnitudes that are very small compared to unity. Further, assume the propagation distances $r_{A_1P_1}$ and $r_{A_2P_2}$ to be very much larger than the mean wavelength $\bar{\lambda} = 2\pi/\bar{k}$ of the radiation. Under these approximations, we have:

$$\cos\theta_1 \approx 1, \quad \cos\theta_2 \approx 1, \quad \frac{1}{r_{A_1P_1}} - i\bar{k} \approx -i\bar{k}, \quad \frac{1}{r_{A_2P_2}} + i\bar{k} \approx i\bar{k}, \tag{1.139}$$

which may be substituted into eqn (1.138) to give (Zernike 1938):

$$J(\mathbf{x}_1, \mathbf{x}_2) = \frac{\bar{k}^2}{(2\pi)^2} \iiiint J(\mathbf{x}'_1, \mathbf{x}'_2) \frac{\exp[i\bar{k}(r_{A_1P_1} - r_{A_2P_2})]}{r_{A_1P_1} r_{A_2P_2}} d\mathbf{x}'_1 d\mathbf{x}'_2. \tag{1.140}$$

The above diffraction integral allows one to propagate the mutual intensity associated with a paraxial quasi-monochromatic field, given the values which the mutual intensity takes at all pairs of points $\mathbf{x}'_1, \mathbf{x}'_2$ lying in the plane $z = 0$. Setting $\mathbf{x}_1 = \mathbf{x}_2$ in the above expression, and making use of the fact that $J(\mathbf{x}_1, \mathbf{x}_1)$ is equal to the time-averaged intensity $I(\mathbf{x}_1)$ at the point $\mathbf{x}_1$ lying in the half-space

$z \geq 0$, we reach the important conclusion that the propagated time-averaged intensity is not simply a function of the intensity $I(\mathbf{x}'_1) = J(\mathbf{x}'_1, \mathbf{x}'_1)$ at each point $\mathbf{x}'_1$ in the plane $z = 0$, but also depends on the 'off diagonal' components of the mutual intensity, namely those pairs of points $\mathbf{x}'_1, \mathbf{x}'_2$ for which the unpropagated mutual intensity is non-zero. Bearing in mind our earlier discussions on the Young interferometer, this state of affairs is rather natural: the existence of non-zero off-diagonal components, in the unpropagated mutual intensity, is indicative of field correlations existing in the plane $z = 0$; in turn, these correlations will lead to interference effects when the disturbances from each of the points are allowed to propagate through free space, before being superposed at a given observation point.

Equation (1.140) is the first of the two approximate propagation formulae, for the mutual intensity, to be considered in the present sub-section. The second such formula, namely the van Cittert–Zernike theorem, specializes the above equation to the case of an incoherent plane quasi-monochromatic source lying in the plane $z = 0$. Such a source may be described by the following mutual intensity:

$$J(\mathbf{x}'_1, \mathbf{x}'_2) = I(\mathbf{x}'_1)\delta(\mathbf{x}'_1 - \mathbf{x}'_2), \tag{1.141}$$

where δ is the two-dimensional Dirac delta. By assumption, therefore, disturbances at distinct points on the source are uncorrelated with one another. Substituting the above expression into eqn (1.140), we arrive at the van Cittert–Zernike theorem (van Cittert 1934; Zernike 1938):

$$J(\mathbf{x}_1, \mathbf{x}_2) = \frac{\overline{k}^2}{(2\pi)^2} \iint I(\mathbf{x}'_1) \frac{\exp[i\overline{k}(r_{A_1P_1} - r_{A_1P_2})]}{r_{A_1P_1} r_{A_1P_2}} d\mathbf{x}'_1. \tag{1.142}$$

1.10 Higher-order correlation functions

In our preceding discussions on partial coherence, we focussed attention on the two-point correlation functions associated with ergodic statistically stationary partially coherent fields. However, as mentioned in Section 1.7.1, two-point correlation functions do not, in general, capture all of the information associated with such random processes. In addition to the two-point correlation function furnished by the mutual coherence, together with its normalized form given by the complex degree of coherence, one can envisage an infinite hierarchy of correlation functions associated with more than two space–time points in a given statistically stationary partially coherent field. Non-zero values for such higher-order correlation functions are then indicative of the correlations that exist between the disturbances at more than two space–time points.

Along this line of development, Glauber (1963) has introduced an infinite hierarchy of normalized correlation functions, the first of which may be associated with the complex degree of coherence. The previously stated condition for coherence—that is, that the modulus of the complex degree of coherence must be unity—is now spoken of as 'second-order coherence'.[30] In order to be fully coherent, according to Glauber's theory, all normalized correlation functions should have modulus unity, not just the two-point correlations. As one can readily show, a strictly monochromatic field obeys this criterion: indeed, since the time-development of such a field is purely harmonic, at each point in space, the disturbances at all sets of space–time points will be perfectly correlated. Strictly monochromatic fields are therefore fully coherent, that is, coherent to all orders in the sense described by Glauber.

Of particular interest are statistically stationary partially coherent scalar fields described by Gaussian random processes. It turns out that such processes are completely characterized by their mutual coherence function, with the corollary that no further information is provided by the previously mentioned higher-order correlation functions. This observation is of particular utility in the context of intensity interferometry using partially coherent electromagnetic wave-fields, which is to be described in our later discussions on the X-ray analogue of the classic series of experiments by Hanbury Brown and Twiss (see Section 4.6.3). Here, intensity correlations—which are related to correlation functions that are fourth order in the field quantities—are used to determine the second-order coherence properties of such a field.

1.11 Summary

The present chapter was devoted to a study of X-ray wave-fields in free space, laying much of the theoretical groundwork that is to be applied in later chapters. Broadly speaking, this chapter was split into the three parts consecutively summarized below.

In the first part, vacuum wave equations were derived for electromagnetic fields in free space, taking the Maxwell equations as a starting point. We saw that the free-space electric and magnetic fields both obey d'Alembert wave equations. Some remarks were then given regarding how one can make a transition from a vector to a scalar theory, with the electromagnetic disturbance described by a complex scalar function of position and time that is known as the complex analytic signal, rather than by the electric and magnetic field vectors at each space–time point. Such a complex scalar disturbance, which also obeys the d'Alembert equation in free space, is comprised only of non-negative frequencies. Each monochromatic component, of a particular instance of such a field, was seen to obey the Helmholtz equation.

The second part of this chapter dealt with the optics of strictly monochromatic scalar electromagnetic fields, based on the Helmholtz equation. We began with a description of the angular-spectrum formalism, for solving plane-to-plane diffraction problems for coherent scalar electromagnetic waves in vacuum. This led to the development of a diffraction operator, which could be used to act upon the coherent disturbance in a given plane, so as to propagate it through

[30]Glauber (1963) denotes this 'first-order coherence' rather than 'second-order coherence'.

a specified distance. The Fresnel and Fraunhofer approximations were outlined, with both being developed as special cases of the angular spectrum formalism. The Kirchhoff and Rayleigh–Sommerfeld diffraction theory was also described.

The third and final part of this chapter dealt with the optics of partially coherent fields. Several key concepts were introduced via the simple example of unfair dice, including the concepts of a random variable, random process, statistical stationarity, correlation functions, ensemble versus time averaging, and ergodicity. The notion of partial coherence, lying somewhere between complete coherence and complete incoherence, was then introduced. Elementary discussions on spatial and temporal coherence were given, before launching into a more formal treatment based on the mutual coherence function. A diffraction theory was developed for the mutual coherence function, together with related two-point correlation functions known as the cross-spectral density and mutual intensity. The last of these is useful for a description of quasi-monochromatic fields, namely those for which the spread of frequencies present is much smaller than the mean frequency. Finally, we made some brief remarks on higher-order correlation functions.

Further reading

Born and Wolf (1999) and Mandel and Wolf (1995) are comprehensive works on the optics of coherent and partially coherent electromagnetic fields, to which the reader can turn for a more detailed exposition of most of the topics given in the present chapter. Nieto-Vesperinas (1991) is a very useful volume on the diffraction theory of coherent fields. Regarding the optics of partially coherent fields, see Beran and Parrent (1964), Papoulis (1968), Peřina (1972), Marathay (1982), and Goodman (1985). The volume edited by Oughstun (1992) provides a very useful compendium of primary literature on scalar wave diffraction, with the two volumes edited by Mandel and Wolf (1990) reprinting many of the classic papers in the theory of partial coherence. For a detailed coverage of Fourier optics, see Bracewell (1965), Goodman (1968), Reynolds *et al.* (1989), and Cowley (1995).

References

B.B. Baker and E.T. Copson, *The mathematical theory of Huygens' principle*, second edition, Oxford University Press, Oxford (1950).

M.J. Beran and G.B. Parrent, *Theory of partial coherence*, Prentice-Hall, Englewood Cliffs (1964).

M. Born and E. Wolf, *Principles of optics*, seventh edition, Cambridge University Press, Cambridge (1999).

R.N. Bracewell, *The Fourier transform and its applications*, McGraw-Hill Book Company, New York (1965).

P.H. van Cittert, *Die wahrscheinliche Schwingungsverteilung in einer von einer Lichtquelle direkt oder mittels einer Linse beleuchteten Ebene*, Physica **1**, 201–210 (1934).

J.M. Cowley, *Diffraction physics*, third revised edition, North-Holland, Amsterdam (1995).

D. Gabor, *Theory of communication*, J. Instn. Elect. Engrs. **93**, 429–457 (1946).

R.J. Glauber, *The quantum theory of optical coherence*, Phys. Rev. **130**, 2529–2539 (1963).

R.H. Good, *Classical electromagnetism*, Saunders College Publishing, Fort Worth (1999).

J.W. Goodman, *Introduction to Fourier optics*, McGraw-Hill, San Francisco, CA (1968).

J.W. Goodman, *Statistical optics*, John Wiley & Sons, New York (1985).

H.S. Green and E. Wolf, *A scalar representation of electromagnetic fields*, Proc. Phys. Soc. A **66**, 1129–1137 (1953).

T.E. Gureyev, Private communication (1994).

E. Hecht, *Optics*, second edition, Addison-Wesley, Reading, MA (1987).

J.D. Jackson, *Classical electrodynamics*, third edition, John Wiley & Sons, New York (1999).

E. Kreysig, *Advanced engineering mathematics*, fifth edition, John Wiley & Sons, New York (1983).

E. Lalor, *Inverse wave propagator*, J. Math. Phys. **9**, 2001–2006 (1968).

S.G. Lipson and H. Lipson, *Optical physics*, second edition, Cambridge University Press, Cambridge (1981).

L. Mandel and E. Wolf, *Coherence properties of optical fields*, Rev. Mod. Phys. **37**, 231–287 (1965).

L. Mandel and E. Wolf (eds), *Selected papers on coherence and fluctuations of light (1850–1966)*, SPIE Milestone Series volume M19 (in two parts), SPIE Optical Engineering Press, Bellingham (1990).

L. Mandel and E. Wolf, *Optical coherence and quantum optics*, Cambridge University Press, Cambridge (1995).

A.S. Marathay, *Elements of optical coherence theory*, John Wiley & Sons, New York (1982).

A.S. Marathay and G.B. Parrent Jr, *Use of a scalar theory in optics*, J. Opt. Soc. Am. **60**, 243–245 (1970).

E.W. Marchand and E. Wolf, *Angular correlation and the far-zone behaviour of partially coherent fields*, J. Opt. Soc. Am. **62**, 379–385 (1972).

W.D. Montgomery, *Algebraic formulation of diffraction applied to self-imaging*, J. Opt. Soc. Am. **58**, 1112–1124 (1968).

W.D. Montgomery, *Distribution formulation of diffraction. The monochromatic case*, J. Opt. Soc. Am. **59**, 136–141 (1969).

W.D. Montgomery, *Unitary operators in the homogeneous wave field*, Opt. Lett. **6**, 314–315 (1981).

M. Nazarathy and J. Shamir, *Fourier optics described by operator algebra*, J. Opt. Soc. Am. **70**, 150–159 (1980).

M. Nieto-Vesperinas, *Scattering and diffraction in physical optics*, John Wiley & Sons, New York (1991).

H.C. Ohanian, *Classical electrodynamics*, Allyn and Bacon, Boston, MA (1988).

K.E. Oughstun (ed.), *Selected papers on scalar wave diffraction*, SPIE Milestone Series volume M51, SPIE Optical Engineering Press, Bellingham (1992).

W.K.H. Panofsky and M. Phillips, *Classical electricity and magnetism*, Addison Wesley, Reading, MA (1962).

A. Papoulis, *Systems and transforms with applications in optics*, McGraw-Hill, New York (1968).

G.B. Parrent, *On the propagation of mutual coherence*, J. Opt. Soc. Am. **49**, 787–793 (1959).

J. Peřina, *Coherence of light*, Van Nostrand Reinhold Company, London (1972).

W.H. Press, S.A. Teukolsky, W.M. Vetterling, and B.P. Flannery, *Numerical recipes in FORTRAN*, second edition, Cambridge University Press, Cambridge (1992).

G.O. Reynolds, J.B. DeVelis, G.B. Parrent, and B.J. Thompson, *The new physical optics notebook: tutorials in Fourier optics*, SPIE Optical Engineering Press, Bellingham (1989).

B.E.A. Saleh and M.C. Teich, *Fundamentals of photonics*, John Wiley & Sons, New York (1991).

G.C. Sherman, *Diffracted wave fields expressible by plane-wave expansions containing only homogeneous components*, Phys. Rev. Lett. **21**, 761–764 (1968).

J.R. Shewell and E. Wolf, *Inverse diffraction and a new reciprocity theorem*, J. Opt. Soc. Am. A **58**, 1596–1603 (1968).

M.R. Spiegel, *Theory and problems of complex variables*, McGraw-Hill Book Company, Singapore (1964).

E. Wolf, *A macroscopic theory of the interference and diffraction of light from finite sources. I. Fields with narrow spectral range*, Proc. Roy. Soc. Ser. A **225**, 96–111 (1954a).

E. Wolf, *Optics in terms of observable quantities*, Nuovo Cim. **12**, 884–888 (1954b).

E. Wolf, *A macroscopic theory of the interference and diffraction of light from finite sources. II. Fields with a spectral range of arbitrary width*, Proc. Roy. Soc. Ser. A **230**, 246–265 (1955).

E. Wolf, *A scalar representation of electromagnetic fields: II*, Proc. Phys. Soc. A **74**, 269–280 (1959).

E. Wolf and J.R. Shewell, *The inverse wave propagator*, Phys. Lett. A **25**, 417–418 (1967); see also E. Wolf and J.R. Shewell, *Errata*, Phys. Lett. A **26**, 104 (1967).

F. Zernike, *The concept of degree of coherence and its application to optical problems*, Physica **5**, 785–795 (1938).

2

X-ray interactions with matter

The manner, in which X-rays interact with matter, is a subject of central importance to coherent X-ray optics. Indeed, it is the coupling of X-ray radiation to matter—a coupling which is often sufficiently strong for the matter to imprint some information regarding its state upon the X-ray wave-field passing through it, while being sufficiently weak for the radiation to probe the full three-dimensional volume occupied by the sample—which makes the former such a powerful tool for probing the latter. Accordingly, the purpose of this chapter is to study the physics behind X-ray interactions with matter, arriving at several key results that will be freely drawn upon in later chapters.

We shall begin by deriving the wave equations governing the evolution of the electric and magnetic fields, in the presence of matter, by taking the Maxwell equations as a starting point. Assuming all scattering materials to be non-magnetic, and to be slowly varying in their material properties over length scales on the order of the wavelength of the illuminating radiation, one can make a transition from a vector theory of the electromagnetic field (in which the field is described by an electric and a magnetic field vector at each point in space, at each instant of time) to a scalar theory (where the field is described by a single complex scalar function of space and time). This leads to a time-dependent scalar wave equation for the electromagnetic disturbance in the presence of matter. We then obtain a corresponding time-independent theory by Fourier decomposing the scalar electromagnetic field, which may correspond to either a polychromatic field or a given realization of a stochastic process describing a partially coherent field, into a superposition of strictly monochromatic fields. Each such monochromatic component is found to obey an 'inhomogeneous' Helmholtz equation in which the scattering term is a specified function of the refractive index of the material. Note that we shall show how the refractive index can be expressed in terms of the electrical permittivity and magnetic permeability of the medium.

The 'inhomogeneous' Helmholtz equation, which is a central equation for describing the interaction of monochromatic X-ray scalar wave-fields with matter, can rarely be solved exactly. Rather, one must resort to various means of approximation, several of which are discussed in this chapter. The first of these is the projection approximation, which, within its domain of applicability, gives a straightforward means for calculating the phase and amplitude shifts that are imparted on a paraxial monochromatic scalar electromagnetic wave traversing a weakly refracting non-magnetic sample. Upon revisiting the notion of a Green function, in the context of X-ray scattering, we are able to study another class of approximations known as the first Born and higher-order Born approximations.

The first Born approximation is closely linked to the notion of the Ewald sphere, which is so often used in the study of X-ray diffraction from crystals. We give a brief treatment of the Ewald sphere, which shows how it emerges naturally from the first Born approximation. Further, we show how the concept of the Ewald sphere remains well defined even when the scatterers are non-crystalline. The multislice and eikonal approximations, for studying the interactions of X-rays with matter, are also discussed. A largely qualitative treatment is given of inelastic scattering, focussing on the Compton and photoelectric effects, including a mention of how these mechanisms are related to X-ray absorption and fluorescence. We close the chapter with some remarks on the information content of scattered wave-fields.

Throughout, some emphasis is placed on the relationships between the various topics under discussion. In this context, we have already mentioned the link between the first Born approximation and the Ewald sphere. As another example, in deriving an approximate expression for the X-ray refractive index in terms of the electron density of a sample, one obtains an appreciation of the link between the scattering and refraction viewpoints of the interaction of X-rays with matter. As a last example, our studies of the eikonal approximation shed some light on the relation between the ray and wave formalisms for X-ray radiation. The multiplicity of viewpoints is related through the fact that they are all ultimately derivable from the Maxwell equations, which govern the evolution of classical electromagnetic fields in the presence of matter.

2.1 Wave equations in the presence of scatterers

The central task of this section is to determine certain differential equations that govern the evolution of electromagnetic fields in the presence of scattering media. Our starting point is the Maxwell equations (see, for example, Jackson (1999)):

$$\nabla \cdot \mathbf{D}(x,y,z,t) = \rho(x,y,z,t), \tag{2.1}$$

$$\nabla \cdot \mathbf{B}(x,y,z,t) = 0, \tag{2.2}$$

$$\nabla \times \mathbf{E}(x,y,z,t) + \frac{\partial}{\partial t}\mathbf{B}(x,y,z,t) = \mathbf{0}, \tag{2.3}$$

$$\nabla \times \mathbf{H}(x,y,z,t) - \frac{\partial}{\partial t}\mathbf{D}(x,y,z,t) = \mathbf{J}(x,y,z,t). \tag{2.4}$$

Here, $\mathbf{D}$ is the electric displacement, $\mathbf{B}$ is the magnetic induction, $\mathbf{E}$ is the electric field, $\mathbf{H}$ is the magnetic field, ρ is the charge density, $\mathbf{J}$ is the current density, (x,y,z) are Cartesian coordinates in three-dimensional space, and t is time. Equation (2.1) is Gauss' Law for the proportionality of the electric displacement flux through a closed surface and the total charge enclosed within that surface. Equation (2.2) asserts the non-existence of magnetic monopoles. Equation (2.3) is Faraday's Law of induction, while eqn (2.4) is Maxwell's modification of Ampère's Law.

In general, both the electric displacement $\mathbf{D}$ and the magnetic induction $\mathbf{B}$ will be functions of $\mathbf{E}$ and $\mathbf{H}$, respectively. We restrict consideration to linear isotropic materials, in which $\mathbf{D} = \varepsilon\mathbf{E}$ and $\mathbf{B} = \mu\mathbf{H}$, where ε and μ respectively denote the electrical permittivity and the magnetic permeability of the material.[31] Notwithstanding the fact that we have thereby excluded a rich variety of nonlinear phenomena, and any account of non-isotropic response, these two approximations allow us to rewrite the Maxwell equations as:

$$\nabla \cdot [\varepsilon(x,y,z,t)\mathbf{E}(x,y,z,t)] = \rho(x,y,z,t), \tag{2.5}$$

$$\nabla \cdot \mathbf{B}(x,y,z,t) = 0, \tag{2.6}$$

$$\nabla \times \mathbf{E}(x,y,z,t) + \frac{\partial}{\partial t}\mathbf{B}(x,y,z,t) = \mathbf{0}, \tag{2.7}$$

$$\nabla \times \left[\frac{\mathbf{B}(x,y,z,t)}{\mu(x,y,z,t)}\right] - \frac{\partial}{\partial t}[\varepsilon(x,y,z,t)\mathbf{E}(x,y,z,t)] = \mathbf{J}(x,y,z,t). \tag{2.8}$$

With a view to determining the wave equation for the electric field, take the curl of eqn (2.7) and then make use of the vector identity:

$$\nabla \times [\nabla \times \mathbf{g}(x,y,z)] = \nabla [\nabla \cdot \mathbf{g}(x,y,z)] - \nabla^2 \mathbf{g}(x,y,z) \tag{2.9}$$

for any suitably well-behaved vector field $\mathbf{g}(x,y,z)$. One thereby obtains:

$$\nabla [\nabla \cdot \mathbf{E}(x,y,z,t)] - \nabla^2 \mathbf{E}(x,y,z,t) + \frac{\partial}{\partial t}\nabla \times \mathbf{B}(x,y,z,t) = \mathbf{0}. \tag{2.10}$$

To proceed further, recall the vector identity:

$$\nabla \times [f(x,y,z)\mathbf{g}(x,y,z)] = f(x,y,z)[\nabla \times \mathbf{g}(x,y,z)] + [\nabla f(x,y,z)] \times \mathbf{g}(x,y,z), \tag{2.11}$$

for suitably well-behaved scalar and vector fields $f(x,y,z)$ and $\mathbf{g}(x,y,z)$. Apply this identity to the first term on the left-hand side of (2.8), with $f \equiv \mu^{-1}$ and $\mathbf{g} \equiv \mathbf{B}$, and then solve the resulting equation for $\nabla \times \mathbf{B}(x,y,z,t)$. Use this expression to eliminate $\nabla \times \mathbf{B}(x,y,z,t)$ from eqn (2.10), leaving:

$$\begin{aligned} &\nabla [\nabla \cdot \mathbf{E}(x,y,z,t)] - \nabla^2 \mathbf{E}(x,y,z,t) + \frac{\partial}{\partial t}[\nabla \log_e \mu(x,y,z,t) \times \mathbf{B}(x,y,z,t)] \\ &+ \frac{\partial}{\partial t}\left\{\mu(x,y,z,t)\frac{\partial}{\partial t}[\varepsilon(x,y,z,t)\mathbf{E}(x,y,z,t)]\right\} \\ &+ \frac{\partial}{\partial t}[\mu(x,y,z,t)\mathbf{J}(x,y,z,t)] = \mathbf{0}. \end{aligned} \tag{2.12}$$

[31] We are excluding, for example, the case of both ferroelectric and ferromagnetic materials, for which the values of $\mathbf{D}$ and $\mathbf{B}$ depend on the previous history of the material, rather than solely upon the applied fields at a given instant of time.

Assume that the material is static, so that both μ and ε are independent of time. Thus the above equation becomes:

$$\begin{aligned}&\left[\varepsilon(x,y,z)\mu(x,y,z)\frac{\partial^2}{\partial t^2}-\nabla^2\right]\mathbf{E}(x,y,z,t)\\&\quad-\left[\nabla\log_e\mu(x,y,z)\right]\times\left[\nabla\times\mathbf{E}(x,y,z,t)\right]\\&\quad=-\nabla\left[\nabla\cdot\mathbf{E}(x,y,z,t)\right]-\mu(x,y,z)\frac{\partial}{\partial t}\mathbf{J}(x,y,z,t),\end{aligned}\tag{2.13}$$

where we have made use of eqn (2.3) to replace $\partial\mathbf{B}/\partial t$ by $-\nabla\times\mathbf{E}$. This is the desired partial differential equation governing the spatial and temporal evolution of the electric field. Note that the individual components of the electric field vector are coupled to one another, through both the last term on the left-hand side, and the first term on the right-hand side. This implies that the three components of the electric field vector are not independent of one another, in general.

Using a similar line of reasoning to that which led to eqn (2.13), and with the same assumption that μ and ε are independent of time, one arrives at the following expression governing the magnetic field $\mathbf{H}$:

$$\begin{aligned}&\left[\varepsilon(x,y,z)\mu(x,y,z)\frac{\partial^2}{\partial t^2}-\nabla^2\right]\mathbf{H}(x,y,z,t)\\&=\nabla\times\mathbf{J}(x,y,z,t)-\nabla\left[\nabla\cdot\mathbf{H}(x,y,z,t)\right]\\&\quad+\frac{1}{\varepsilon(x,y,z)}\left\{\nabla\varepsilon(x,y,z)\times\left[\nabla\times\mathbf{H}(x,y,z,t)\right]-\nabla\varepsilon(x,y,z)\times\mathbf{J}(x,y,z,t)\right\}.\end{aligned}\tag{2.14}$$

Evidently, the various components of the magnetic field vector are in general coupled to one another.

In addition to this, if we now consider eqns (2.13) and (2.14) as a pair, we see that the electric and magnetic fields are coupled to one another through the sources of these fields, namely the current and charge densities.[32] These various couplings—between different Cartesian components of the electric field, between different Cartesian components of the magnetic field, and between the electric and magnetic fields—are lifted *in vacuo*, with eqns (2.13) and (2.14) then reducing to the d'Alembert vacuum wave equations for the electric and magnetic field (eqns (1.7) and (1.8)):

[32]Note that the charge density appears implicitly in one of the equations given here, as the term containing the divergence of the electric field can be re-expressed in terms of the electric charge density.

$$\left(\varepsilon_0\mu_0\frac{\partial^2}{\partial t^2}-\nabla^2\right)\mathbf{E}(x,y,z,t)=\mathbf{0}, \tag{2.15}$$

$$\left(\varepsilon_0\mu_0\frac{\partial^2}{\partial t^2}-\nabla^2\right)\mathbf{H}(x,y,z,t)=\mathbf{0}. \tag{2.16}$$

We remind the reader that ε_0 and μ_0 respectively denote the electrical permittivity of free space, and the magnetic permeability of free space. In the above pair of equations, the separate components of the electric and magnetic fields are decoupled from one another. As we saw, this decoupling motivates the introduction of a single scalar function $\Psi(x,y,z,t)$ to describe the vacuum disturbance due to an electromagnetic field, which satisfies the wave equation (see eqns (1.10) and (1.13)):

$$\left(\frac{1}{c^2}\frac{\partial^2}{\partial t^2}-\nabla^2\right)\Psi(x,y,z,t)=0. \tag{2.17}$$

Now, the purpose of this chapter is to discuss the interactions of X-rays with matter. Accordingly, we seek to generalize the above scalar theory of vacuum electromagnetic fields so as to incorporate interactions of these fields with matter. To this end let us return to eqns (2.13) and (2.14), restricting ourselves to non-magnetic materials,[33] so that $\mu(x,y,z)=\mu_0$. Further, we will assume that neither current densities nor charge densities are present; thus we take $\rho(x,y,z,t)=0$ and $\mathbf{J}(x,y,z,t)=\mathbf{0}$. Under these assumptions, eqns (2.13) and (2.14) become:

$$\left[\varepsilon(x,y,z)\mu_0\frac{\partial^2}{\partial t^2}-\nabla^2\right]\mathbf{E}(x,y,z,t)=-\nabla\left[\nabla\cdot\mathbf{E}(x,y,z,t)\right] \tag{2.18}$$

$$\left[\varepsilon(x,y,z)\mu_0\frac{\partial^2}{\partial t^2}-\nabla^2\right]\mathbf{H}(x,y,z,t)=\frac{1}{\varepsilon(x,y,z)}\nabla\varepsilon(x,y,z)\times\left[\nabla\times\mathbf{H}(x,y,z,t)\right]. \tag{2.19}$$

While this pair of equations uncouples the electric from the magnetic fields, individual components of each of these fields remain coupled to one another through the right-hand sides of these equations. This is indicative of depolarization, namely the process where interaction with matter may mix different linear polarization states of the electric and magnetic fields.

To proceed further we make another assumption, that the scatterers are sufficiently slowly varying over length scales comparable to the wavelength of the X-ray radiation, such that one may neglect the right-hand sides of the above pairs of equations, leading to:

[33]In this book, we make no mention of the important subject of magnetic scattering, which exploits the interaction of X-ray photons with the magnetic properties of scattering materials. Treatments on this subject, together with references to the primary literature, include Altarelli (1993), Lovesey and Collins (1996) and Gibbs *et al.* (2002).

$$\left[\varepsilon(x,y,z)\mu_0\frac{\partial^2}{\partial t^2}-\nabla^2\right]\mathbf{E}(x,y,z,t)=\mathbf{0}, \tag{2.20}$$

$$\left[\varepsilon(x,y,z)\mu_0\frac{\partial^2}{\partial t^2}-\nabla^2\right]\mathbf{H}(x,y,z,t)=\mathbf{0}. \tag{2.21}$$

Since there is now no mixing between any of components of the electric and magnetic field vectors, one can immediately make a transition to a scalar theory, with this scalar being equal to any single component of the electromagnetic field.[34] This scalar theory models the electromagnetic field by the disturbance $\Psi(x,y,z,t)$, which obeys the differential equation:

$$\left[\varepsilon(x,y,z)\mu_0\frac{\partial^2}{\partial t^2}-\nabla^2\right]\Psi(x,y,z,t)=0. \tag{2.22}$$

While the quantity $\Psi(x,y,z,t)$ is intrinsically real, it is nevertheless convenient to treat this as a complex disturbance—namely, the analytic signal discussed in Section 1.2—whose real part corresponds to the physical field. Subject to the approximations catalogued earlier, eqn (2.22) then describes the spatial and temporal evolution of the complex scalar electromagnetic wave $\Psi(x,y,z,t)$. This complex disturbance may be expressed as a continuous superposition of monochromatic components, via a Fourier integral over non-negative angular frequencies ω (see eqn (1.14)):

$$\Psi(x,y,z,t)=\frac{1}{\sqrt{2\pi}}\int_0^\infty \psi_\omega(x,y,z)\exp(-i\omega t)d\omega. \tag{2.23}$$

Physically, this amounts to decomposing $\Psi(x,y,z,t)$ as a continuous sum of infinitely many strictly monochromatic fields $\psi_\omega(x,y,z)\exp(-i\omega t)$. Substitute the above decomposition into eqn (2.22), bring the differential operator inside the integral sign, replace $\partial^2/\partial t^2$ by $(-i\omega)^2=-\omega^2$ and then cancel a common factor of -1, to give:

$$\int_0^\infty \left\{\left[\nabla^2+\varepsilon(x,y,z)\mu_0\omega^2\right]\psi_\omega(x,y,z)\right\}\exp(-i\omega t)d\omega=0. \tag{2.24}$$

The quantity in braces must vanish. Therefore:

$$\left[\nabla^2+\varepsilon_\omega(x,y,z)\mu_0 c^2k^2\right]\psi_\omega(x,y,z)=0, \tag{2.25}$$

where we have made use of the fact that $\omega=ck$, and put an ω subscript on the electrical permittivity so as to indicate that this quantity will in general depend on frequency.

[34] One need not be so restrictive—see references cited at the end of Section 1.1.

Next, we identify $\varepsilon_\omega(x,y,z)\mu_0 c^2$ with the square of the frequency-dependent refractive index $n_\omega(x,y,z)$, so that:

$$n_\omega(x,y,z) = c\sqrt{\varepsilon_\omega(x,y,z)\mu_0} = \sqrt{\frac{\varepsilon_\omega(x,y,z)}{\varepsilon_0}}. \tag{2.26}$$

To motivate this identification, consider the form that eqn (2.25) takes in a medium with a uniform value for ε, and then write $k = 2\pi/\lambda$, where λ is the vacuum wavelength of the radiation, to give:

$$\left[\nabla^2 + \left(\frac{2\pi}{\lambda/\{c\sqrt{\varepsilon_\omega(x,y,z)\mu_0}\}}\right)^2\right]\psi_\omega(x,y,z) = 0. \tag{2.27}$$

Since we require the wavelength in the medium to be equal to the vacuum wavelength divided by the refractive index of the medium, we immediately see that the quantity in braces above is equal to the refractive index of the uniform medium. Generalizing this notion to non-magnetic media whose electrical permittivity varies slowly over length scales comparable to the wavelength of the monochromatic X-ray field, we obtain the expression for the position-dependent refractive index given in eqn (2.26).

Using expression (2.26) for the refractive index, eqn (2.25) becomes the following 'inhomogeneous' Helmholtz equation[35]:

$$\left[\nabla^2 + k^2 n_\omega^2(x,y,z)\right]\psi_\omega(x,y,z) = 0. \tag{2.28}$$

This equation, which governs the scattering of the spatial part $\psi_\omega(x,y,z)$ of a particular monochromatic component $\psi_\omega(x,y,z)\exp(-i\omega t)$ in the spectral decomposition of the scalar field $\Psi(x,y,z,t)$ given in eqn (2.23), is a central result of the present chapter. Since the governing equation (2.22) is linear, one may use a Fourier integral to express a given polychromatic disturbance as a sum of strictly monochromatic disturbances,[36] each of which solve eqn (2.28). Accordingly, most of the remainder of the chapter we will concentrate on various means for solving this equation.

[35]Here, and elsewhere in the text, an 'inhomogeneous' field equation is often used to refer to a field equation in the presence of an inhomogeneous medium. This differs from the usual usage of the term 'inhomogeneous', in the theory of partial differential equations, where the term is used to refer to the presence of sources. With this understanding, we will henceforth drop the inverted commas from the term 'inhomogeneous'.

[36]In addition to this advantage of linearity, there is an associated disadvantage: eqn (2.22) cannot describe inelastic scattering. This is because we can express any solution to eqn (2.22) as a linear superposition of strictly monochromatic solutions to eqn (2.28), each of which has a specified angular frequency. Linearity implies that distinct angular frequencies are uncoupled from one another, which therefore precludes the possibility of inelastic scattering.

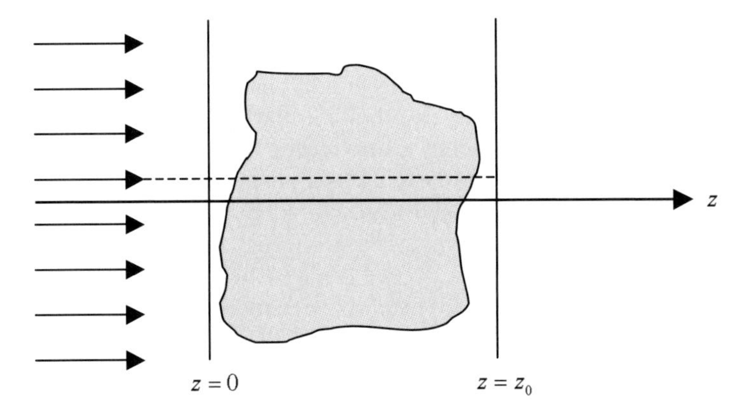

Fig. 2.1. Incident upon the plane $z = 0$ is the z-directed monochromatic plane wave indicated by arrows at the left of the diagram. All scatterers, denoted by the grey region, are assumed to lie between the planes $z = 0$ and $z = z_0$. We seek an approximate expression for the field over the exit surface $z = z_0$. The dotted line shows the path that an incident 'ray' would take, in the absence of the scatterer.

2.2 The projection approximation

Consider Fig. 2.1. Here, we see a monochromatic plane wave which is incident upon a scattering volume from the left, as indicated by the series of parallel arrows. Align the direction of propagation of this plane wave with the z-axis of a Cartesian coordinate system. Assume that the regions $z \leq 0$ and $z \geq z_0$ consist of vacuum, with all scatterers contained within the slab of space $0 < z < z_0$. These scatterers are indicated by the shaded grey region in the figure.

Despite the fact that we are working within the context of wave optics, one can nevertheless visualize 'ray paths' such as that which is indicated by a dotted line in the figure. If we were working within the framework of geometric optics, this dotted line would represent the trajectory of a certain ray in the absence of the scatterer. Within the framework of scalar wave optics, however, this path may be defined such that the phase gradient of the unscattered field is everywhere parallel to the trajectory.[37] The essence of the projection approximation is to assume that the scatterers are sufficiently weak so as to negligibly perturb the ray paths which would have existed in the volume occupied by the scatterer had the scatterer been absent; in this situation the phase and amplitude of the disturbance at the exit surface $z = z_0$ (see Fig. 2.1) can be expressed in terms of the phase and amplitude shifts accumulated as the disturbance traverses a given ray path connecting the entrance and exit surfaces, with the ray paths

[37]This is closely related to the notion of streamlines in fluid dynamics, with the wave-field phase being replaced by the so-called velocity potential for the fluid. See Section 2.8, for a further exploration of this point.

corresponding to those that would have existed if the scattering volume were replaced by vacuum. This idea is explored, in more detail, in the remainder of this section.

Consider the inhomogeneous Helmholtz equation (2.28). In the context of studying the scattering of incident plane waves by a localized distribution of scatterers, express its solution $\psi_\omega(x,y,z)$ as a perturbed plane wave, by writing this solution as the product of the unscattered plane wave $\exp(ikz)$ with an envelope $\tilde{\psi}_\omega(x,y,z)$:

$$\psi_\omega(x,y,z) \equiv \tilde{\psi}_\omega(x,y,z)\exp(ikz). \tag{2.29}$$

Conveniently, $|\tilde{\psi}_\omega(x,y,z)|^2 = |\psi_\omega(x,y,z)|^2$, so that both the field and its envelope have the same distribution of intensity in three-dimensional space. Substitute the above ansatz into eqn (2.28), and then make use of the identity:

$$\begin{aligned}\nabla^2[A(x,y,z)B(x,y,z)] = {} & A(x,y,z)\nabla^2 B(x,y,z) + B(x,y,z)\nabla^2 A(x,y,z) \\ & + 2\nabla A(x,y,z)\cdot\nabla B(x,y,z),\end{aligned} \tag{2.30}$$

for suitably well-behaved functions $A(x,y,z)$ and $B(x,y,z)$, to arrive at:

$$\left\{2ik\frac{\partial}{\partial z} + \nabla_\perp^2 + \frac{\partial^2}{\partial z^2} + k^2[n_\omega^2(x,y,z) - 1]\right\}\tilde{\psi}_\omega(x,y,z) = 0. \tag{2.31}$$

In the above equation, we have written $\nabla^2 = \nabla_\perp^2 + \partial^2/\partial z^2$, where the Laplacian in the xy plane (i.e. the 'transverse Laplacian') is given by:

$$\nabla_\perp^2 \equiv \frac{\partial^2}{\partial x^2} + \frac{\partial^2}{\partial y^2}. \tag{2.32}$$

The paraxial approximation amounts to neglecting the second derivative with respect to z in eqn (2.31). This will be a good approximation if the envelope $\tilde{\psi}_\omega(x,y,z)$ is 'beamlike', so that it is much more strongly varying in the x and y directions than in the z direction. We thereby arrive at the inhomogeneous paraxial equation:

$$\left(2ik\frac{\partial}{\partial z} + \nabla_\perp^2 + k^2[n_\omega^2(x,y,z) - 1]\right)\tilde{\psi}_\omega(x,y,z) = 0. \tag{2.33}$$

Note that, in vacuum, this equation reduces to the homogeneous paraxial equation (hereafter termed the 'paraxial equation'):

$$\left(2ik\frac{\partial}{\partial z} + \nabla_\perp^2\right)\tilde{\psi}_\omega(x,y,z) = 0, \tag{2.34}$$

for which an exact solution is given by the Fresnel diffraction theory outlined in the previous chapter.[38] Note that we will henceforth refer to the envelope

[38] In establishing this connection, the following calculation may be useful. Take the Fourier transform of the paraxial equation with respect to x and y, denoting the associated operator by

$\tilde{\psi}_\omega(x,y,z)$ as the 'wave-field', even though it differs from the true wave-field $\psi_\omega(x,y,z)$ by the linear phase factor $\exp(ikz)$.

Returning to the core problem of this section, consider once again the scenario sketched in Fig. 2.1, assuming the scattering to be sufficiently weak for the projection approximation to hold. As mentioned earlier, this is equivalent to assuming that the value, of the wave-field at the exit-surface $z = z_0$, is entirely determined by the phase and amplitude shifts that are accumulated along streamlines of the unscattered beam. Since we have assumed a normally incident z-directed plane wave as the unscattered beam, and since the transverse Laplacian is the only portion of the inhomogeneous paraxial equation that couples neighbouring ray trajectories, neglecting this term amounts to the projection approximation. One can then solve the boundary value problem for the resulting partial differential equation:

$$\frac{\partial}{\partial z}\tilde{\psi}_\omega(x,y,z) \approx \frac{k}{2i}[1 - n_\omega^2(x,y,z)]\tilde{\psi}_\omega(x,y,z), \tag{2.35}$$

to obtain the following approximate expression, which relates the wave-field $\tilde{\psi}_\omega(x,y,z=0)$ at the entrance surface $z=0$ of the slab to the wave-field $\tilde{\psi}_\omega(x,y,z=z_0)$ at the exit surface $z=z_0$ of the slab:

$$\tilde{\psi}_\omega(x,y,z=z_0) \approx \exp\left\{\frac{k}{2i}\int_{z=0}^{z=z_0}[1-n_\omega^2(x,y,z)]dz\right\}\tilde{\psi}_\omega(x,y,z=0). \tag{2.36}$$

Since the refractive index for X-rays is typically very close to unity, this index is often expressed in the form:

$$n_\omega = 1 - \delta_\omega + i\beta_\omega, \tag{2.37}$$

where δ_ω and β_ω are real numbers that are both very much less than unity in magnitude. Using this form to evaluate the quantity $1 - n_\omega^2(x,y,z)$ in eqn (2.36), keeping only terms to first order in δ_ω and β_ω, we have:

$$1 - n_\omega^2(x,y,z) \approx 2[\delta_\omega(x,y,z) - i\beta_\omega(x,y,z)]. \tag{2.38}$$

Under this approximation, eqn (2.36) becomes:

$\mathcal{F}$ and using the convention given by eqn (A.5). Making use of the Fourier derivative theorem, and recalling that k_x and k_y are defined to be the Fourier-space variables reciprocal to the position coordinates x and y, we see that $[2ik\partial/\partial z - (k_x^2+k_y^2)]\mathcal{F}\{\tilde{\psi}_\omega(x,y,z)\} = 0$. Upon rewriting this differential equation as $(\partial/\partial z)\mathcal{F}\{\tilde{\psi}_\omega(x,y,z)\} = -[i/(2k)](k_x^2+k_y^2)\mathcal{F}\{\tilde{\psi}_\omega(x,y,z)\}$, we may immediately write down the plane-to-plane diffraction integral $\mathcal{F}\{\tilde{\psi}_\omega(x,y,z\geq 0)\} = \exp\{-[iz/(2k)](k_x^2+k_y^2)\}\mathcal{F}\{\tilde{\psi}_\omega(x,y,z=0)\}$. Apply $\mathcal{F}^{-1}$ to both sides of this diffraction integral, and then make use of eqn (2.29) to replace the field envelope with the field itself. This yields the operator form of the Fresnel diffraction integral, namely $\psi_\omega(x,y,z\geq 0) = \exp(ikz)\mathcal{F}^{-1}\exp\{-[iz/(2k)](k_x^2+k_y^2)\}\mathcal{F}\{\psi_\omega(x,y,z=0)\}$, which was given in eqn (1.28) of the previous chapter.

$$\tilde{\psi}_\omega(x,y,z=z_0) \approx \exp\left\{-ik\int_{z=0}^{z=z_0}[\delta_\omega(x,y,z)-i\beta_\omega(x,y,z)]dz\right\}\tilde{\psi}_\omega(x,y,z=0). \tag{2.39}$$

This equation is the main result of the present section.

Prior to passing onto some interpretive statements regarding the physics underlying this result, it may be helpful to separately consider the predictions of this formula for the phase and amplitude shifts imparted on a wave-field traversing the scattering object, including the simple special case of a single-material scatterer. (i) According to eqn (2.39), valid in the projection approximation, the phase shift $\Delta\phi(x,y)$, imparted on the modulated plane wave as it passes from the entrance to the exit plane of the slab of space containing the weakly scattering object, is:

$$\Delta\phi(x,y) = -k\int \delta_\omega(x,y,z)dz. \tag{2.40}$$

For the case of a scatterer which is composed of a single material, of projected thickness $T(x,y)$ along the z direction, this reduces to:

$$\Delta\phi(x,y) = -k\delta_\omega T(x,y). \tag{2.41}$$

(ii) By taking the squared modulus of eqn (2.39), and identifying $|\tilde{\psi}_\omega(x,y,z)|^2$ with the intensity $I_\omega(x,y,z)$ of the wave-field, we obtain:

$$I_\omega(x,y,z=z_0) = \exp\left[-2k\int \beta_\omega(x,y,z)dz\right] I_\omega(x,y,z=0). \tag{2.42}$$

For the case of a scatterer composed of a single material of projected thickness $T(x,y)$, this becomes the familiar Beer's law of absorption:

$$I_\omega(x,y,z=z_0) = \exp\left[-\mu_\omega T(x,y)\right] I_\omega(x,y,z=0), \tag{2.43}$$

where:

$$\mu_\omega = 2k\beta_\omega \tag{2.44}$$

is the linear attenuation coefficient (linear absorption coefficient) of the material.

While we have ultimately derived eqns (2.40) and (2.41) by taking the Maxwell equations as a starting point and then introducing a succession of approximations, they can also be derived in a rather elementary way using a modified form of geometric optics in which one can associate a phase with a ray-path.[39] Such an argument, which we now outline, serves the three purposes of: (i) aiding in a

[39] Here, there is a strong analogy with certain semi-classical approximations in quantum mechanics, such as the Wentzel–Kramers–Brillouin (WKB) approximation. See, for example, Messiah (1961).

physical understanding of these equations; (ii) giving a simple means of arriving at these equations, without needing to take the scalar wave equation as a starting point; (iii) providing a grounding in some of the notions which will be used in the later section on the eikonal approximation. Note that a similar argument can be made for Beer's law of absorption, with this case being left as an exercise for the reader.

Consider Fig. 2.2. Here we see z-directed monochromatic scalar plane waves $\exp(ikz)$, of wave-number k and angular frequency ω, that are incident from the left. The half-space $z < 0$ is taken to be vacuum. In the first instance consider the scatterer to be of uniform density and comprised of a single material, with projected thickness equal to $T(x, y)$, this projection being taken along the positive z-axis. The real part $\mathrm{Re}(n_\omega)$ of the refractive index of the material is taken to be $1-\delta_\omega$, in accord with eqn (2.37). Within the framework of geometric optics, the rays associated with the unscattered beam are given by a series of parallel lines aligned with the z-axis. Two such lines are given as AB and CD in the diagram. The line AB pierces the plane $z = 0$ at the point $(x_{AB}, y_{AB}, 0)$, and pierces the plane $z = z_0$ at the point (x_{AB}, y_{AB}, z_0). Similarly, the line CD pierces the plane $z = 0$ at the point $(x_{CD}, y_{CD}, 0)$, and pierces the plane $z = z_0$ at the point (x_{CD}, y_{CD}, z_0). The line AB passes through the scatterer, while the line CD does not pass through the scatterer.

We now use the projection approximation to estimate the phase of the wavefield over the plane $z = z_0$. At the level of geometric optics, we are ignoring any transverse deviations of the rays within the scattering volume that lies between the planes $z = 0$ and $z = z_0$. Therefore, to estimate the phase at a given point (x, y, z_0) at the exit surface, we count up the number of wavelengths which fit along the line joining $(x, y, 0)$ to (x, y, z_0), and then multiply by 2π. For example, consider the phase accumulated at the point B. In travelling from the point A in the plane $z = 0$ to the point B in the plane $z = z_0$, the wave has travelled through a thickness $T(x_{AB}, y_{AB})$ of material, and through a thickness $z_0 - T(x_{AB}, y_{AB})$ of vacuum. Since the wavelength in the material is given by $\lambda/\mathrm{Re}(n_\omega)$, where λ is the vacuum wavelength, the phase shift $\phi_{\mathrm{mat}}(x_{AB}, y_{AB})$ accumulated in passing through the material is equal to 2π multiplied by the number of wavelengths N which fit into the projected thickness, so that:

$$\begin{aligned}\phi_{\mathrm{mat}}(x_{AB}, y_{AB}) &= 2\pi N = 2\pi \frac{T(x_{AB}, y_{AB})}{\lambda/\mathrm{Re}(n_\omega)} \\ &= \frac{2\pi \mathrm{Re}(n_\omega) T(x_{AB}, y_{AB})}{\lambda}. \end{aligned} \tag{2.45}$$

Using a similar argument, the phase shift $\phi_{\mathrm{vac}}(x_{AB}, y_{AB})$ accumulated in traversing a thickness $z_0 - T(x_{AB}, y_{AB})$ of vacuum is given by:

$$\phi_{\mathrm{vac}}(x_{AB}, y_{AB}) = 2\pi \frac{z_0 - T(x_{AB}, y_{AB})}{\lambda}. \tag{2.46}$$

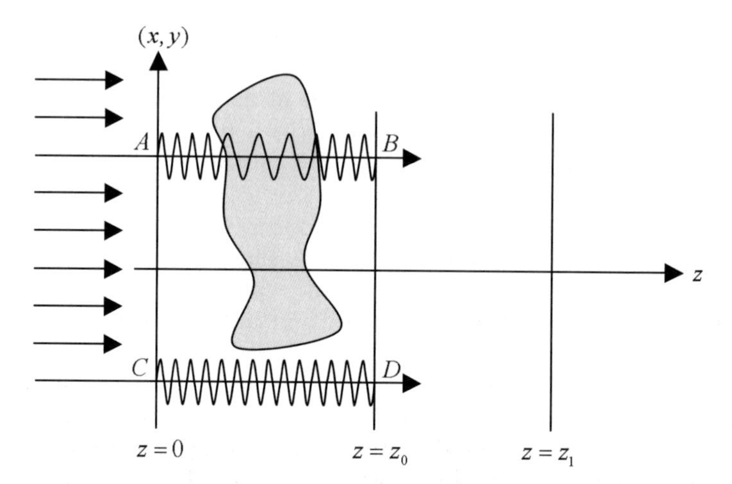

Fig. 2.2. Monochromatic plane waves, travelling along the z direction, are incident from the left. All scatterers are assumed to lie between the planes $z = 0$ and $z = z_0$. Two ray paths, associated with the unscattered beam, are given by the lines AB and CD. We wish to determine the intensity and phase of the wave at the exit-surface $z = z_0$, using the projection approximation. Under this approximation, the ray paths in the presence of the scatterer are taken to be equal to their unscattered counterparts. Thus the phase and amplitude at the point B is considered to be accumulated along the line AB, with a similar procedure used to approximate the amplitude and phase at any point (x, y, z_0) over the exit surface.

Therefore, according to the projection approximation, the phase $\phi(x_{AB}, y_{AB})$ of the wave at the point B is given by:

$$\phi(x_{AB}, y_{AB}) = \phi_{\mathrm{mat}}(x_{AB}, y_{AB}) + \phi_{\mathrm{vac}}(x_{AB}, y_{AB}) = \frac{2\pi}{\lambda}[z_0 - T(x_{AB}, y_{AB})\delta_\omega]. \tag{2.47}$$

Note that, in writing the last equality above, we have made use of the fact that $\mathrm{Re}(n_\omega) = 1 - \delta_\omega$.

With reference to the line CD in Fig. 2.2, the phase $\phi(x_{CD}, y_{CD})$ of the wave at the point D is given by setting T to zero in the above formula, so that:

$$\phi(x_{CD}, y_{CD}) = \frac{2\pi}{\lambda} z_0. \tag{2.48}$$

Note that we could have written this formula on inspection, since it is equal to 2π multiplied by the number z_0/λ of vacuum wavelengths λ that fit into a distance of z_0. Now, since the phase of a monochromatic field is only defined up to an arbitrary additive constant, we can consider the phase shift $\Delta\phi(x, y, z_0)$ over the plane $z = z_0$ to be relative to the vacuum phase shift $\phi(x_{CD}, y_{CD})$, so that $\Delta\phi(x, y, z_0) \equiv \phi(x, y, z_0) - (2\pi/\lambda)z_0$. Making use of eqns (2.47) and (2.48), with (x_{AB}, y_{AB}, z_0) being replaced with an arbitrary point (x, y, z_0) over the exit surface $z = z_0$, and writing $2\pi/\lambda = k$, we arrive at eqn (2.41). We leave it as an exercise to the reader, to generalize the above geometric argument so as to derive the more general form given by eqn (2.40).

We close this section by noting that once the intensity and phase of the scattered wave-field have been estimated over the plane $z = z_0$ using the projection approximation, this information can subsequently be used to calculate the wave-field which results when this exit-surface field is propagated from $z = z_0$ to $z = z_1 > z_0$ (see Fig. 2.2). This propagation could be effected using an appropriate diffraction integral, including any of the means for propagating coherent wave-fields which were introduced in the previous chapter. Alternatively, one can compute the image produced by a well characterized imaging system which takes, as input, the exit field over the surface $z = z_0$ (cf. Chapter 4).

2.3 Point scatterers and the outgoing Green function

Green functions, already encountered in the previous chapter's discussions on X-ray wave-fields in free space, are a powerful and widely applicable tool in mathematical physics. Here we use them to study the problem of a single point scatterer in the presence of an incident monochromatic scalar electromagnetic field, defining the Green function to be the field that is scattered from such a point. In subsequent sections, we extend this formalism to study the elastic scattering of X-rays by an extended distribution of matter.

Consider the scattering scenario sketched in Fig. 2.3. Here, the entire volume is taken to be vacuum, with the exception of a point scatterer located at (x', y', z'). We assume a z-directed monochromatic plane wave to be incident upon this scatterer. One can mentally decompose the total field, in the presence of the point scatterer, as the sum of two fields: the field which would have existed in the absence of the scatterer, and the scattered field. This decomposition is suggested in the figure, with the unscattered field indicated by a succession of parallel wavefronts, and the scattered field denoted by spherical wavefronts emanating from the point scatterer. The outgoing Green function $G_\omega(x, y, z, x', y', z')$, for this problem, is by definition equal to the scattered field due to a point scatterer located at (x', y', z'), as a function of the position coordinate (x, y, z) in three-dimensional space.

We have seen that the Maxwell equations imply the inhomogeneous Helmholtz equation (2.28) for a scalar field in the presence of a scattering medium described by the time-independent complex refractive index $n_\omega(x, y, z)$, subject to the approximations that: (i) the medium is non-magnetic; (ii) the electrical permittivity, and therefore the refractive index, is time-independent and slowly varying over spatial length scales comparable with the wavelength of the radiation with which it interacts; (iii) we may work with a single complex quantity to quantify

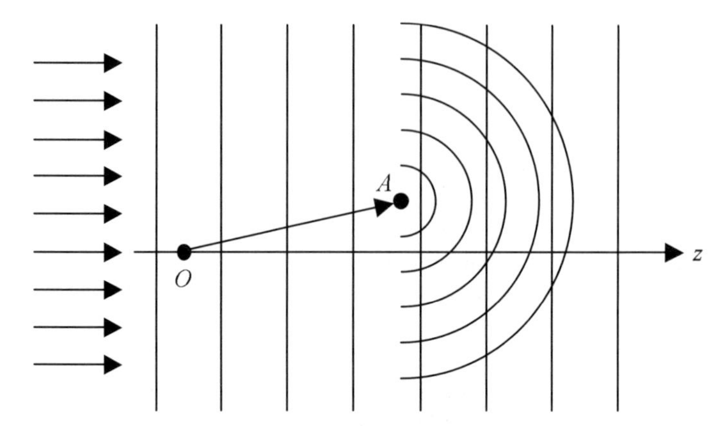

Fig. 2.3. Monochromatic plane waves, travelling along the z direction, are incident from the left. The three-dimensional space is taken to be vacuum, with the exception of a single point scatterer located at $A \equiv (x', y', z')$ with respect to an origin O.

the disturbance due to the monochromatic electromagnetic field; (iv) the effects of inelastic scattering may be ignored. This inhomogeneous equation, namely eqn (2.28), may be written in the form:

$$\left(\nabla^2 + k^2\right) \psi_\omega(x, y, z) = k^2 \left[1 - n_\omega^2(x, y, z)\right] \psi_\omega(x, y, z). \tag{2.49}$$

In vacuum, the right-hand side of this equation vanishes, leaving the homogeneous Helmholtz equation for coherent scalar fields, which was studied in Chapter 1. Here, however, we are interested in the case of a point scatterer, in which case the right-hand side of the equation is proportional to a three-dimensional Dirac delta $\delta(x - x', y - y', z - z')$ located at the point (x', y', z'). Bearing in mind our earlier definition of the Green function $G_\omega(x, y, z, x', y', z')$ as the field scattered from a point scatterer located at (x', y', z'), we conclude that the Green function obeys the equation (cf. eqn (1.56)):

$$\left(\nabla^2 + k^2\right) G_\omega(x, y, z, x', y', z') = -4\pi\delta(x - x', y - y', z - z'). \tag{2.50}$$

Note that the factor of -4π, on the right side of this equation, is introduced by convention.

Since free space is homogeneous and isotropic, $G_\omega(x, y, z, x', y', z') = G_\omega(x - x', y - y', z - z')$, amounting to the statement that the scattered field observed at a given point depends only on the relative position of the observation point with respect to the point scatterer. Making this simplification to the Green function in eqn (2.50), and then shifting the origin of coordinates to coincide with the location of the point scatterer, we arrive at:

$$\left(\nabla^2 + k^2\right) G_\omega(x, y, z) = -4\pi\delta(x, y, z). \tag{2.51}$$

We now outline two popular methods for solving the above equation, to yield the desired free-space Green function for the inhomogeneous Helmholtz equation. The first of these methods has the virtue of greater simplicity, while the second method[40]—which makes use of complex analysis, and which may be skipped without loss of continuity—gives us the opportunity to introduce some mathematical results that will be of occasional use in later discussions.

2.3.1 *First method for obtaining free-space Green function*

Introduce spherical polar coordinates centred at $(x, y, z) = (0, 0, 0)$, with radial distance from this origin denoted by r. Making use of the form which the three-dimensional Laplacian takes in spherical polar coordinates, together with the fact that the rotationally symmetric Green function depends only on r, eqn (2.51) becomes:

$$\frac{1}{r^2}\frac{d}{dr}\left[r^2 \frac{d}{dr} G_\omega(r)\right] + k^2 G_\omega(r) = -4\pi\delta(r). \tag{2.52}$$

Multiply both sides by r, thereby eliminating the Dirac delta, so that:

$$\frac{1}{r}\frac{d}{dr}\left[r^2 \frac{d}{dr} G_\omega(r)\right] = -k^2 r G_\omega(r). \tag{2.53}$$

This may be rewritten as:

$$\frac{d^2}{dr^2}\left[r G_\omega(r)\right] = -k^2 r G_\omega(r), \tag{2.54}$$

indicating that $rG_\omega(r)$ is an oscillatory eigenfunction of d^2/dr^2, so that $rG_\omega(r) = \exp(\pm ikr)$. This directly yields the desired explicit form for the free-space Green function:

$$G_\omega(r) = \frac{\exp(\pm ikr)}{r}. \tag{2.55}$$

Choosing the positive sign in the exponent yields the outgoing spherical wave, which is required in the present scattering context. This outgoing spherical wave, as its name implies, is so called because it describes energy flowing radially outwards from the point scatterer. On the other hand, if one were to choose the negative sign in the above exponent, one would arrive at an incoming spherical wave, with energy flowing towards the point scatterer. While this solution is to be rejected as unphysical in the present context, the reader will recall that such an incoming Green function was required in the previous chapter's discussions on the propagation of mutual coherence (see Section 1.9.3).

[40]Note that our treatment of this second method is after Merzbacher (1970).

2.3.2 *Second method for obtaining free-space Green function*

Express both the Green function and the Dirac delta, appearing in eqn (2.51), as three-dimensional Fourier integrals:

$$G_\omega(x,y,z) = \frac{1}{(2\pi)^{3/2}} \iiint \breve{G}_\omega(k_x,k_y,k_z)\exp[i(k_x x + k_y y + k_z z)]dk_x dk_y dk_z, \tag{2.56}$$

$$\delta(x,y,z) = \frac{1}{(2\pi)^3} \iiint \exp[i(k_x x + k_y y + k_z z)]dk_x dk_y dk_z. \tag{2.57}$$

Here, $\breve{G}_\omega(k_x,k_y,k_z)$ denotes the Fourier transform of $G_\omega(x,y,z)$ with respect to x, y and z, and (k_x,k_y,k_z) are the Fourier coordinates corresponding to the real-space coordinates (x,y,z). Note that eqn (2.57) expresses the fact that the Dirac delta is proportional to the inverse Fourier transform of a constant function.

Substitute the above pair of equations into the defining equation (2.51) for the Green function, and then bring the operator $(\nabla^2 + k^2)$ inside the integrals. In what amounts to a use of the Fourier derivative theorem, one can then replace ∇^2 with $-(k_x^2 + k_y^2 + k_z^2)$, since $\nabla^2 \exp[i(k_x x + k_y y + k_z z)] = -(k_x^2 + k_y^2 + k_z^2)\exp[i(k_x x + k_y y + k_z z)]$. Collecting together the resulting equation under a single triple integral, we obtain:

$$0 = \iiint \left\{ \left[k^2 - (k_x^2 + k_y^2 + k_z^2)\right] \breve{G}_\omega(k_x,k_y,k_z) + \sqrt{\frac{2}{\pi}} \right\} \\ \times \exp[i(k_x x + k_y y + k_z z)]dk_x dk_y dk_z. \tag{2.58}$$

The quantity in braces must vanish. Therefore:

$$\breve{G}_\omega(k_x,k_y,k_z) = \frac{\sqrt{2/\pi}}{k_x^2 + k_y^2 + k_z^2 - k^2}. \tag{2.59}$$

This can now be substituted into eqn (2.56), to give the following Fourier integral representation for the Green function:

$$G_\omega(x,y,z) = \frac{1}{2\pi^2} \iiint \frac{\exp[i(k_x x + k_y y + k_z z)]}{k_x^2 + k_y^2 + k_z^2 - k^2} dk_x dk_y dk_z. \tag{2.60}$$

Transform to spherical polar coordinates (r,θ,ϕ) in real space, and spherical polar coordinates (k_r,k_θ,k_ϕ) in Fourier space. Align the latter coordinate system such that k_θ is the angle between (r,θ,ϕ) and (k_r,k_θ,k_ϕ). Therefore the above integral becomes:

$$G_\omega(r,\theta,\phi) = \frac{1}{2\pi^2}\int_{k_r=0}^{\infty}\int_{k_\theta=0}^{\pi}\int_{k_\phi=0}^{2\pi}\frac{\exp(ik_r r\cos k_\theta)}{k_r^2-k^2}k_r^2\sin k_\theta dk_r dk_\theta dk_\phi. \quad (2.61)$$

Evaluate the integral with respect to k_ϕ, which gives a multiplicative factor of 2π, and then re-order the remaining integrals to obtain:

$$G_\omega(r,\theta,\phi) = \frac{1}{\pi}\int_{k_r=0}^{\infty}\left\{\frac{k_r^2}{k_r^2-k^2}\left[\int_{k_\theta=0}^{\pi}\exp(ik_r r\cos k_\theta)\sin k_\theta dk_\theta\right]\right\}dk_r. \quad (2.62)$$

By making the change of variables $\eta = \cos k_\theta$, the integral in square brackets is readily evaluated, leaving:

$$G_\omega(r,\theta,\phi) = \frac{2}{\pi r}\int_{k_r=0}^{\infty}\frac{k_r\sin(k_r r)}{k_r^2-k^2}dk_r = -\frac{2}{\pi r}\frac{d}{dr}\int_{k_r=0}^{\infty}\frac{\cos(k_r r)}{k_r^2-k^2}dk_r. \quad (2.63)$$

Since the final integrand is even in k_r, we may extend the lower limit to $-\infty$ and divide the result by 2. Further, we may then replace $\cos(k_r r)$ with $\exp(ik_r r)$, since this amounts to adding an odd function to the integrand, thereby making no change to the resulting integral. Thus we obtain:

$$G_\omega(r,\theta,\phi) = -\frac{1}{\pi r}\frac{d}{dr}\int_{-\infty}^{\infty}\frac{\exp(ik_r r)}{k_r^2-k^2}dk_r = -\frac{1}{\pi r}\frac{d}{dr}\int_{-\infty}^{\infty}\frac{\exp(ik_r r)}{(k_r-k)(k_r+k)}dk_r. \quad (2.64)$$

To proceed, we shall make use of some of the central results of complex analysis, this being the calculus of complex-valued functions of a single complex variable. Accordingly, for the remainder of this sub-section, it will be assumed that the reader is familiar with the following: complex poles, residues, and the residue theorem. Excellent introductions to this material can be found in many texts, such as those by Spiegel (1964) and Kreysig (1983).

In the integrand of the final integral in eqn (2.64), the variable k_r takes on all real values between $-\infty$ and ∞. Suppose that we allow k_r to take on complex values, so that possible values for k_r can be plotted as points in the complex plane. Such a complex plane is shown in Fig. 2.4. In this plane, the horizontal coordinate of a specified point is given by the real value of k_r, with the vertical coordinate being given by the imaginary part of k_r.

Consider the closed contour Γ shown in Fig. 2.4. This closed contour can be expressed as the sum of six segments: $\Gamma = \Gamma_1 + \Gamma_2 + \Gamma_3 + \Gamma_4 + \Gamma_5 + \Gamma_6$. The segment Γ_1 is an anti-clockwise traversed semi-circle in the upper half of the complex plane, centred on the origin and with a radius R that will be made to tend to infinity. Γ_3 is a clockwise-traversed semi-circle in the upper-half plane, centred at $k_r = -k$. Similarly, Γ_5 is an anticlockwise-traversed semi-circle in the lower-half plane, centred at $k_r = +k$. The radius ε of the contours Γ_3 and Γ_5

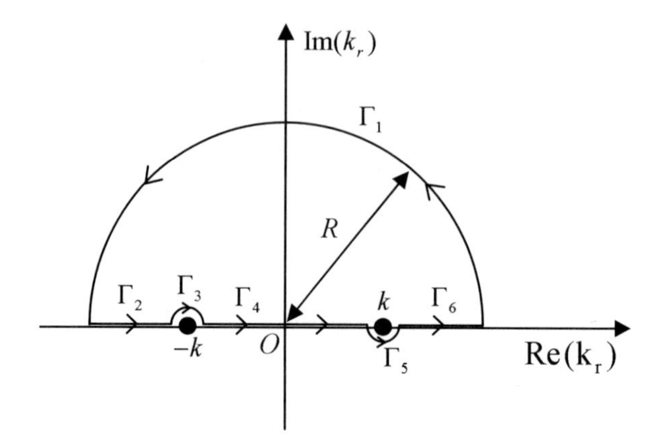

Fig. 2.4. Contour integral used during the derivation of the outgoing Green function for the inhomogeneous Helmholtz equation. Details in main text.

will be taken to tend to zero from above. Finally, the contours Γ_2, Γ_4, and Γ_6 all lie along the real axis; in the limit as R tends to infinity from below, and ε tends to zero from above, this triple of contours covers every point on the real axis, with the exception of the points located at $k_r = \pm k$.

Rather than directly evaluating the final integral in eqn (2.64), let us evaluate a related integral, which is obtained by allowing k_r to be complex and then integrating with respect to this variable, over the closed contour Γ described in the previous paragraph. Note that the 'detours' due to paths Γ_3 and Γ_5 serve to avoid the simple poles, of the integrand, located at the points $k_r = \pm k$. One thereby arrives at the expression on the left side of the equation below, with the right side being given by the residue theorem of complex analysis:

$$\oint_\Gamma \frac{\exp(ik_r r)}{(k_r - k)(k_r + k)} dk_r = 2\pi i \, \mathrm{Res}(k_r = k). \tag{2.65}$$

Here, $\mathrm{Res}(k_r = k)$ denotes the residue of the simple pole at $k_r = k$. Note that only poles inside the contour Γ will contribute to the integral over this contour. The residue is evaluated, in the usual manner, to give:

$$\mathrm{Res}(k_r = k) = \lim_{k_r \to k} (k_r - k) \times \frac{\exp(ik_r r)}{(k_r - k)(k_r + k)} = \frac{\exp(ikr)}{2k}. \tag{2.66}$$

Make use of this result, together with the fact that $\Gamma = \Gamma_1 + \Gamma_2 + \Gamma_3 + \Gamma_4 + \Gamma_5 + \Gamma_6$, to rewrite eqn (2.65) as:

$$\int_{\Gamma_1} \frac{\exp(ik_r r) dk_r}{(k_r - k)(k_r + k)} + \int_{\Gamma_2 + \Gamma_3 + \Gamma_4 + \Gamma_5 + \Gamma_6} \frac{\exp(ik_r r) dk_r}{(k_r - k)(k_r + k)} = \frac{\pi i}{k} \exp(ikr). \tag{2.67}$$

The first integral, in the above equation, will vanish. To see this, note that the modulus of the integrand decreases as $|k_r|^{-2}$ as $|k_r| \to \infty$, while the length of the contour increases as $|k_r|$. Regarding the second integrand: as R tends to infinity from below, and ε tends to zero from above, this becomes the very integral we wished to evaluate, namely the integral on the right-hand side of eqn (2.64). Thus we can return to this equation, thereby completing our derivation to arrive once again at the outgoing Green function:

$$G_\omega(r,\theta,\phi) = -\frac{1}{\pi r}\frac{d}{dr}\frac{\pi i}{k}\exp(ikr) = \frac{\exp(ikr)}{r}. \tag{2.68}$$

Expressed in terms of Cartesian coordinates, we may write down the free-space outgoing Green function for the inhomogeneous Helmholtz equation (2.28), as:

$$G_\omega(x,y,z) = \frac{\exp(ik\sqrt{x^2+y^2+z^2})}{\sqrt{x^2+y^2+z^2}} \equiv \frac{\exp(ik|\mathbf{x}|)}{|\mathbf{x}|}, \quad \mathbf{x} \equiv (x,y,z). \tag{2.69}$$

We close this sub-section by noting that there is a certain freedom in the choice of contour Γ that was adopted in Fig. 2.4, this contour being such that we 'hopped around' the singularities of the integrand in a particular way, including one pole but not the other, within the contour. The Green function, thus obtained, was outgoing at infinity. This outgoing/retarded Green function may be compared to the incoming/advanced Green function, which would have been arrived at if we had chosen to include only the pole at $k_r = -k$, as the reader can verify.

2.4 Integral-equation formulation of scattering

In the previous section, we considered the interaction of an X-ray plane wave with a point scatterer. This culminated in a derivation of the outgoing Green function, given by eqn (2.69), representing the spherical wave which is scattered from the point scatterer when it is located at the origin of coordinates. Here, we build upon this result, allowing us to rewrite the differential equation (2.28) as an integral equation. Note that this integral equation is more restrictive than the differential equation from which it is obtained, in the sense that the latter must be supplemented by appropriate boundary values/conditions in order to be solved, whereas the former has the boundary values/conditions 'built in'. This integral equation will be used in subsequent sections to derive various approximate expressions describing the scattering of X-rays by matter.

Rather than deriving the desired integral equation, for the field $\psi_\omega(x,y,z)$ in the presence of a localized non-magnetic scattering distribution described by the refractive index $n_\omega(x,y,z)$, we merely write this integral equation down and then verify that it indeed leads to the differential equation (2.28). Accordingly, we assert that the required integral equation is given by:

$$\psi_\omega(\mathbf{x}) = \psi_\omega^0(\mathbf{x}) - \frac{k^2}{4\pi}\iiint G_\omega(\mathbf{x}-\mathbf{x}')\left[1-n_\omega^2(\mathbf{x}')\right]\psi_\omega(\mathbf{x}')d\mathbf{x}', \qquad (2.70)$$

where $\mathbf{x} \equiv (x, y, z)$, $\mathbf{x}' \equiv (x', y', z')$, $d\mathbf{x}' \equiv dx'dy'dz'$, $\psi_\omega^0(\mathbf{x})$ represents the unscattered wave and is therefore a solution to eqn (2.28) in the absence of scatterers, and $G_\omega(\mathbf{x}-\mathbf{x}') = \exp(ik|\mathbf{x}-\mathbf{x}'|)/|\mathbf{x}-\mathbf{x}'|$ is the outgoing Green function derived in the previous section. Note that the desired field $\psi_\omega(\mathbf{x})$ appears both on the left side of this expression, and under the integral sign on the right-hand side, which is why we speak of it as an integral equation rather than an integral expression for $\psi_\omega(\mathbf{x})$.

Let us show that $\psi_\omega(\mathbf{x})$ indeed obeys eqn (2.28), in the form given by eqn (2.49). To this end, apply the operator $(\nabla^2 + k^2)$ to the left side of eqn (2.70), to give:

$$\begin{aligned}(\nabla^2 + k^2)\psi_\omega(\mathbf{x}) &= (\nabla^2 + k^2)\psi_\omega^0(\mathbf{x}) \\ &\quad - \frac{k^2}{4\pi}(\nabla^2 + k^2)\iiint G_\omega(\mathbf{x}-\mathbf{x}')\left[1-n_\omega^2(\mathbf{x}')\right]\psi_\omega(\mathbf{x}')d\mathbf{x}'.\end{aligned} \qquad (2.71)$$

The first term on the right-hand side vanishes, since $\psi_\omega^0(\mathbf{x})$ is the unscattered wave and therefore obeys the form of eqn (2.28) which is obtained when $n_\omega(\mathbf{x}) = 1$. Regarding the second term on the right-hand side, bring the operator (∇^2+k^2) inside the integral sign, noting that it acts only on the variable $\mathbf{x}$, and then make use of the defining equation (2.51). Thus:

$$\begin{aligned}(\nabla^2 + k^2)\psi_\omega(\mathbf{x}) &= -\frac{k^2}{4\pi}\iiint \left[(\nabla^2 + k^2)G_\omega(\mathbf{x}-\mathbf{x}')\right]\left[1-n_\omega^2(\mathbf{x}')\right]\psi_\omega(\mathbf{x}')d\mathbf{x}' \\ &= -\frac{k^2}{4\pi}\iiint \left[-4\pi\delta(\mathbf{x}-\mathbf{x}')\right]\left[1-n_\omega^2(\mathbf{x}')\right]\psi_\omega(\mathbf{x}')d\mathbf{x}' \\ &= k^2\left[1-n_\omega^2(\mathbf{x})\right]\psi_\omega(\mathbf{x}),\end{aligned} \qquad (2.72)$$

and so we see that solutions to the integral equation (2.70) are indeed solutions to the differential equation (2.28) for scattering of monochromatic scalar electromagnetic waves by localized non-magnetic scatterers characterized by their distribution of refractive index.

We close this section by noting that the integral equation contains more information—and is therefore less general—than the corresponding differential equation, since: (i) in the complete absence of any scatterers, the integral in eqn (2.70) will vanish, and so we recover an expression for the unscattered field $\psi_\omega^0(\mathbf{x})$; (ii) in choosing to use the outgoing Green function in eqn (2.70), we automatically guarantee that the scattered field represents an outgoing wave when one is infinitely far from all scatterers.

2.5 First Born approximation for kinematical scattering

Consider the integral equation (2.70), for the elastic scattering of incident monochromatic scalar electromagnetic waves from a non-magnetic material described by the time-independent distribution of refractive index $n_\omega(\mathbf{x})$. Note that the integrand is only non-zero within the volume of the scatterer, as the refractive index of vacuum is unity (cf. eqn (2.26)). Therefore, if one can assume that the X-ray disturbance inside the scattering volume is only slightly different from the disturbance that would have existed at each point $\mathbf{x}'$ in that volume in the absence of the scatterer, then one can replace $\psi_\omega(\mathbf{x}')$ by the unscattered disturbance $\psi_\omega^0(\mathbf{x}')$, inside the integral. The resulting expression, namely:

$$\psi_\omega(\mathbf{x}) = \psi_\omega^0(\mathbf{x}) - \frac{k^2}{4\pi} \iiint G_\omega(\mathbf{x}-\mathbf{x}') \left[1 - n_\omega^2(\mathbf{x}')\right] \psi_\omega^0(\mathbf{x}') d\mathbf{x}', \tag{2.73}$$

is no longer an integral equation, but rather an approximate expression for the total wave-field. This approximate expression is known as the 'first Born approximation'.[41]

It will prove convenient, in the following discussions, to take eqn (2.73), use eqn (2.69) to write the outgoing Green function explicitly as $G_\omega(\mathbf{x}-\mathbf{x}') = \exp(ik|\mathbf{x}-\mathbf{x}'|)/|\mathbf{x}-\mathbf{x}'|$, and take the incident wave-field to be a monochromatic plane wave $\exp(i\mathbf{k}_0 \cdot \mathbf{x})$. Thus we write down:

$$\psi_\omega(\mathbf{x}) = \exp(i\mathbf{k}_0 \cdot \mathbf{x}) - \frac{k^2}{4\pi} \iiint \frac{\exp(ik|\mathbf{x}-\mathbf{x}'|)}{|\mathbf{x}-\mathbf{x}'|} \left[1 - n_\omega^2(\mathbf{x}')\right] \exp(i\mathbf{k}_0 \cdot \mathbf{x}') d\mathbf{x}'. \tag{2.74}$$

This form of the first Born approximation corresponds to the scattering scenario sketched in Fig. 2.5, which we now describe. In this figure, we see an incident plane wave-field, with its surfaces of constant phase indicated by a series of parallel vertical lines. This incident field is considered to fill the whole space, as we choose to view the field everywhere as a sum of the unscattered field, together with the field scattered by the sample. Regarding the field scattered by the sample, the sketch is intended to suggest a so-called 'single scattering approximation', in which each point within the volume of the scatterer is taken to emanate spherical waves, the strength of which is proportional to the wave which is incident upon such a point. The scatterer is taken to be sufficiently weak that, for each scattering point, we can take the field incident upon it to be equal to the unscattered field. The total field is then given by the totality of spherical waves (free-space outgoing Green functions) which emanate from each point in the scatterer, together with the unscattered wave-field, according to eqn (2.74).

Evidently, the first Born approximation corresponds to a single-scattering scenario in which the incident wave-field is either not scattered at all, or scattered

[41] The second-order and higher-order Born approximations are considered in Section 2.6.

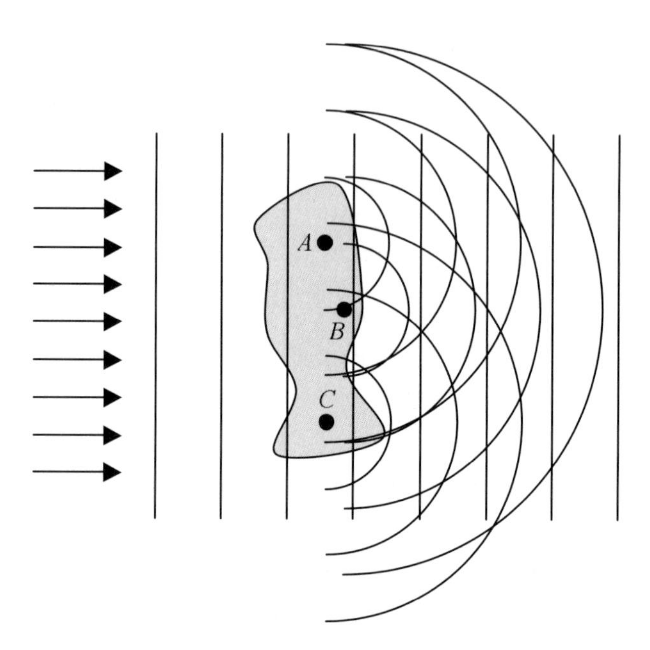

Fig. 2.5. Plane monochromatic scalar electromagnetic waves are incident, from the left, upon a localized distribution of static non-magnetic scatterers. Under the first Born approximation, the resulting wave-field is considered to be the sum of an unscattered wave-field, together with the wave-field that results from single-scattering events between the incident wave and any given point in the medium. The unscattered wave is represented by the plane waves on the right side of the sample, together with the scattered wave emanating from each point within the sample. Three such scattering points, labelled A, B, and C, are shown in the diagram.

only once by a single point within the sample. Such single-scattering approximations are known as 'kinematical' theories. They may be compared to the so-called 'dynamical' theories, in which the incident wave-field may be scattered multiple times (see Section 2.6).

The remainder of this section is broken into three parts. The first part considers an asymptotic form of the first Born approximation, corresponding to scattering through a distance that is sufficiently large to place one in the far field. This first case will be shown to reduce to the Fraunhofer diffraction integral when the scatterer is a two-dimensional screen. The second part considers an angular-spectrum expression for the diffracted field due to a three-dimensional scattering distribution, under the first Born approximation, over any plane which is downstream of the source. The third part shows how the first Born approximation may be related to the famous Ewald sphere.

2.5.1 *Fraunhofer diffraction and the first Born approximation*

Here, we treat the special case of the first Born approximation that corresponds to the scattered radiation being detected very far from all scatterers. The treatment given here is a standard one, which is in direct mathematical correspondence to the treatment of non-relativistic potential scattering that is given in many elementary quantum-mechanics texts[42].

Set up a common origin of coordinates for both $\mathbf{x}$ and $\mathbf{x}'$, that lies somewhere within the volume occupied by the scatterers. Assume the position vector of the observation point $\mathbf{x}$, with respect to this origin, to have a magnitude that is much greater than the size of the scatterer. Since the integrand in eqn (2.74) will vanish whenever $\mathbf{x}'$ does not lie within the scattering volume, as the term in square brackets vanishes in vacuum, this allows us to take $|\mathbf{x}'| \ll |\mathbf{x}|$ in this integrand. Hence we obtain the following approximate form for the spherical wave scattered from a point $\mathbf{x}'$ within the scattering volume:

$$\begin{aligned}\frac{\exp(ik|\mathbf{x}-\mathbf{x}'|)}{|\mathbf{x}-\mathbf{x}'|} &= \frac{\exp[ik\sqrt{(\mathbf{x}-\mathbf{x}')\cdot(\mathbf{x}-\mathbf{x}')}]}{|\mathbf{x}-\mathbf{x}'|} \\ &= \frac{\exp(ik\sqrt{|\mathbf{x}|^2 - 2\mathbf{x}\cdot\mathbf{x}' + |\mathbf{x}'|^2})}{|\mathbf{x}-\mathbf{x}'|} \\ &\approx \frac{\exp(ik\sqrt{|\mathbf{x}|^2 - 2\mathbf{x}\cdot\mathbf{x}'})}{r} \\ &= \frac{\exp(ikr\sqrt{1-2r^{-2}\mathbf{x}\cdot\mathbf{x}'})}{r}. \end{aligned} \tag{2.75}$$

Here, $r \equiv |\mathbf{x}|$ is the distance from the origin of coordinates, located within the scattering volume, to the observation point $\mathbf{x}$.

Make the binomial approximation $\sqrt{1-2r^{-2}\mathbf{x}\cdot\mathbf{x}'} \approx 1 - r^{-2}\mathbf{x}\cdot\mathbf{x}'$ in the above expression, substituting the result into eqn (2.74), to give:

$$\begin{aligned}\psi_\omega(\mathbf{x}) &= \exp(i\mathbf{k}_0\cdot\mathbf{x}) \\ &\quad - \frac{k^2\exp(ikr)}{4\pi r}\iiint \exp(-ikr^{-1}\mathbf{x}\cdot\mathbf{x}')\left[1-n_\omega^2(\mathbf{x}')\right]\exp(i\mathbf{k}_0\cdot\mathbf{x}')d\mathbf{x}'.\end{aligned} \tag{2.76}$$

Introduce the unit vector $\hat{\mathbf{x}}$, pointing in the $\mathbf{x}$ direction, defined by $\hat{\mathbf{x}} \equiv \mathbf{x}/|\mathbf{x}| \equiv \mathbf{x}/r$. Thus the above equation becomes:

[42] This correspondence arises because eqns (2.28) and (2.49) are mathematically identical in form to the time-independent Schrödinger equation for spinless non-relativistic particles in the presence of a scalar potential. Accordingly, the treatment given here is influenced by the discussions on the corresponding quantum-mechanical potential scattering problem, which are given in standard quantum-mechanics texts such as Messiah (1961) and Merzbacher (1970).

$$\psi_\omega(\mathbf{x}) = \exp(i\mathbf{k}_0 \cdot \mathbf{x}) + \frac{\exp(ikr)}{r} f(\Delta\mathbf{k}), \tag{2.77}$$

$$f(\Delta\mathbf{k}) \equiv \frac{k^2}{4\pi} \iiint \left[n_\omega^2(\mathbf{x}') - 1\right] \exp(-i\Delta\mathbf{k} \cdot \mathbf{x}') d\mathbf{x}', \tag{2.78}$$

$$\Delta\mathbf{k} \equiv k\hat{\mathbf{x}} - \mathbf{k}_0. \tag{2.79}$$

We have arrived at a rather simple picture for the wave-field in the presence of the scattering potential, when one is sufficiently far from all scatterers, and under the first Born approximation. According to the second term on the right side of eqn (2.77), the scattered radiation is given by a distorted spherical wave emanating from the point at which the scattering distribution is located. The form of the distortion, of this spherical wave, is quantified by the envelope f. We shall henceforth speak of f as the 'scattering amplitude', which multiplies the undistorted outgoing spherical wave $\exp(ik)/r$ in order to form the distorted spherical wave emerging from the scatterer. The scattering amplitude, according to eqn (2.78), is proportional to the three-dimensional Fourier transform of $[n_\omega^2(\mathbf{x}) - 1]$, where $n_\omega(\mathbf{x})$ is the position-dependent refractive index of the scatterer. To determine the value which the scattering amplitude f takes in the far field, one must evaluate it as a function of $\Delta\mathbf{k}$, this being the vector difference between the incident wave-vector $\mathbf{k}_0$ and the scattered wave-vector $k\hat{\mathbf{x}}$ of the X-ray wave-field.

We close this sub-section by showing how eqns (2.77) through (2.79) relate to the Fraunhofer diffraction integral. With this end in mind, assume all scatterers to lie between the planes $z = 0$ and $z = z_0 > 0$, with the incident plane wave having a wave-vector $\mathbf{k}_0$ aligned with the positive z-axis. This is the same as the scenario sketched in Fig. 2.1. Making use of eqn (2.38), which is a good approximation for X-rays since refractive indices are always close to unity in this regime, and writing $\Delta\mathbf{k} = (\Delta k_x, \Delta k_y, \Delta k_z)$, expression (2.78) becomes:

$$\begin{aligned} &f(\Delta k_x, \Delta k_y, \Delta k_z) \\ &= -\frac{k^2}{2\pi} \iiint [\delta_\omega(x', y', z') - i\beta_\omega(x', y', z')] \\ &\quad \times \exp[-i(\Delta k_x x' + \Delta k_y y' + \Delta k_z z'] dx' dy' dz' \\ &= \frac{-k^2}{2\pi} \int_{x'=-\infty}^{\infty} \int_{y'=-\infty}^{\infty} \exp[-i(\Delta k_x x' + \Delta k_y y'] \\ &\quad \times \left\{ \int_{z'=0}^{z_0} [\delta_\omega(x', y', z') - i\beta_\omega(x', y', z')] \exp(-i\Delta k_z z') dz' \right\} dx' dy'. \end{aligned} \tag{2.80}$$

Under the projection approximation, the quantity in braces above is proportional to the wave-field formed at the exit surface $z = z_0$ of the scattering distribution.[43]

[43] Assume that the measured scattered wave-vectors only make small angles with respect to the direction of the incident plane wave. As stated in the main text, the incident plane wave is

The far-field diffraction pattern is then calculated in the usual way, as described in the previous chapter, via the Fraunhofer prescription of taking the Fourier transform with respect to x and y (see Section 1.5).

2.5.2 *Angular spectrum and the first Born approximation*

In this sub-section, we show how the first Born approximation may be used to develop a formula for the propagated field over any plane downstream of a localized scattering volume (Wolf 1969). Our development will split into two parts. First, we will explain how to derive a certain integral representation of the spherical wave, known as the 'Weyl expansion' (Weyl 1919; reprinted in Oughstun 1992). Second, we will make use of this expansion to obtain the desired formula for the propagated field downstream of the scattering volume.

Let us begin with a derivation of the Weyl expansion. Consider the Fourier integral representation for the Green function given in eqn (2.60), which we write in the form:

$$G_\omega(x,y,z) = \frac{1}{2\pi^2}\iint \exp[i(k_x x + k_y y)]\left\{\int \frac{\exp(ik_z z)dk_z}{k_z^2 - (k^2 - k_x^2 - k_y^2)}\right\}dk_x dk_y. \tag{2.81}$$

Next we compute the integral contained in braces. To this end we can recycle the result obtained in eqn (2.67): (i) in this equation, make the replacements $k_r \to k_z$, $r \to z$, and $k \to (k^2 - k_x^2 - k_y^2)^{1/2}$; (ii) then note that the first integral in eqn (2.67) vanishes, with the second integral being equal to the integral in the braces of the equation above. Hence we see that:

$$\int \frac{\exp(ik_z z)dk_z}{k_z^2 - (k^2 - k_x^2 - k_y^2)} = \pi i \frac{\exp(iz\sqrt{k^2 - k_x^2 - k_y^2})}{\sqrt{k^2 - k_x^2 - k_y^2}}, \tag{2.82}$$

where use has been made of eqn (1.19) in the exponent on the right-hand side. Substitute the above result into eqn (2.81), and replace $G_\omega(x,y,z)$ with the explicit expression for an expanding spherical wave given in eqn (2.69), to obtain the Weyl expansion:

travelling along the z direction. These two assumptions imply that the term $\exp(-i\Delta k_z z')$, in the last line of eqn (2.80), can be replaced by unity. The term in braces, in this equation, then becomes $\int_{z=0}^{z_0}[\delta_\omega(x',y',z') - i\beta_\omega(x',y',z')]dz'$. This expression is proportional to that which is obtained if one takes the projection approximation given by eqn (2.39), assumes a sufficiently weak scatterer to allow one to linearize the exponent that appears there, sets $\tilde{\psi}_\omega(x,y,z=0)$ equal to unity since the normally incident plane wave $\exp(ikz)$ has this value in the plane $z=0$, and then discards the unscattered component of the resulting wave-field.

$$\frac{\exp(ik\sqrt{x^2+y^2+z^2})}{\sqrt{x^2+y^2+z^2}} = \frac{i}{2\pi}\iint \frac{\exp[i(k_x x + k_y y + z\sqrt{k^2-k_x^2-k_y^2})]}{\sqrt{k^2-k_x^2-k_y^2}} dk_x dk_y,$$
$$z > 0. \quad (2.83)$$

Now that we have obtained the above two-dimensional Fourier decomposition for an outgoing spherical wave, we can substitute it into eqn (2.74), to give:

$$\begin{aligned}\psi_\omega(\mathbf{x}) &= \exp(i\mathbf{k}_0\cdot\mathbf{x}) \qquad (2.84)\\ &-\frac{ik^2}{8\pi^2}\iint \frac{dk_x dk_y \exp\left[i\left(k_x x + k_y y + z\sqrt{k^2-k_x^2-k_y^2}\right)\right]\Gamma(\mathbf{k}_0,k_x,k_y)}{\sqrt{k^2-k_x^2-k_y^2}},\end{aligned}$$

$$\begin{aligned}\Gamma(\mathbf{k}_0,k_x,k_y) \equiv & \iiint [1-n_\omega^2(\mathbf{x}')]\exp(i\mathbf{k}_0\cdot\mathbf{x}')\\ &\times\exp\left[-i\left(k_x x' + k_y y' + z'\sqrt{k^2-k_x^2-k_y^2}\right)\right]d\mathbf{x}', \quad z > z_0. \quad (2.85)\end{aligned}$$

This expression allows one to evaluate the scattered field over any plane at a distance $z > z_0$ downstream of a set of scatterers, which are sufficiently weakly scattering for the first Born approximation to hold, and which are contained in the slab of space between the planes $z = 0$ and $z = z_0 > 0$. Note, in this context, that $\Gamma(\mathbf{k}_0, k_x, k_y)$ comprises a certain two-dimensional section through the Fourier transform of $[1 - n_\omega^2(\mathbf{x})]$ with respect to $\mathbf{x}$. Thus—notwithstanding the fact that the three-dimensional Fourier transform of $\left[1 - n_\omega^2(\mathbf{x})\right]$ has three dimensions of information contained within it—for a particular scattering experiment, and under the first Born approximation, it is only the value of this function over a particular two-dimensional surface, which determines the form of the scattered field. This surface, known as the Ewald sphere, will be discussed in the next section. In the present context, however, we may immediately conclude that only a subset of information regarding the scatterer is 'imprinted' upon the field scattered by it, in a given scattering scenario under the first Born approximation. Note that this observation will be explored in more detail, in Section 2.11.

2.5.3 *The Ewald sphere*

The Ewald sphere is a very commonly used construction in the study of kinematical X-ray diffraction from crystals. However, the notion of the Ewald sphere may also be fruitfully applied to scattering from non-crystalline specimens. It is in this latter context that we will place our emphasis, on account of the vast literature that exists on kinematical X-ray diffraction from periodic assemblies of atoms.

The present sub-section is broken up into two parts. In the first part, we will show how the Ewald sphere is a direct consequence of the kinematical theory of diffraction based on the first Born approximation. Second, we will examine the important special case of this construct, for scattering by crystals.

2.5.3.1 *From the first Born approximation to the Ewald sphere* In this subsection, we develop a direct link between the first Born approximation, and the concept of the Ewald sphere. Interestingly, we shall need no further mathematical development in order to make this connection—rather, we merely need to look at some of our earlier equations in a different light.

Accordingly, let us briefly revise the physics which underlies eqns (2.77) through (2.79). Under the single-scattering scenario implied by the first Born approximation, these equations give an expression for the coherent scalar X-ray wave-field $\psi_\omega(x, y, z)$, which results when a plane wave $\exp(i\mathbf{k}_0 \cdot \mathbf{x})$ is incident upon a weak non-magnetic localized scatterer characterized by a given spatial distribution of time-independent refractive index $n_\omega(x, y, z)$. According to eqn (2.77), the resulting wave-field can be written as a sum of the unscattered wave-field, together with a distorted spherical wave $f \exp(ikr)/r$. The distortion on the spherical wave $\exp(ikr)/r$, which emanates from the scattering volume, is quantified by the scattering amplitude f which modulates the otherwise perfect outgoing spherical wave. Determination of the scattering amplitude f, which is the central task to be performed in determining the scattered wave-field given by eqn (2.77), proceeds in two stages. First, one evaluates eqn (2.78) for all values of $\Delta\mathbf{k}$, to give a function proportional to the Fourier transform of $[n_\omega^2(x, y, z) - 1]$ with respect to x, y, and z. Second, one notes from eqn (2.79) that only a certain two-dimensional subset of points, in the three-dimensional Fourier transform of $[n_\omega^2(x, y, z) - 1]$, contributes to diffracted wave-field.

Indeed, according to eqn (2.79), to evaluate the radiation wave-field which results from the scattering scenario described above, one must evaluate the Fourier transform of $[n_\omega^2(x, y, z) - 1]$ at the series of points $k\hat{\mathbf{x}} - \mathbf{k}_0$ in Fourier space. Since all scattering directions $\hat{\mathbf{x}}$ are possible, this series of points forms a spherical surface in Fourier space.

Known as the Ewald sphere, this Fourier-space construction is sketched in Fig. 2.6(a). The incident wave-vector $\mathbf{k}_0$ is denoted by the vector pointing vertically downwards. The tip of this vector is identified with the origin O of Fourier space. The scattered wave-vector $k\hat{\mathbf{x}}$ has a modulus which is equal to that of the incident wave-vector—that is, the scattering is elastic—since $\hat{\mathbf{x}}$ is a unit vector. The totality of all possible scattered wave-vectors $k\hat{\mathbf{x}}$, associated with the set of all possible scattering directions $\hat{\mathbf{x}}$, have tips which all lie on the surface of the sphere indicated in part (a) of the figure, provided that the tails of all such scattered wave-vectors are made to coincide with the tail of the incident wave-vector. Now imagine that the Fourier space, in which the Ewald sphere is located, is also occupied by a function equal to the Fourier transform of $[n_\omega^2(x, y, z) - 1]$. In general, this function will be smeared throughout the space, although for localized scatterers it will typically decay rapidly to zero for Fourier-space radii whose magnitude is much larger than the reciprocal of the smallest length scale over

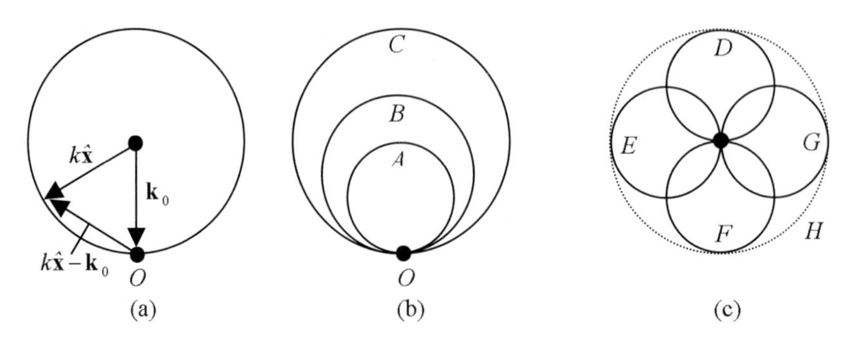

Fig. 2.6. (a) Ewald sphere for elastic scattering of incident plane waves under the first Born approximation. Incident and scattered wave-vectors are respectively denoted by $\mathbf{k}_0$ and $k\hat{\mathbf{x}}$. (b) Series of Ewald spheres, for three different incident beam energies, all of which have the scatterer in the same orientation relative to the incident plane wave. (c) Series of Ewald spheres, for four different beam angles of incidence, all of which have the illuminating beam at the same energy.

which the refractive index varies appreciably. Since the Fourier transformation is an invertible operation, all of the information, contained in $[n_\omega^2(x, y, z) - 1]$, is contained in its Fourier transform with respect to x, y, and z. However, it is only the values of this transform over points on the surface of the Ewald sphere, which will contribute to the scattered radiation wave-field in a particular experiment, with plane wave illumination and under the first Born approximation.

Suppose one wishes to probe values of the Fourier transform of $[n_\omega^2(x, y, z)-1]$, for values other than those which lie at the surface of a given Ewald sphere, for a particular plane-wave scattering experiment. One means of doing this, as sketched in Fig. 2.6(b), is to vary the energy of the incident plane wave while leaving unchanged both the orientation of the object and the direction of the incident plane wave. One will then have a series of Ewald spheres of different radii, all of which have a common point O. Three such spheres, denoted A, B and C, are sketched in the figure. Another strategy, as sketched in Fig. 2.6(c), is to leave the energy of the incident beam unchanged, while rotating either the direction of the incident beam or the orientation of the sample. Regardless of how this second strategy is effected, we can view the Ewald sphere as thereby rotating about the origin of Fourier space. This will sweep out a spherical volume, centred on this point and with a radius equal to $2k = 4\pi/\lambda$, as indicated by the dotted sphere H (known as the limiting Ewald sphere) in this figure. Within this limiting sphere, we see four Ewald spheres corresponding to four different experiments, these being labelled D, E, F, and G.

2.5.3.2 *The Ewald sphere for scattering from crystals* Notwithstanding the importance of the Ewald sphere in the context of kinematical scattering from crystals, we shall only give a cursory treatment of this topic here. There is a vast literature on the subject, to which the interested reader is referred. Excellent introductory accounts are given by Ashcroft and Mermin (1976), Kittel (1986), and Hammond (2001), with advanced treatments including the texts of Azaroff (1968), Warren (1969), Pinsker (1978), Cullity (1978), Woolfson (1997), and Authier (2001).

Here, we confine ourselves to a brief look at how the first Born approximation, for kinematical X-ray diffraction, is specialized to the case where the scattering potential is periodic (i.e. where the scatterer is a crystal). Introduce the notation:

$$n_\omega^2(\mathbf{x}) - 1 \equiv q(\mathbf{x})S(\mathbf{x}) \tag{2.86}$$

for these periodic variations in refractive index associated with a non-magnetic crystal. Here, $q(\mathbf{x})$ is periodic in three dimensions, and therefore infinite in extent. The so-called shape function $S(\mathbf{x})$ has been introduced to account for the fact that real crystals are finite in size. This shape function, whose form will be left arbitrary and which will be not be considered again until later in this subsection, is defined to have a value of unity inside the volume of space occupied by a finite-sized but otherwise perfect crystal, and to have a value of zero outside that crystal.

Note that, in writing down the above idealized equation, we have ignored the following features of real crystals: (i) real crystals typically contain defects—namely deviations from strict periodicity—which include but are not limited to stacking faults, vacancies, interstitials, screw dislocations, and twins (see, for example, Ashcroft and Mermin (1976) and Kittel (1986)); (ii) at all temperatures, each atom will oscillate about its equilibrium position in the crystal, making the refractive index time dependent, and thereby influencing the X-ray diffraction pattern recorded from such a structure (see, for example, Warren (1969)); (iii) we have neglected the possibility of relaxation effects, whereby the structure near the surface of a crystal may differ from that within the bulk of the crystal.

Notwithstanding our neglect of the above features of real crystals, eqn (2.86)—which models a crystal as a perfectly periodic three-dimensional distribution of refractive index that is truncated to the volume of space in which the shape function $S(\mathbf{x})$ is equal to unity—is an idealization that is sufficient for our purposes. Focussing upon $q(\mathbf{x})$ for the present, we note that since this quantity is periodic, there exist three linearly independent vectors $\mathbf{a}, \mathbf{b}, \mathbf{c}$ which are such that:

$$q(\mathbf{x} + u\mathbf{a} + v\mathbf{b} + w\mathbf{c}) = q(\mathbf{x}), \tag{2.87}$$

for any integers u, v, and w. Note that the vectors $\mathbf{a}, \mathbf{b}, \mathbf{c}$ will not, in general, be orthogonal to one another. Their choice is not unique, but for a given series of three directions, one can demand that each member of this set is as short as possible. The parallelepiped formed by the triple of vectors $\mathbf{a}, \mathbf{b}, \mathbf{c}$ can then be associated with a so-called 'unit cell' of the lattice. Associated with each of these unit cells is a particular distribution of scatterers within the volume of the cell, with the full scattering distribution $q(\mathbf{x})$ being built up by stacking these parallelepiped 'bricks' to fill all of space. All possible linear combinations of the vectors $\mathbf{a}, \mathbf{b}, \mathbf{c}$, with integer coefficients in front of each vector, results in the so-called 'direct lattice'. This is a periodic lattice of points in three dimensions, at each point of which is placed the same atom or group of atoms.

Since $q(\mathbf{x})$ is periodic, it can be expanded as a Fourier series. As we shall justify in a moment, the required Fourier series is:

$$q(\mathbf{x}) = \sum_h \sum_k \sum_l q_{hkl} \exp\left(i\mathbf{g}_{hkl} \cdot \mathbf{x}\right), \quad \mathbf{g}_{hkl} \equiv h\mathbf{a}^\star + k\mathbf{b}^\star + l\mathbf{c}^\star, \tag{2.88}$$

where h, k, and l can each take any integer value, q_{hkl} are the Fourier coefficients of $q(\mathbf{x}) \equiv q(x, y, z)$, and:

$$\mathbf{a}^\star = \frac{2\pi\mathbf{b} \times \mathbf{c}}{\mathbf{a} \cdot (\mathbf{b} \times \mathbf{c})}, \qquad \mathbf{b}^\star = \frac{2\pi\mathbf{c} \times \mathbf{a}}{\mathbf{a} \cdot (\mathbf{b} \times \mathbf{c})}, \qquad \mathbf{c}^\star = \frac{2\pi\mathbf{a} \times \mathbf{b}}{\mathbf{a} \cdot (\mathbf{b} \times \mathbf{c})}. \tag{2.89}$$

The vectors $\mathbf{a}^\star, \mathbf{b}^\star, \mathbf{c}^\star$ are the basis vectors for the so-called 'reciprocal lattice' $\mathbf{g}_{hkl}$, a concept of central importance in the theory of X-ray diffraction from crystals. As the reader can readily show, all possible dot products between any member of the direct-lattice basis $\mathbf{a}, \mathbf{b}, \mathbf{c}$, and any member of the reciprocal-lattice basis $\mathbf{a}^\star, \mathbf{b}^\star, \mathbf{c}^\star$, are given by:

$$\begin{array}{lll} \mathbf{a}^\star \cdot \mathbf{a} = 2\pi, & \mathbf{a}^\star \cdot \mathbf{b} = 0, & \mathbf{a}^\star \cdot \mathbf{c} = 0, \\ \mathbf{b}^\star \cdot \mathbf{a} = 0, & \mathbf{b}^\star \cdot \mathbf{b} = 2\pi, & \mathbf{b}^\star \cdot \mathbf{c} = 0, \\ \mathbf{c}^\star \cdot \mathbf{a} = 0, & \mathbf{c}^\star \cdot \mathbf{b} = 0, & \mathbf{c}^\star \cdot \mathbf{c} = 2\pi. \end{array} \tag{2.90}$$

We are now ready to verify the correctness of the Fourier-series decomposition given in eqn (2.88), by showing that the left side of this equation indeed obeys the condition for periodicity given in eqn (2.87). Accordingly, let us prove that eqn (2.88) is invariant under the replacement of $\mathbf{x}$ with $\mathbf{x} + u\mathbf{a} + v\mathbf{b} + w\mathbf{c}$, as stipulated in eqn (2.87), for any triple of integers (u, v, w):

$$\begin{aligned} & q(\mathbf{x} + u\mathbf{a} + v\mathbf{b} + w\mathbf{c}) \\ & = \sum_h \sum_k \sum_l q_{hkl} \exp\left[i\mathbf{g}_{hkl} \cdot (\mathbf{x} + u\mathbf{a} + v\mathbf{b} + w\mathbf{c})\right] \\ & = \sum_h \sum_k \sum_l q_{hkl} \exp\left(i\mathbf{g}_{hkl} \cdot \mathbf{x}\right) \exp[i(h\mathbf{a}^\star + k\mathbf{b}^\star + l\mathbf{c}^\star) \cdot (u\mathbf{a} + v\mathbf{b} + w\mathbf{c})] \\ & = \sum_h \sum_k \sum_l q_{hkl} \exp\left(i\mathbf{g}_{hkl} \cdot \mathbf{x}\right) \exp[i(hu\mathbf{a}^\star \cdot \mathbf{a} + kv\mathbf{b}^\star \cdot \mathbf{b} + lw\mathbf{c}^\star \cdot \mathbf{c})] \\ & = \sum_h \sum_k \sum_l q_{hkl} \exp\left(i\mathbf{g}_{hkl} \cdot \mathbf{x}\right) \exp[2\pi i(hu + kv + lw)] \\ & = q(\mathbf{x}). \end{aligned} \tag{2.91}$$

Note that, in obtaining the above equations, we have made use of the definition for $\mathbf{g}_{hkl}$ given in eqn (2.88), the dot products given in eqn (2.90), and the fact that $hu + kv + lw$ is an integer. This completes our proof, that the Fourier decomposition given in eqn (2.88), indeed has the periodicity required by eqn (2.87).

We are now ready to consider the problem of X-ray diffraction from the truncated but otherwise perfect crystal having the distribution of refractive index given by eqn (2.86), under the first Born approximation. Evidently, this is a special case of the theory presented earlier, which culminated in eqns (2.77) through (2.79). Accordingly, let us calculate the quantity $f(\Delta\mathbf{k})$ in eqn (2.78). In the first instance, assume the shape function $S(\mathbf{x})$ in eqn (2.86) to be equal to unity, which amounts to ignoring effects in the diffraction pattern that are due to the finite size of the crystal. If we then substitute the Fourier-series representation (2.88) into eqn (2.78) and interchange the order of the integrations and summations, we obtain:

$$f(\Delta\mathbf{k}) = \frac{k^2}{4\pi}\sum_h\sum_k\sum_l q_{hkl}\iiint \exp\left[i(\mathbf{g}_{hkl} - \Delta\mathbf{k})\cdot\mathbf{x}'\right] d\mathbf{x}'. \tag{2.92}$$

The integral, on the right side of this equation, is proportional to the Fourier-integral representation of the three-dimensional Dirac delta:

$$\delta(\mathbf{k}) = \frac{1}{(2\pi)^3}\iiint \exp(i\mathbf{k}\cdot\mathbf{x})d\mathbf{x}. \tag{2.93}$$

Hence eqn (2.92) becomes:

$$f(\Delta\mathbf{k}) = 2\pi^2 k^2 \sum_h\sum_k\sum_l q_{hkl}\delta(\mathbf{g}_{hkl} - \Delta\mathbf{k}). \tag{2.94}$$

Substitute this into eqn (2.77), then make use of eqn (2.79), to arrive at the following expression for kinematical X-ray diffraction from a crystal:

$$\psi_\omega(\mathbf{x}) = \exp(i\mathbf{k}_0\cdot\mathbf{x}) + 2\pi^2 k^2 \frac{\exp(ikr)}{r}\sum_h\sum_k\sum_l q_{hkl}\delta(\mathbf{g}_{hkl} - k\hat{\mathbf{x}} + \mathbf{k}_0). \tag{2.95}$$

The interpretation of this equation is similar to that which was given earlier, in the context of the first Born approximation for scattering from non-crystalline samples. Once again, we see that the wave-field $\psi_\omega(\mathbf{x})$ far from the scatterer is given by superposing the unscattered plane wave $\exp(i\mathbf{k}_0\cdot\mathbf{x})$ with a distorted expanding spherical wave. Once again, the envelope on the distorted spherical wave is related to the Fourier transform of the scattering distribution. However, it is at this point that we reach the central point of difference between kinematical scattering by crystalline and non-crystalline samples. For the case of a crystal considered here, and for an incident plane wave with wave-vector $\mathbf{k}_0$, the presence of the Dirac delta in the above equation implies that scattered waves will only be observed in directions $\hat{\mathbf{x}}$ which are such that $\mathbf{g}_{hkl} - k\hat{\mathbf{x}} + \mathbf{k}_0 = 0$. Recalling eqn (2.79), this condition becomes the 'von Laue diffraction condition':

$$k\hat{\mathbf{x}} - \mathbf{k}_0 = \Delta\mathbf{k} = \mathbf{g}_{hkl}. \tag{2.96}$$

One of the most-used equations in the study of kinematical X-ray diffraction from crystals, this result—which can readily be shown to be equivalent to the famous Bragg Law (see Section 3.2.3)—may be visualized in terms of the Ewald sphere sketched in Fig. 2.6(a). As was the case in our earlier discussions of this construction, the incident and scattered wave-vectors are respectively denoted by $\mathbf{k}_0$ and $k\hat{\mathbf{x}}$, with k being the wave-number of the radiation and $\hat{\mathbf{x}}$ being a unit vector pointing along the direction of the scattered radiation. These two vectors are placed such that the former has its tip coinciding with the origin O of the reciprocal lattice. As mentioned earlier, this reciprocal lattice consists of the set of points with position vectors $\mathbf{g}_{hkl} \equiv h\mathbf{a}^\star + k\mathbf{b}^\star + l\mathbf{c}^\star$, where h, k, l are all possible triplets of integers, and $\mathbf{a}^\star, \mathbf{b}^\star, \mathbf{c}^\star$ are the basis vectors for the reciprocal lattice, which may be obtained from the basis vectors $\mathbf{a}, \mathbf{b}, \mathbf{c}$ for the direct lattice, using eqn (2.89). The origin O of the reciprocal lattice is then given by the point $\mathbf{g}_{000}$.

Equipped with these preliminaries, we are ready to interpret eqn (2.96) in terms of the Ewald sphere. With reference to Fig. 2.6(a), scattering will only occur in directions $\hat{\mathbf{x}}$ which are such that the vector $k\hat{\mathbf{x}}$ points to a reciprocal lattice point—that is, diffraction occurs when a reciprocal lattice point intersects the surface of the Ewald sphere. Stated differently, diffraction will only occur if incident and scattered wave-vectors differ by a reciprocal lattice vector.

Since the Ewald sphere is finite in size and the spacing between points in the reciprocal lattice is non-zero, this implies that the diffracted beam will only be non-zero in a finite number of directions. This corresponds to the well-known fact that the X-ray diffraction pattern of a crystal, which is sufficiently weakly scattering for the assumption of single-scattering to be a good one, will consist of a series of diffraction spots. Such a series of spots is shown in Fig. 2.7, this being the diffraction pattern of zinc sulphide included in a 1912 paper by Friedrich, Knipping, and von Laue. The unscattered beam is evident as the bright spot at the centre of the image, with a number of diffraction spots surrounding this undiffracted beam.

Thus far, we have ignored the effects of the finite size of the crystal, on the kinematical diffraction of X-rays from such a structure. At a first level of approximation, the effects of this finite size may be included via the shape function $S(\mathbf{x})$, which was introduced in eqn (2.86). We leave it as an exercise to the reader to modify the calculation, leading to eqn (2.95), so as to include the effects of this shape function. One thereby concludes that the points in reciprocal space become smeared out so as to occupy a non-zero volume, with the corollary that

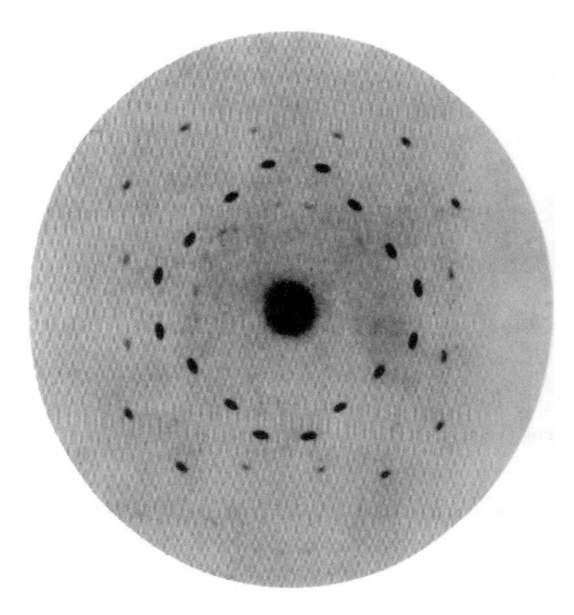

Fig. 2.7. X-ray diffraction image of zinc sulphide, taken from a 1912 paper by Friedrich, Knipping, and von Laue. Image reproduced from Hammond (2001).

the associated diffraction spots become decorated with fine structure indicative of the shape of the diffracting crystal.

We close this section by mentioning the Rytov approximation, which is closely related to the first Born approximation. We shall not give an account of this theory in the present book, nor its relationship with the first Born approximation. The interested reader may consult Tatarskii (1967), Nieto-Vesperinas (1991), and Born and Wolf (1999), together with primary references therein. As an entry point to the X-ray literature on this subject, we refer the reader to the paper by Davis (1994), which uses the Rytov approximation to present a unified account of X-ray scattering, absorption, and refraction.

2.6 Born series and dynamical scattering

Thus far, we have spoken at length regarding the first Born approximation, this being an approximate means of solving the integral equation for scattering given by eqn (2.70). With a view to obtaining the second and higher-order Born approximations, we write this integral equation in operator form as:

$$\psi_\omega(\mathbf{x}) = \psi_\omega^0(\mathbf{x}) + \mathcal{G}_\omega v_\omega(\mathbf{x})\psi_\omega(\mathbf{x}), \tag{2.97}$$

where the distribution of refractive index in the scatterer has been written as:

$$v_\omega(\mathbf{x}) \equiv \frac{k^2}{4\pi}\left[n_\omega^2(\mathbf{x}) - 1\right] \tag{2.98}$$

and the Green operator $\mathcal{G}_\omega$ is defined to be the operator which acts upon a given input function of $\mathbf{x}$, say $f(\mathbf{x})$, and returns as output the convolution of the input function with the outgoing Green function of angular frequency ω:

$$\mathcal{G}_\omega f(\mathbf{x}) \equiv \iiint G_\omega(\mathbf{x} - \mathbf{x}')f(\mathbf{x}')d\mathbf{x}'. \tag{2.99}$$

In writing down eqn (2.97), and subsequently, we assume that all operators act from right to left, and that the Green operator acts to everything on its right. For example, we have:

$$\mathcal{G}_\omega v_\omega(\mathbf{x})\psi_\omega(\mathbf{x}) \equiv \mathcal{G}_\omega\left[v_\omega(\mathbf{x})\psi_\omega(\mathbf{x})\right]. \tag{2.100}$$

Having established the above notation, we see that the first Born approximation—obtained, as before, by replacing the wave-field on the right side of eqn (2.97) by its unscattered value—is now written as:

$$\psi_\omega^{(1)}(\mathbf{x}) = \psi_\omega^0(\mathbf{x}) + \mathcal{G}_\omega v_\omega(\mathbf{x})\psi_\omega^0(\mathbf{x}) = \left[1 + \mathcal{G}_\omega v_\omega(\mathbf{x})\right]\psi_\omega^0(\mathbf{x}). \tag{2.101}$$

Note that we have introduced a superscript on the left side of this equation, to indicate that $\psi_\omega^{(1)}(\mathbf{x})$ is the approximation to $\psi_\omega(\mathbf{x})$ that is obtained under the first Born approximation. Similarly, we denote by $\psi_\omega^{(m)}(\mathbf{x})$ the approximation to $\psi_\omega(\mathbf{x})$ under the mth Born approximation, where m is an integer.

Now that we have eqn (2.101), this being the first Born approximation's estimate for $\psi_\omega(\mathbf{x})$, we may obtain a better estimate for $\psi_\omega(\mathbf{x})$ by replacing the wave-field on the right side of eqn (2.97) by the estimate for this function, which is given by the first Born approximation. This results in the second Born approximation:

$$\psi_\omega^{(2)}(\mathbf{x}) = \psi_\omega^0(\mathbf{x}) + \mathcal{G}_\omega v_\omega(\mathbf{x})\psi_\omega^{(1)}(\mathbf{x}) = \left\{1 + \mathcal{G}_\omega v_\omega(\mathbf{x}) + \left[\mathcal{G}_\omega v_\omega(\mathbf{x})\right]^2\right\}\psi_\omega^0(\mathbf{x}). \tag{2.102}$$

This procedure may be iterated, giving an infinite hierarchy of approximations known as the 'Born series'. The mth member of this series, namely the mth Born approximation, is given by:

$$\psi_\omega^{(m)}(\mathbf{x}) = \left\{1 + \mathcal{G}_\omega v_\omega(\mathbf{x}) + \left[\mathcal{G}_\omega v_\omega(\mathbf{x})\right]^2 + \cdots + \left[\mathcal{G}_\omega v_\omega(\mathbf{x})\right]^m\right\}\psi_\omega^0(\mathbf{x}). \tag{2.103}$$

Each term, in the braces of the equation above, may be interpreted in terms of the degree of scattering to which it corresponds. The first term, which is unity, corresponds to no scattering. The second term corresponds to the singly scattered wave-field predicted by the first Born approximation. The third term corresponds to doubly scattered radiation, and so on. Evidently, in the limit as m tends to infinity, the Born series decomposes the wave-field in the presence of a scatterer, as a sum of wave-fields, each of which has been scattered a given number of times in interacting with the scatterer.

Now, we have previously used the term 'kinematical' to refer to scattering theories that assume a single-scattering approximation. As has already been stated, scattering theories that go beyond the single-scattering approximation, such as the Born series outlined here, are known as 'dynamical' theories of X-ray diffraction.

In considering a particular X-ray scattering scenario, one can make use of second and higher-order Born approximations if the first Born approximation is insufficient. In practice, however, it is often the case that the Born series is poorly converging, in which case alternative means of approximation are needed to treat X-ray diffraction which is dynamical in character. Note that the multislice and eikonal approximations, discussed in the following two sections, together with the projection approximation treated earlier in this chapter, are examples of dynamical diffraction theories. An important omission, from our accounts, is the Bloch-wave approach to dynamical X-ray scattering. The interested reader is referred to the book on dynamical X-ray diffraction by Authier (2001), and references therein, for more detail on the vast and fascinating subject of dynamical X-ray diffraction.

2.7 Multislice approximation

The multi-slice method is an intuitive procedure for treating dynamical scattering by a localized scattering potential. This methodology, of which there are many variants, was originally developed in the context of electron diffraction by Cowley and Moodie (1957)—see also Cowley (1995). Our outline, of one such variant, will make use of both the angular spectrum formalism (see Section 1.3) and the projection approximation (see Section 2.2). Using these tools, we will construct a multislice algorithm for the scattering of coherent scalar X-rays from a given potential, which is sufficiently weakly scattering for us to be able to neglect backscatter.

Consider the scenario shown in Fig. 2.8, but with the scatterer absent. Suppose that one was given the value of a given coherent scalar X-ray disturbance, over the plane A. Suppose, further, that one knew the wave-field to be forward propagating, that is, there are no points on the plane A at which there is a flow of energy from right to left. Given knowledge of the forward-propagating wave-field over the plane A, we saw in Section 1.3 that the angular-spectrum formalism allows us to construct an operator $\mathcal{D}_\Delta$ that may be applied to this disturbance, yielding the propagated disturbance over any parallel plane which lies at a non-negative distance Δ downstream of A. We saw that this operator has the explicit form given by eqn (1.25) of the previous chapter, namely:

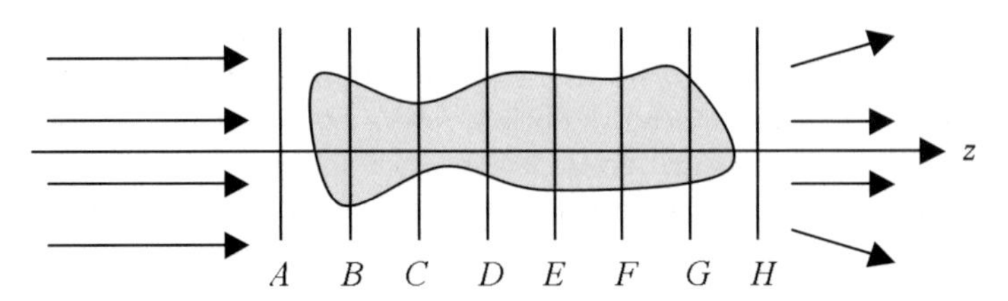

Fig. 2.8. Coherent paraxial X-ray radiation is incident from the left, with optic axis z. Multislice algorithms consider the scatterer, here shown in grey, to be sliced into a number of slabs. Here, these slabs are delineated by the series of equally spaced parallel planes marked A, B, C, etc. After passing through all slices of the scatterer, one has the exit-surface wave-function over the plane marked H. This exit-surface wave-function then propagates into the half-space downstream of the exit surface.

$$\mathcal{D}_\Delta = \mathcal{F}^{-1} \exp\left[i\Delta\sqrt{k^2 - k_x^2 - k_y^2} \right] \mathcal{F}. \tag{2.104}$$

Consider, once again, the scattering scenario sketched in Fig. 2.8, but this time with the scatterer present. Here, we see a forward-propagating paraxial wave-field incident from the left, which creates a known disturbance over the plane marked A. All scatterers, denoted by the irregular grey shape in the figure, are assumed to lie between the planes marked A and H, respectively. The object is assumed to be sufficiently weakly scattering such that the wave-field is forward propagating, over any of the sequence of parallel planes marked A through H.

In order to obtain the wave-field over the exit surface H, multislice algorithms split the problem into that of propagating from plane A to B, then from B to C, and so on until one reaches the plane H. Accordingly, let us consider the question of how to propagate the wave-field from the plane $z = z_0$ to the plane $z = z_0 + \Delta$, the thickness of the 'slice' being given by $\Delta > 0$. The plane $z = z_0$ is assumed to lie somewhere between the planes A and H, as marked in the figure. Within the slab, between the planes $z = z_0$ and $z = z_0 + \Delta$, the scatterers are described by their three-dimensional distribution of refractive index $n_\omega(x, y, z)$, with $z_0 \leq z \leq z_0 + \Delta$.

According to the projection approximation, the wave-field $\psi_\omega(x, y, z_0)$ at the entrance surface of the slice is related to the wave-field $\psi_\omega(x, y, z_0 + \Delta)$ at the exit-surface of the slice by the appropriate form of eqn (2.36). Thus, bearing eqn (2.29) in mind, we write down:

$$\begin{aligned} &\psi_\omega(x, y, z = z_0 + \Delta) \\ &\approx \exp(ik\Delta) \exp\left\{ \frac{k}{2i} \int_{z=z_0}^{z=z_0+\Delta} [1 - n_\omega^2(x, y, z)] dz \right\} \psi_\omega(x, y, z = z_0). \end{aligned} \tag{2.105}$$

Since it was obtained under the projection approximation, the above expression ignores the effects of free-space propagation between the planes $z = z_0$ and $z = z_0 + \Delta$. The essence of the multislice approximation is to separately consider the effects of free-space propagation within the slab, from the phase and amplitude shifts obtained under the projection approximation and quantified in the above equation. Thus the multislice approximation simply applies the scaled free-space diffraction operator $\exp(-ik\Delta)\mathcal{D}_\Delta$ (see eqn (1.25))[44] to the right side of eqn (2.105), to obtain the following approximate expression for the wave-field at the exit-surface $z = z_0 + \Delta$ of the slab, from the wave-field at the entrance surface $z = z_0$ of the slab:

$$\begin{aligned}
&\psi_\omega(x, y, z = z_0 + \Delta) \qquad (2.106)\\
&\quad \approx \mathcal{D}_\Delta \left(\exp\left\{ \frac{k}{2i} \int_{z=z_0}^{z=z_0+\Delta} [1 - n_\omega^2(x, y, z)]dz \right\} \psi_\omega(x, y, z = z_0) \right).
\end{aligned}$$

This expression acquires a simpler form if we note that the paraxiality of the radiation allows us to apply the binomial approximation (1.26) to the explicit form of the free space diffraction operator given in eqn (1.25). This results in:

$$\begin{aligned}
\psi_\omega(x, y, z = z_0 + \Delta) &= \exp(ik\Delta)\mathcal{F}^{-1} \exp\left(-i\Delta \frac{k_x^2 + k_y^2}{2k} \right) \\
&\times \mathcal{F} \exp\left\{ \frac{k}{2i} \int_{z=z_0}^{z=z_0+\Delta} [1 - n_\omega^2(x, y, z)]dz \right\} \psi_\omega(x, y, z = z_0). \qquad (2.107)
\end{aligned}$$

In the above equation, as has been our practice for cascaded operators, all operators are assumed to act from right to left (cf. eqn (2.100)).

By recursively applying the above equation, one can propagate the wave-field from $z = z_0$ to $z = z_0 + \Delta$, then from $z = z_0 + \Delta$ to $z = z_0 + 2\Delta$, and so forth, until one arrives at the wave-field at the exit-surface of the scattering volume. Evidently, the slice thickness Δ must be chosen to be sufficiently small. In practice, a rule of thumb is to continue reducing the slice thickness until further reduction in this thickness produces no change in the calculated exit-surface wave-function, to within the numerical accuracy desired for this quantity.

2.8 Eikonal approximation and geometrical optics

Here, we introduce the eikonal approximation for treating the interactions of coherent X-rays with matter. Closely related to the projection approximation introduced in Section 2.2, this formalism serves to forge a link between the

[44]Note that the free-space propagator has been scaled by the phase factor $\exp(-ik\Delta)$, since the complex conjugate of this phase factor has already been accounted for in making the projection approximation.

wave and ray theories of X-ray radiation, by showing how the latter emerges as a limiting case of the former.[45] Our treatment begins with a derivation of the so-called hydrodynamic formulation of the scalar wave optics of coherent X-ray wave-fields. One of the resulting equations is a continuity equation, being an expression for the local conservation of optical energy. The other equation in the hydrodynamic formulation, known as the eikonal equation, is of more importance in the present context. This equation passes, in the limit of infinitely small wavelength, to a central equation of geometric optics.[46] It therefore forms a conceptual bridge—one of many—between the wave and ray theories of the X-ray wave-field. At this point, we shall make contact with the formalism for the projection approximation which was outlined in Section 2.2, by rederiving one of its central results within the context of the eikonal approximation. Here, one can visualize the X-ray disturbance in terms of rays that continuously evolve as they travel through a refractive medium. In the absence of a medium, and in the geometric optics limit of zero wavelength, these rays travel in straight lines through free space. However, if the geometric optics limit is relaxed, a so-called 'diffraction term' is present in the eikonal equation, which serves to bend the rays even when they are travelling through free space. This bending of rays, which the formalism also indicates to be present within a material medium, is associated with the phenomenon of diffraction.

With a view to obtaining a hydrodynamic formulation of coherent scalar X-ray wave-fields, return to the inhomogeneous Helmholtz equation (2.28). Write the coherent wave-field $\psi_\omega(x,y,z)$ in terms of its intensity $|\psi_\omega(x,y,z)|^2 \equiv I_\omega(x,y,z)$ and phase $\arg[\psi_\omega(x,y,z)] \equiv \phi_\omega(x,y,z)$, so that eqn (2.28) becomes:

$$\left[\nabla^2 + k^2 n_\omega^2(x,y,z)\right]\left\{\sqrt{I_\omega(x,y,z)}\exp\left[i\phi_\omega(x,y,z)\right]\right\} = 0. \qquad (2.108)$$

Expanding out the Laplacian of the term in braces, we obtain[47]:

[45]Note, also, that these calculations have direct parallels with the so-called 'hydrodynamic' formulation of quantum mechanics. Due to Madelung (1926), this formulation of quantum mechanics is discussed in standard texts such as that of Messiah (1961). Note that one speaks of a 'hydrodynamic' formulation, on account of the evident parallels between such a formulation—which works with currents and energy densities rather than a complex wave-field—and the equations of fluid mechanics.

[46]Rather confusingly, this limiting case is also known as the eikonal equation—for clarity, we will always refer to this as the eikonal equation of geometric optics.

[47]In the second line of eqn (2.109), we have used the identity $\nabla^2 = \nabla \cdot \nabla$. In the third line, we have used a form of the product rule, namely $\nabla(AB) = A\nabla B + B\nabla A$, for suitably well-behaved functions A and B. By 'suitably well behaved', we mean that these functions must be continuous, single-valued, and differentiable. Note that these conditions will not be fulfilled at points in space where the intensity vanishes, together with points where the phase possesses singularities of the form discussed in Chapter 5.

$$\begin{aligned}
&\nabla^2 \left\{ \sqrt{I_\omega(x,y,z)} \exp\left[i\phi_\omega(x,y,z)\right] \right\} \\
&= \nabla \cdot \nabla \left\{ \sqrt{I_\omega(x,y,z)} \exp[i\phi_\omega(x,y,z)] \right\} \\
&= \nabla \cdot \left\{ \sqrt{I_\omega(x,y,z)} \nabla \exp[i\phi_\omega(x,y,z)] + \exp[i\phi_\omega(x,y,z)] \nabla \sqrt{I_\omega(x,y,z)} \right\} \\
&= \nabla \cdot \left\{ \sqrt{I_\omega(x,y,z)} \mathrm{e}^{i\phi_\omega(x,y,z)} i \nabla \phi_\omega(x,y,z) + \frac{\mathrm{e}^{i\phi_\omega(x,y,z)} \nabla I_\omega(x,y,z)}{2\sqrt{I_\omega(x,y,z)}} \right\}.
\end{aligned} \tag{2.109}$$

Each term, inside the braces of the last line, is equal to a scalar function multiplied by the gradient of another scalar function. With a view to evaluating the divergence of each of these terms, recall the following vector identity for two suitably well-behaved functions $A(x,y,z)$ and $B(x,y,z)$:

$$\nabla \cdot [A(x,y,z) \nabla B(x,y,z)] = A(x,y,z) \nabla^2 B(x,y,z) + \nabla A(x,y,z) \cdot \nabla B(x,y,z). \tag{2.110}$$

Making use of this identity, one can expand the last line of eqn (2.109). Having done so, substitute the result into eqn (2.108), and then cancel a common factor of $\sqrt{I_\omega(x,y,z)} \exp[i\phi_\omega(x,y,z)]$, to arrive at:

$$\begin{aligned}
&k^2 n_\omega^2(x,y,z) + i\nabla^2 \phi_\omega(x,y,z) + \frac{\nabla^2 I_\omega(x,y,z)}{2I_\omega(x,y,z)} + \frac{i\nabla I_\omega(x,y,z) \cdot \nabla \phi_\omega(x,y,z)}{I_\omega(x,y,z)} \\
&\quad - |\nabla \phi_\omega(x,y,z)|^2 - \frac{|\nabla I_\omega(x,y,z)|^2}{4[I_\omega(x,y,z)]^2} = 0.
\end{aligned} \tag{2.111}$$

The real part of this expression, and $I_\omega(x,y,z)$ multiplied by the imaginary part of this expression, respectively lead to the following pair of equations:

$$k^2 n_\omega^2(x,y,z) + \frac{\nabla^2 I_\omega(x,y,z)}{2I_\omega(x,y,z)} - |\nabla \phi_\omega(x,y,z)|^2 - \frac{|\nabla I_\omega(x,y,z)|^2}{4[I_\omega(x,y,z)]^2} = 0, \tag{2.112}$$

$$I_\omega(x,y,z) \nabla^2 \phi_\omega(x,y,z) + \nabla I_\omega(x,y,z) \cdot \nabla \phi_\omega(x,y,z) = 0. \tag{2.113}$$

The first of these equations may be simplified by noting that:

$$\frac{\nabla^2 \sqrt{I_\omega(x,y,z)}}{\sqrt{I_\omega(x,y,z)}} = \frac{\nabla^2 I_\omega(x,y,z)}{2I_\omega(x,y,z)} - \frac{|\nabla I_\omega(x,y,z)|^2}{4I_\omega^2(x,y,z)}, \tag{2.114}$$

with the second equation being simplified by recognizing its left-hand side to be the divergence of $I_\omega(x,y,z) \nabla \phi_\omega(x,y,z)$. We thereby arrive at two key results of the present section, namely a hydrodynamic formulation of the inhomogeneous Helmholtz equation (2.28):

$$|\nabla\phi_\omega(x,y,z)|^2 = k^2 n_\omega^2(x,y,z) + \frac{\nabla^2\sqrt{I_\omega(x,y,z)}}{\sqrt{I_\omega(x,y,z)}}, \tag{2.115}$$

$$\nabla \cdot [I_\omega(x,y,z)\nabla\phi_\omega(x,y,z)] = 0. \tag{2.116}$$

Equation (2.115) is known as the eikonal equation. In the first instance, let us consider the limiting case of this equation which results when the length scale, over which the intensity changes appreciably, is much larger than the wavelength of the radiation. In this case, which corresponds to the geometric optics limit, the second term on the right side will be negligible in comparison to the other terms.[48] Thus, in the geometric-optics limit, eqn (2.115) becomes the eikonal equation of geometric optics:

$$|\nabla\phi_\omega(x,y,z)| = k n_\omega(x,y,z), \quad k \to \infty. \tag{2.117}$$

With a view to interpreting this equation, and the caution that this interpretation will take a few paragraphs to reach, consider Fig. 2.9(a). Here we depict coherent scalar X-ray waves incident from the left, passing though a medium in which the refractive index varies with position. For clarity, this three-dimensional distribution of refractive index has not been included in the diagram. Locally, and in the absence of topological defects such as the phase vortices discussed in Chapter 5, the wave-field is planar. This is suggested in the diagram, by having the wave-fronts being close to parallel to one another, over sufficiently small regions. Since they are identified with the surfaces of constant phase, the normal to the wave-fronts is given by the phase gradient $\nabla\phi_\omega(\mathbf{x})$. One such vector is shown in the diagram. Subtleties regarding gauge freedom[49] aside, we may identify this vector with the direction of the energy flow at a given point in the field.

In steady-state fluid flow, the notion of 'streamlines' is often useful. These streamlines may be defined as the set of lines, passing through the fluid, which are such that the local current vector is tangent to every point on the line. As a corollary to this construction, a small test particle, placed at one point on a given streamline, will move along the streamline as it is carried along by the fluid. This notion has an analogue in Fig. 2.9(a), with the line BAC being one

[48]To see this, note that the phase of the wave-field will change by approximately 2π over distances on the order of one wavelength. Therefore the phase gradient scales as the inverse of the wavelength, that is, it scales with the wave-number k. Therefore, the left side of eqn (2.115) scales as k^2, as does the first term on the right side. The second term on the right side scales as the inverse square of the characteristic length scale over which the intensity changes appreciably. Therefore, this term is negligible compared to the other terms in eqn (2.115), in the geometrical optics limit, where k tends to infinity.

[49]The gauge freedom associated with the electromagnetic field, and its relevance to the ambiguities inherent in the concept of energy flow and energy density at a point, will not be dealt with in this book. For an entry point into the literature on this subject, see the references cited in Nieto-Vesperinas (1991).

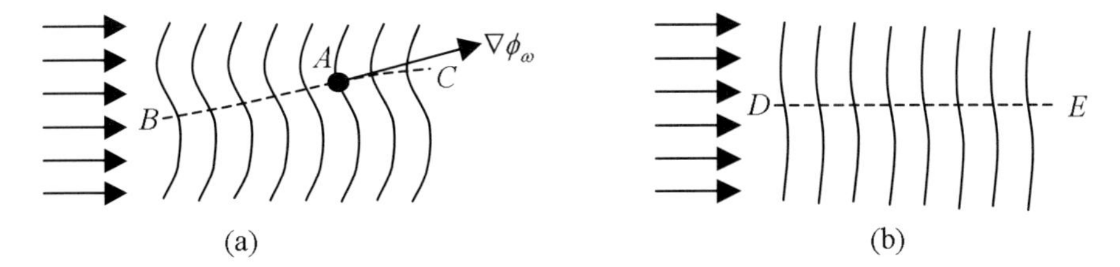

Fig. 2.9. (a) Coherent X-ray waves are incident from the left, passing through a space with a given distribution of refractive index (not shown). Surfaces of constant phase are indicated by wavy lines. The phase gradient $\nabla\phi_\omega$ at a point A is shown, with the dashed curve BC corresponding to a 'streamline' of the field. (b) Corresponding diagram in which plane waves pass through a non-uniform medium (not shown) which is sufficiently weakly scattering for the phase gradient, at each point in the field, to point in almost the same direction as that of the incident field. Thus the streamline DE, of the unscattered field, is very close to being a streamline of the field in the presence of the scattering medium.

such X-ray 'streamline'. Streamlines, of the optical flow, are lines which are such that the phase gradient is tangent to every point on the streamline.

Introduce the arc length s measured along a particular streamline, with respect to a convenient origin on that streamline. Since the phase gradient is tangent to every point on the streamline, and with the agreement that the direction of increasing s is chosen to be parallel rather than antiparallel to the local phase gradient, the modulus of eqn (2.117) implies that:

$$\frac{\partial\phi_\omega[x(s),y(s),z(s)]}{\partial s}=kn_\omega[x(s),y(s),z(s)],\quad k\rightarrow\infty. \tag{2.118}$$

Integrating with respect to s, we arrive at:

$$\begin{aligned}\phi_\omega[x(s),y(s),z(s)]&=\int_{s_0}^{s}\frac{\partial\phi_\omega[x(s),y(s),z(s)]}{\partial s}ds\\&=2\pi\int_{s_0}^{s}\frac{ds}{\lambda/\{n_\omega[x(s),y(s),z(s)]\}},\quad k\rightarrow\infty.\end{aligned} \tag{2.119}$$

Here, the phase origin is chosen to be the point $[x(s_0),y(s_0),z(s_0)]$ on the streamline, at $s=s_0$, with an arbitrary point on the streamline being denoted by $[x(s),y(s),z(s)]$.

From a physical point of view, the eikonal equation specifies how the spacing between locally planar wave-fronts is altered by the presence of a refractive medium. In so distorting the phase of the radiation wave-field, the energy flow of the field is altered, that is, the wave-fronts are refracted. This distortion of the phase is quantified by eqn (2.119).

To deepen our understanding of this result, note that one may re-derive it using the following geometric-optics argument: (i) At a given point on the streamline, the local wavelength is equal to λ/n_ω, where λ is the vacuum wavelength and n_ω is the value of the refractive index at the specified point on the streamline. (ii) Therefore, when one moves through a distance ds on the streamline, this corresponds to moving through a total of $ds/(\lambda/n_\omega)$ wavelengths. (iii) Since 2π radians of phase are accumulated when one moves through a distance of one local wavelength along the streamline, this implies that the phase accumulated in moving through a distance ds is equal to $2\pi ds/(\lambda/n_\omega)$. (iv) Integrating along the streamline, from a convenient origin, one arrives at eqn (2.119). As a corollary to this geometric-optics view, note that the rays travel in straight lines in the absence of any medium, but that they do not, in general, travel in straight lines in the presence of a medium.

As a special case of eqn (2.119), consider Fig. 2.9(b). Once again, a three-dimensional refractive medium is considered to be present, but is not explicitly indicated for the sake of clarity. The medium is assumed to be sufficiently weakly scattering that the streamlines, of the scattered field, do not differ appreciably from the streamlines of the unscattered field. Further, the incident field is assumed to be a coherent scalar z-directed plane wave. Under these assumptions, eqn (2.119) reduces to:

$$\phi_\omega(x,y,z) = 2\pi \int \frac{dz}{\lambda/n_\omega(x,y,z)} = 2\pi \int \frac{dz}{\lambda/[1-\delta_\omega(x,y,z)]}, \qquad (2.120)$$

which is identical to the corresponding result obtained under the projection approximation of Section 2.2. Indeed, there is evidently a rather close correspondence between the two theories.

Having investigated the eikonal equation (2.117) of geometrical optics, let us relax the geometric optics limit and return to the wave-optical eikonal equation (2.115). Since it is the only term which vanishes in the limit of zero wavelength, the term $I_\omega^{-1/2}\nabla^2 I_\omega^{1/2}$ must embody the effects of diffraction. Further, since eqn (2.116) remains unchanged in the geometric-optics limit, and the pair of equations (2.115) and (2.116) are fully equivalent to the inhomogeneous Helmholtz equation from which they were derived (at points with non-zero intensity), we conclude that all diffraction effects are accounted for by the term $I_\omega^{-1/2}\nabla^2 I_\omega^{1/2}$. We shall therefore speak of this as the 'diffraction term'.[50] Taking this diffraction term into account, and realizing that the previously discussed notions of X-ray

[50]As an interesting historical aside, note that there is a direct parallel between the diffraction term and the so-called quantum potential of David Bohm's famous hidden-variable theory of non-relativistic quantum mechanics (Bohm 1952). Now, we have already remarked on the identical mathematical form of the inhomogeneous Helmholtz equation (2.28), and the time-independent Schrödinger equation for spinless non-relativistic particles in the presence of a scalar potential. Accordingly, the hydrodynamic formulation of coherent scalar wave optics, presented in the main text, has a direct parallel with Madelung's hydrodynamic formulation of quantum mechanics (Madelung 1926). Independently of this work, Bohm arrived at the

streamlines still hold good when this term is not neglected, we write down the following generalized form of eqn (2.119):

$$\begin{aligned}\phi_\omega[x(s), y(s), z(s)] &= \int_{s_0}^{s} \frac{\partial \phi_\omega}{\partial s} ds \\ &= \int_{s_0}^{s} ds \sqrt{k^2 n_\omega^2 + \frac{\nabla^2 \sqrt{I_\omega}}{\sqrt{I_\omega}}} \\ &= 2\pi \int_{s_0}^{s} \frac{ds}{\lambda / n_\omega} \sqrt{1 + \frac{\lambda^2}{4\pi^2 n_\omega^2} \frac{\nabla^2 \sqrt{I_\omega}}{\sqrt{I_\omega}}}. \end{aligned} \tag{2.121}$$

In the last line, the geometric optics limit corresponds to replacing the square root with unity. We see that the rate, at which phase is accumulated as one travels along an X-ray streamline, is influenced by the diffraction term. From a wave-optics point of view, this diffraction term can be thought of as providing an additional distortion to the shape of the locally planar wave-fronts of the coherent field, which exists in addition to the distortion provided by the local value and spatial rate of change of the refractive index of the medium through which the wave is travelling. From a ray optics point of view, the diffraction term can be thought of as serving to alter the refractive index of the medium, making this quantity a function of both the field intensity and its spatial derivatives, via the prescription:

$$n_\omega \to n_\omega \sqrt{1 + \frac{\lambda^2}{4\pi^2 n_\omega^2} \frac{\nabla^2 \sqrt{I_\omega}}{\sqrt{I_\omega}}}. \tag{2.122}$$

Regardless of the adopted viewpoint, it is evident that the diffraction term is greatest where the intensity of the wave-field is most rapidly changing. This is intuitively sensible, since it harmonizes with the knowledge that increasingly rapid spatial fluctuations, in the intensity of a given coherent field in a given region, typically leads to an increase in the rate at which the field diffracts as it propagates. In general this diffraction occurs whether or not the X-rays are travelling through a material medium. In particular, the vacuum form of eqns (2.121) or (2.122) implies that rays do not in general travel along straight lines

same equations, namely the quantum analogues of eqns (2.115) and (2.116). What we call the geometric optics limit (corresponding to zero radiation wavelength) is considered to be the classical limit in the case of quantum mechanics (corresponding to zero de Broglie wavelength). Bohm's contribution was to interpret the diffraction term as providing an additional potential for deviating the particle trajectories, allowing one to maintain a particle viewpoint even when the diffraction term is non-negligible. Thus, when finite de Broglie wavelengths are allowed, the particle viewpoint was considered to be preserved, provided that the potential was appropriately modified by addition of the diffraction term. In turn, this idea has a parallel in the main text, where we shall briefly investigate the effects of the diffraction term on ray trajectories. The resulting theory is equivalent to the wave theory, at points of non-zero intensity, but works in terms of rays.

when propagating through free space, with these free-space trajectories becoming progressively straighter as the wavelength is reduced.

As already mentioned, eqns (2.115) and (2.116) are fully equivalent to the inhomogeneous Helmholtz equation (2.28) from which they were derived, except at points where the intensity vanishes or the phase becomes non-differentiable. We close this section with a very brief return to the second member of this pair of equations, namely the continuity equation (2.116). The continuity equation is so named because it expresses local conservation of optical energy: application of the Gauss divergence theorem converts this equation into the statement that the flux of the energy-flow vector of the coherent scalar field, through any closed surface, is equal to zero. This conservation law remains unchanged in the geometric optics limit. Further, precisely because it is a local conservation law for optical energy, the continuity equation is more general than the inhomogeneous Helmholtz equation (2.28) from which it was derived. For example, if one allows the refractive index to be an arbitrary real function of intensity, in the inhomogeneous Helmholtz equation, the continuity equation will remain unchanged. This amounts to the natural statement that the inclusion, of an arbitrary non-dissipative nonlinearity into an isotropic non-magnetic refractive medium, is a complexity which does not alter local conservation of optical energy.

2.9 Scattering, refractive index, and electron density

Under the single-scattering viewpoint of the first Born approximation, the radiation downstream of an illuminated scatterer is considered to be the superposition of an unscattered wave-field, together with the sum total of all spherical wave-fields that originate from single-scattering events at each point in the sample. This viewpoint is epitomized by Fig. 2.5. A different viewpoint, which does not decompose the wave-field into unscattered and spherical waves, is furnished by the projection approximation. With reference to Fig. 2.1, the projection approximation regards the wave-field to be a single entity, which continuously evolves as the wave proceeds from the entrance surface to the exit surface of the scatterer.

Of course, the first Born and projection approximations are different approximations, with different domains of validity. Having said this, there exists at least one case—that of a weakly forward-scattering thin uniform slab of material irradiated by spherical waves—in which the two methods yield identical results. In this section, this case will be separately treated under the first Born and projection approximations, thereby forging an instructive link between the two methods. More importantly, by equating the results obtained using each method, we will arrive at a formula that expresses the X-ray refractive index in terms of the electron density of the sample. Our presentation of this lovely calculation is adapted from the treatments of Sakurai (1967), Als-Nielsen (1993), and Als-Nielsen and McMorrow (2001).

Begin by considering the elastic scattering of a planar coherent X-ray wave-field by a single free stationary electron, of charge e and rest mass m_e, as sketched in Fig. 2.10. Here we see a z-directed planar incident wave-field ψ_0 which illu-

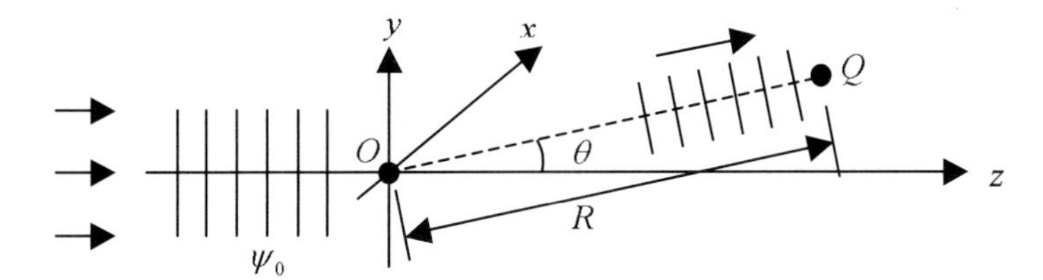

Fig. 2.10. A free electron is located at the origin O of a Cartesian coordinate system (x, y, z). Incident upon this electron is the coherent z-directed X-ray plane wave ψ_0. An observation point Q is located at a distance R, far from the electron. The line from O to Q is assumed to make a small angle θ, with respect to the positive z-axis, allowing us to neglect the dependence of the scattering amplitude on the polarization of the incident plane wave.

minates a stationary electron located at the origin O of a Cartesian coordinate system $(x, y, z) \equiv \mathbf{x}$. We consider an observation point Q, which is sufficiently far from the origin for the asymptotic form of the scattered field to hold, and sufficiently close to the positive z-axis for the magnitude of the angle θ to be much smaller than unity.[51] Under these conditions, a standard calculation based on classical electrodynamics leads to the following form for the scattered field[52]:

$$\psi(\mathbf{x}) \sim \psi_0(\mathbf{x}) - r_0 \psi_0^{(0)} \frac{\exp(ik|\mathbf{x}|)}{|\mathbf{x}|}, \quad r_0 = \frac{e^2}{4\pi\varepsilon_0 m_e c^2}. \tag{2.123}$$

Here, $\psi(\mathbf{x})$ is the asymptotic form of the complex scalar wave-field which results when the incident wave $\psi_0(\mathbf{x})$ is scattered through a small angle by an electron located at the origin of coordinates. The quantity r_0, known variously as the Thomson scattering length or the classical electron radius, quantifies the strength of scattering due to the electron. Lastly, $\psi_0^{(0)}$ is the value of the incident disturbance at the location of the electron. We see that the scattered wave-field

[51]Note, in this context, that the assumption of a small scattering angle allows us to work within the framework of a scalar theory (which ignores polarization). The scattering properties of the electron are polarization dependent, if this condition of small scattering angle is relaxed.

[52]This scattering scenario, without the approximation of small scattering angles, is known as Thomson scattering. The resulting polarization-dependent scattering amplitude leads to the polarization-independent eqn (2.123), when the scattering angles are small. Derivation of the Thomson scattering amplitude can proceed via at least three levels of sophistication. At the simplest of these levels, one can use classical electrodynamics—as embodied in the Maxwell equations—to obtain this scattering amplitude. Such a derivation is presented in many standard texts on electrodynamics, such as that of Jackson (1999). A more sophisticated treatment may be based on the non-covariant form of quantum electrodynamics, in which the electromagnetic field is quantized and the electron is described using the Schrödinger equation. The most sophisticated treatment, of the three that are described here, uses the covariant formulation of quantum electrodynamics. In this formulation, both the photon and electron fields are quantized, being described in a fully covariant manner. Each of the two quantum treatments of Thomson scattering are described in the text by Sakurai (1967). See also Heitler (1954) and Mandl and Shaw (1993).

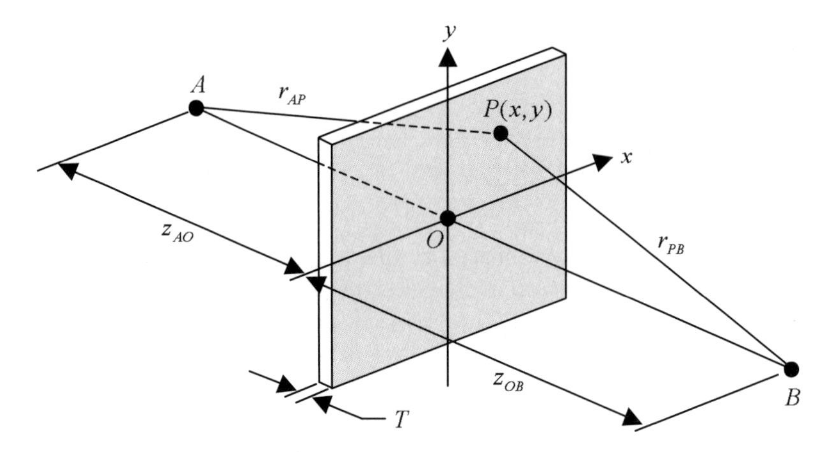

Fig. 2.11. A source A emits coherent spherical X-ray waves that are incident upon a thin scattering screen, of thickness T. The entrance surface of the screen coincides with the xy plane of a Cartesian coordinate system, with the source A lying on the negative z-axis, and the observation point B lying on the positive z-axis. The distances from the source to the origin, and from the origin to the observation point, are respectively denoted z_{AO} and z_{OB}. A point $P(x, y)$, within the volume of the scatterer, is indicated. The distances from the source A to P, and from P to the observation point B, are respectively denoted by r_{AP} and r_{PB}. After Als-Nielsen (1993) and Als-Nielsen and McMorrow (2001).

is decomposed into two parts, an unscattered wave-field and a spherical wave emanating from the scattering electron. Note that when the incident wave-field is considered to be a plane wave, the above formula is a special case of that given in eqn (2.77), in the context of the first Born approximation.

Equipped with the Thomson formula (2.123) for the scattering of a coherent scalar X-ray wave by a free electron, through a small angle, we are ready to consider the scenario sketched in Fig. 2.11. Here, we see a source of spherical waves A, which pass through a thin scattering screen whose entrance surface coincides with the xy plane of a Cartesian coordinate system. Both the source A, and the observation point B, are taken to lie on the z-axis. The screen is taken to be an infinitely extended uniform-density single-material slab of thickness T. This slab is considered to be a continuum of scattering points, all of which have the same scattering strength, and all of which are uniformly distributed through the volume of the material. The thickness of the slab is taken to be sufficiently small for there to be negligible phase differences between scattered waves emanating from a point $(x, y, 0)$ on the entrance surface of the slab, and the corresponding point (x, y, T) on the exit surface of the slab.

The scattering due to nuclear matter will be neglected—we consider only the scattering of the incident X-ray radiation by the electron density present in the slab. Further, we will assume that the energy of the incident radiation is much greater than the binding energies of all electrons present in the material,[53] so that the motion of the electrons may be neglected, and that all scattering angles are small. Further, we take the photon energy to be much smaller than the rest-mass energy of the electrons. Lastly, we make the first Born approximation, by considering the wave-field incident upon any given scatterer to be approximately equal to the incident wave-field. With all of these assumption in place, we may make use of the small-angle version of the Thomson scattering amplitude in eqn (2.123), for each individual electron in the scattering screen. The total wave-field, downstream of the scattering screen, can then be considered to be a sum of the unscattered incident spherical wave emanating from A, and the sum total of all Thomson-scattered wave-fields due to each electron present in the screen.

Given the assumptions listed above, let us calculate the contribution to the scattered radiation at the observation point B, by an infinitesimal column of the scatterer that passes through the point $P(x, y)$. This column of scatterers has a length T perpendicular to the slab, and an infinitesimal cross-sectional area given by the areal element $dxdy$. The total number of electrons in this scattering column is given by $\rho T dxdy$, where ρ is the number of electrons per unit volume in the uniform material. In accord with one of the assumptions listed earlier, we will neglect the phase differences due to electrons being at different positions within the scattering column, which will be the case if the thickness T of the screen is taken to be sufficiently small.

At the entrance surface of the scattering column the incident disturbance, due to the spherical waves emitted from the source A, is given by $\mathscr{A} \exp(ikr_{AP})/r_{AP}$. Here, r_{AP} is the distance from the source to the entrance surface of the column, and $\mathscr{A}$ is a complex constant that specifies the strength of the source. According to the Thomson scattering formula (2.123), in order to obtain the contribution $d\psi_{\mathrm{S},B}^{(P)}$ of the scattering column at P to the scattered disturbance at B, we must multiply the incident disturbance $\mathscr{A} \exp(ikr_{AP})/r_{AP}$ by three factors: (i) the negative[54] of the Thomson scattering length r_0; (ii) the number $\rho T dxdy$ of electrons in the column; (iii) the spherical wave $\exp(ikr_{PB})/r_{PB}$ emanating from the scattering volume. We therefore obtain:

$$d\psi_{\mathrm{S},B}^{(P)} = \mathscr{A} \frac{\exp(ikr_{AP})}{r_{AP}} \times (-r_0) \times \rho T dxdy \times \frac{\exp(ikr_{PB})}{r_{PB}}. \tag{2.124}$$

To continue, let us assume that both z_{AO} and z_{OB} are much greater than either $|x|$ or $|y|$. This allows us to make a binomial approximation for the distance r_{AP}:

[53]This allows us to consider the electrons to be approximately free.

[54]Since $\exp(i\pi) = -1$, we see that the presence of this negative sign, in formula (2.123), indicates that the scattered wave is π radians out of phase with the incident radiation. As we shall show, this fact implies the real part of the refractive index to be less than unity.

$$r_{AP} = \sqrt{z_{AO}^2 + x^2 + y^2} = z_{AO}\left(1 + \frac{x^2+y^2}{z_{AO}^2}\right)^{1/2} \approx z_{AO} + \frac{x^2+y^2}{2z_{AO}}, \quad (2.125)$$

and similarly for the distance r_{PB}:

$$r_{PB} \approx z_{OB} + \frac{x^2+y^2}{2z_{OB}}. \quad (2.126)$$

Substitute the above pair of binomial approximations into eqn (2.124), and then replace the denominator $r_{AP}r_{PB}$ by $z_{AO}z_{OB}$, to give:

$$d\psi_{\mathrm{S},B}^{(P)} \approx -\frac{r_0\rho T\mathscr{A}\,dxdy}{z_{AO}z_{OB}}\exp\left[ik(z_{AO}+z_{OB})\right] \times \exp\left[\frac{1}{2}ik(x^2+y^2)\left(\frac{1}{z_{AO}}+\frac{1}{z_{OB}}\right)\right]. \quad (2.127)$$

This is the contribution to the scattered radiation at B, due to the scattering column at $P(x,y)$. To obtain the total scattered disturbance $\psi_{\mathrm{S},B}$ at B, we integrate over the whole screen:

$$\psi_{\mathrm{S},B} \approx -\frac{r_0\rho T\mathscr{A}}{z_{AO}z_{OB}}\exp\left[ik(z_{AO}+z_{OB})\right] \times \iint \exp\left[\frac{1}{2}ik(x^2+y^2)\left(\frac{1}{z_{AO}}+\frac{1}{z_{OB}}\right)\right]dxdy. \quad (2.128)$$

Transform to plane polar coordinates (r,θ). Performing the trivial angular integration gives a multiplicative factor of 2π, leaving:

$$\psi_{\mathrm{S},B} \approx -\frac{2\pi r_0\rho T\mathscr{A}}{z_{AO}z_{OB}}\exp\left[ik(z_{AO}+z_{OB})\right] \times \lim_{R\to\infty}\int_{r=0}^{r=R}\exp\left[\frac{1}{2}ikr^2\left(\frac{1}{z_{AO}}+\frac{1}{z_{OB}}\right)\right]rdr. \quad (2.129)$$

Next, we need to evaluate the integral appearing in the above equation. We use a simple albeit non-rigorous method for computing this integral,[55] by making the change of variables $u=r^2$ and then replacing k with $k+i\varepsilon$, where ε is a small real constant that is taken to tend to zero from above. The resulting integral is elementary, and so we are led to the following chain of reasoning:

[55]For the reader requiring greater rigour, express the integral in terms of the Fresnel integrals (see Section 1.4.2), and then make use of their asymptotic form. Other alternatives include the method of stationary phase, and a method based on contour integration.

$$\begin{aligned}
&\lim_{R\to\infty}\int_{r=0}^{r=R}\exp\left[\frac{1}{2}ikr^2\left(\frac{1}{z_{AO}}+\frac{1}{z_{OB}}\right)\right]rdr \\
&=\frac{1}{2}\lim_{R\to\infty}\lim_{\varepsilon\to 0^+}\int_{u=0}^{u=R^2}\exp\left[\frac{1}{2}iu(k+i\varepsilon)\left(\frac{1}{z_{AO}}+\frac{1}{z_{OB}}\right)\right]du \\
&=\lim_{R\to\infty}\lim_{\varepsilon\to 0^+}\frac{\exp\left[(1/2)i(k+i\varepsilon)R^2\left(z_{AO}^{-1}+z_{OB}^{-1}\right)\right]-1}{i(k+i\varepsilon)\left(z_{AO}^{-1}+z_{OB}^{-1}\right)} \\
&=\frac{i}{k\left(z_{AO}^{-1}+z_{OB}^{-1}\right)}. \qquad (2.130)
\end{aligned}$$

Note that, in writing down the last equality, the penultimate line's term in square brackets is assumed to vanish. This will be the case if the term $\exp[i^2\varepsilon R^2(z_{AO}^{-1}+z_{OB}^{-1})/2]=\exp[-\varepsilon R^2(z_{AO}^{-1}+z_{OB}^{-1})/2]$ is exponentially damped in the limits as ε tends to zero from above, and R tends to infinity from below. This will be the case if the limits are taken such that εR^2 is arbitrarily large, so that R^2 tends to infinity faster than ε tends to zero. One means of ensuring this is to choose $R=1/\varepsilon$.

Having computed the integral in eqn (2.129), we re-write this equation as:

$$\psi_{\mathrm{S},B}\approx-\frac{2\pi i r_0\rho T\mathscr{A}\exp\left[ik(z_{AO}+z_{OB})\right]}{k(z_{AO}+z_{OB})}. \qquad (2.131)$$

This expression for the scattered wave-field $\psi_{\mathrm{S},B}$ may then be used to determine the total wave-field ψ_B at the observation point B. Recalling that the incident disturbance is a spherical wave, so that the unscattered wave-field has an amplitude of $\mathscr{A}\exp\left[ik(z_{AO}+z_{OB})\right]/(z_{AO}+z_{OB})$ at the observation point B, we obtain:

$$\begin{aligned}
\psi_B &\sim \mathscr{A}\frac{\exp\left[ik(z_{AO}+z_{OB})\right]}{z_{AO}+z_{OB}}+\psi_{\mathrm{S},B} \\
&=\mathscr{A}\frac{\exp\left[ik(z_{AO}+z_{OB})\right]}{z_{AO}+z_{OB}}\left\{1-\frac{2\pi i r_0\rho T}{k}\right\} \\
&\approx\mathscr{A}\frac{\exp\left[ik(z_{AO}+z_{OB})\right]}{z_{AO}+z_{OB}}\exp\left(-\frac{2\pi i r_0\rho T}{k}\right). \qquad (2.132)
\end{aligned}$$

Note that, in passing from the second to the third line of this equation, we have used the fact that $|2\pi i r_0\rho T/k|\ll 1$, allowing us to write $1-2\pi i r_0\rho T/k$ as the exponential of $-2\pi i r_0\rho T/k$. This can always be arranged, by taking the scattering slab to be sufficiently thin.

Equation (2.132) is our final result for the on-axis X-ray disturbance ψ_B, at the observation point B downstream of the uniform scattering slab shown in Fig. 2.11. We now evaluate this same quantity, using the projection approximation outlined in Section 2.2. Under this approximation, consider the slab to have a real refractive index $n_\omega = 1 - \delta_\omega$, with the usual dependence of the refractive index on angular frequency ω being explicitly indicated. Now, the unscattered disturbance at B is given by $\mathscr{A} \exp[ik(z_{AO} + z_{OB})] / (z_{AO} + z_{OB})$, this being the spherical wave emanating from the source at A. Under the projection approximation, the scattering screen serves to shift the phase of the disturbance at B, by an amount given by eqn (2.41). Therefore, according to the projection approximation, the disturbance at the observation point B is given by:

$$\psi_B \approx \mathscr{A} \frac{\exp[ik(z_{AO} + z_{OB})]}{z_{AO} + z_{OB}} \exp(-ik\delta_\omega T). \tag{2.133}$$

Equating results (2.132) and (2.133), which were respectively obtained under the first Born and projection approximations, we obtain the following expression relating the refractive index decrement δ_ω to the number density ρ of electrons in the scatterer:

$$\delta_\omega = \frac{2\pi r_0 \rho}{k^2} = \frac{e^2 \rho}{2\varepsilon_0 m_e c^2 k^2}. \tag{2.134}$$

In writing the last equality, we have used the expression for the Thomson scattering length in eqn (2.123).

We close this section with four remarks: (i) In deriving the above result, the incident photon energy was assumed to be sufficiently large for us to be able to assume the electron to be essentially free. If the incident photon energy is sufficiently close to an absorption edge of the scattering atoms, this assumption will break down and correction terms must be added to the above result. (ii) By substituting in typical values for all quantities appearing in the above equation, the reader will conclude that δ is typically on the order of 10^{-5} or 10^{-6}, for X-rays in the medium-to-high energy range. The refractive effects of matter upon X-rays is therefore very slight, although we shall see in later chapters that there are many means by which these refractive effects can be visualized, and that there are many circumstances in which refractive effects dominate over absorptive effects. (iii) The refractive index for X-rays is slightly less than unity. Among many other things, this implies that matter imposes a phase retardation rather than a phase advance on weakly scattered radiation passing through it, that the phenomenon of total internal reflection (for visible light) has the counterpart of total external reflection for X-rays, and that concave rather than convex lenses are the appropriate refractive optical element for focussing X-rays. (iv) While the above equation was obtained under the approximation of a thin uniform scattering screen, it also holds for irregularly shaped scatterers with non-uniform density and composition, provided that the material properties do not vary too rapidly with position.

2.10 Inelastic scattering and absorption

Thus far we have restricted our considerations, on the interactions of X-rays with matter, to elastic scattering. In such processes, the energy of the X-ray is unchanged by the scattering event. In realistic scattering scenarios, such 'coherent scatter' is only one component of the radiation that emerges when X-rays illuminate a given sample. In addition to the coherently scattered radiation, there will in general exist an inelastically scattered component ('incoherent scatter'). Such inelastic photon scatter, together with its relationship to the absorption of X-rays, will be discussed in the present section.

Two major mechanisms, for inelastic photon scatter in the X-ray regime, are the Compton effect and the photoelectric effect. The Compton effect, which will be treated in the first sub-section, occurs when an X-ray photon is inelastically scattered from a nearly free electron. The photoelectric effect, which refers to the absorption of an X-ray photon by an atom with the associated emission of an electron, will be treated in the second sub-section. Together with the coherent scattering treated earlier, the Compton and photoelectric effects form the major mechanisms of X-ray scattering with which we shall be primarily concerned.[56]

2.10.1 *Compton scattering*

Consider the inelastic interaction of an incident X-ray with an initially-stationary electron, which is either free or sufficiently weakly bound so as to be considered essentially free. This process, known as Compton scattering, cannot be treated in terms of the classical picture of X-rays that has been employed for the majority of this chapter. Rather, a photon description of the field is required. Regarding such a description, we recall the words of Einstein, from his famous 1905 paper on the photoelectric effect in which the quantum of the electromagnetic field was hypothesized (Einstein 1905): '... the energy of a light ray spreading out from a point source is not continuously distributed over an increasing space but consists of a finite number of energy quanta which are localized at points in space, which move without dividing, and which can only be produced and absorbed as complete units.' In Compton's analysis of X-ray scattering by free electrons, the incident X-ray light is treated as one of Einstein's photons rather than as a classical electromagnetic wave (Compton 1923).

The process of Compton scattering is sketched in Fig. 2.12. Since the initially-stationary electron will in general recoil after interacting with the incident pho-

[56]Two further mechanisms for inelastic photon scattering, namely pair production and photonuclear absorption, will not be treated here. (i) Pair production occurs when the incident photon possesses sufficient energy for the creation of an electron–positron pair. A crude lower bound, on the photon energy required for pair production, is given by doubling the rest-mass energy of the electron (2×511 keV $= 1022$ keV). Thus pair production will not occur if the photon energy is less than 1022 keV, as will always be the case for the X-ray regimes considered in this book. (ii) Photonuclear absorption occurs when the energy of the incident photon is sufficiently high to produce excitations in the atomic nucleus. Since excited nuclear energy levels typically have energies of 1 MeV or more relative to the nuclear ground state, photonuclear absorption may also be safely ignored in our discussions.

Fig. 2.12. Compton scattering of a photon from an electron. (a) A photon γ is incident upon a stationary free electron e^-. (b) After the interaction the electron recoils, with the photon being scattered through an angle θ.

ton, the energy of the electron will evidently increase. Therefore, the energy of the scattered photon will be less than that of the incident photon, leading to the conclusion that the wavelength of the scattered photon is greater than that of the incident photon.[57] In the following paragraphs, we will obtain expressions for the change in both the wavelength and energy of the scattered photon, as a function of the deflection angle θ.

Consider, once again, the scenario sketched in Fig. 2.12. Let the energy and momentum of the incident photon be respectively denoted by $E_\gamma^{(\mathrm{i})}$ and $\mathbf{p}_\gamma^{(\mathrm{i})}$, with the initial energy and momentum of the electron being given by $E_\mathrm{e}^{(\mathrm{i})}$ and $\mathbf{p}_\mathrm{e}^{(\mathrm{i})} = \mathbf{0}$. After the collision these quantities become $E_\gamma^{(\mathrm{f})}, \mathbf{p}_\gamma^{(\mathrm{f})}, E_\mathrm{e}^{(\mathrm{f})}$, and $\mathbf{p}_\mathrm{e}^{(\mathrm{f})}$, respectively. The above properties are readily related to one another, by separately invoking energy and momentum conservation, yielding the following pair of equations:

$$E_\gamma^{(\mathrm{i})} + E_\mathrm{e}^{(\mathrm{i})} = E_\gamma^{(\mathrm{f})} + E_\mathrm{e}^{(\mathrm{f})}, \tag{2.135}$$

$$\mathbf{p}_\gamma^{(\mathrm{i})} = \mathbf{p}_\gamma^{(\mathrm{f})} + \mathbf{p}_\mathrm{e}^{(\mathrm{f})}. \tag{2.136}$$

To proceed further, we make use of the following three observations: (i) Since the electron is initially stationary, its energy will initially be equal to its rest-mass energy, so that:

$$E_\mathrm{e}^{(\mathrm{i})} = m_\mathrm{e} c^2, \tag{2.137}$$

where m_e is the rest mass of the electron. (ii) The energy E_γ of a photon is related to its wavelength λ_γ by $E_\gamma = hc/\lambda_\gamma$, where h is Planck's constant, so that:

[57] If the initial state of the electron does not correspond to it being at rest, then one speaks of 'inverse Compton scattering'. Evidently, it will no longer be true that energy is not transferred from the electron to the photon. Indeed, in inverse Compton scattering a significant fraction of energy may be transferred from the electron to the photon. This is believed to be a mechanism for the astrophysical generation of X-rays (see, for example, Rindler (1991)). On a less cosmic note, inverse Compton scattering may be important in terrestrial X-ray optics if the energy of the initial electron is not sufficiently small, with respect to the energy of the incident photon, for it to be considered free.

$$E_{\gamma}^{(\mathrm{i})} = \frac{hc}{\lambda_{\gamma}^{(\mathrm{i})}}, \qquad E_{\gamma}^{(\mathrm{f})} = \frac{hc}{\lambda_{\gamma}^{(\mathrm{f})}}. \tag{2.138}$$

(iii) The energy E, momentum $\mathbf{p}$, and rest mass m_0, of a material particle in free space, obey the relativistic energy–momentum–mass relationship[58]:

$$E^2 = m_0^2 c^4 + |\mathbf{p}|^2 c^2. \tag{2.139}$$

For the case of the scattered electron, the square root of this expression gives:

$$E_{\mathrm{e}}^{(\mathrm{f})} = \sqrt{m_{\mathrm{e}}^2 c^4 + \left|\mathbf{p}_{\mathrm{e}}^{(\mathrm{f})}\right|^2 c^2}. \tag{2.140}$$

Next, substitute eqns (2.137), (2.138), and (2.140) into the energy-conservation expression (2.135). Isolating the square root in the resulting equation, we then have:

$$m_{\mathrm{e}} c^2 + hc\left(\frac{1}{\lambda_{\gamma}^{(\mathrm{i})}} - \frac{1}{\lambda_{\gamma}^{(\mathrm{f})}}\right) = \sqrt{m_{\mathrm{e}}^2 c^4 + \left|\mathbf{p}_{\mathrm{e}}^{(\mathrm{f})}\right|^2 c^2}. \tag{2.141}$$

Take the square of the above equation, before cancelling the square of the electron rest-mass energy from both sides, and then dividing through by c^2. Thus:

$$h^2\left(\frac{1}{\lambda_{\gamma}^{(\mathrm{i})}} - \frac{1}{\lambda_{\gamma}^{(\mathrm{f})}}\right)^2 + 2m_{\mathrm{e}} hc\left(\frac{1}{\lambda_{\gamma}^{(\mathrm{i})}} - \frac{1}{\lambda_{\gamma}^{(\mathrm{f})}}\right) = \left|\mathbf{p}_{\mathrm{e}}^{(\mathrm{f})}\right|^2. \tag{2.142}$$

Next, we wish to eliminate the electron's recoil momentum $\mathbf{p}_{\mathrm{e}}^{(\mathrm{f})}$, as we are primarily interested in the properties of the Compton-scattered photon. To this end, isolate the electron recoil momentum on one side of eqn (2.136), and then take the squared modulus of the result. This yields:

[58]This expression is derived in many standard texts on special relativity, such as those of French (1968) and Rindler (1991). Here, we note three special cases of the above formula: (i) When the rest mass of the particle is taken to zero, the relativistic energy–momentum–mass relation reduces to $E = |\mathbf{p}|c$, which is the familiar relationship between the energy and momentum of a photon of well-defined momentum. (ii) When the momentum is taken to zero, one obtains what is perhaps the most well-known formula in physics, namely Einstein's relation $E = m_0 c^2$ for the rest-mass energy of a stationary free particle of rest mass m_0. (iii) In the non-relativistic limit, where the first term on the right side of (2.139) is much greater than the second, one can perform a binomial approximation on the square root of this relation, to yield $E \approx m_0 c^2 + |\mathbf{p}|^2/(2m_0)$. This expresses the total energy of a non-relativistic material particle, of rest mass m_0, as the sum of its rest mass energy $m_0 c^2$ and the Newtonian expression $|\mathbf{p}|^2/(2m_0)$ for its kinetic energy.

$$\left|\mathbf{p}_{\mathrm{e}}^{(\mathrm{f})}\right|^2 = \left|\mathbf{p}_\gamma^{(\mathrm{i})} - \mathbf{p}_\gamma^{(\mathrm{f})}\right|^2 = \left|\mathbf{p}_\gamma^{(\mathrm{i})}\right|^2 + \left|\mathbf{p}_\gamma^{(\mathrm{f})}\right|^2 - 2\mathbf{p}_\gamma^{(\mathrm{i})} \cdot \mathbf{p}_\gamma^{(\mathrm{f})}. \tag{2.143}$$

Regarding the first and second terms on the right-hand side of this equation, we note the de Broglie relation $\lambda_\gamma = h/|\mathbf{p}_\gamma|$ for the wavelength λ_γ and momentum $\mathbf{p}_\gamma$ of a photon, so that:

$$\left|\mathbf{p}_\gamma^{(\mathrm{i})}\right| = \frac{h}{\lambda_\gamma^{(\mathrm{i})}}, \qquad \left|\mathbf{p}_\gamma^{(\mathrm{f})}\right| = \frac{h}{\lambda_\gamma^{(\mathrm{f})}}. \tag{2.144}$$

Further, regarding the third term on the right side of eqn (2.143), we have:

$$\mathbf{p}_\gamma^{(\mathrm{i})} \cdot \mathbf{p}_\gamma^{(\mathrm{f})} = \left|\mathbf{p}_\gamma^{(\mathrm{i})}\right| \left|\mathbf{p}_\gamma^{(\mathrm{f})}\right| \cos\theta = \frac{h^2 \cos\theta}{\lambda_\gamma^{(\mathrm{i})} \lambda_\gamma^{(\mathrm{f})}}, \tag{2.145}$$

where use has been made of eqn (2.144), and θ is the angle through which the incident photon has been scattered (see Fig. 2.12(b)).

Substitute eqns (2.144) and (2.145) into the right side of expression (2.143). The resulting formula, for the square of the recoil momentum of the electron, may then be used to eliminate this quantity from eqn (2.142), to give:

$$\begin{aligned} h^2 \left(\frac{1}{\lambda_\gamma^{(\mathrm{i})}} - \frac{1}{\lambda_\gamma^{(\mathrm{f})}} \right)^2 &+ 2m_{\mathrm{e}}hc \left(\frac{1}{\lambda_\gamma^{(\mathrm{i})}} - \frac{1}{\lambda_\gamma^{(\mathrm{f})}} \right) \\ &= \frac{h^2}{\left[\lambda_\gamma^{(\mathrm{i})}\right]^2} + \frac{h^2}{\left[\lambda_\gamma^{(\mathrm{f})}\right]^2} - \frac{2h^2 \cos\theta}{\lambda_\gamma^{(\mathrm{i})} \lambda_\gamma^{(\mathrm{f})}}. \end{aligned} \tag{2.146}$$

After a little algebra, together with the trigonometric formula:

$$\sin^2\left(\frac{\theta}{2}\right) = \frac{1}{2}(1 - \cos\theta), \tag{2.147}$$

one arrives at Compton's formula for photon scattering from a stationary free electron:

$$\lambda_\gamma^{(\mathrm{f})} - \lambda_\gamma^{(\mathrm{i})} = 2\frac{h}{m_{\mathrm{e}}c} \sin^2\left(\frac{\theta}{2}\right). \tag{2.148}$$

This gives an expression for the change in wavelength, between the incident and scattered photons, as a function of the angle θ through which the incident photon is scattered by the stationary electron. This change in wavelength is independent of the energy of the incident photon, and is proportional to the so-called Compton wavelength $h/(m_{\mathrm{e}}c)$. Further, we note that $\lambda_\gamma^{(\mathrm{f})} - \lambda_\gamma^{(\mathrm{i})}$ is evidently positive, consistent with the previous verbal argument that the scattered photon must have a longer wavelength than the incident photon, on account of the energy transferred to the electron upon setting it in motion.

Rather than considering the change of photon wavelength in the process of Compton scattering, one may consider the change in photon energy. Accordingly, divide both sides of eqn (2.148) by $\lambda_\gamma^{(\mathrm{i})}$, before making use of eqn (2.138) to translate all photon wavelengths into their corresponding energies. This leads to:

$$\frac{E_\gamma^{(\mathrm{f})}}{E_\gamma^{(\mathrm{i})}} = \left[1 + \frac{2E_\gamma^{(\mathrm{i})}\sin^2(\theta/2)}{m_\mathrm{e}c^2}\right]^{-1}. \tag{2.149}$$

We see that the ratio, of the final and initial energies of the Compton-scattered photon, is a function of the energy of the incident photon. Indeed, we see from the right-hand side of the above expression, that the ratio $E_\gamma^{(\mathrm{f})}/E_\gamma^{(\mathrm{i})}$ can only be appreciably different from unity if $E_\gamma^{(\mathrm{i})}/(m_\mathrm{e}c^2)$ is non-negligible. Thus the energy of the incident photon should be some non-negligible fraction of the rest-mass energy of the electron. Noting that the rest-mass energy of an electron is 511 keV, one crudely estimates that the energy shifts due to Compton scattering will begin to become non-negligible at photon energies of a few keV or higher, corresponding to the hard X-ray regime and to the harder end of the soft X-ray regime. The energy shifts increase with both the scattering angle θ and the incident photon energy $E_\gamma^{(\mathrm{i})}$, according to eqn (2.149). In Fig. 2.13, we have used this result to plot the ratio of the scattered to the incident photon energies, as a function of scattering angle θ, for incident photon energies of 5, 10, 20, 40, 60, 80, and 100 keV.

We close this section with a few qualitative remarks regarding the angular distribution of Compton-scattered photons. To properly study such a problem requires the use of relativistic quantum mechanics, leading to the so-called Klein–Nishina formula for photon-electron scattering. This may be obtained using both the non-covariant and covariant formulations of quantum electrodynamics, and is covered in many advanced texts on quantum mechanics (see, for example, Sakurai (1967)). In many X-ray imaging contexts, the incoherent scatter resulting from the Compton effect contributes to a gently-varying background signal in the recorded intensity distribution. In the regime from 1 to 100 keV, the contribution to incoherent scatter from the Compton effect is typically an increasing function of energy.

2.10.2 *Photoelectric absorption and fluorescence*

Consider the plot shown in Fig. 2.14. Here, we see a logarithmic curve of the linear attenuation coefficient μ of copper (normalized against density ρ, in units of $\mathrm{cm}^{-1} \div (\mathrm{g/cm}^3) = \mathrm{cm}^2/\mathrm{g}$), as a function of photon energy in keV. Two salient features are immediately evident. First, the normalized linear attenuation coefficient shows a general decrease with increasing photon energy, indicating that copper becomes progressively more transparent as the photon energy is increased. Second, there is a region having a sharp rise in normalized linear attenuation coefficient, at a photon energy of 8.98 keV. Such features are known

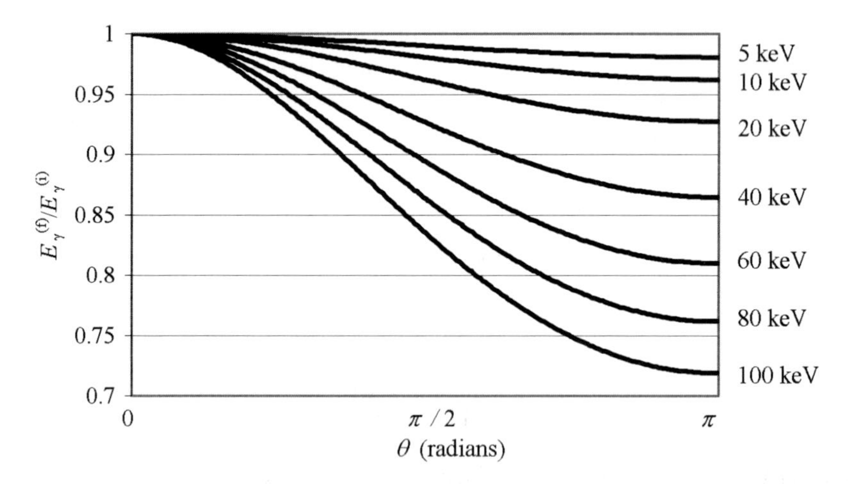

Fig. 2.13. Ratio of scattered to incident photon energies, for Compton scattering of a photon from a free stationary electron. This ratio is plotted as a function of the angle θ through which the photon is scattered, for photon energies of 5, 10, 20, 40, 60, 80, and 100 keV. Curves based on eqn (2.149). After Als-Nielsen and McMorrow (2001).

as absorption edges, the structure of which is intimately related to the concept of photoelectric absorption. It is to this subject that we now turn.

Consider a free atom, in its ground state. The electrons surrounding the atom are considered to be arranged in atomic orbitals, which are grouped together to form shells. These shells are denoted by the letters K, L, M, and so forth, with K corresponding to the electron shell that is most tightly bound to the atom, and sequential letters describing shells of electrons that are progressively more weakly bound. The notion of 'electron binding energy', which may be defined as the minimum energy required to liberate a given electron from the atom, is a convenient means of quantifying the degree of binding of an electron to an atom.

Next, suppose our atom to be irradiated by X-ray photons of a given wavelength. For the purposes of this argument, suppose the X-ray energy to be intermediate between the binding energy of the innermost (and therefore most tightly bound) shell, that is, the K shell, and the L shell. The incident photon does not have enough energy to liberate the K shell electron from the atom, but it does have sufficient energy to liberate electrons from the L and higher shells. When, for example, an L electron is liberated from the atom, the X-ray is considered to have been absorbed by the atom. The kinetic energy, of the liberated electron, will then be equal to the difference between the energy of the absorbed photon, and the binding energy of the electron prior to its ejection (Einstein 1905).

Such a mechanism, for photon absorption, is known as photoelectric absorption, with the ejected electron being known as a photoelectron. As was the case with our earlier discussions on the Compton effect, photoelectric absorp-

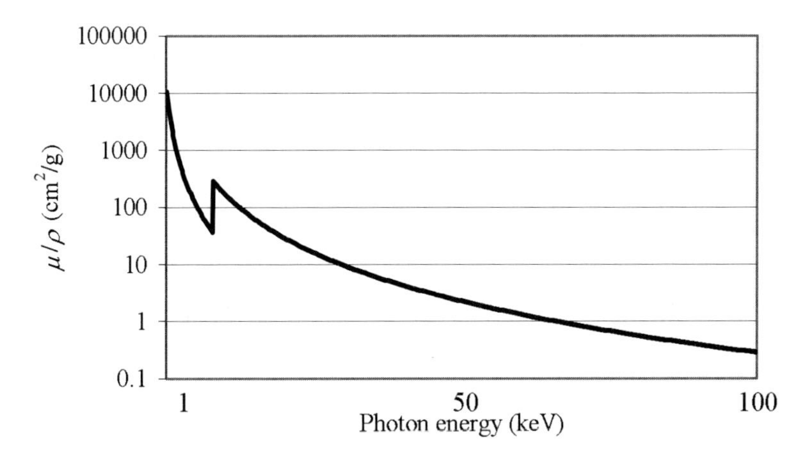

Fig. 2.14. Linear attenuation coefficient μ for free atoms of copper, normalized against density ρ, as a function of photon energy in keV. An absorption edge is apparent for a photon energy of 8.98 keV, corresponding to the binding energy of the 1s electron in the K shell of copper.

tion requires a photon description of the electromagnetic field, and cannot be adequately described using classical electrodynamics. Note, further, that photoelectric absorption is one factor which contributes to the linear attenuation coefficient of a material, with coherent scattering and Compton scattering providing additional contributions. Further contributions to the linear attenuation coefficient are also possible (inverse Compton scattering, Rayleigh scattering, non-resonant and resonant fluorescence, photonuclear absorption, pair production, etc.).

Returning to the example of a free atom illuminated by photons of insufficient energy to eject a K shell electron, but sufficient energy to liberate an L electron, suppose that one gradually increases the photon energy until it just coincides with the binding energy of the K electrons. At such an energy, a new channel for absorption will become available to the incident photons, with a corresponding sharp increase in the linear attenuation coefficient of the material. This yields an example of the previously mentioned absorption edge of a material, making the connection between such absorption edges and the photoelectric effect. As an example we note that the binding energy of a K electron in free copper is equal to 8.98 keV, with a corresponding absorption edge, at this photon energy, being evident in the plot of Fig. 2.14.

After an electron has been photo-ejected from a free atom, several relaxation processes may occur. (i) One such mechanism is for one or more of the electrons to 'drop down', from a higher-energy orbital into the vacant electron orbital, liberating a photon in the process. This mechanism is known as fluorescence. If the liberated photon is in the X-ray range, then one speaks of X-ray fluorescence. Note that X-ray fluorescence may also be produced when an ionized electron recombines with a given ion. (ii) Another mechanism, for relaxation of an atom from which a photoelectron has been ejected, is for an electron to 'drop down' from a higher-energy orbital into the vacant orbital, with the resulting energy loss being transferred to another electron in the atom, which is then ejected. Such ejected electrons are known as Auger electrons.

In this sub-section, we have often spoken of free atoms. Typically, the atoms in a given sample will not be free and the various mechanisms outlined above will display some sensitivity to the chemical and structural state of the atoms. As one example of this sensitivity, plots of the linear attenuation coefficient versus photon energy often display fine oscillations in the vicinity of an absorption edge. Such oscillations are known as extended X-ray absorption fine structure (EXAFS), the structure of which depends on the chemical state of the atom in question, together with structural parameters such as the orientation of neighbouring atoms. Such oscillations are typically confined to photon energies within a few hundred electron volts of the absorption edge. Further, we note that fine oscillations, within a range of a few tens of electron volts near an absorption edge, are termed either near-edge extended X-ray absorption fine structure (NEXAFS) or X-ray absorption near-edge structure (XANES).

2.11 Information content of scattered fields

We close this chapter with a discussion of the information content of scattered scalar electromagnetic fields. In general, this information is imprinted in both the modulus and phase of the scattered field, with only the former quantity being directly measurable in experiment (at optical and higher frequencies). Nevertheless there are means for inferring the phase of a scattered radiation wave-field from intensity measurements, such as phase retrieval and interferometry.[59]

Here, we adopt a viewpoint that is more concerned with questions of principle, assuming the scattered wave-fields to be known in both modulus and phase. What is the nature of the information content of these fields? What information can be inferred regarding a sample, given the wave-field or wave-fields scattered by this sample in a given experiment or series of experiments? These questions will be discussed in the present section. Our outline is split into two parts, which respectively consider the information contained within monochromatic and polychromatic scattered fields.

2.11.1 *Information content of scattered monochromatic fields*

Consider, once more, the scattering scenario sketched in Fig. 2.8. Allow the incident wave-field to be any monochromatic scalar wave, which is forward propagating with respect to some nominal optic axis z, with all scatterers lying between the planes $z = z_0$ and $z = z_0 + \Delta$. These planes will respectively be denoted as the entrance and exit surfaces of the scattering volume. For the moment, we

[59]See Chapter 4, for a discussion of some methods for inferring wave-field phase from intensity measurements.

make no approximations regarding the weakness of the scatterer, other than that which is implicit in working with a scalar theory. In particular, we will now allow the incident field to be back-scattered from the scattering volume, so that the scattered wave-field may be non-zero over both the entrance and the exit surfaces of this volume. Elastic scattering is assumed throughout.

Over the exit surface $z = z_0 + \Delta$ of the scattering volume, which forms a boundary for the half-space $z \geq z_0 + \Delta$ in which the scattered field obeys the Helmholtz equation (1.16), the forward-propagating component $\psi_\omega^{\rm S}(x, y, z)$ of the scattered field admits an angular-spectrum representation given by (see eqn (1.23)):

$$\psi_\omega^{\rm S}(x,y,z) = \frac{1}{2\pi}\iint \breve{\psi}_\omega^{\rm S}(k_x, k_y, z = z_0 + \Delta) \exp\left[i(z - z_0 - \Delta)\sqrt{k^2 - k_x^2 - k_y^2}\right] \times \exp\left[i(k_x x + k_y y)\right] dk_x dk_y, \quad z \geq z_0 + \Delta. \qquad (2.150)$$

By taking the case $z = z_0 + \Delta$ of this formula, we see that $\breve{\psi}_\omega^{\rm S}(k_x, k_y, z = z_0 + \Delta)$ is equal to the Fourier transform, with respect to x and y, of the forward scattered field over the exit surface.

A first consequence, of the above equation, is the two-dimensional nature of the information contained within the forward-scattered field. This is so, because the above integral expresses the forward-scattered field, over the entire three-dimensional half-space $z \geq z_0 + \Delta$, in terms of the function $\breve{\psi}_\omega^{\rm S}(k_x, k_y, z = z_0 + \Delta)$ of two real variables k_x and k_y. Note that this statement is rather obvious from a physical point of view, since the forward-scattered field is comprised of a two-dimensional continuum of forward-propagating plane waves, the propagation vectors of which all have the same magnitude. These propagation vectors are parameterized by two variables, namely their x and y projections.

A second consequence, of the above equation, flows from the fact that the free-space propagator $\exp\left[i(z - z_0 - \Delta)(k^2 - k_x^2 - k_y^2)^{1/2}\right]$ gives an exponential damping[60] with propagation distance $z - z_0 - \Delta$, for all two-dimensional Fourier components $\breve{\psi}_\omega^{\rm S}(k_x, k_y, z = z_0 + \Delta)$ of the exit-surface forward-scattered field, which are such that $k_x^2 + k_y^2 > k^2$. When the propagation distance is greater than a few wavelengths, such information is essentially lost by the act of free-space propagation. This corresponds to the loss of any sub-wavelength information encoded in the exit-surface scattered wave-field. Evidently, free space acts as a low-pass filter for coherent scalar wave-fields, disallowing the transmission of sub-wavelength structural information over distances much greater than a few wavelengths.[61]

[60] As mentioned in Section 1.3, such exponentially damped plane waves are rather poetically known as 'evanescent waves'.

[61] When evanescent waves are excluded from the analysis, eqn (2.150) expresses the field downstream of the scatterer as a sum of propagating plane waves, all of which have a real propagation vector which has a non-negative projection onto the optic axis z. If all of these

From the above two consequences, we conclude that the information content of the forward-scattered field is essentially two-dimensional in nature, being essentially limited to those two-dimensional Fourier components $\breve{\psi}^{\rm S}_{\omega}(k_x, k_y, z = z_0+\Delta)$ of the exit-surface forward-scattered field that lie within the Fourier-space disc $k_x^2 + k_y^2 \leq k^2$. Via a simple extension of the above arguments, a similar conclusion applies to the back-scattered wave-field over the entrance surface $z = z_0$ of the scattering volume, with the back-scattered field being propagated into the half-space $z < z_0$ lying upstream of the scattering volume.

This two-dimensional nature, of the information contained within the scattered monochromatic scalar field, may be contrasted with the three-dimensional nature of the information which will in general be contained within the scattering volume. This suggests that not all information, associated with the distribution of matter in the scattering volume, is contained within the scattered field that is obtained in a given experiment.

Let us now examine, in a little more detail, the connection between the information content of elastically scattered monochromatic fields, and the distribution of matter that results in such scattered fields. In particular, how much information regarding the scatterer can one obtain from the scattered field? We separately treat this question, under the projection and first Born approximations.

2.11.1.1 *Projection approximation* Regarding the information content of elastically scattered monochromatic scalar electromagnetic fields under the projection approximation, return to eqn (2.105). This gives an approximate expression for the wave-field over the exit-surface of the scattering volume, with all information regarding the scatterer being imprinted into the exit-surface wave-field via the projection $\int_{z=z_0}^{z=z_0+\Delta}[1 - n_{\omega}^2(x, y, z)]dz$.[62] In accord with the discussions earlier in this section, the information encoded, by the matter upon the scattered wave-field, is essentially two-dimensional in nature. Thus the exit-surface wave-field, for a single scattering experiment, only contains partial knowledge regarding the scatterer, under the projection approximation.

The methods of computed tomography may be used to obtain three-dimensional information regarding the scatterer, provided that the assumption of the projection approximation is valid, and provided that one is prepared to take a series of measurements of the scattered radiation. Two of the most widely used such methods are known as the 'summation back-projection algorithm', and an algorithm based on the so-called 'central slice theorem'. Here, we give a very brief description of the latter method.

allowed propagation vectors are arranged so that their tails occupy a common point, the tips of these wave-vectors will all lie on the surface of a hemisphere whose axis points in the z direction.

[62] Recall that $n_{\omega}(x, y, z)$ denotes the three-dimensional distribution of refractive index in the scatterer, as a function of the spatial coordinates x, y, z, for a given angular frequency ω of the complex scalar X-ray wave-field.

Let us take the Fourier transform of $1 - n_\omega^2(x, y, z)$ with respect to x, y, and z, denoting this transform by $\breve{g}(k_x, k_y, k_z)$:

$$\breve{g}(k_x, k_y, k_z) = \frac{1}{(2\pi)^{3/2}} \iiint_{-\infty}^{\infty} \left[1 - n_\omega^2(x, y, z)\right] \exp[-i(k_x x + k_y y + k_z z)] dx dy dz. \tag{2.151}$$

Set $k_z = 0$ in the above expression, and then make use of the fact that $1 - n_\omega^2(x, y, z)$ vanishes unless $z_0 \leq z \leq z_0 + \Delta$, to see that:

$$\begin{aligned} &\sqrt{2\pi}\, \breve{g}(k_x, k_y, k_z = 0) \qquad (2.152) \\ &= \frac{1}{2\pi} \iint_{-\infty}^{\infty} \left\{ \int_{z=z_0}^{z_0+\Delta} \left[1 - n_\omega^2(x, y, z)\right] dz \right\} \exp[-i(k_x x + k_y y)] dx dy. \end{aligned}$$

The quantity in braces is the z projection of $1 - n_\omega^2(x, y, z)$, which features in eqn (2.105) for the exit-surface wave-field, under the projection approximation. The right-side of the above equation is equal to the two-dimensional Fourier transform of this z projection, with respect to x and y. Bearing the left side of the above equation in mind, we have now proven the central-slice theorem: the Fourier transform (with respect to x and y) of the z projection of a given localized three-dimensional function, is equal to the three-dimensional Fourier transform of that function, evaluated over a plane ('central slice') $k_z = 0$ in the three-dimensional Fourier space, multiplied by $\sqrt{2\pi}$.

Once again, we see the two-dimensional nature of the wave-field at the exit surface of the scattering volume. However, armed with the central-slice theorem, we can now construct a means for obtaining three-dimensional information regarding the scattering volume. Consider a series of scattering experiments, in which the sample is rotated through a large number of equally spaced angles between 0 and 180 degrees, with respect to a given axis (say, the y axis). Via the central-slice theorem, the sum total of the scattered wave-fields, from this series of experiments, furnishes information along a series of planes, each of which pass through the k_y axis of the Fourier space. Provided that the number of experiments is sufficiently large, the volume of Fourier space swept out by these planes can be used to build up an estimate for the three-dimensional Fourier transform of the scattering potential. By inverse Fourier transformation, this can then be used to estimate the three-dimensional distribution of refractive index in the scatterer. For further information on tomographic imaging, see, for example, Natterer (1986).

2.11.1.2 *First Born approximation* Having briefly discussed the information content of elastically scattered monochromatic scalar electromagnetic fields under the projection approximation, let us consider this same question in the context of the first Born approximation. Unlike the former case, this latter case admits the possibility of a back-scattered field, namely a non-zero scattered field over the entrance surface of the scattering volume.[63]

Consider the Ewald sphere construction in Fig. 2.6(a). Developed under the first Born approximation for scattered wave-fields with a planar incident wave, this construction states that the difference between incident and scattered wave-vectors must connect two points located on the Ewald sphere. More precisely, the origin O of reciprocal space is placed at the tip of the incident wave-vector $\mathbf{k}_0$, with the tail of the scattered wave-vector $k\hat{\mathbf{x}}$ coinciding with the tail of the incident vector. With the scattered wave-vector so placed in reciprocal space, scattering will only occur if the vector lies on the surface of the Ewald sphere. When incident and allowed scattered wave-vectors have an angle between them that is less than 90 degrees in magnitude, the scattered radiation will be forward scattered into the half-space downstream of the scattering volume.[64] Otherwise, allowed scattered wave-vectors correspond to back-scattered radiation.

Evidently, there is a two-dimensional continuum of allowed scattered wave-vectors, in the first Born approximation. Once again, we note the two-dimensional information contained within the scattered wave-field. Bearing eqns (2.77) through (2.79) in mind, and for a single scattering experiment with a given incident plane wave, only a two-dimensional closed surface of the Fourier representation of the scattering potential is probed, under the first Born approximation for planar incident radiation.

As was the case in our earlier discussions on three-dimensional imaging under the projection approximation, one may rotate the object in order to obtain three-dimensional information regarding the scattering potential. This possibility is illustrated in Fig. 2.6(c). Rotation of the object evidently corresponds to rotating the Ewald sphere, with four such Ewald spheres—marked D, E, F, and G—being sketched in the figure. If one rotates the object through all possible orientations, a three-dimensional volume of Fourier space will be swept out. This volume, bounded by a surface known as the 'limiting Ewald sphere', is denoted by the dotted sphere H shown in the figure. This sphere has a radius equal to twice that of any given Ewald sphere.

The finite size of the limiting Ewald sphere implies a resolution limit to the three-dimensional information regarding the scatterer, which is contained in the totality of all possible elastic scattering experiments that can be performed, using incident plane waves of a given frequency, under the first Born approximation.[65] The question of how to recover the three-dimensional scattering potential, from

[63]As an example of this possibility, we cite the so-called Laue back-reflection method, in which the back-scattered X-ray field from a small crystal is measured in an X-ray diffractometer. For examples of X-ray diffraction patterns obtained using this method, see, for example, Warren (1969) or Hammond (2001).

[64]Note that our use of the term 'forward scattering' is relative to an optic axis aligned with the propagation direction of the incident plane wave.

[65]See, however, the use of the method of analytic extension discussed in both Habashy and Wolf (1994) and Born and Wolf (1999).

such a series of images, is the subject of diffraction tomography.[66] Following on from the seminal paper by Wolf (1969), this subject has generated a vast literature. For more information on diffraction tomography, we refer the reader to the excellent accounts in Nieto-Vesperinas (1991) and Born and Wolf (1999), together with references to the primary literature which are contained therein.

2.11.2 *Information content of scattered polychromatic fields*

We have discussed the essentially two-dimensional nature of the information that is contained within coherent scalar wave-fields, irrespective of whether or not they are scattered by a given sample of interest. As a corollary to these discussions, we have seen that—under both the projection and first Born approximations—only a subset of the information, contained within a given three-dimensional distribution of scattering potential, can be obtained from a single scattering experiment involving a plane wave incident on the scatterer. This observation naturally led to the question of strategies for reconstructing three-dimensional information regarding the scatterer. One such strategy, that of performing scattering experiments with a single frequency of incident plane waves but for a number of different orientations of the object, was discussed in the previous sub-section. Here we discuss an alternative strategy, which keeps the orientation of the object fixed, but uses a range of incident wavelengths in the incident planar radiation.

Consider the scattering geometry described in the first paragraph of Section 2.11.1, with the exception that the incident z-directed monochromatic scalar plane wave is replaced with an incident z-directed polychromatic plane wave. Bearing the spectral decomposition (2.23) in mind, the incident polychromatic plane wave $\Psi^{(\mathrm{PW})}(x, y, z, t)$ is given by the following linear combination of monochromatic plane waves:

$$\Psi^{(\mathrm{PW})}(x, y, z \le z_0, t) = \frac{1}{\sqrt{2\pi}} \int s(\omega) \exp[i(kz - \omega t)] d\omega, \quad z \le z_0. \qquad (2.153)$$

Here, $s(\omega)$ is a specified complex function of angular frequency ω, the squared modulus of which is proportional to the intensity of a given monochromatic component of the incident polychromatic plane wave. The phase of $s(\omega)$ specifies the phase relationship between different frequencies of the incident plane wave. Note that the temporal development of the incident polychromatic plane wave will in general be rather complicated, due to the 'beating' of different plane-wave frequencies against one another. Note, also, that only non-negative angular frequencies ω are included in the above integral (see Section 1.2).

When this field is incident upon an arbitrary scatterer contained within the slab $z_0 < z < z_0 + \Delta$, the different frequencies will in general be mixed with

[66] The term 'diffraction tomography' distinguishes this field from that of computed tomography, which makes use of the projection approximation, as sketched in the previous sub-section.

one another.[67] This possibility, namely that of inelastic scattering, will be allowed for the moment. The time-dependent forward-scattered scalar disturbance $\Psi^{\mathrm{S}}(x,y,z,t)$, in the half-space $z \geq z_0 + \Delta$ downstream of the exit-surface of the scattering volume, is given by the following polychromatic generalization of eqn (2.150):

$$\begin{aligned}\Psi^{\mathrm{S}}(x,y,z,t) = & \frac{1}{(2\pi)^{3/2}} \iiint \breve{\psi}^{\mathrm{S}}_{\omega}(k_x,k_y,z=z_0+\Delta) \\ & \times \exp\left[i(z-z_0-\Delta)\sqrt{k^2-k_x^2-k_y^2}\right] \\ & \times \exp\left[i(k_x x + k_y y - \omega t)\right] dk_x dk_y d\omega, \quad z \geq z_0 + \Delta. \end{aligned} \tag{2.154}$$

Here, $\breve{\psi}^{\mathrm{S}}_{\omega}(k_x,k_y,z=z_0+\Delta)$ is the Fourier transform, with respect to x and y, of the spatial part of a given monochromatic component of the forward scattered field over the exit surface $z = z_0 + \Delta$. As was the case for the corresponding analysis of monochromatic fields, the information content of each monochromatic component is: (i) two-dimensional in nature; (ii) unable to propagate sub-wavelength information through distances of more than a few wavelengths of the given monochromatic component. Similar considerations apply to the angular-spectrum decomposition of the scattered polychromatic field over the entrance surface $z = z_0$ of the scattering volume, with the disturbance propagating into the half-space $z < z_0$.

However, there is a third dimension of information present in the scattered polychromatic field, given that there is a continuum of frequencies ω present.[68] Following Gabor (1961), each monochromatic field may be viewed as a separate information channel, each of which will in general contain different information regarding the scatterer. This observation leads us to consider the question of how much three-dimensional information, regarding the scatterer, is contained within the scattered polychromatic field. To parallel the treatment given for the same question in the case of monochromatic incident fields, we will separately treat this question under the projection and first Born approximations.

Before launching into a discussion of each of these sub-cases, in the following sub-sections, we note that the question is considerably complicated by the fact that the refractive index $n_{\omega}(x,y,z)$ will in general have a complicated dependence on the angular frequency ω of the incident radiation. Accordingly, for the sake of simplicity in the presentation, we will make the additional strong assumption that the scattering medium is composed of a known single material of variable density. Further, we will also assume that the X-ray refractive index is very close to unity, so that eqn (2.38) holds.

[67] That is, a single frequency of incident wave may lead to a range of frequencies in the scattered wave.

[68] If the polychromatic field exists for all time, this statement is correct. If the field only exists for a finite time, however, then there will in general be a countable infinity of different monochromatic components in the polychromatic field.

2.11.2.1 *Projection approximation* Since we have made the assumption that the scatterer is composed of a single known material, we can write both the real and imaginary parts of the refractive index distribution, at any angular frequency, as the product of the unknown position-dependent density $\rho(x,y,z)$ of the material, and known functions of angular frequency. Thus we have the scaling relations:

$$\delta_\omega(x,y,z) = D(\omega)\rho(x,y,z), \qquad \beta_\omega(x,y,z) = B(\omega)\rho(x,y,z), \tag{2.155}$$

where $D(\omega)$ and $B(\omega)$ are known material-dependent real functions of ω. Therefore the three-dimensional distribution of complex refractive index, at any angular frequency, has a one-to-one relationship with the density $\rho(x,y,z)$ of the sample. Since the refractive index has been taken to be very close to unity, we may make use of both (2.38) and the above scaling relations, so that:

$$\begin{aligned} 1 - n_\omega^2(x,y,z) &\approx 2[\delta_\omega(x,y,z) - i\beta_\omega(x,y,z)] \\ &\approx 2[D(\omega) - iB(\omega)]\rho(x,y,z). \end{aligned} \tag{2.156}$$

Substituting into eqn (2.105), and performing some simple algebraic manipulations, we see that:

$$\int_{z=z_0}^{z=z_0+\Delta} \rho(x,y,z)dz \approx \frac{(i/k)\log_e\left[\frac{\psi_\omega(x,y,z=z_0+\Delta)}{\psi_\omega(x,y,z=z_0)}\right] + \Delta}{D(\omega) - iB(\omega)}. \tag{2.157}$$

If one knows the monochromatic components of the entrance and exit-surface wave-fields at a particular angular frequency, together with the functions $D(\omega)$ and $B(\omega)$ and the distance Δ between the entrance and exit surfaces of the slab containing the single-material scattering sample, then the right side of the above expression will be known. One is thereby able to determine the left side of the above equation, yielding two-dimensional (projection) information regarding the three-dimensional distribution of density in the sample. No additional information is provided by the entrance-surface and exit-surface wave-fields at other angular frequencies, under the projection approximation. We conclude that, for the present case of a static single-material sample, which is illuminated by a polychromatic scalar plane wave under the projection approximation, the use of polychromatic rather than monochromatic incident plane waves provides no additional information regarding the three-dimensional distribution of density in the scatterer.

2.11.2.2 *First Born approximation* Consider the generalization of the Ewald sphere construction to the case of polychromatic scalar plane-wave illumination of a static non-magnetic sample. In expression (2.153) for the incident polychromatic plane wave, let us assume that $|s(\omega)|$ is non-negligible only for angular frequencies ω that lie in a given range, and that the first Born approximation is valid for all of these frequencies. Associated with each monochromatic component, of the incident polychromatic plane wave, will be an Ewald sphere of a given radius. See Fig. 2.6(b), where three such Ewald spheres are sketched. Spheres A and C are respectively associated with the minimum and maximum angular frequency for which $|s(\omega)|$ is non-negligible, with sphere B corresponding to a frequency lying between these extremes. The set of Ewald spheres, corresponding to the non-negligible spectral components of the incident polychromatic plane wave, occupies the volume of Fourier space given by the intersection of the exterior to sphere A with the interior to sphere C. This 'onion-like' set of Ewald spheres is sometimes spoken of as a 'nest' of Ewald spheres. Note, also, that as was the case in the monochromatic version of the present analysis, both forward-scattered and back-scattered radiation is being considered.

If the refractive index of the scatterer could be taken to be independent of the angular frequency of the incident radiation, then we could conclude that the scattered polychromatic field contains information regarding the Fourier coefficients of the scattering potential that lie between the Ewald spheres A and C. The refractive index does depend on frequency, however, so let us make use of the single-material assumption, as embodied in eqns (2.155) and (2.156). Thus the scattering amplitude f in eqn (2.78) obeys:

$$\frac{-2\pi f(\Delta\mathbf{k})}{k^2[D(\omega) - iB(\omega)]} = \iiint \rho(\mathbf{x}') \exp(-i\Delta\mathbf{k}\cdot\mathbf{x}')d\mathbf{x}'. \tag{2.158}$$

One can therefore determine the three-dimensional Fourier transform $f(\Delta\mathbf{k})$ of the density $\rho(x, y, z)$ of the single-material sample, as a function of all Fourier-space vectors $\Delta\mathbf{k}$ within the Ewald nest. This is done using a single polychromatic plane wave illuminating the single-material sample of interest, under the first Born approximation. The Ewald nest covers a certain volume of reciprocal space, this being indicative of the three-dimensional nature of the information contained within the scattered polychromatic field. The thinness of the Ewald nest, in the forward-scattering direction, limits the depth information in samples which may be obtained using radiation scattered through small angles. Again, one may rotate the sample, in order to 'fill out' the missing values of $f(\Delta\mathbf{k})$. Note also that the previously discussed notion, of a limiting Ewald sphere, has a direct analogue in the present context.

2.12 Summary

The present chapter was devoted to a study of the interactions of X-rays with matter, developing techniques that will be built upon in later chapters.

We opened with a derivation of certain wave equations governing the evolution of electromagnetic fields in the presence of scatterers. Such wave equations constitute a generalization of the vacuum wave equations obtained in the opening pages of the first chapter. As was the case there, we then effected a transition from a vector to a scalar theory of the electromagnetic field, this time in the presence of scatterers. The resulting complex scalar field (analytic signal) was seen to obey an inhomogeneous d'Alembert equation, with each spectral component of the field being governed by an inhomogeneous form of the Helmholtz equation. This last-mentioned equation, in which the effects of scatterers is quantified through a frequency-dependent refractive index that varies with spatial position, was seen to be a central result of the present chapter. Further, we note that the just-summarized discussions elucidated the manner in which the notion of refractive index arises from the Maxwell equations, by making an explicit connection between this quantity and the electrical permittivity and magnetic permeability.

We discussed several means of approximately treating the interactions of x-rays with matter, taking the inhomogeneous Helmholtz equation as a starting point. The first of these, namely the projection approximation, allowed us to deduce rather simple expressions for the field at the exit surface of a scattering volume, in terms of line integrals of the complex refractive index. These line integrals are carried out along paths corresponding to the classical ray paths of the unscattered radiation. When this approximation is valid, which will be the case when the classical ray paths within the volume of the scatterer are only slightly different from those which would have existed in the same volume in the absence of the scatterer, it provides a very simple means for quantifying the interactions of X-rays with a sample. Accordingly, the projection approximation will be used on many occasions in later chapters.

We then developed an integral form of the inhomogeneous Helmholtz equation. As we saw, this integral equation is equivalent to the original differential equation, together with both (i) knowledge of the incident wave-field and (ii) incorporation of the boundary condition that the scattered field be outgoing at infinity. In the context of developing this integral equation, we gave two explicit derivations of the free-space Green function for the complex scalar electromagnetic field. We were then able to derive an approximate solution, known as the first Born approximation, to the scattering problem based on the integral equation mentioned earlier. This was seen to be a single-scattering (kinematical) approximation. The first Born approximation was then related to our earlier studies on both the Fraunhofer approximation and the angular spectrum. Having done this, we then proceeded to demonstrate how the famous Ewald sphere naturally emerges from the first Born approximation, whether or not the scattering distributions are crystalline. Second and higher-order Born approximations were briefly considered, leading to the notion of multiple (dynamical) scattering.

We then made another point of connection, between the single-scattering picture of the first Born approximation, and the notion of refractive index which was developed earlier in the chapter. In making this connection, we obtained a useful approximate expression relating the real part of the refractive index of a material to its electron number density. We argued that this expression is often a good approximation, if one is sufficiently far from absorption edges of the material.

Further methods of approximation, regarding the elastic scattering of coherent scalar X-rays from static non-magnetic media, included the multislice and eikonal approximations. This latter approximation also served to elucidate some of the connections between the wave and ray pictures of electromagnetic radiation.

We closed the chapter with a discussion of various mechanisms for the inelastic scattering of X-rays, of which Compton and photoelectric scattering were considered to be the most important, together with some discussions on the information content of scattered fields.

Further reading

For further information on the scattering theory outlined in this chapter, see Nieto-Vesperinas (1991) and Born and Wolf (1999). Note, in this regard, that the integral-equation treatment of scattering, together with the first and higher-order Born approximations, is very closely related to the problem of potential scattering in non-relativistic quantum mechanics. Useful references, in this respect, include Messiah (1961) and Merzbacher (1970). Scattering by magnetic materials, which was not covered here, is treated in Lovesy and Collins (1996). Regarding the use of complex analysis made in text, see, for example, Morse and Feshbach (1953), Spiegel (1964), and Markusevich (1983). For further information on the Ewald sphere, the reader may consult Warren (1969), Ashcroft and Mermin (1976), Kittel (1986), and Cowley (1995). Many important topics, on the dynamical diffraction of X-rays, have been omitted from this chapter. For an authoritative treatment, see Authier (2001). Elements of this chapter have made use of the theory of special relativity, excellent accounts of which are given in the texts of French (1968) and Rindler (1991). Tabulations of quantities, which may be related to the real and imaginary parts of the complex refractive index used in the present chapter, together with descriptions of the underlying physics, are given in Henke *et al.* (1993) and Chantler (1995, 2000). Note, also, that much of this information is now available online—see, for example, the internet website of the National Institute of Standards and Technology (NIST). Lastly, we note the very useful 'X-ray data booklet', the most recent version of which is available through the internet website of the Lawrence Berkeley National Laboratory.

References

J. Als-Nielsen, *Diffraction, refraction and absorption of X-rays and neutrons: a comparative study*, in J. Baruchel, J.-L. Hodeau, M.S. Lehmann, J.-R. Regnard, and C. Schlenker (eds), *Neutron and synchrotron radiation for condensed matter studies, volume 1: theory, instruments and methods*, Springer-Verlag, Berlin (1993), pp. 3–33.

J. Als-Nielsen and D. McMorrow, *Elements of modern X-ray physics*, John Wiley & Sons, New York (2001).

M. Altarelli, *Polarized X-rays*, in J. Baruchel, J.-L. Hodeau, M.S. Lehmann, J.-R. Regnard, and C. Schlenker (eds), *Neutron and synchrotron radiation for condensed matter studies, volume 1: theory, instruments and methods*, Springer-Verlag, Berlin (1993), pp. 261–269.

N.W. Aschroft and N.D. Mermin, *Solid state physics*, Thomson Learning (1976).

A. Authier, *Dynamical theory of X-ray diffraction*, Oxford University Press, Oxford (2001).

L.V. Azaroff, *Elements of X-ray crystallography*, McGraw-Hill, New York (1968).

L.V. Azaroff, R. Kaplow, N. Kato, R.J. Weiss, A.J.C. Wilson, and R.A. Young, *X-ray diffraction*, McGraw-Hill, Sydney (1974).

D. Bohm, *A suggested interpretation of the quantum theory in terms of 'hidden' variables, I and II*, Phys. Rev. **85**, 166–193 (1952).

M. Born and E. Wolf, *Principles of optics*, seventh edition, Cambridge University Press, Cambridge (1999).

C.T. Chantler, *Theoretical form factor, attenuation and scattering tabulation for Z=1–92 from E=1–10 eV to E=0.4–1.0 MeV*, J. Phys. Chem. Ref. Data **24**, 71–643 (1995).

C. T. Chantler, *Detailed tabulation of atomic form factors, photoelectric absorption and scattering cross section, and mass attenuation coefficients in the vicinity of absorption edges in the soft X-ray (Z=30–36, Z=60–89, E=0.1 keV–10 keV), addressing convergence issues of earlier work*, J. Phys. Chem. Ref. Data **29**, 597–1048 (2000).

A.H. Compton, *A quantum theory of the scattering of X-rays by light elements*, Phys. Rev. **21**, 483–502 (1923).

J.M. Cowley, *Diffraction physics*, third revised edition, North-Holland, Amsterdam (1995).

J.M. Cowley and A.F. Moodie, *The scattering of electrons by atoms and crystals. I. A new theoretical approach.*, Acta. Cryst. **10**, 609–619 (1957).

B.D. Cullity, *Elements of X-ray diffraction*, second edition, Addison-Wesley, Reading, MA (1978).

T.J. Davis, *A unified treatment of small-angle X-ray scattering, X-ray refraction and absorption using the Rytov approximation*, Acta Cryst. **A50**, 686–690 (1994).

A. Einstein, *Über einen die Erzeugung und Verwandlung des Lichtes betreffenden heuristischen Gesichtspunkt*, Ann. Phys. **17**, 132–148 (1905). The English translation, cited in the text, is given in A.B. Arons and M.B. Peppard (trans.), *Einstein's proposal of the photon concept—a translation of the Annalen der Physik paper of 1905*, Am. J. Phys. **33**, 367–374 (1965).

A.P. French, *Special relativity*, Chapman and Hall, London (1968).

D. Gabor, *Light and information*, in E. Wolf (ed.), *Progress in Optics*, volume 1, North-Holland, Amsterdam (1961), pp. 109–153.

D. Gibbs, J.P. Hill, and C. Vettier, *New directions in X-ray magnetic scattering*, in D.M. Mills (ed.), *Third-generation hard X-ray synchrotron radiation sources: source properties, optics and experimental techniques*, John Wiley &

Sons, New York, pp. 267–310 (2002).

T. Habashy and E. Wolf, *Reconstruction of scattering potentials from incomplete data*, J. Mod. Opt. **41**, 1679–1685 (1994).

C. Hammond, *The basics of crystallography and diffraction*, second edition, Oxford University Press, Oxford (2001).

W. Heitler, *The quantum theory of radiation*, third edition, Oxford University Press, Oxford (1954).

B.L. Henke, E.M. Gullikson, and J.C. Davis, *X-ray interactions: photoabsorption, scattering, transmission and reflection at E=50–30,000 eV, Z=1–92*, At. Data Nucl. Data Tables **54**, 181–342 (1993).

J.D. Jackson, *Classical electrodynamics*, third edition, John Wiley & Sons, New York (1999).

C. Kittel, *Introduction to solid state physics*, sixth edition, John Wiley & Sons, New York (1986).

E. Kreysig, *Advanced engineering mathematics*, fifth edition, John Wiley & Sons, New York (1983).

S.W. Lovesy and S.P. Collins, *Scattering and absorption by magnetic materials*, Oxford University Press, Oxford (1996).

L. Mandel and E. Wolf, *Optical coherence and quantum optics*, Cambridge University Press, Cambridge (1995).

F. Mandl and G. Shaw, *Quantum field theory*, revised edition, John Wiley & Sons, Chichester (1993).

E. Madelung, *Quantentheorie in Hydrodynamischer form*, Z. Phys. **40**, 322–326 (1926).

A.I. Markusevich, *The theory of analytic functions: a brief course*, Mir Publishers, Moscow (1983).

E. Merzbacher, *Quantum mechanics*, second edition, John Wiley & Sons, New York (1970).

A. Messiah, *Quantum mechanics*, volumes 1 and 2, North-Holland, Amsterdam (1961).

P.M. Morse and H. Feshbach, *Methods of theoretical physics, volume 1*, McGraw-Hill, New York (1953).

F. Natterer, *The mathematics of computerized tomography*, John Wiley & Sons, New York (1986).

M. Nieto-Vesperinas, *Scattering and diffraction in physical optics*, John Wiley & Sons, New York (1991).

K.E. Oughstun (ed.), *Selected papers on scalar wave diffraction*, SPIE Milestone Series volume M51, SPIE Optical Engineering Press, Bellingham (1992).

Z.G. Pinsker, *Dynamical theory of X-ray scattering in ideal crystals*, Springer-Verlag, Berlin (1978).

W. Rindler, *Introduction to special relativity*, Oxford University Press, Oxford (1991).

J.J. Sakurai, *Advanced quantum mechanics*, Addison-Wesley, Reading, MA (1967).

M.R. Spiegel, *Theory and problems of complex variables*, McGraw-Hill Book Company, Singapore (1964).

V.I. Tatarskii, *Wave propagation in a turbulent medium*, Dover Publications, New York (1967).

B.E. Warren, *X-ray diffraction*, Addison-Wesley, Reading, MA (1969).

H. Weyl, *Ausbreitung elektromagnetischer Wellen über einem ebenen Leiter*, Ann. d. Physik **60**, 481–500 (1919).

E. Wolf, *Three-dimensional structure determination of semi-transparent objects from holographic data*, Opt. Commun. **1**, 153–156 (1969).

M.M. Woolfson, *An introduction to X-ray crystallography*, second edition, Cambridge University Press, Cambridge (1997).

3

X-ray sources, optical elements, and detectors

X-ray sources, optical elements, and detectors are the subject of this chapter. Each of these three topics will be treated in turn, with some emphasis placed on their role in the science of coherent X-ray optics.

The first part of the chapter treats X-ray sources. As a simple means for quantifying the quality of radiation from such sources, the notions of brightness and emittance are introduced. We then outline several classes of X-ray source. The first such class, which includes X-ray tubes and rotating anode sources, is the least coherent of those to be treated. Synchrotron sources occupy the majority of our discussion on the means of X-ray generation, as the bulk of present-day work in coherent X-ray optics is undertaken at such facilities. Finally, we consider three newer classes of X-ray source, whose properties are in many respects complementary to those of synchrotron radiation: free-electron lasers, energy-recovering linear accelerator sources, and soft X-ray lasers. Such newer sources, all of which are at an advanced stage of development, are likely to assume increased prominence in future research involving coherent X-ray optics.

In the second part of the chapter we turn our attention to X-ray optical elements, many of which are rather dissimilar to their visible-light counterparts. Such optics may be classified as 'absorptive', 'diffractive', 'reflective' and 'refractive'. We will not discuss the simple absorptive optical elements such as shutters, slits, pinholes, knife edges, and anti-scatter grids. Optical elements to be considered include diffraction gratings, Fresnel zone plates, monochromators, analyser crystals, interferometers, Bragg–Fresnel crystal optics, capillaries, mirrors, square-channel arrays, prisms, and compound refractive lenses. The role of virtual optical elements, in which the computer forms an integral part of an optical imaging system, is also treated.

In the third and final part of the chapter, we treat X-ray detectors. Their key parameters are discussed, before outlining several commonly used varieties of detector. Broadly, these devices may be split into two categories: integrating detectors and counting detectors. The former, which typically have better spatial resolution but poorer dynamic range, measure a signal integrated over many X-ray photons. Conversely, the latter are capable of counting individual photons. A comparison will be given, of the relative merits of counting and integrating detectors. Lastly, we offer some comments on the relationship between detectors and coherence.

3.1 Sources

3.1.1 *Brightness and emittance*

Simple means for quantifying the quality of an X-ray source, in the context of coherent X-ray optics, are often useful. Such simple measures include the notions of 'brightness' and 'emittance', to which we now turn.

Erect a Cartesian xy coordinate system in a plane perpendicular to a given optic axis z, with x and y respectively coinciding with the horizontal and vertical directions. Let σ_x and σ_y denote the root-mean-square width and height of a given X-ray beam (in millimetres), with σ_{θ_x} and σ_{θ_x} respectively denoting the divergence in the horizontal and vertical directions (measured in milliradians). Finally, let $\Phi(E)$ denote the X-ray flux (in photons per second) at a specified energy E, within a 0.1% bandwidth centred at this energy.

Given the above definitions, we may write down the following formula for spectral brightness $\mathscr{B}(E)$ (see, for example, Kim (2001)):

$$\mathscr{B}(E) \equiv \frac{\Phi(E)}{4\pi^2 \sigma_x \sigma_y \sigma_{\theta_x} \sigma_{\theta_y}}. \tag{3.1}$$

Brightness—which is maximized by making the beam size and divergence as small as possible, and the photon flux as large as possible—is the first of the promised simple measures of source quality in the context of coherent X-ray optics.

The second such measure is known as the emittance, which may defined in both the horizontal and vertical directions by the respective equations (see, for example, Kim (2001)):

$$\varepsilon_x \equiv \sigma_x \sigma_{\theta_x}, \qquad \varepsilon_y \equiv \sigma_y \sigma_{\theta_y}. \tag{3.2}$$

This allows us to write eqn (3.1) in the alternative form:

$$\mathscr{B}(E) \equiv \frac{\Phi(E)}{4\pi^2 \varepsilon_x \varepsilon_y}. \tag{3.3}$$

We can now give a simple answer to the question of what constitutes a high-quality source for experiments in coherent X-ray optics: high brightness and low emittance. For a more detailed description in the context of a particular optical scenario, one may choose to work with the mutual coherence function and its variants, as introduced in the first chapter.[69] Notwithstanding this, for the purposes of the present chapter we shall work with the simpler measures of brightness and emittance, in order to quantify the quality of an X-ray source in the context of experiments in coherent X-ray optics.

[69] Note that the brightness—considered as a function of both real and reciprocal space coordinates, rather than as a single number—may be obtained from the mutual coherence function (see, for example, Kim (1986)). This brightness function is closely related to the Wigner distribution. See also the review by Wolf (1978), and references therein.

3.1.2 *Fixed-anode and rotating-anode sources*

Consider a free electron which is travelling through vacuum at a given speed, corresponding to a particular momentum $\mathbf{p}$. This electron has kinetic energy K, this being the difference between its total energy E and its rest mass energy $m_e c^2$. Here, m_e is the rest mass of the electron and c is the speed of light in vacuum. Invoking the relativistic energy–momentum–mass relation, as given in eqn (2.139) of the previous chapter, we see that the kinetic energy of the electron is[70]:

$$K = E - m_e c^2 = \sqrt{m_e^2 c^4 + |\mathbf{p}|^2 c^2} - m_e c^2. \tag{3.4}$$

Now suppose that our electron is abruptly decelerated to rest. The electron's kinetic energy K is thereby transferred to the energy of the electromagnetic field, which is radiated by the electron in the process of coming to rest. Such radiation is known as 'bremsstrahlung', after the German for 'braking radiation'. The theory of bremsstrahlung will not be given here, as it is treated in many advanced texts on electrodynamics, such as that of Jackson (1999). For relativistic electrons such braking radiation has a broad spectrum, ranging from the lowest energies through to a cutoff wavelength whose energy corresponds to the initial kinetic energy K of the electron.

A convenient means of decelerating the electron is to cause it to strike a suitable fixed target, which is typically made of materials such as copper or molybdenum. In decelerating within the target material, bremsstrahlung is emitted. Upon observing the spectrum of X-rays that is thereby produced, one typically notes the presence of spikes at energies which depend upon the material used for the target. Such spikes, which overlay the broad bremsstrahlung spectrum, are known as 'characteristic lines'. They arise due to the fact that the incident electrons may knock out core electrons from the material of the target, with 'fluorescent' X-rays being emitted when these energy levels are filled as the perturbed atom relaxes to its ground state. The energy of the fluorescent X-rays is therefore a characteristic of the atomic energy levels of the material used to construct the target.[71]

The above two processes, namely bremsstrahlung and the production of characteristic lines, are the two dominant mechanisms which result in X-ray production by fixed-anode and rotating-anode sources. Fixed-anode sources, such as the

[70]Note that this reduces to the familiar Newtonian expression for the kinetic energy, in the non-relativistic limit where the total energy is dominated by the rest-mass energy. To see this, remove a factor of $m_e c^2$ from the square root, and then perform a first-order binomial expansion: $K = m_e c^2 (1 + |\mathbf{p}|^2 m_e^{-2} c^{-2})^{1/2} - m_e c^2 \approx m_e c^2 (1 + \frac{1}{2}|\mathbf{p}|^2 m_e^{-2} c^{-2}) - m_e c^2 = |\mathbf{p}|^2/(2m_e)$.

[71]If the fluoresced X-ray is due to the relaxation of an atomic electron from the L to the K shell, then the associated peak in the X-ray spectrum is termed a 'K_α peak'. Similarly, if the fluoresced X-ray is due to the relaxation of an atomic electron from the M to the K shell, then the associated peak in the X-ray spectrum is termed a 'K_β peak'. The L_α peaks, and so on, are similarly defined.

Hittorf, Lenard, and Crookes tubes used by Röntgen in his discovery of X-ray radiation (Röntgen 1896), are so called because the target material comprises the stationary ('fixed') anode of a pair of electrodes through which an electrical discharge is passed. When electrons from the discharge strike the anode they are rapidly decelerated, yielding both bremsstrahlung and characteristic radiation. The modern Coolidge X-ray tube still works on this principle, albeit with a hot cathode and a cooled anode.

Fixed-anode sources have the lowest brightness of those that shall be considered in this text. While it is difficult to give specific numbers on account of the variability which exists between sources, together with the conditions under which they operate, the peak brightnesses of bremsstrahlung from modern X-ray tubes are on the order of 10^5 photons $s^{-1}mm^{-2}mrad^{-2}$ per 0.1% bandwidth (see, for example, Kim (2001)). The strongest characteristic lines, from targets such as copper and molybdenum, have a brightness which is two to three orders of magnitude larger than that of the bremsstrahlung.

One major limitation, on the brightness achievable with fixed-anode sources, is the rate at which heat can be removed from the anode. The greater the heat load that can be dissipated from the target, per unit time, the greater will be the electron currents which can be tolerated and the greater the flux of X-ray photons which are emitted. Rotating-anode sources, in which a cooled anode is made to rotate, are able to more efficiently remove heat and thereby sustain higher fluxes of X-ray photons. Such sources gain roughly an order of magnitude in brightness, compared to their fixed-anode counterparts, for both the characteristic and bremsstrahlung components of the X-ray spectrum.

A further order-of-magnitude gain in brightness can be achieved with the so-called 'microfocus' rotating-anode sources, in which a focused electron beam can be used to achieve a smaller source size for the emitted X-ray radiation. Enhanced brightness can also be achieved in fixed-anode machines, by using the focusing technology of an electron microscope to achieve very small spot sizes on the target (von Ardenne 1939; Marton 1939; von Ardenne 1940; Cosslett and Nixon 1951, 1952, 1953, 1960; see also Mayo *et al.* 2002, and references therein).

3.1.3 *Synchrotron sources*

'Synchrotron light' may be defined as the electromagnetic radiation emitted by relativistic charged particles that are accelerated along curved trajectories. Such radiation predates human technology, being naturally produced by astrophysical sources in which highly energetic electrons are accelerated along helical paths in the presence of magnetic fields. Our emphasis, of course, is on the development of terrestrial sources of synchrotron radiation. We begin our account with a sketch of some of the earliest theoretical and experimental investigations in this area, before passing onto a description of modern synchrotron facilities.

3.1.3.1 *Early researches on synchrotron radiation* One of the earliest investigations on the theory of synchrotron radiation is due to Liénard (1898), who developed an expression for the electromagnetic field radiated by a point charge moving along an arbitrary path. As a special case, Liénard derived an expression for the energy loss of a point charge due to the synchrotron radiation it gives off when following a circular path. Liénard's work was extended by Schott (1912), who obtained formulae for the spectral, polarization, and angular properties of synchrotron radiation. Later theoretical works include those of Pomeranchuk (1940), Iwanenko and Pomeranchuk (1944), Arzimovich and Pomeranchuk (1945), and Schwinger (1949). These investigations led to the following key results: (i) the synchrotron radiation, emitted by a relativistic classical point charge which travels around a circle at constant speed, has a radiated energy per orbit that is proportional to both the square of the electric charge of the particle and the fourth power of the energy of the particle, while being inversely proportional to the fourth power of the rest mass of the particle; (ii) the synchrotron radiation is strongly concentrated in the direction of the charge's velocity $\mathbf{v}$, in a forward-pointing cone of half-angle

$$\theta \approx \sqrt{1 - \frac{|\mathbf{v}|^2}{c^2}}, \tag{3.5}$$

which is equal to the ratio of the rest-mass energy of the charged particle, to the total energy E of the particle; (iii) the broad spectrum of synchrotron radiation has a critical wavelength[72] of:

$$\lambda = \frac{4\pi R}{3} \left(\frac{m_0 c^2}{E} \right)^3, \tag{3.6}$$

where R is the radius of curvature of the circle traversed by the point charge of rest mass m_0.

The above theoretical developments overlap with the history of experimental efforts which resulted in the production of synchrotron radiation. While electron accelerators were developed in the early 1920s, it was not until much later that consideration was given to the production of synchrotron radiation by such devices. Indeed, such radiation was considered to be a nuisance, resulting as it did in energy losses from the accelerated electrons.[73] Following the near miss described by Blewett (1946), synchrotron radiation was first observed in 1947 (Elder *et al.* 1947a,b). This synchrotron light was in the visible range of the spectrum, being obtained using the 70 MeV synchrotron at the General Electric Research Laboratory in Schenectady, New York. Note that the radius, of the circular electron

[72]By definition, half of the emitted power lies at wavelengths below the critical wavelength.

[73]Three examples epitomizing this negative attitude: (i) The paper by Iwanenko and Pomeranchuk (1944) concerns the 'limitation for maximal energy attainable' in a circular accelerator for relativistic electrons, due to 'radiative dissipation of energy of electrons moving in a magnetic field'. (ii) Similarly, the paper by Blewett (1946) is firmly focussed on the detrimental effect of radiation losses on electron accelerators due to synchrotron radiation. (iii) The opening paragraph, of Schwinger's famous paper (Schwinger 1949), speaks of 'limitations to the attainment of high energy electrons imposed by the radiative energy loss of the accelerated electrons'. None of these papers mention the possible positive use of synchrotron radiation.

orbit in this machine, was rather small by most modern standards, at 29.3 cm. Further researches, during these earliest years, include those reported by Elder *et al.* (1948) and Tomboulian and Hartman (1956). For further information regarding the history of the development of synchrotron radiation, together with references to the primary literature, we refer the reader to the review article by Lea (1978). For a first-hand account of the early days of synchrotron radiation, see Blewett (1998).

3.1.3.2 *The three generations of synchrotron light source* Broadly speaking, the history of synchrotron radiation sources may be classified into three phases, commonly termed 'generations'. We examine each of these in turn.

'First-generation' sources, which include those employed in the experimental work outlined in the previous sub-section, may be defined as those facilities that produce synchrotron radiation as a by product, but which are not specifically designed for the production of such radiation. The primary purpose of first-generation sources was as accelerators for experiments in high-energy particle physics.

'Second-generation' sources refer to the first machines purpose built for, and solely dedicated to, the production and use of synchrotron radiation. The first such facility, the Daresbury Synchrotron Radiation Source, became operational in 1981. Rather than being based on machines (e.g. betatrons and synchrotrons) whose function was to accelerate electrons or positrons, the second generation of sources were based on 'storage rings', these being designed to store electrons or positrons in a closed circular orbit at a fixed energy. Such fixed-energy charged-particle storage is maintained by having suitable radio-frequency cavities, at appropriate points along the orbit of the storage ring, which serve to replenish the energy that is lost through synchrotron radiation. Rather than being optimized for the acceleration of particles, as was the case with the first generation of synchrotron sources, second-generation sources were optimized with respect to the production of synchrotron light.

The primary source of radiation, in a second-generation source, arises from 'bending magnets'—see Fig. 3.1. Here, we see an electron bunch[74] at A which is travelling along a straight section of an electron storage ring. Upon passing through the bending magnet B, within which there is an approximately uniform magnetic field pointing into the plane of the page, the associated Lorentz force will cause the electrons to move in a circle of radius R. This circle is about the point O, which lies in the plane of the storage ring. In accelerating the electrons within the bending magnet, a tightly collimated cone C of synchrotron radiation is produced. Finally, after traversing the bending magnet, the electrons will have been deflected, subsequently travelling along the straight segment D of the storage ring. The peak brightness of the radiation, produced by a bending magnet, is on the order of 10^{15} photons $\mathrm{s}^{-1}\mathrm{mm}^{-2}\mathrm{mrad}^{-2}$ per 0.1% bandwidth.

[74]This term reflects the fact that the electrons are localized in small clumps about the circumference of the ring.

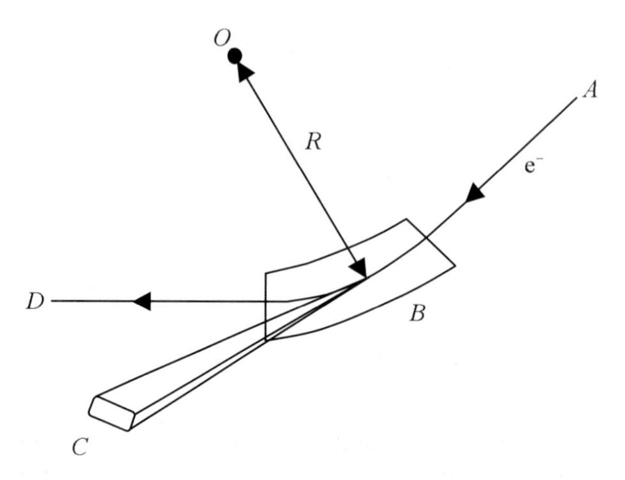

Fig. 3.1. Bending magnet as a source of synchrotron radiation. A relativistic electron bunch at A passes through a bending magnet B with approximately uniform magnetic field pointing into the plane of the page. This causes the electrons to traverse an arc with radius R. At each point within the bending magnet, the accelerated electrons emit a narrow cone of synchrotron radiation C, before emerging from the magnet to travel along the straight path D.

This is some ten orders of magnitude larger than the peak brightness associated with fixed-anode X-ray sources (see, for example, Kim (2001)).

This remarkable increase in brightness is further improved upon, in the 'third-generation' synchrotron facilities to which we now turn. These differ from their predecessors in two principal respects. First, a good deal of experience was gained from the development and operation of second-generation sources, which was incorporated into improved machine designs with a view to increasing brightness and decreasing emittance. Second, and perhaps more distinctly, the bending magnets were augmented with 'insertion devices' as the main means of producing synchrotron radiation.

Modern insertion devices may be broken into two classes, whimsically known as 'wigglers' and 'undulators'. These devices are inserted into the straight sections of the storage ring. They have the common feature of producing synchrotron radiation by passing relativistic electron bunches through periodic magnetic structures (Motz 1951), such as that shown in Fig. 3.2. Here, we see an electron bunch at A entering a periodic array of magnetic dipoles, producing a periodic magnetic field that causes the electrons to follow an approximately sinusoidal path. This may be contrasted with the circular arc traced by electrons traversing bending magnets.

An insertion device may be spoken of as either a 'wiggler' or an 'undulator', depending upon the parameter regime in which it operates. With a view to elucidating this distinction, consider Fig. 3.3. Here, we see a sinusoidal electron path

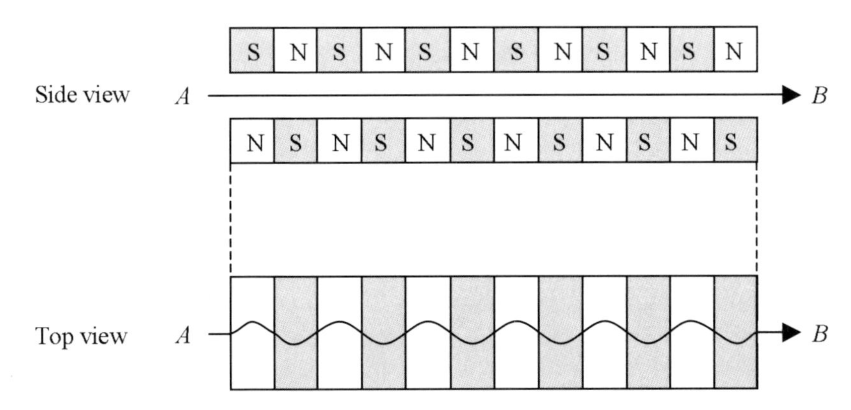

Fig. 3.2. Schematic of an insertion device, which may be a wiggler or an undulator. A relativistic electron bunch at A enters an array of magnetic dipoles, causing it to follow an approximately sinusoidal trajectory in the plane of the storage ring. North and south ends, of the magnetic dipoles, are respectively denoted by 'N' and 'S'.

corresponding to the wiggler regime. The narrow cones of synchrotron radiation, emitted from the points B, C, and D on the electron trajectory, are respectively indicated by E, F, and G. As suggested by the diagram, the defining characteristic of the wiggler regime is that the various synchrotron-radiation cones do not all have a significant degree of overlap. Notwithstanding this fact, there will evidently be a series of forward-pointing cones (such as F) which will overlap, the number of which depends on the number of periods in the insertion device. The fields of each overlapping forward-pointing cone (even if they are due to the same electron traversing subsequent periods of the magnetic lattice of the wiggler) add incoherently, with a resulting spectrum which is smooth and broad. The more periods in the wiggler, the greater will be the number of forward-directed X-ray cones which overlap, and the more intense will be the resulting synchrotron radiation. Typically, the brightness achievable with a wiggler is one to two orders of magnitude larger than that obtained using bending magnets (see, for example, Kim (2001)).

Next, suppose that the peak magnetic field of the wiggler is chosen to be sufficiently weak, and the period of the magnetic array sufficiently small, that the cones E, F, and G, in Fig. 3.3, have a significant degree of overlap. The insertion device would now be referred to as an 'undulator'. The associated X-ray spectrum is no longer smooth, being sharply peaked at the harmonics of the device, which are approximately equally spaced with respect to energy. The radiation from a single electron, which radiates from consecutive periods of the undulator's magnetic lattice, adds coherently for radiation wavelengths corresponding to one of the previously mentioned harmonics; however, the radiation from different electrons adds incoherently, as does radiation from any given electron which

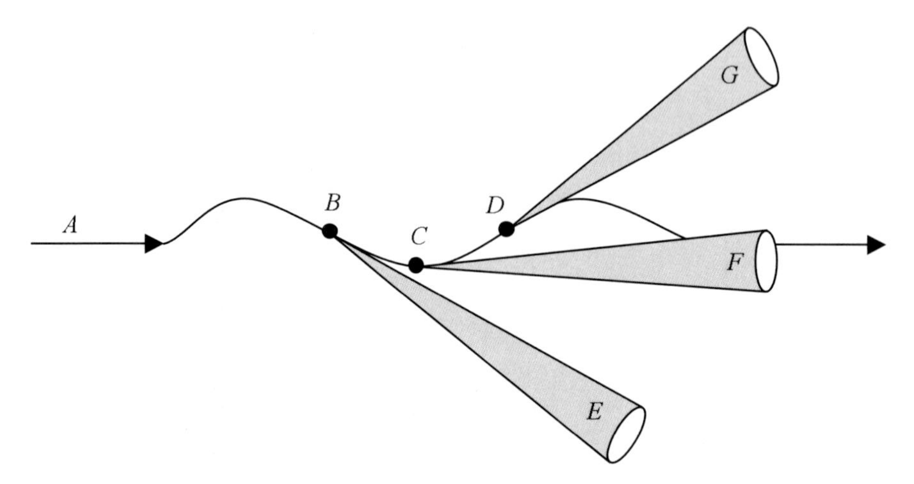

Fig. 3.3. Wiggler as a source of synchrotron radiation. A relativistic electron bunch at A passes through an array of magnets of alternating polarity (not shown), causing it to travel along an approximately sinusoidal path. The narrow cones of synchrotron radiation, emitted from the points B, C, and D, are respectively indicated by E, F, and G. After Margaritondo (2002).

does not correspond to one of the harmonics of the device. Due to the constructive interference mentioned above, the peak brightness of undulator radiation is typically three to four orders of magnitude greater than that of bending magnets (see, for example, Kim (2001)).

On account of their high brightness and low emittance, undulators are the workhorses of third-generation storage-ring facilities for the production of synchrotron radiation. A simplified schematic, of such a facility, is given in Fig. 3.4. Here, energetic electron bunches enter the storage ring via an injector A. Note that these injected electrons are typically accelerated, prior to injection, using both a linear accelerator and a so-called booster synchrotron (not shown). In order to pass from the injector to the storage ring, a bending magnet B is required. Bending magnets C and D are shown, leading to bending-magnet radiation such as that denoted by E. However, as stated earlier, third-generation sources are designed to have a large number of straight sections, into which insertion devices such as F may be placed. While these insertion devices may be either wigglers or undulators, the latter are usually preferred, on account of the superior brightness of the radiation that they produce. A beam G of insertion-device radiation is indicated in the diagram, together with a radio-frequency cavity H used to replenish the electron energy lost due to synchrotron radiation.

Associated with each of the synchrotron beams produced by the storage-ring facility, is a so-called 'beamline'. These typically contain a variety of specialized equipment suited to a particular class of experiments. For example, one often encounters beamlines dedicated to crystallography, diffraction from powdered

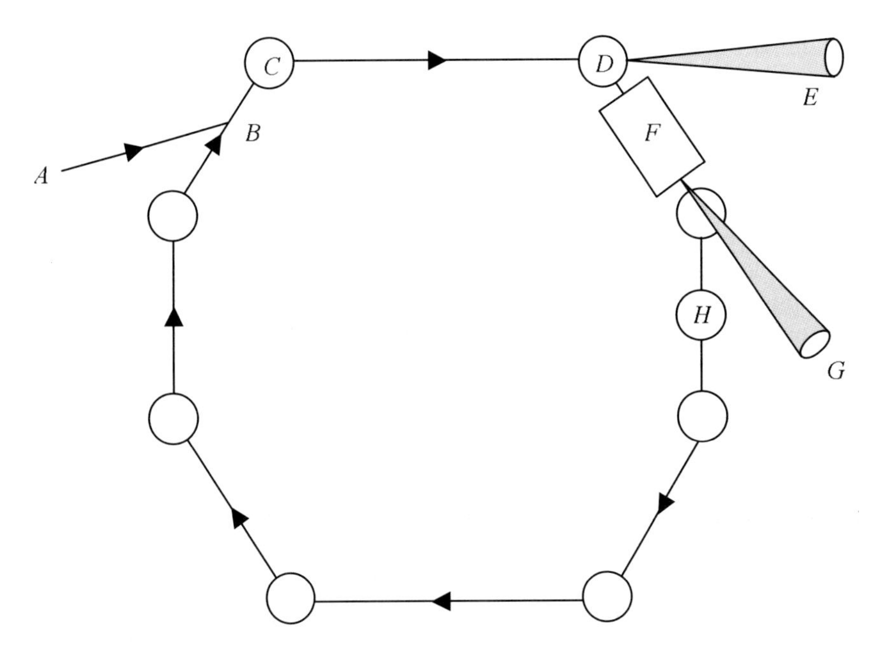

Fig. 3.4. Schematic of a third-generation storage ring for the production of synchrotron radiation. Injector, A; injector bending magnet, B; bending magnets, C, D; synchrotron radiation from bending magnet, E; insertion device, F; synchrotron radiation from insertion device, G; radio-frequency cavity, H. After Lewis (1997).

crystals, surface studies, infrared spectroscopy, X-ray microscopy, topography, lithography, small-angle scattering, and X-ray absorption fine structure, to name but a few.

3.1.4 *Free-electron lasers*

Recall the discussions of Section 2.10.2 on photon absorption, in which we saw that an absorbed photon's energy may promote an atomic electron from a lower-energy state to a higher-energy state (possibly in the continuum, resulting in ionization of the atom). If a plane-wave beam of photons is normally incident upon a single-material slab of constant thickness T, then, in the X-ray regime, such a mechanism for photon absorption ('photo-absorption') typically gives a dominant contribution to the linear attenuation coefficient μ, which results in the intensity of the exit beam being attenuated with respect to that of the incident beam, by a factor of $\exp(-\mu T)$ (Beer's Law, cf. eqn (2.43)).

Now suppose that at least some of the atoms, in the illuminated slab, are in an excited state. In addition to the photon-absorption mechanism described above, there will now be the possibility for photons to be emitted by atoms in the slab. Two such mechanisms, known as 'spontaneous emission' and 'stimulated emission', will now be described. (i) Spontaneous emission occurs when an excited atomic state emits a photon, in the absence of other photons of the same type, as an atomic electron relaxes from a higher-energy state to a lower-energy state. Such spontaneously emitted photons bear no fixed phase relationship with respect to one another, and therefore constitute an incoherent source of light. The characteristic lines, discussed in Section 3.1.2, are an example of such a source of X-ray photons. (ii) Stimulated emission (Einstein 1917) refers to photon emission which is induced by virtue of the excited atom being immersed in a photon field. From a classical point of view, one might view the pre-existing electromagnetic field as causing the electron to oscillate, and thereby emit electromagnetic radiation at the same frequency as that which stimulated the emission. Using an argument based on the quantum-mechanical first-order time-dependent perturbation theory of the photon–atom interaction (see, for example, Sakurai (1967)), one can show that the probability amplitude, for an excited atom to emit a photon of a given type (i.e. with a particular wavelength, propagation direction, and polarization), increases with the number of photons, of the same type, which are already present. Thus, the more photons of a given type are present, the greater is the chance that more such photons will be created through the mechanism of stimulated emission. Further, when stimulated photon emission is effected by pre-existing photons of a given type, the emitted photons have the same wavelength, propagation direction, and polarization as the photon or photons which stimulated the emission.

Stimulated emission is the process which underpins the conventional laser, this being an acronym for 'light amplification by stimulated emission of radiation'. Such light amplification may be contrasted with the previously discussed light attenuation via photon absorption. Rather than the photons giving energy to the medium via processes such as photo-absorption, the mechanism of stimulated emission raises the possibility that the medium may transfer energy to the light field in which it is immersed. This process underpins visible-light lasers, the generic features of which are sketched in Fig. 3.5. Light amplification by stimulated emission is achieved by creating a so-called 'population inversion' in the lasing medium. This refers to a state of the lasing medium in which an excited atomic level (often a meta-stable excited state) is more greatly populated than a lower-energy level. The injection of energy, required to create such a population inversion in the lasing medium, is termed 'pumping'. The population inversion serves to increase the probability of stimulated emission, while reducing the probability of photon absorption. The condition for lasing is that the light amplification, by stimulated emission of radiation, be greater than the light attenuation by absorption. Conditions for lasing are typically strengthened by placing the lasing medium C in an optical cavity, which often comprises a fully silvered mirror A at one end and a partially silvered mirror B at the other end. Particular modes of the cavity may be excited by the process of spontaneous emission, with one or more such modes subsequently being amplified, until saturation is reached, by the process of stimulated emission. Under these conditions,

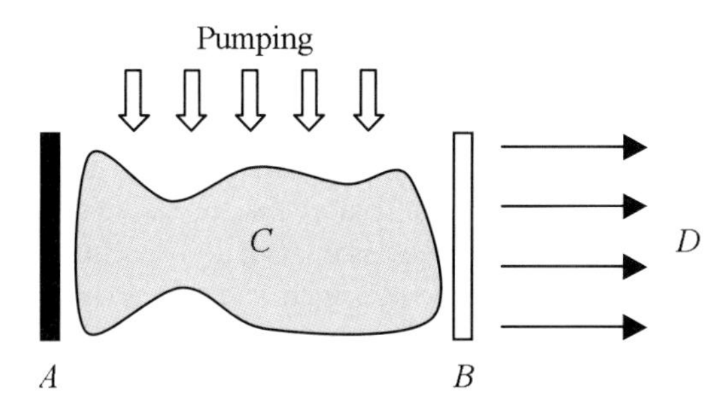

Fig. 3.5. Schematic of features common to many visible-light lasers. An optical cavity is formed by a fully silvered mirror A and a partially silvered mirror B. Within this cavity is a lasing medium C, in which population inversion may be achieved by pumping with a suitable energy source. If the lasing condition is fulfilled, laser light D will exit from the cavity.

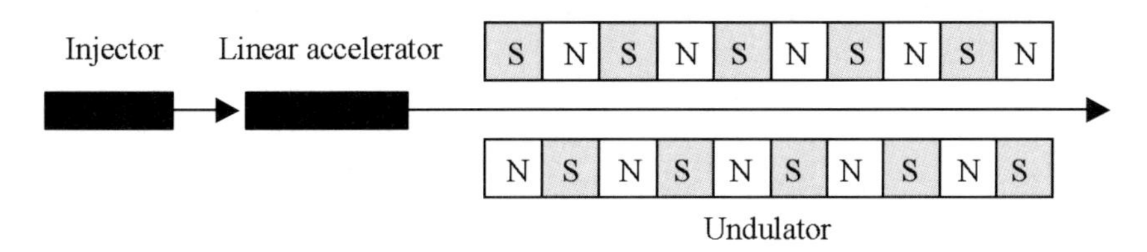

Fig. 3.6. Generic setup for a free-electron laser. 'Free' electrons, so called because they are not bound to atoms in a lasing medium, are injected into a linear accelerator. This provides a low-emittance source of energetic electrons which subsequently enters an undulator.

a laser beam D may be caused to exit from the cavity.

Having covered the above preliminaries, we are now ready to consider the free-electron laser (Madey 1971). Figure 3.6 shows an injector-fed linear accelerator, which provides a source of energetic electrons that subsequently pass through an undulator. Evidently, this setup can function as an 'ordinary' undulator, as described in the previous sub-section. In this undulator regime, photons emitted by different electrons, via spontaneous emission of synchrotron radiation, add incoherently. However, if the emittance of the electrons from the linear accelerator is sufficiently small, and the undulator sufficiently long, then appropriate choices for the remaining experimental parameters (such as the undulator period and field strength) allow a qualitatively new regime of undulator operation. In this 'free-electron laser' regime one can realize coherent addition of photons emitted from a large number of electrons within the electron bunch passing through the undulator.

Let us explore this idea in a little more detail. Suppose a linear accelerator injects a low-emittance relativistic electron bunch into an undulator, as shown in Fig. 3.6. As the bunch travels through the periods of the magnetic lattice at the start of the undulator, photons will be spontaneously emitted as synchrotron radiation. The electromagnetic field, felt by the electrons in the bunch, will then be a combination of that due to the undulator, that due to any spontaneously emitted photons which are present, and that due to the electrons themselves.[75] With this electromagnetic field may be associated a potential, known as the 'ponderomotive potential',[76] the presence of which leads to a collective instability of the electron–photon system (Bonifacio *et al.* 1984). This instability causes the electrons to form a one-dimensional crystalline structure within the bunch, transversely slicing it up into a number of thin discs of equal thickness ('microbunching'). This structure has a period that is equal to the wavelength of a certain longitudinally travelling photon mode, which may be amplified through stimulated emission. For this amplification to occur, a resonance condition must be met—namely that the microbunched electron pulse should fall one microbunch-period behind the z-directed photon plane-wave, when one period of the magnetic array is traversed. Under this condition, there can be a net transfer of energy, from the microbunched electrons to the photon beam. This net transfer occurs through the process of stimulated emission, as described earlier. The resulting photon beam becomes exponentially amplified as it passes through the undulator, which is why we mentioned earlier that the undulator must be sufficiently

[75]In the laboratory reference frame, the undulator creates a static magnetic field. However, in a frame of reference which moves with the mean velocity of the electrons in the undulator, the undulator field is both electric and magnetic in nature. Note in this context that neither the electric nor the magnetic fields have any absolute meaning in the context of special relativity, the axioms of which are consistent with the Maxwell equations (see, for example, Jackson (1999)). Note, also, that since the photon field is identified with the electromagnetic field in quantum mechanics (see, for example, Sakurai (1967)), we can meaningfully speak of the net electromagnetic field due to the photons, the undulator, and the electron bunch.

[76]The existence of such a potential is rather natural, as we shall now argue. In the previous chapter, on the interactions of X-rays with matter, we discussed how light is influenced by its interactions with matter (e.g. via scattering, diffraction, or refraction). The existence, of such an interaction, is indicative of a coupling between light and matter. On a classical level, this coupling is quantified by the presence of source terms (charge and current densities) in the Maxwell equations, together with macroscopic properties of material media such as electrical permittivity and magnetic permeability. On a quantum-mechanical level, the coupling of electromagnetic and matter fields is quantified by the coupling of the Dirac and Maxwell equations in quantum electrodynamics. Now, if the electron and photon fields are coupled to one another, thereby allowing matter to 'push light around', then the converse must also be true: light 'pushes matter around'. Rather than viewing the photon field as being manipulated by the potential created by the matter field, which is the starting point for radiation-field optics in which radiation is manipulated using optical elements made from matter, one can view the matter field as being manipulated by the potential created by the photon field. This is the starting point for matter-field optics in which matter fields are manipulated using optics made from photons. A common example of this is the magnetic lenses in an electron microscope. Another example is the Kapitza-Dirac effect, in which a beam of electrons is diffracted from the ponderomotive potential due to a 'grating' of light which is composed of the interference pattern produced by counter-propagating electromagnetic plane waves. A third example is the effect of the photon field, on the electron bunch travelling through the undulator of a free-electron laser, as described in the main text.

long for free-electron lasing to occur. After the period of exponential growth in intensity, saturation of the photon beam is reached, and the free-electron laser beam emerges from the end of the undulator. The process, described above, is known as 'self amplified spontaneous emission' (SASE) (see, for example, Pellegrini (2001), and references therein).

For an early experimental demonstration of the free-electron laser, in the infrared regime, see Elias *et al.* (1976). In the infrared to visible regions of the electromagnetic spectrum, SASE free-electron lasing has been experimentally demonstrated for some time (see, for example, references 36–46 of Pellegrini (2001)). Indeed, many free-electron laser facilities are currently operational at infrared wavelengths. Further, SASE free-electron lasing has been achieved at vacuum ultraviolet wavelengths (see, for example, Andruszkow *et al.* (2000)). At the time of this writing no SASE free-electron lasers are operational in the hard X-ray regime, although there are such facilities currently at an advanced stage of development. The Linac[77] Coherent Light Source (LCLS), at the Stanford Linear Accelerator Center (SLAC), is likely to be one of the first hard X-ray free-electron lasers to come online (Tatchyn *et al.* 1996; Cornacchia *et al.* 2004). Scheduled to be operational by 2009, this SASE facility plans to lase at wavelengths down to 0.15 nm, yielding X-ray pulses down to 230 fs in duration. Another future facility is the European X-Ray Laser Project (XFEL). This SASE facility, scheduled to be operating by 2012, will be housed at the Deutsches Elektronen-Synchrotron (DESY) in Hamburg. It is planned to lase at wavelengths down to 0.085 nm, with pulse lengths of 100 fs or less. Other X-ray SASE free-electron lasers are also under development, including the SPring–8[78] Compact SASE Source (SCSS) in Japan (Shintake *et al.* 2004), and the Fourth Generation Light Source (4GLS) at Daresbury Laboratory in the United Kingdom (Quinn *et al.* 2004).

The peak brightness, of such SASE X-ray free-electron laser facilities as LCLS and XFEL, is expected to be some ten orders of magnitude larger than that of undulator radiation at a contemporary third-generation source. The average brightness is expected to be roughly two orders of magnitude larger, and the emittance to be around two orders of magnitude smaller (Pellegrini and Stöhr 2003). It is to be anticipated that these remarkable advances, in source coherence, will open up many new avenues for scientific research which cannot be realized at present facilities.

3.1.5 *Energy-recovering linear accelerators*

Storage rings remain the dominant contemporary source of high-brightness X-ray radiation. Such rings are typically designed to contain electron bunches which circulate for the order of tens of hours, or more, this being the expected duration of continuous beam-time for X-ray users.

However, this design criterion restricts the smallness of the electron-bunch emittance (and therefore the greatness of the source brightness) that can be

[77]Note that this is a contraction of 'linear accelerator'.

[78]This is a contraction of 'Super Photon Ring – 8 GeV'.

sustained by the facility. Indeed, consider the various vicissitudes which the electron bunch encounters as it repeatedly circulates within the storage ring: insertion devices, bending and focusing magnets, aberration-correcting magnets, electron–electron interactions, dispersion and perturbation due to the radio-frequency cavities which replenish lost energy, and emission of synchrotron radiation. Such factors both determine and limit the equilibrium properties of the electron bunch in the storage ring. Typical electron-bunch equilibration times are on the order of milliseconds, corresponding to thousands of orbits around the storage ring (Gruner 2004).

In the previous sub-section, we discussed what can now be re-interpreted as one means of addressing this limitation of electron-bunch equilibration: the free-electron laser. Since this is a single-pass device, bunch equilibration does not have time to eventuate, permitting electron emittances which are sufficiently high for SASE to occur. Having said this, we note that the process is inherently inefficient, as the spent electrons are passed into a beam dump once they have traversed the undulator. This wastes a large amount of energy, as less than 10% (and often less than 5%) of the energy of the electrons is converted into photons in the free-electron laser (Kurennoy *et al.* 2004).

There is an evident tradeoff here: the storage ring achieves beam longevity, energy efficiency, and high repetition rates at the expense of brightness, while the single-pass free-electron laser compromises beam longevity, efficiency, and high repetition rates in the pursuit of brightness.

Between these extremes lies the notion of an energy-recovering linear accelerator (ERL), developed by Tigner (1965) in the context of high-energy physics applications. The photon-producing variant of this idea, with which we are here concerned, is sketched in Fig. 3.7. This shows an injector A that feeds low-emittance electron bunches into a linear accelerator B, which energizes the bunches without overly compromising their emittance. The energized bunches then pass around a ring, containing both bending magnets C and insertion devices D, thereby producing synchrotron radiation for the users of the facility. Having passed around the ring once, the ERL is designed such that the 'used' bunches re-enter the linear accelerator 180° out of phase with the 'fresh' bunches from the injector. Thus, while the linear accelerator serves to accelerate the fresh bunches, it serves to decelerate the used bunches. The used bunches thereby transfer much of their energy back to the radio-frequency cavities in the accelerator, with this energy being available for the acceleration of fresh electron bunches from the injector. Having returned a substantial fraction of their energy for subsequent re-use, the spent low-energy electrons are then directed to a beam dump E.

The central idea of the ERL, therefore, is to recycle the energy of the electron bunches, rather than recycling the electron bunches themselves. Since individual electron bunches are used only once, the previously mentioned equilibration effects (whose millisecond timescales are typically around three orders of magnitude longer than the ring orbit time) do not pose a limitation. Notwithstanding this, the brightness available from these machines is not likely to be as high as

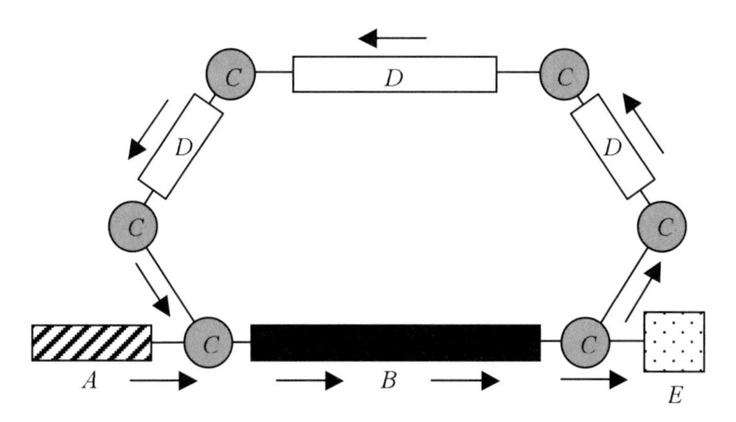

Fig. 3.7. ERL facility. Injector, A; linear accelerator, B; bending magnets, C; insertion devices, D; beam dump, E.

that obtained with single-pass free-electron lasers, at least in existing designs. However, ERLs have the advantage of higher repetition rates than single-pass free-electron lasers—in this respect, they are rather closer to modern storage rings than single-pass free-electron lasers. Further, we note that in many respects the planned and proposed ERLs will be able to support many of the experiments that are currently undertaken at third-generation sources, while providing the higher brightness and shorter pulses that will enable new experiments not feasible at the current facilities.

Early experimental demonstrations, of the ERL concept at sub-optical frequencies, include those of Smith *et al.* (1987) and Neil *et al.* (2000). As is the case with hard-X-ray free-electron lasers, however, hard-X-ray energy-recirculating technology has yet to become operational, at the time of this writing. Current ERL plans and proposals in the X-ray regime include a facility at Cornell University (Bilderback *et al.* 2003), the 4GLS suite of sources at Daresbury Laboratory (Quinn *et al.* 2004), the Multiturn Accelerator–Recuperator Source (MARS) at the Budker Institute of Nuclear Physics at Novosibirsk (Kulipanov *et al.* 1998), together with ERL sources at the Photon Factory in Tsukuba and the University of Erlangen.

3.1.6 *Soft X-ray lasers*

We close this section by very briefly treating soft X-ray lasers. As is the case for the free-electron lasers described in Section 3.1.4, soft X-ray lasers transfer energy from matter to photons via stimulated emission. A point of difference, however, is that the soft X-ray sources considered here are 'bound-electron lasers', in the sense that their electrons are contained within a lasing medium.

As is the case with visible-light lasers containing a lasing medium, the X-ray lasing medium must be 'pumped' in order to achieve a population inversion. Such inversion is a necessary condition for lasing to occur, as described in the opening paragraphs of Section 3.1.4. In soft X-ray lasers, pumping to population inversion may be achieved in a number of ways. These include: (i) illuminating the lasing medium with a sufficiently powerful pulsed laser operating at a longer wavelength (e.g. infrared or optical wavelengths) than that desired of the resulting laser beam; (ii) bombarding the medium with a rapid electric discharge.

Typically, such pumping ionizes the atoms in the lasing medium, yielding a plasma. Note that this ionization is a consequence of the facts that: (i) the excitation energy of the electrons should be on the order of the energy corresponding to the lasing wavelength; (ii) X-ray wavelengths are often sufficiently strong to ionize many atoms. This has led to some difficulty in allowing lasing-medium-based lasers to pass the threshold from the soft-X-ray to the hard-X-ray regime. Notwithstanding this, soft X-ray lasers are the subject of much ongoing research and development, with many researchers striving to push this technology into the hard X-ray domain.

3.2 Diffractive optical elements

Here, we discuss several commonly used diffractive optical elements for coherent X-ray optics. We sequentially treat diffraction gratings, Fresnel zone plates, analyser crystals, monochromators, crystal beam-splitters, interferometers, and Bragg–Fresnel crystal optics. We also consider slabs of free space to comprise a diffractive optical element.

3.2.1 *Diffraction gratings*

Consider the binary transmission grating, sketched in cross section in Fig. 3.8. The maximum and minimum projected thicknesses of the grating are taken to be A and B, respectively, with the grating period being equal to L. Note that, for simplicity, we have taken the width of the grooves to be equal to the spacing between the grooves (i.e. both quantities are equal to $\frac{1}{2}L$). With these assumptions, and adopting (x, y) Cartesian coordinates in the plane of the grating, we can write its projected thickness $T(x, y)$ as:

$$T(x, y) = \begin{cases} A, & \text{if } \sin(2\pi x/L) \geq 0, \\ B, & \text{otherwise.} \end{cases} \tag{3.7}$$

Assume the grating to be made of a single homogeneous isotropic non-magnetic material of known complex refractive index n_ω, where (see eqn (2.37)):

$$n_\omega = 1 - \delta_\omega + i\beta_\omega. \tag{3.8}$$

The real numbers δ_ω and β_ω respectively quantify the refractive and absorptive properties of the material, as a function of the angular frequency ω of the X-rays. Note that β_ω is related to the linear attenuation coefficient μ_ω which appears in Beer's law of absorption (eqn (2.43)), by the relation $\mu_\omega = 2k\beta_\omega$ (see eqn (2.44)).

Next, suppose that a z-directed monochromatic scalar X-ray plane wave $\mathscr{D}\exp(ikz)$ is normally incident upon our single-material grating, with z being a nominal optic axis perpendicular to the xy plane. Here, $\mathscr{D}$ is a complex

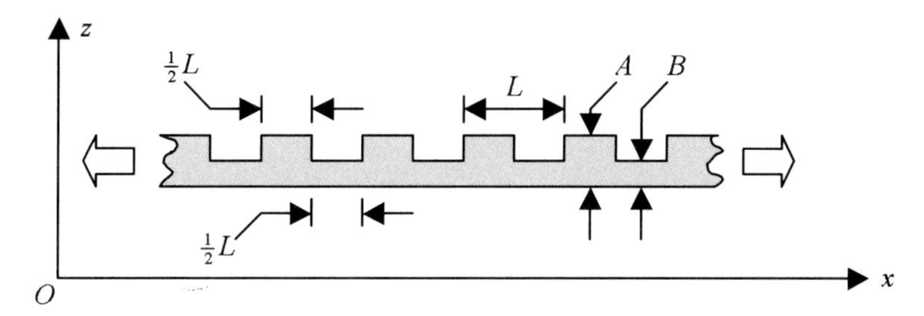

Fig. 3.8. Simple two-level binary diffraction grating, in side view. Maximum and minimum projected thicknesses are given by A and B, respectively. The grating period is L, with the width of grooves and peaks being equal.

constant whose squared modulus is equal to the intensity of the incident plane wave, and $k = 2\pi/\lambda$ is the usual wave-number associated with monochromatic scalar radiation of wavelength λ. Let the plane $z = 0$ coincide with the exit surface of the grating. Assume the grating to be sufficiently thin, and the grating period to be sufficiently long with respect to the wavelength of the radiation, for the projection approximation to be valid (see Section 2.2). Then the spatial part $\psi_\omega(x, y, z = 0)$ of the complex disturbance, at the exit surface of the grating, will have both an amplitude and phase that is proportional to the projected thickness $T(x, y)$:

$$\psi_\omega(x, y, z = 0) = \mathscr{D}\sqrt{\exp[-\mu_\omega T(x, y)]}\exp[-ik\delta_\omega T(x, y)]. \tag{3.9}$$

The right side, of the above equation, is a product of three quantities: (i) $\mathscr{D}$ reflects the fact that the exit-surface wave-field is proportional to the complex amplitude of the incident plane wave; (ii) $\sqrt{\exp[-\mu_\omega T(x, y)]}$ is indicative of Beer's law of absorption, in the projection approximation, with the square root being present because we are here working with amplitudes rather than with intensities (the reader will recall that the latter is identified with the squared modulus of the former); (iii) the phase $-k\delta_\omega T(x, y)$ of the third term, which for convenience is taken relative to the uniform exit-surface phase that would have existed had no grating been present, is given by eqn (2.41).

Making use of eqns (2.44) and (3.7), we see that eqn (3.9) becomes:

$$\psi_\omega(x, y, z = 0) = \begin{cases} \mathscr{D}\exp[-kA(\beta_\omega + i\delta_\omega)], & \text{if } \sin(2\pi x/L) \geq 0, \\ \mathscr{D}\exp[-kB(\beta_\omega + i\delta_\omega)], & \text{otherwise.} \end{cases} \tag{3.10}$$

At this point, it serves us to recall the Fourier-series decomposition of a square wave. This square wave $\Upsilon(x)$, of period L, may be defined by:

$$\Upsilon(x) = \text{sgn}\left[\sin\left(\frac{2\pi x}{L}\right)\right], \tag{3.11}$$

where the sign function 'sgn' returns a value of +1 if its argument is positive or zero, and −1 if its argument is negative. The Fourier-series decomposition of this square wave, derived in most elementary texts on Fourier analysis, is:

$$\Upsilon(x) = \frac{4}{\pi}\sum_{m=1}^{\infty}\frac{\sin[2\pi(2m-1)x/L]}{2m-1}. \tag{3.12}$$

The above Fourier series takes on the value +1 when $\sin(2\pi x/L) \geq 0$, being equal to −1 otherwise. Now, the ordered pair $\{-1, 1\}$ is changed into the ordered pair $\{c_1, c_2\}$, where c_1 and c_2 are arbitrary complex numbers, if one applies the following sequence of operations to each member of the first pair: add 1, multiply by $\frac{1}{2}(c_2 - c_1)$, then add c_1. We conclude that:

$$\begin{aligned}\left\{[\Upsilon(x)+1]\times\frac{1}{2}(c_2-c_1)\right\}+c_1 &= \frac{1}{2}\left[(c_2-c_1)\Upsilon(x)+c_1+c_2\right]\\ &= \begin{cases} c_2, & \text{if } \sin(2\pi x/L)\geq 0,\\ c_1, & \text{otherwise.}\end{cases}\end{aligned} \tag{3.13}$$

After our diversion on the Fourier series of square waves, we are ready to return to the main thread of the argument. To this end, note that if we choose:

$$c_1 = \mathscr{D}\exp[-kB(\beta_\omega + i\delta_\omega)], \qquad c_2 = \mathscr{D}\exp[-kA(\beta_\omega + i\delta_\omega)], \tag{3.14}$$

then the right-hand sides of eqns (3.10) and (3.13) are identical. Therefore, their left sides are identical, so that:

$$\psi_\omega(x,y,z=0) = \frac{1}{2}[(c_2-c_1)\Upsilon(x)+c_1+c_2]. \tag{3.15}$$

Making use of eqn (3.12), the above equation becomes the following Fourier-series decomposition for the wave-field at the exit surface of our grating:

$$\psi_\omega(x,y,z=0) = \frac{c_1+c_2}{2} + \frac{2(c_2-c_1)}{\pi}\sum_{m=1}^{\infty}\frac{\sin[2\pi(2m-1)x/L]}{2m-1}, \tag{3.16}$$

with c_1 and c_2 being given by eqn (3.14). Next, recall the complex representation of the sine function:

$$\sin(x) = \frac{1}{2i}[\exp(ix) - \exp(-ix)], \tag{3.17}$$

allowing us to re-write eqn (3.16) as:

$$\psi_\omega(x,y,z=0) = \frac{c_1+c_2}{2} + \frac{c_2-c_1}{\pi i}\sum_{m=1}^{\infty}\frac{\exp[2i\pi(2m-1)x/L]}{2m-1}$$
$$- \frac{c_2-c_1}{\pi i}\sum_{m=1}^{\infty}\frac{\exp[-2i\pi(2m-1)x/L]}{2m-1}. \quad (3.18)$$

Having obtained the above expression, for the complex amplitude of the optical wave-field at the exit surface $z=0$ of the grating, our next task is to propagate this solution into the vacuum-filled half-space $z \geq 0$ downstream of the grating. This propagated solution must obey the Helmholtz equation (1.16), in that half space. To evaluate the propagated field, we use a method that is directly related to the discussions of Section 1.3, in which we treated the angular-spectrum formalism for the propagation of monochromatic complex scalar wave-fields.

The elementary plane wave $\exp[i(k_x x + k_z z)]$, with wave-vector $(k_x, k_y = 0, k_z)$, solves the Helmholtz equation if:

$$k_x^2 + k_z^2 = k^2. \quad (3.19)$$

This equation can be solved for k_z, with the positive square root being chosen on account of the fact that there are no sources in the half space into which the field propagates. Therefore:

$$k_z = \sqrt{k^2 - k_x^2}, \quad (3.20)$$

so that our elementary plane wave becomes $\exp[i(k_x x + z\sqrt{k^2-k_x^2})]$. The boundary value of this elementary plane wave, over the surface $z=0$, is $\exp(ik_x x)$; to propagate the boundary value into the half-space $z \geq 0$, one need only multiply by the free-space propagator[79]:

$$\exp(ik_z z) = \exp[iz\sqrt{k^2-k_x^2}]. \quad (3.21)$$

Now, eqn (3.18) expresses the exit-surface wave-field of the grating as a linear superposition of such boundary values (i.e. over the surface $z=0$) of the elementary plane waves $\exp[i(k_x x + k_z z)]$. Since the Helmholtz equation is linear, we may propagate $\psi_\omega(x,y,z=0)$ into the vacuum-filled half-space $z \geq 0$ by multiplying each of the summed boundary values by the appropriate propagator. Bearing all of this in mind, one may immediately write down the following formula for the propagated field $\psi_\omega(x,y,z\geq 0)$ in the half-space downstream of the illuminated diffraction grating:

[79]See the lines of text immediately after eqn (1.21), noting that $k_y = 0$ in the present context.

$$\psi_\omega(x,y,z\geq 0)=\frac{(c_1+c_2)\exp(ikz)}{2}$$
$$+\frac{c_2-c_1}{\pi i}\sum_{m=1}^{\infty}\frac{\exp[2i\pi(2m-1)x/L]}{2m-1}\exp\left[iz\sqrt{k^2-4\pi^2\left(\frac{2m-1}{L}\right)^2}\right]$$
$$-\frac{c_2-c_1}{\pi i}\sum_{m=1}^{\infty}\frac{\exp[-2i\pi(2m-1)x/L]}{2m-1}\exp\left[iz\sqrt{k^2-4\pi^2\left(\frac{2m-1}{L}\right)^2}\right].\tag{3.22}$$

Known as the 'zeroth-order diffracted beam', the first term on the right side of this equation represents an attenuated and phase-delayed form of the incident plane wave. Note that this zeroth-order beam will vanish if $c_1+c_2=0$, a point we shall return to later in the context of grating efficiency.

The second term, on the right side of eqn (3.22), represents a series of plane waves which are mutually inclined with respect to one another. Each such plane wave corresponds to a particular positive diffracted order, labelled by one of the integers $m=1,2,3,\cdots$ in the summation.

Consider the $m=1$ term, namely the first diffracted order. This is given by the plane wave:

$$\frac{c_2-c_1}{\pi i}\exp\left(\frac{2i\pi x}{L}\right)\exp\left(2\pi iz\sqrt{\frac{1}{\lambda^2}-\frac{1}{L^2}}\right),\tag{3.23}$$

where we have made use of the fact that $k=2\pi/\lambda$, with λ equal to the radiation wavelength. This $m=1$ order will be a propagating (rather than an evanescent) plane wave if the argument of the above square root is positive, which will be the case if the grating period L is longer than the wavelength of the radiation. Since we assumed $L\gg\lambda$ in making the projection approximation near the beginning of this calculation, the $m=1$ order will be propagating. This diffracted order propagates with respect to the optic axis at an angle $\theta_{m=1}$ given by:

$$\theta_{m=1}=\tan^{-1}\left(\frac{k_x}{k_z}\right)=\tan^{-1}\left(\frac{2\pi/L}{2\pi\sqrt{\lambda^{-2}-L^{-2}}}\right)=\tan^{-1}\left(\frac{\lambda}{\sqrt{L^2-\lambda^2}}\right).\tag{3.24}$$

Note, also, that $L\gg\lambda$ implies that eqn (3.24) has the approximate form:

$$\theta_{m=1}\approx\frac{\lambda}{L},\quad L\gg\lambda.\tag{3.25}$$

Thus, the diffraction angle of the first order is very small compared to unity, if the period of the diffraction grating is very much longer than the wavelength of the illuminating radiation.

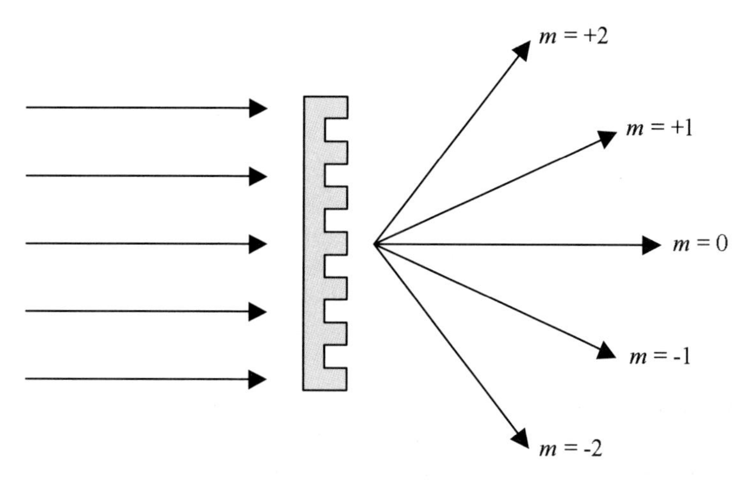

Fig. 3.9. When monochromatic plane waves are incident upon a ruled transmission grating, a series of diffracted orders results.

The angles $\theta_{m=2}, \theta_{m=3}, \cdots$, made by the other propagating positive-order diffracted beams with respect to the optic axis, can be determined in a similar manner to that employed in the previous paragraph. As the reader can easily show, the propagating negative diffracted orders—as given by each of the summed terms in the bottom line of eqn (3.22), with m being sufficiently small in modulus for the argument of the square root to be real—have propagation angles which are the negative of those for the corresponding positive diffracted orders, so that $\theta_{-m} = -\theta_m$. Specifically, one has the following generalization of eqn (3.24), for the angles θ_m made by the various diffracted orders with respect to the zeroth-order beam:

$$\theta_m = \operatorname{sgn}(m) \tan^{-1}\left(\frac{(2|m|-1)\lambda}{\sqrt{L^2 - (2|m|-1)^2\lambda^2}}\right), \quad m = \pm 1, \pm 2, \cdots, \quad L > (2|m|-1)\lambda. \quad (3.26)$$

In light of the preceding analysis, it is clear that the diffraction grating can function as a beam splitter, as indicated in Fig. 3.9. Here, we see the normally incident monochromatic beam being split into the various diffracted orders, upon passage through the grating.

In addition to functioning as a beam splitter, eqn (3.26) implies that the transmission grating can function as a spectrometer. This is because the angle, which a given order of diffraction makes with the optic axis, depends on the wavelength (and therefore upon the energy) of the radiation diffracted into that order. Note that such grating-based X-ray spectroscopy is typically restricted to the soft end of the X-ray spectrum, using reflection rather than transmission geometries, with crystals typically being used for spectroscopy at the harder energies.

Next, let us examine the relative intensities of the various diffracted orders, produced when monochromatic scalar plane waves normally illuminate the simple transmission grating sketched in Fig. 3.8. In this context, we introduce the efficiency χ_m of the grating in diffracting the incident plane wave into the mth order. This efficiency is defined as the ratio of the squared modulus of the complex amplitude of the mth diffracted order, to the squared modulus of the complex amplitude of the incident beam. For our grating, eqn (3.22) allows us to write down the efficiency:

$$\chi_m = \left| \frac{c_2 - c_1}{(2|m|-1)\pi i} \div \mathscr{D} \right|^2 = \frac{|c_2 - c_1|^2}{\pi^2 (2|m|-1)^2 |\mathscr{D}|^2}, \tag{3.27}$$

where $m = \pm 1, \pm 2, \cdots$, and the condition $L > (2|m|-1)\lambda$ restricts the above formula to propagating diffracted orders. Making use of eqn (3.14), the above result becomes:

$$\chi_m = \frac{|\exp[-kA(\beta_\omega + i\delta_\omega)] - \exp[-kB(\beta_\omega + i\delta_\omega)]|^2}{\pi^2(2|m|-1)^2}, \\ m = \pm 1, \pm 2, \cdots, \quad L > (2|m|-1)\lambda. \tag{3.28}$$

For our simple two-level ruled grating, the most intense of the diffracted orders occurs for $m = \pm 1$.

Having introduced the notion of diffraction efficiency χ_m, and analytically evaluated it for the simple grating model being considered here, it is natural to enquire as to how this quantity can be maximized. With reference to eqn (3.28), we see that the relevant variables in the required optimization are A, B, δ_ω, and β_ω.[80] Let us give a geometric argument for how this optimization may be achieved.

For any given m, the numerator of eqn (3.28) tells us that we need to maximize the modulus of the difference between the complex numbers $\exp[-kA(\beta_\omega + i\delta_\omega)]$ and $\exp[-kB(\beta_\omega + i\delta_\omega)]$. Note that each of these complex numbers has a modulus which is less than or equal to unity. As can be readily seen by drawing these complex numbers as points in the complex plane, each of which must lie

[80]The grating period L is only of implicit relevance, insofar as it must be sufficiently large with respect to the radiation wavelength λ, for a given diffracted order to exist as a propagating rather than as an evanescent wave. Since the $m = \pm 1$ diffracted orders contain the most energy in the model being considered here, it suffices to ensure that L is sufficiently large that these first orders be propagating rather than evanescent. The condition for this, as we saw earlier, is that the grating period be longer than the wavelength of the illuminating radiation. This is consistent with the requirement that $L \gg \lambda$, upon which was predicated the projection-approximation expression (3.9) for the exit-surface wave-field of the normally illuminated grating.

within a circle of unit radius, the modulus of the difference will be maximized if the following two conditions simultaneously hold: (i) the modulus of both $\exp[-kA(\beta_\omega + i\delta_\omega)]$ and $\exp[-kB(\beta_\omega + i\delta_\omega)]$ should be maximal, and therefore each of these moduli should be equal to unity; (ii) the phase of $\exp[-kA(\beta_\omega+i\delta_\omega)]$ and $\exp[-kB(\beta_\omega + i\delta_\omega)]$ should differ by an odd integer multiple of π, so that the corresponding position vectors (phasors) in the complex plane are pointing in opposite directions. The first condition implies that:

$$kA\beta_\omega = kB\beta_\omega = 0. \tag{3.29}$$

Since we cannot have $A = B = 0$, which would imply the absence of grooves in the grating, we must have:

$$\beta_\omega = 0. \tag{3.30}$$

This implies the grating to be non-absorbing. Such a grating is known as a 'phase grating'. The second condition for maximum efficiency can be written as:

$$-kA\delta_\omega = -kB\delta_\omega + (2n+1)\pi, \tag{3.31}$$

where n is some integer. Bearing eqn (2.41) in mind, we see that this has a simple physical interpretation: the phase shift of the X-rays on passing through the thin part of the grating, and the phase shift on passing through the thick part of the grating, should differ by an odd multiple of π. With both of the above conditions met, the numerator of eqn (3.28) is equal to $|1-(-1)|^2 = 4$, yielding the maximal efficiency:

$$\chi_m^{(\text{MAX})} = \frac{4}{\pi^2(2|m|-1)^2}, \quad m = \pm 1, \pm 2, \cdots, \quad L > (2|m|-1)\lambda. \tag{3.32}$$

For the $m = \pm 1$ orders, this efficiency obtains its maximum value of $4/\pi^2$, which is about 41%. Further, the zeroth order vanishes when this maximum efficiency is attained, as can be seen by substituting eqns (3.30) and (3.31) into (3.14), and then inserting the result into (3.22).

Having seen that maximal efficiency is obtained for the case of a phase grating, we briefly consider the maximal efficiency which can be obtained for an absorptive grating, under the simple model being considered here. Such a grating is, by definition, one in which $B = 0$, with A being sufficiently thick—and the material being sufficiently absorbing—that we can approximate it as being completely absorbing. Using a simple variation on the geometric argument given above, one can show that the maximal efficiency is one-quarter of that for the corresponding phase grating case. In particular, the maximal efficiency, for diffraction into the $m = \pm 1$ orders, is now $1/\pi^2 \approx 10\%$.

Some remarks: (i) We have considered the question of maximal efficiency within the simple model for a single-material two-level grating as sketched in

Fig. 3.8, under the projection approximation. Less-restrictive models permit efficiencies in excess of those discussed here. For example, the efficiency can be improved by having more than two levels in the grating. As a popular example of this, due to Rayleigh, one can have so-called 'blazed' gratings whose projected thickness has a saw-tooth shape rather than that of a square wave. (ii) For normally incident plane-wave illumination, the location of the various diffracted orders is unchanged by changing the profile of the periodic grating, while keeping the period fixed, whether or not the projection approximation is valid. (iii) Gratings can also be operated in reflection rather than transmission, using glancing angles of incidence. (iv) For non-plane-wave monochromatic incident waves, the above formulae can be suitably modified by first considering their generalization to the case of non-normal plane-wave incidence, and then expressing the incident field as a suitable Fourier integral. The case of polychromatic illumination, including pulsed beams which are sufficiently weak to not appreciably modify the structure of the grating, can also be treated upon the introduction of suitable spectral sums. (v) For a recent study of a soft X-ray transmission grating, which may be read as an experimental case study highlighting several of the concepts discussed in this section, see, for example, Desauté *et al.* (2000).

3.2.2 *Fresnel zone plates*

With a view to designing a diffractive lens for focussing X-rays, consider the construction shown in Fig. 3.10. Here, we see a thin diffractive optical element which is illuminated by normally incident monochromatic scalar plane waves. We wish this device, known as a Fresnel zone plate, to focus the incident radiation to a point C that lies at a distance f downstream of it.

Before proceeding with the analysis, let us make an educated guess as to what this device should look like. Under normal plane-wave incidence, we wish the diffracted field to have rotational symmetry; therefore, the element itself should be rotationally symmetric.[81] Also, in the simplest incarnation of these devices, let us assume them to be: (i) composed of a single non-magnetic isotropic material of constant density; (ii) binary in structure, so that only two values for the projected thickness are allowed. Thus, we may take our simple zone plate to be characterized by its projected thickness $T(r)$, which is a binary function of the radial distance r from the centre of the plate.

At this point, let us recall the discussions of the previous sub-section. There we saw that the simple linear diffraction grating, sketched in Fig. 3.9, was able to diffract incident plane waves through a variety of different angles. The diffraction angle $\theta_{m=1}$, of the first diffracted order under illumination by a normally incident plane-wave beam of wavelength λ, was seen to increase as the period L of the grating decreases; specifically, we saw that $\theta_{m=1} \approx \lambda/L$, when $L \gg \lambda$. Returning to Fig. 3.10 once more, we see that the zone plate deflects X-rays through different angles, depending on their distance r from the optic axis. This immediately suggests that the zone plate will be a rotationally symmetric grating whose

[81] Compare, however, the discussion on 'photon sieves' near the end of this sub-section.

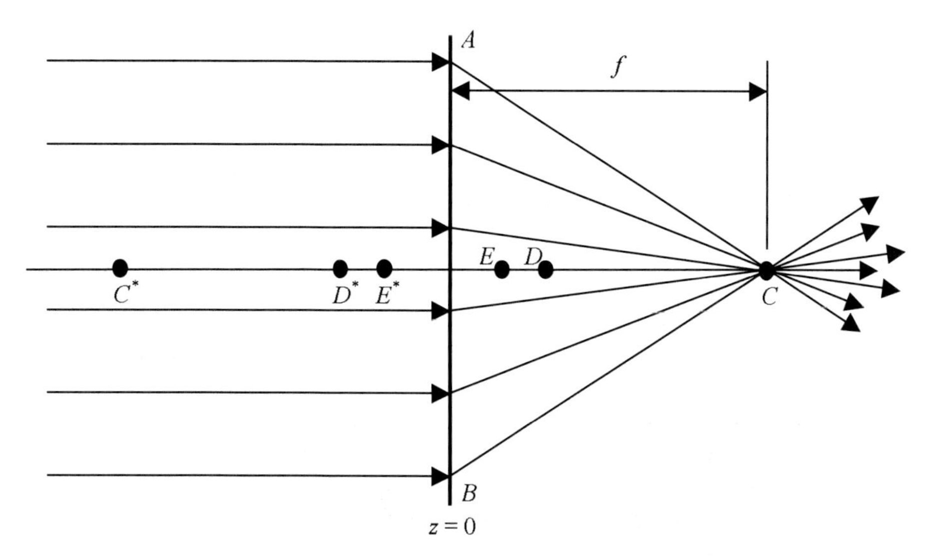

Fig. 3.10. A Fresnel zone plate AB lies in the plane $z = 0$. This thin diffractive optical element is rotationally symmetric about the optic axis z. It is designed to focus X-rays to a point C, when illuminated with normally incident monochromatic plane waves. The focal length of the zone plate is the distance f between C and the centre of the zone plate. Additional foci lie at distances $\pm f/3$, $\pm f/5$, etc. Two such real foci are shown at D and E, together with three virtual foci at C^*, D^*, and E^*.

period decreases with increasing r, so that rays that are further from the optic axis are diffracted through larger angles. These diffraction angles should be such that X-rays are brought to a focus at C. Further, this circular diffraction grating should be such that the focussed radiation constructively interferes at C.

The above essential features of zone plates—namely that they are rotationally symmetric circular diffraction gratings whose period decreases with increasing radial distance from the centre of the device, such that they are able to function as a focussing element for electromagnetic radiation—should be borne in mind during the development which follows. Our analysis is somewhat unorthodox, insofar as it relies on what is essentially a holographic argument (cf. Rogers (1950)), although we shall not assume any prior knowledge of in-line holography.[82] This treatment may be contrasted with the more usual development based on the notion of Fresnel zones, about which more will be said later.

Consider Fig. 3.10 once more, but with the zone plate removed. Suppose that a point scatterer is placed at C^*, this being the mirror image of the desired focal point C, obtained by reflecting the latter point through the plane $z = 0$. With the exception of the scatterer at C^*, the space is considered to be filled with vacuum.

[82]Having said this, we note that in-line holography is treated in Section 4.3.1.

When z-directed plane waves $\mathscr{D}\exp(ikz)$ illuminate the point scatterer, spherical waves emerging from C^* will result (cf. Fig. 2.3). Denote these scattered waves by $\mathscr{E}\exp(ikR)/R$, where $\mathscr{E}$ is a complex constant whose modulus is indicative of the strength of the scatterer, R denotes radial distance from the point C^*, and k is the usual wave-number corresponding to X-ray radiation of wavelength λ and angular frequency ω. The resulting wave-field, $\psi_\omega(x,y,z)$, is given by the coherent superposition of the incident plane wave and the scattered spherical wave (cf. eqn (2.123)):

$$\psi_\omega(x,y,z) = \mathscr{D}\exp(ikz) + |\mathscr{E}|\frac{\exp\{ik[R(x,y,z)+\varpi(\mathscr{E})]\}}{R(x,y,z)}. \tag{3.33}$$

Here, we have written the scattering amplitude $\mathscr{E}$ as:

$$\mathscr{E} = |\mathscr{E}|\exp[ik\varpi(\mathscr{E})], \tag{3.34}$$

serving to define $\varpi(\mathscr{E})$ as the phase of $\mathscr{E}$, divided by k.

Let $r = \sqrt{x^2+y^2}$ denote radial distance in cylindrical polar coordinates, measured from an optic axis that pierces the centre of the zone plate. In the plane $z=0$ the disturbance may then be written as:

$$\psi_\omega(r, z=0) = \mathscr{D} + |\mathscr{E}|\frac{\exp\{ik[\sqrt{f^2+r^2}+\varpi(\mathscr{E})]\}}{\sqrt{f^2+r^2}}. \tag{3.35}$$

The corresponding intensity $I_\omega(r, z=0)$, over the plane $z=0$, is obtained by taking the squared modulus of the above expression. Thus:

$$I_\omega(r, z=0) = |\mathscr{D}|^2 + \frac{|\mathscr{E}|^2}{f^2+r^2} + \frac{2|\mathscr{D}\mathscr{E}|}{\sqrt{f^2+r^2}}\cos[k(\sqrt{f^2+r^2}+\varpi)], \tag{3.36}$$

where we have introduced the following constant, which is indicative of the phase difference between the incident plane wave and the scattered spherical wave:

$$\varpi \equiv \varpi(\mathscr{E}) - \frac{\mathrm{Arg}(\mathscr{D})}{k}. \tag{3.37}$$

The function $I_\omega(r)$ may be interpreted as the interference pattern formed due to the coherent superposition of the incident plane wave $\mathscr{D}\exp(ikz)$ and the scattered spherical wave $\mathscr{E}\exp(ikR)/R$. In the language of holography, we would speak of this intensity distribution as the 'in-line hologram' which results when the planar 'reference wave' interferes with the 'object wave' scattered by the point lying at C^*. For more on this subject, we refer the reader to the discussion on holography in the next chapter.

We may now note the following remarkable fact, which shall emerge from the analysis of the following paragraphs: if one creates a thin transmissive mask, based on a binarized version of the interferogram in eqn (3.36), subsequently illuminating this mask with normally incident plane waves of the same wavelength as was used to compute the interferogram, then the resulting wave-field will form a real image of the scatterer at the mirror-image point C. This real image—of the spherical wave expanding away from the point C^* in Fig. 3.10—is evidently a spherical wave collapsing upon the point C. Stated differently, the aforementioned transmissive mask is a diffractive lens that focuses incident plane waves to the point C. It therefore coincides with the desired Fresnel zone plate, the profile of which it is the object of these discussions to determine.

Bearing the above ideas in mind, let us binarize the interferogram in eqn (3.36). The first two terms, on the right side of this expression, will be considered to be sufficiently slowly varying that they may be safely ignored. We therefore need only binarize the third term on the right side of this equation, so that the projected thickness $T(r)$ of the desired zone plate is equal to either A or B, according to the following prescription:

$$T(r) = \begin{cases} A, & \text{if } \cos[k(\sqrt{f^2+r^2}+\varpi)] \geq 0, \\ B, & \text{otherwise.} \end{cases} \tag{3.38}$$

Next, we suppose this binarized zone plate to lie in the plane $z = 0$ in Fig. 3.10, being illuminated by the normally incident plane wave $\mathscr{D}\exp(ikz)$. Assume the zone plate to be sufficiently thin and slowly varying for the projection approximation (see Section 2.2) to hold good. Then the complex disturbance $\psi_\omega(x, y, z = 0)$, at the exit surface of the illuminated zone plate, will have both an amplitude and phase which is proportional to its projected thickness $T(x, y)$:

$$\psi_\omega(r, z=0) = \begin{cases} \mathscr{D}\exp[-kA(\beta_\omega + i\delta_\omega)] \equiv c_2, & \text{if } \cos[k(\sqrt{f^2+r^2}+\varpi)] \geq 0, \\ \mathscr{D}\exp[-kB(\beta_\omega + i\delta_\omega)] \equiv c_1, & \text{otherwise.} \end{cases} \tag{3.39}$$

Note that the definitions for c_1 and c_2 are the same as those that appear in eqn (3.14), in the context of linear diffraction gratings.

Our next step is to decompose the exit-surface wave-field in the above expression, in a manner corresponding to the various diffracted orders which emerge from the zone plate. To this end, compare the above formula to eqn (3.10) of the previous sub-section, to see that the latter may be converted into the former via the replacement:

$$x \rightarrow \frac{kL(\sqrt{f^2+r^2}+\varpi)}{2\pi} + \frac{L}{4}. \tag{3.40}$$

Taking eqn (3.18), and making the above replacement on the right hand side, we obtain the desired decomposition of the wave-field $\psi_\omega(r, z = 0)$, at the exit-surface of the illuminated zone plate:

$$\begin{aligned}\psi_\omega(r, z=0) =& \mathscr{A}(r)\frac{c_1+c_2}{2} \\ &+ \mathscr{A}(r)\frac{c_2-c_1}{\pi i}\sum_{m=1}^{\infty}\frac{\exp\{i(2m-1)[k(\sqrt{f^2+r^2}+\varpi)+(\pi/2)]\}}{2m-1} \\ &- \mathscr{A}(r)\frac{c_2-c_1}{\pi i}\sum_{m=1}^{\infty}\frac{\exp\{-i(2m-1)[k(\sqrt{f^2+r^2}+\varpi)+(\pi/2)]\}}{2m-1}.\end{aligned} \tag{3.41}$$

In writing down this expression, we have introduced the aperture function $\mathscr{A}(r)$, to take into account the finite radius r_{MAX} of the zone plate. By definition, this aperture function is equal to unity if $r \leq r_{\text{MAX}}$, being equal to zero otherwise.

We now separately consider each of the three terms on the right side of the above equation. (i) The first such term is the zeroth diffracted order of the zone plate. This corresponds to an attenuated version of the plane wave which is incident upon it. (ii) The second term consists of a sum of diffracted orders. The first of these, namely:

$$\mathscr{A}(r)\frac{(c_2-c_1)\exp(ik\varpi)}{\pi}\exp(ik\sqrt{f^2+r^2}), \tag{3.42}$$

is an expanding truncated spherical wavefront, evaluated over the plane $z = 0$ and apparently emanating from the point C^* in Fig. 3.10. The remaining terms, in the first summation on the right-hand side of eqn (3.41), correspond to distorted truncated spherical waves apparently emerging from the points D^*, E^* etc., in the figure. (iii) Like the second term, various diffracted orders are present in the third term on the right side of eqn (3.41). The first of these is:

$$\mathscr{A}(r)\frac{(c_2-c_1)\exp(-ik\varpi)}{\pi}\exp(-ik\sqrt{f^2+r^2}), \tag{3.43}$$

this being an apertured collapsing spherical wavefront converging towards the point C. It is this diffracted order which indicates that the zone plate performs its desired function as a diffractive X-ray lens of focal length f. The remaining diffracted orders, in the third term on the right-hand side of eqn (3.41), correspond to higher-order foci, namely distorted apertured spherical waves converging on the points D, E, etc.

In most contexts involving X-ray focussing by zone plates, the focal length f will be much larger than the radius r_{MAX} of the plate. This permits the following binomial approximation, to the complex exponentials appearing in eqn (3.41):

$$\sqrt{f^2+r^2}+\varpi = f\sqrt{1+\left(\frac{r}{f}\right)^2}+\varpi \approx f\left[1+\frac{1}{2}\left(\frac{r}{f}\right)^2\right]+\varpi = \frac{r^2}{2f}+f+\varpi. \tag{3.44}$$

Now, there is evidently some freedom in the choice for ϖ. To keep our formulae as simple as possible, we take:

$$\varpi = -f. \tag{3.45}$$

Inserting this choice into the binomial approximation (3.44), we see that the paraxial form of eqn (3.41) is:

$$\begin{aligned}\psi_\omega(r, z=0) =& \mathscr{A}(r)\frac{c_1+c_2}{2} + \mathscr{A}(r)\frac{c_2-c_1}{\pi i}\sum_{m=1}^{\infty}\frac{i^{2m-1}}{2m-1}\exp\left\{i\frac{kr^2}{2[f/(2m-1)]}\right\}\\ &- \mathscr{A}(r)\frac{c_2-c_1}{\pi i}\sum_{m=1}^{\infty}\frac{(-i)^{2m-1}}{2m-1}\exp\left\{-i\frac{kr^2}{2[f/(2m-1)]}\right\},\\ &f \gg r_{\mathrm{MAX}}.\end{aligned} \tag{3.46}$$

Thus, the zone plate has foci at odd-integer multiples of the first-order focal length f, at the following points on the optic axis:

$$z = \pm f, \pm\frac{f}{3}, \pm\frac{f}{5}, \cdots. \tag{3.47}$$

Next we pass onto questions of efficiency, in a discussion which very closely parallels that given in our earlier treatment of linear diffraction gratings. As was the case there, we define the efficiency χ_m, of the mth order of the zone plate, to be the ratio of the squared modulus of the mth diffracted order of the beam, to the squared modulus of the incident beam. In the calculation of such efficiencies, the squared moduli of the diffracted beams are calculated over the exit surface of the zone plate. The resulting efficiencies may be written down upon inspection of eqn (3.41), these being identical to those which were obtained earlier, based on eqn (3.27). In particular, we see that the efficiency, of diffraction into the first order ($m = 1$) of the zone plate, can be maximized using an identical geometric argument to that given previously. The efficiency again attains the maximum value of $4/\pi^2 \approx 40\%$, if the following two conditions hold: (i) the zone-plate is non-absorbing; (ii) the phase difference, between radiation traversing the thick and thin parts of the zone plate, is an odd integer multiple of π radians.[83] For such zone plates, which are known as 'Rayleigh–Wood' zone plates, the diffraction efficiency for the mth order is $4/[\pi^2(2|m|-1)^2]$ (see eqn (3.32)). All

[83] This observation is closely related to the notion of 'Fresnel zones'. Such zones are formed by the concentric rings of the zone plate, together with the central disc. This central disc is the first zone, the surrounding ring is the second zone, and so forth. Note that many workers consider the thick and thin parts of the zone plate to be counted separately, as will be the case here. As can be readily shown using a simple ray-based picture, the second condition (as stated in the main text) ensures that there is no destructive interference between the radiation emerging from the various zones, at the location of the desired focal spot.

of these efficiencies are reduced by a factor of 4, for two-level 'absorption' zone plates whose minimum thickness is zero, and whose maximum thickness may be taken to completely absorb the incident radiation. In particular, for absorption zone plates the efficiency, for diffraction into the first-order focus, is $1/\pi^2 \approx 10\%$.

These efficiencies can be further improved if one goes beyond the simple two-level model adopted here, for example, via multi-level or blazed zone plates. Indeed, it is rather simple—in principle, but not in practice—to design a zone plate with perfect efficiency, under the projection approximation.[84] Such a 'kinoform' zone plate[85] (see, for example, Nazmov *et al.* (2004), and references therein) is composed of a negligibly absorbing material, of projected thickness $T(r)$ given by (cf. eqn (3.97)):

$$T(r) = \left(\frac{\sqrt{f^2 + r^2} - f}{\delta_\omega} \right)_{\mathrm{mod T}_{2\pi}} . \tag{3.48}$$

Here, 'mod $T_{2\pi}$' indicates that the thickness should be taken modulo the thickness of material $T_{2\pi}$, which imparts a phase shift of 2π on monochromatic X-rays passing through it.[86] Indeed, the Rayleigh–Wood zone plate may be seen as a two-level approximation to the above formula, an approximation which carries the twin corollaries of introducing additional diffracted orders, and reducing the efficiency with which radiation is directed into the nominal focus of the device.

X-ray zone plates may be fabricated using a variety of materials including gold, nickel, carbon, silicon, molybdenum, and aluminium. As an example of a modern device, consider the gold Fresnel zone plate shown in Fig. 3.11, taken from the study by Di Fabrizio and colleagues (Di Fabrizio *et al.* 1999). Fabricated using electron-beam lithography, this four-level zone plate focuses X-rays in the range from 5–8 keV. It has a diameter of 150 μm, with the width of the outermost ring being 500 nm. At 8 keV, the focal length of the zone plate is 1 m. The efficiency, for diffraction into the first order by the gold four-level zone plate, was measured to be 38%. Using a nickel four-level zone plate with identical diameter and zone boundaries to the gold plate, an efficiency of 55% was measured, for diffraction into the first-order focus. Note that the latter figure is higher than the theoretical maximum of $4/\pi^2 \approx 40\%$, which we calculated for the two-level phase zone plate.

Some remarks: (i) Fresnel zone plates, as described here, are evidently designed to work with radiation whose spectral bandwidth is sufficiently narrow. If the incident radiation is polychromatic then these devices will suffer from chromatic aberration, that is, the focal length will depend upon wavelength. A

[84]Note that the projection approximation will break down at the infinitely sharp edges of the mask given in eqn (3.48).

[85]Note that this is also known as a 'Fresnel lens'. See, for example, Miyamoto (1961).

[86]If one removes the 'modulo $T_{2\pi}$' from the above equation, it becomes the projected thickness distribution for a simple refractive lens. Unfortunately, this is very often unsatisfactory for the focusing of X-rays, as the resulting projected thicknesses are typically too large. Compare, however, our later discussions on compound refractive lenses in Section 3.4.2.

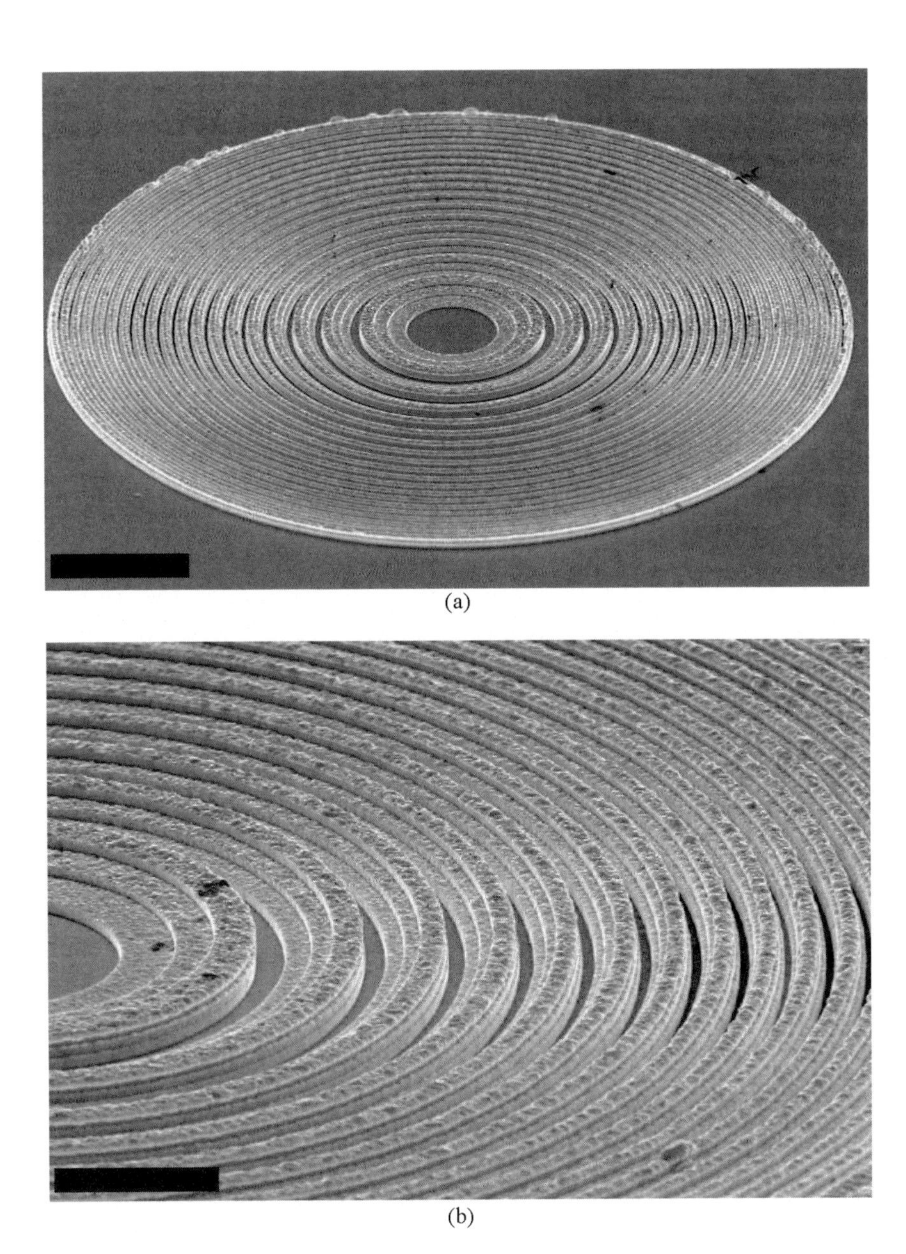
(a)

(b)

Fig. 3.11. Two views of a four-level gold Fresnel zone plate for focussing X-rays in the range 5–8 keV. Scale bar in (a) is 25 μm, with scale bar in (b) at 10 μm. Images taken from Di Fabrizio *et al.* (1999). Width of outermost level is 500 nm. Used with permission.

means, for considerably increasing the spectral bandwidth which can be efficiently focused by Fresnel zone-plate optics, has been developed by Wang *et al.* (2003). The essence of this idea is to have a thin uniform-thickness membrane, on one side of which is a conventional two-level zone plate, with a Fresnel lens (see eqn (3.48)) on the other side. The zone plate and the Fresnel lens, which are co-axial with one another, should be made of different materials. At the mean wavelength to be focussed, one of the materials should have a refractive index which increases with increasing X-ray energy,[87] with the other material having a refractive index which decreases with increasing X-ray energy. If the relevant parameters are chosen properly, then the resulting optical element has a focal length that has a much weaker dependence on the wavelength of the illuminating radiation, when compared to the simpler zone plate described earlier in the text. (ii) The analytical calculations, of this sub-section, are all predicated on the projection approximation. When the Fresnel zone plate has very fine zones, and/or if these zones have a sufficiently high aspect ratio, then this approximation will begin to break down. One will then need to invoke a wave theory for calculating the wave-field at the exit surface of the zone plate, which explicitly accounts for the diffraction that occurs within the zone plate. Such a wave theory may be based, for example, on the inhomogeneous paraxial equation (2.33) discussed in the previous chapter (see, for example, Kurokhtin and Popov (2002), and references therein, for the solution of this equation in the context of the theory of zone plates). If the scattering is sufficiently strong for paraxiality to be violated, then the analysis may instead proceed via the inhomogeneous Helmholtz equation (see eqn (2.28)). Finally, if the scattering within the element is so strong that the scalar theory breaks down, a treatment based on Maxwell's equations is required. (iii) In the preceding development we used the interference pattern, between a plane wave and a spherical wave, as part of the logic leading to the design of two-level zone plates. More generally, one can consider the interference pattern between spherical waves in designing Fresnel zone plates (see, for example, Erko *et al.* (1996)). Such an argument is not restricted to the theory of zone-plate design—it can also be actively used in their fabrication. Indeed, the interference of two spherical waves—which may themselves be generated by zone plates, although this need not be the case—may be used in the holographic production of Fresnel zone plates. For an entry point into the literature on this subject, see, for example, Solak *et al.* (2004). (iv) In coherent X-ray imaging, it is often the case that zone-plate-like structures appear in one's images. These may be produced by interference between the incident wave, and the distorted spherical wave scattered from small particles upstream of the image plane. Again, we note that this observation is consistent with the notion of a zone plate as an in-line hologram of a point scatterer. (v) To obtain a Fresnel zone plate with a point focus, one may binarize the interference pattern formed by superposing a

[87]This often occurs in the so-called 'anomalous dispersion' regime in the vicinity of a photoelectric absorption edge.

plane wave and the wave scattered by a point (i.e. a spherical wave). To obtain a Fresnel zone plate with a line focus, one may binarize the interference pattern formed by superposing a plane wave and the wave scattered by a line (i.e. a cylindrical wave). To obtain a Fresnel zone plate with an arbitrary focus (which may be two- or three-dimensional), one may binarize the interference pattern formed by superposing a plane wave and the wave scattered from a scatterer with a density distribution of the same functional form as the desired focal distribution. Under the first Born approximation, this scenario could be modelled using eqn (2.84). Alternatively, see Di Fabrizio *et al.* (2003) for a different means of designing Fresnel zone plates with custom-designed focal-plane patterns. (vi) Fresnel zone plates can work in reflection, as well as in transmission (see, for example, Basov *et al.* (1995)). Such reflective zone plates may be designed by considering the interference pattern between a spherical wave, and a plane wave which is tilted with respect to a nominal optic axis (see, for example, Erko *et al.* (1996)). (vii) Thus far, we have assumed our single-level and multi-level on-axis zone plates to possess rotational symmetry, on account of the evident rotational symmetry of the optical setup sketched in Fig. 3.10. This symmetry is broken in the so-called 'photon sieves', proposed by Kipp *et al.* (2001). These are a variant on the absorptive Fresnel zone plate, in which an element of randomness is introduced into the mask, which consists of a series of randomly sized pinholes whose locations are clustered about the transparent zones of a traditional absorptive zone plate. While these 'sieves' have the disadvantage of being more absorptive than traditional zone plates, they have the advantage of a tighter first-order focus (in both the transverse and longitudinal directions), suppression of side lobes in the first-order focal plane, and suppression of higher-order foci.

3.2.3 *Analyser crystals*

In the previous pair of sub-sections, we treated two-dimensional gratings as diffractive optical elements for X-rays. The next four sub-sections hop up a dimension, considering the use of certain three-dimensional diffraction gratings as X-ray optical elements. If the periodicity of such gratings exists at the atomic level, then these structures are crystalline, by definition. This is fortuitous, from a practical point of view—rather than attempting to piecemeal construct three-dimensional gratings to serve as diffractive optical elements for X-rays, one may exploit the remarkable self-organizing process of crystal formation to the same end. The associated length scales must be appropriate, of course—in order for crystals to diffract X-rays through appreciable angles, their periods (lattice constants) should be comparable to the wavelength of the radiation. This is indeed the case for hard X-rays, provided that the crystal's unit cell does not contain too many atoms, as both length scales (i.e. the crystal's lattice constants and the radiation wavelength) will then typically be on the order of angstroms.

In the present sub-section, we focus on what it perhaps the simplest crystal-based X-ray optical element—a strain-free slab of near-perfect crystal. Such slabs are known as 'analyser crystals', for reasons to be revealed several paragraphs hence. These crystal optics are typically made from silicon, but may also be fabricated from other materials such as diamond or germanium.

Before proceeding with a discussion of analyser crystals, it may be useful to briefly summarize certain salient results, from the previous chapter's discussions on both (i) crystals and (ii) the scattering of X-rays by crystals (see Section 2.5.3.2). (i) In a perfect crystal, any microscopic scalar material property $\Pi(\mathbf{x})$ will be a periodic function of position $\mathbf{x}$ in three-dimensional space, so that (see eqn (2.87)):

$$\Pi(\mathbf{x} + u\mathbf{a} + v\mathbf{b} + w\mathbf{c}) = \Pi(\mathbf{x}). \tag{3.49}$$

Here, $\mathbf{a}, \mathbf{b}$, and $\mathbf{c}$ are the three linearly independent vectors which constitute the axes for the unit cell of the crystal, and u, v, w are integers. Given the above condition on three-dimensional periodicity, $\Pi(\mathbf{x})$ admits the following Fourier-series decomposition (see eqn (2.88)):

$$\Pi(\mathbf{x}) = \sum_h \sum_k \sum_l \Pi_{hkl} \exp\left(i\mathbf{g}_{hkl} \cdot \mathbf{x}\right), \quad \mathbf{g}_{hkl} \equiv h\mathbf{a}^\star + k\mathbf{b}^\star + l\mathbf{c}^\star. \tag{3.50}$$

The Fourier coefficients Π_{hkl} are indexed by the integers h, k, l, with $\mathbf{g}_{hkl}$ being reciprocal-lattice vectors dual to the real-space lattice of the crystal. As indicated above, the reciprocal-lattice vectors $\mathbf{g}_{hkl}$ may be written in terms of the reciprocal-lattice basis vectors $\mathbf{a}^\star, \mathbf{b}^\star, \mathbf{c}^\star$, which are related to the real-space-lattice basis $\mathbf{a}, \mathbf{b}, \mathbf{c}$ via eqn (2.89):

$$\mathbf{a}^\star = \frac{2\pi\mathbf{b} \times \mathbf{c}}{\mathbf{a} \cdot (\mathbf{b} \times \mathbf{c})}, \qquad \mathbf{b}^\star = \frac{2\pi\mathbf{c} \times \mathbf{a}}{\mathbf{a} \cdot (\mathbf{b} \times \mathbf{c})}, \qquad \mathbf{c}^\star = \frac{2\pi\mathbf{a} \times \mathbf{b}}{\mathbf{a} \cdot (\mathbf{b} \times \mathbf{c})}. \tag{3.51}$$

(ii) If one is sufficiently far from an illuminated crystalline scatterer so as to be in the far-field (Fraunhofer) regime, with the crystal assumed to be sufficiently weak for the first Born approximation to hold, then an incident monochromatic scalar plane wave, with wave-vector $\mathbf{k}_0$, will only be scattered in the direction specified by the unit vector $\hat{\mathbf{x}}$, if the von Laue diffraction condition holds (see eqn (2.96)):

$$k\hat{\mathbf{x}} - \mathbf{k}_0 \equiv \Delta\mathbf{k} = \mathbf{g}_{hkl}. \tag{3.52}$$

Here, $\Delta\mathbf{k}$ is defined to be the difference between the wave-vectors of the incident and scattered plane waves, $k = |\mathbf{k}_0|$ ensures that the scattering is elastic, and $\mathbf{g}_{hkl}$ is any reciprocal-lattice vector associated with the crystal.

Next, recall the Fourier-series decomposition of the one-dimensional square-wave grating, based on eqn (3.12). This decomposition may be viewed as expressing the square-wave grating, of period L, as a sum of sinusoidal gratings of periods $L, \frac{1}{3}L, \frac{1}{5}L$, etc. More generally, the Fourier-series expression for the projected thickness of an arbitrary one-dimensional periodic grating, of period

L, would comprise a linear superposition of sinusoidal gratings with periods $L, \frac{1}{2}L, \frac{1}{3}L$, etc. Thus, while the arbitrary one-dimensional grating has period L, one can mentally decompose it into a sinusoidal grating with the same period, together with a sum of sinusoidal gratings with periods equal to L divided by any of the integers $2, 3, \cdots$.

This notion of a 'sum of sinusoidal gratings' generalizes to three-dimensional gratings such as perfect crystals. With this in mind, we again consider the Fourier-series expansion in eqn (3.50), together with the sketch in Fig. 3.12. In this figure, the regular series of dots is a two-dimensional representation of the real-space lattice of a given three-dimensional crystal. Three sets of equally spaced parallel planes, each of which are considered to extend through the entirety of the near-perfect crystal, are shown in the diagram. The first and second sets of such crystal planes, labelled A and B, both have the property that: (i) each lattice point is pierced by one crystal plane; (ii) each crystal plane passes through the same number of lattice points. However, the planes C do not fulfil the second property above. Rather, it is only every second plane that contains lattice points. More generally, one may consider sets of parallel planes for which it is only every mth plane that contains lattice points, where m is an integer; the set of planes C is then the special case for $m = 2$. Any given set of such planes may be described by a vector whose magnitude is equal to 2π divided by the spacing d between the crystal planes, and whose direction is perpendicular to the crystal planes.[88] This vector may be identified with the reciprocal lattice vector $\mathbf{g}_{hkl}$. Conceptually, this amounts to decomposing the crystal as a sum of sinusoidal three-dimensional gratings, each of which is characterized by its period, together with the normal to its planar 'surfaces of constant phase'. Mathematically, each sinusoidal grating corresponds to a particular Fourier harmonic in the decomposition of eqn (3.50). The period d_{hkl} of a given sinusoidal grating, corresponding to the reciprocal lattice vector $\mathbf{g}_{hkl}$ in eqn (3.50), is:

$$d_{hkl} = \frac{2\pi}{|\mathbf{g}_{hkl}|}. \tag{3.53}$$

We are now ready to recast the von Laue diffraction condition (3.52) in an alternative form. To this end, take the squared modulus of this condition and then make use of the above equation. This gives:

$$(k\hat{\mathbf{x}} - \mathbf{k}_0) \cdot (k\hat{\mathbf{x}} - \mathbf{k}_0) = |\mathbf{g}_{hkl}|^2 = \left(\frac{2\pi}{d_{hkl}}\right)^2. \tag{3.54}$$

Expand the scalar product on the left side, and then make use of the trigonometric identity:

[88] This construction is analogous to the wave-vector $\mathbf{k}$ associated with a coherent plane wave, if one replaces parallel crystal planes with surfaces of constant specified phase (modulo 2π).

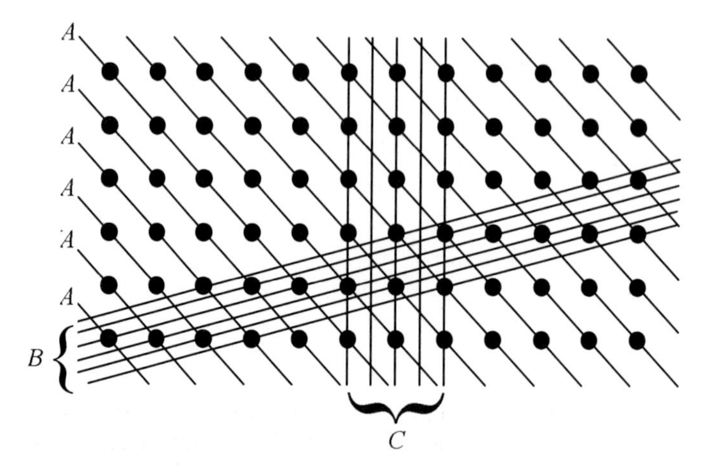

Fig. 3.12. Three sets of crystal planes, labelled A, B, and C. These planes are associated with the points in a crystal lattice, which are represented by dots.

$$1 - \cos(2\theta_{\rm B}) = 2\sin^2\theta_{\rm B}. \tag{3.55}$$

One thereby arrives at:

$$2d_{hkl}\sin\theta_{\rm B} = \lambda. \tag{3.56}$$

Here, as shown in Fig. 3.13, $\theta_{\rm B}$ denotes the angle between the wave-vector $\mathbf{k}_0$ of the incident plane wave, and the set of equally spaced planes (DD', EE', FF', GG', etc.) corresponding to the reciprocal-lattice vector $\mathbf{g}_{hkl}$ through which the incident wave is scattered.[89] The spacing between these planes, namely d_{hkl}, is given by eqn (3.53).

Now, for reasons outlined earlier, one may need to multiply d_{hkl} by an integer m, in order to obtain the spacing d between the corresponding atomic planes in the crystal. Substituting $md_{hkl} = d$ into the previous equation, we obtain the famous Bragg Law:

$$2d\sin\theta_{\rm B} = m\lambda. \tag{3.57}$$

With a view to interpreting this equation, which was obtained as a direct consequence of the von Laue diffraction condition in eqn (3.52), consider the construction sketched in Fig. 3.14. This is a real-space construction corresponding to the Fourier-space construction shown in the previous figure. We see two streamlines of the incident plane wave, labelled HI and JK. Two streamlines of the scattered plane wave, labelled IL and KM, are also shown. The parallel planes $DD', EE', FF', GG', \cdots$, associated with the reciprocal lattice vector

[89]This angle is known as the 'Bragg angle', in view of the Bragg interpretation of the scattering process which is given later in the present sub-section. Hence the subscript on $\theta_{\rm B}$.

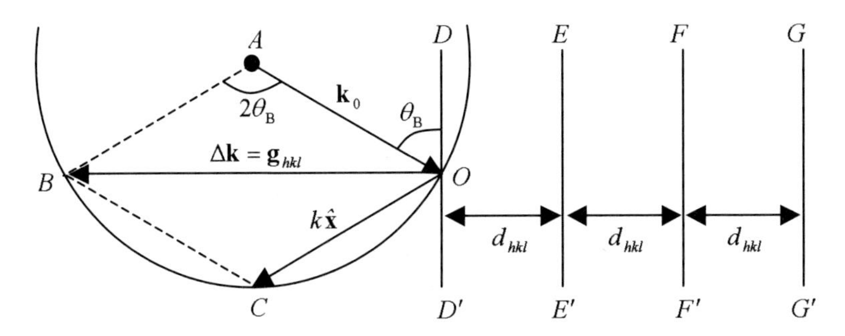

Fig. 3.13. Reciprocal-space picture corresponding to the von Laue diffraction condition in eqn (3.52). An incident plane wave, with wave-vector $\mathbf{k}_0$, is scattered by a crystalline sample so as to yield the scattered plane wave with wave-vector $k\hat{\mathbf{x}}$. Here, $\hat{\mathbf{x}}$ is a unit vector in the direction of the scattered plane wave, and $k = |\mathbf{k}_0|$ is the wave-number of both incident and scattered plane waves. Associated with the reciprocal-lattice vector $\mathbf{g}_{hkl}$, is the set of parallel planes DD', EE', FF', GG', etc., each of which are equally spaced by a distance $d_{hkl} = 2\pi/|\mathbf{g}_{hkl}|$ (note that these planes exist in real space, rather than reciprocal space). The angle, between the incident wave-vector $\mathbf{k}_0$ and the set of planes corresponding to the reciprocal-lattice vector $\mathbf{g}_{hkl}$, is denoted by θ_B. Lastly, the circle centred at A is the Ewald sphere.

$\mathbf{g}_{hkl}$, are considered to be planes from which the incident beam is reflected (with $m = 1$) in order to form the diffracted beam. Remembering that the wave-fronts are perpendicular to the streamlines, we require that the path difference $2x$, between the streamlines JKM and HIL, be equal to the wavelength λ of the incident radiation. Thus:

$$2x = \lambda. \tag{3.58}$$

To proceed further, we note from the triangle KNI that:

$$\cos\left(\frac{\pi}{2} - \theta_B\right) = \frac{x}{d_{hkl}} \Longrightarrow x = d_{hkl}\cos\left(\frac{\pi}{2} - \theta_B\right) = \frac{d}{m}\sin\theta_B, \tag{3.59}$$

which may be substituted into eqn (3.58) to recover the Bragg Law (3.57).

The interpretation, of X-ray scattering as occurring from so-called Bragg planes within the crystal, is a powerful real-space conceptualization which complements the reciprocal-space picture of kinematical crystal scattering based on the Ewald sphere. While we have derived the associated Bragg Law (eqn (3.57)) through a chain of argument which proceeds from the first Born approximation via the reciprocal-space equation (3.52), we note that the Braggs' derivation of their famous law proceeded using the real-space geometric argument of the previous paragraph.

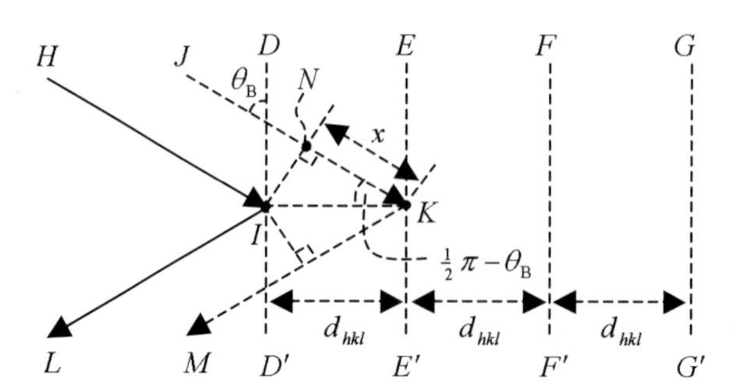

Fig. 3.14. Real-space picture corresponding to the diffraction condition in eqn (3.52). Two streamlines of the incident wave-field are shown as HI and JK; IL and KM are two streamlines of the scattered plane wave. Associated with the reciprocal-lattice vector $\mathbf{g}_{hkl}$, namely the difference between the scattered and incident wave-vectors, is the set of parallel planes DD', EE', FF', GG', etc. These planes are equally spaced by a distance $d_{hkl} = 2\pi/|\mathbf{g}_{hkl}|$. The angle, between the incident wave-vector and any member of this set of parallel planes, is denoted by θ_B. The path difference, between the streamline JKM and HIL, is equal to twice the distance x from N to K.

Both the reciprocal-space and real-space viewpoints have been used to go beyond the kinematical theory of scattering, upon which the analyses of the preceding paragraphs are predicated. Indeed, the kinematical approximation is often inadequate when one considers X-ray reflections[90] from large perfect crystals, such as the analyser crystals with which we are here concerned. One then needs to invoke a dynamical diffraction theory in which the possibility of multiple scattering is accounted for (cf. Section 2.6). Notwithstanding this, in the dynamical theory of X-ray diffraction from perfect crystals many of the concepts of the kinematic theory retain their utility. For example, in Darwin's real-space theory of dynamical diffraction (Darwin 1914a,b; see also Warren 1969), the simple Bragg notion of mirror-like reflecting atomic planes is retained. The resulting theory is perhaps the simplest variant of dynamical diffraction theory for X-rays, but is limited in scope. A more sophisticated dynamical diffraction theory is furnished by Ewald's reciprocal-space formulation (Ewald 1916a,b, 1917), as extended by von Laue (1931) (see also Authier 2001). In this formulation the reciprocal lattice remains of central importance, as does the von Laue diffraction condition, with the Ewald sphere being considered as a limiting case of the so-called 'dispersion surface'.

We shall not go into the details of dynamical crystal diffraction theory for X-rays. Rather, we refer the reader to the excellent texts by Pinsker (1978) and Authier (2001), together with the review by Batterman and Cole (1964). In the

[90]Strictly speaking, we should speak of 'diffracting' rather than 'reflecting' crystal planes. However, the latter terminology is in common use.

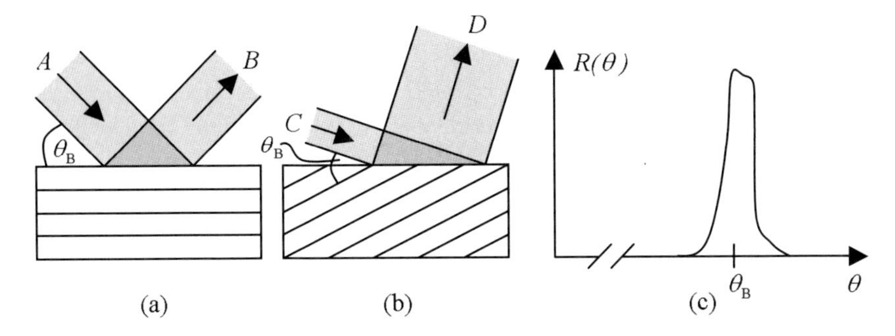

Fig. 3.15. (a) Symmetric reflection of an incident plane wave A from an analyser crystal, yielding the diffracted plane wave B. The angle $\theta_{\rm B}$ corresponds to a Bragg angle, with respect to the Bragg planes indicated in the diagram. Both incident and diffracted plane waves make the same angle with respect to the surface of the crystal, which is parallel to the active Bragg planes. (b) Asymmetric reflection of incident plane wave C from an analyser crystal, yielding the scattered plane wave D. Incident and scattered plane waves no longer make the same angle with respect to the surface of the crystal, which is not parallel to the active set of Bragg planes. However, incident and scattered plane waves do make the same angle with respect to the active Bragg planes. (c) The reflectivity $R(\theta)$, of a plane wave incident on an analyser crystal at angle θ, is sharply peaked at the Bragg angle $\theta_{\rm B}$ corresponding to a given reflection. Such a plot is known as a 'rocking curve'.

present context of analyser crystals, we merely note that a dynamical theory is required to correctly model the shape of the so-called 'rocking curve' associated with such a crystal. It is to this subject that we now turn.

Consider the sketch in Fig. 3.15. Both (a) and (b) show our analyser, this being a slab of near-perfect crystal that is used to filter ('analyse') the X-ray radiation incident upon it. In each of these figures, a certain set of active crystal/Bragg planes has been indicated by a series of parallel lines. Fig. 3.15(a) shows a 'symmetric reflection' from these Bragg planes, so termed because the angle of incidence $\theta_{\rm B}$, of the incident plane wave A with respect to the surface of the crystal, is the same as the angle made by the diffracted plane wave B with respect to the surface of the crystal. This is a consequence of the fact that the crystal surface is parallel to the Bragg planes contributing to the reflection. This scenario may be contrasted with the 'asymmetric reflection', shown in Fig. 3.15(b). Here, the angles of incidence and diffraction differ, on account of the fact that the Bragg planes are not parallel to the surface of the crystal. Note, however, that the incident and diffracted plane waves make the same angle $\theta_{\rm B}$ with respect to the reflecting crystal planes.

According to the von Laue diffraction condition (3.52), together with its alternative form given by the Bragg Law (3.57), the analyser crystal will only reflect if the angle of incidence (between the wave-vector of the incident plane wave and a specified set of Bragg planes) lies at exactly the Bragg angle θ_B. However, what happens if the angle of incidence, of a given plane wave striking the surface of the analyser crystal, is very close to a Bragg angle? Evidently, there will be some non-trivial structure to the degree to which the analyser reflects plane waves, in the vicinity of the Bragg angle. To more precisely quantify this notion, define the reflectivity R of the crystal to be the ratio of the squared modulus of the reflected plane wave, to the squared modulus of the incident plane wave. This reflectivity will be a function of the polarization of the incident plane wave, but we shall ignore this complication in the present simple discussion.

An indicative plot, of the reflectivity versus the angle of incidence θ, is shown in Fig. 3.15(c). This is known as the 'rocking curve' of the crystal, for a given Bragg reflection. We see that the reflectivity is sharply peaked about the Bragg angle θ_B, but does not have the zero width predicted by the simple kinematic theory. As mentioned earlier, the dynamical theory of X-ray diffraction must be invoked to correctly model the rocking curve for large near-perfect crystals.

Noting that the width of a hard-X-ray perfect-crystal rocking curve is typically on the order of seconds of arc, it is evident that it may be used as an exquisitely sensitive angular filter for X-rays. As such, it can 'analyse' an incident wavefront by filtering all but those plane waves whose propagation directions lie in a tightly constrained range of directions passed by the crystal.

Crystal analysers can be split into two classes, known as 'Bragg analysers' and 'Laue analysers'. The former refers to the case where the incident and diffracted X-rays pass through the same surface of the analyser (see, for example, Fig. 3.15(a)). On the other hand, one could have the plane wave incident upon one surface of a slab of analyser crystal, with both transmitted and diffracted beams emerging from the opposite face of the analyser. In such a case, which will be considered in our later discussions on X-ray interferometers, one speaks of 'Laue analysers'. Similarly, one may refer to the distinction between 'Bragg reflections' and 'Laue reflections', together with 'Bragg geometry' and 'Laue geometry'. For example, the reflections in Figs 3.15(a) and (b) would be respectively referred to as symmetric and asymmetric Bragg reflections from the crystal analyser.

3.2.4 *Crystal monochromators*

Rather than considering a slab of near-perfect crystal as an angular filter, as was the case in the previous sub-section, we here consider the crystal as a spectral filter—that is, as a 'monochromator'. As its name implies, the purpose of a monochromator crystal is to only allow a small range of X-ray wavelengths to pass through it, thereby increasing the degree of monochromaticity of the transmitted beam in comparison to the incident beam. Such spectrum-narrowing elements are crucial in many experiments in coherent X-ray optics, very often serving as the first X-ray optic encountered by a synchrotron beam as it passes from the storage ring to the experimental station.

There are a great many different designs for monochromator crystals. As a

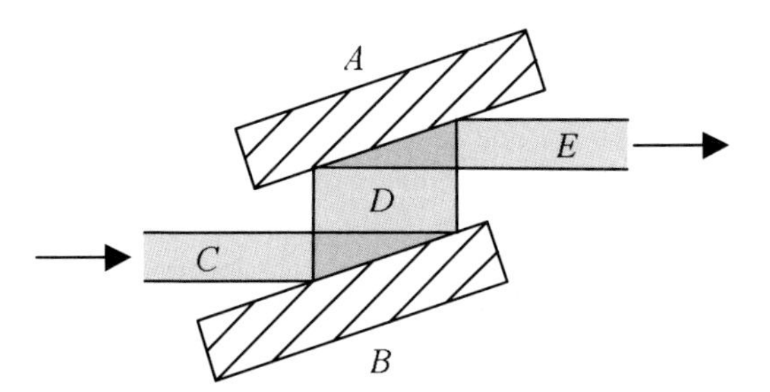

Fig. 3.16. Double-bounce monochromator, comprised of a pair of crystals A and B. These two slabs are such that the Bragg planes of each are aligned as indicated. Polychromatic radiation C has its spectral spread reduced upon being successively Bragg diffracted into the beams D and E.

commonly encountered example, consider the 'double-bounce' monochromator sketched in Fig. 3.16. Here, we see two crystals A and B, upon which is incident a polychromatic beam C. As suggested by the sets of parallel lines running through both A and B, the same Bragg planes are active in both crystals, which are aligned such that the corresponding crystal planes are parallel to one another.

Note that the incident beam C is often the 'white' (i.e. broad-spectrum polychromatic) beam of synchrotron radiation emitted by electrons in a storage ring. In modern third-generation facilities, such white synchrotron beams typically have power densities of tens or hundreds of watts per square millimetre. Such power densities are enormous, and pose a significant challenge for the engineering of crystal monochromators.

The heat-load problem is most severe for the first crystal encountering the white beam from the synchrotron. Any non-negligible thermally induced distortions of this first crystal (e.g. crystal B in the figure) will alter its rocking curve, adversely affecting the performance of the monochromator. This naturally leads to the question of how extreme heat loads may be accounted for in monochromator design.

The performance of a monochromator can be improved via one or more of the following: (i) a decrease in the coefficient of thermal expansion, of the crystal from which the monochromator is composed; (ii) a means of removing energy that is deposited by the X-rays in the monochromator; (iii) an increase in the thermal conductivity of the monochromator material; (iv) reducing the rate at which energy is absorbed by the monochromator (see, for example, Mills (2002a)). There are many means by which the issues raised in this list may be addressed. For example, points (i) and (ii) can be simultaneously effected by cryogenically cooling a silicon monochromator with liquid nitrogen (Bilderback 1986). Such cooling may be either indirect (e.g. by attaching the crystal to a metal block that is cooled by liquid nitrogen), or direct (by having channels cut in the monochromator crystal itself, through which the cooling fluid passes). Alternatively, points (i), (iii), and (iv) may be addressed by replacing silicon with diamond—at room temperature, diamond has a coefficient of thermal expansion that is about 40% of that of silicon, with a thermal conductivity that is approximately 6–13 times higher (depending on any impurities which may be present), augmented by a reduction in linear attenuation coefficient on account of the fact that carbon is lighter than silicon (see, for example, Mills (2002a)).

In addition to the problem of heat load, the broad-spectrum polychromatic nature of an incident X-ray beam may lead to what is known as 'harmonic contamination' in the beam emerging from the monochromator. This term refers to the presence of frequency pass-bands that are equal to an integer multiple of the radiation frequency which one would like to be passed by the monochromator. While dynamical diffraction theory is required for a quantitative analysis of the problem of harmonic contamination, the geometric origin of the problem may be more readily appreciated using the simple kinematic argument to which we now turn.

Consider, once more, the von Laue diffraction condition in eqn (3.52). Multiply this equation by any integer m, so that:

$$mk\hat{\mathbf{x}} - m\mathbf{k}_0 = m\mathbf{g}_{hkl}. \tag{3.60}$$

Suppose, further, that the $m = 1$ case of the above equation holds, so that the kinematic theory predicts a non-forbidden reflection[91] from the set of Bragg planes corresponding to the reciprocal-lattice vector $\mathbf{g}_{hkl}$. Noting that $m\mathbf{g}_{hkl}$ will be a reciprocal-lattice vector if $\mathbf{g}_{hkl}$ is a reciprocal-lattice vector, we reach the following conclusion: if the $m = 1$ case of the above equation holds, then this equation holds for any m. Thus, if the radiation wavelength λ corresponds to reflection from the Bragg planes described by $\mathbf{g}_{hkl}$, then the radiation wavelength λ/m will be reflected from the Bragg planes corresponding to $m\mathbf{g}_{hkl}$ (provided that this higher-order reflection is not forbidden). If such wavelengths are present in the incident polychromatic beam, they will be observed as higher-order 'harmonics' in the beam passed by the monochromator.

There are many means for suppressing such higher-order harmonics—for example, in the double-bounce monochromator sketched in Fig. 3.16, one may slightly rotate crystal A with respect to crystal B, leading to strong suppression of the higher-order harmonics at the expense of reducing the throughput of the monochromator for the desired lowest-order harmonic.

3.2.5 *Crystal beam-splitters and interferometers*

As we have already seen, near-perfect crystalline slabs can function as both angular filters (crystal analysers) and spectral filters (monochromators) for X-ray radiation. These two applications head a long list of other means in which such slabs may be used as optical elements for X-ray radiation. Here, we outline

[91] When the von Laue diffraction condition is fulfilled, the kinematical theory may nevertheless predict that no X-ray reflection is present. Such reflections are known as 'forbidden reflections'.

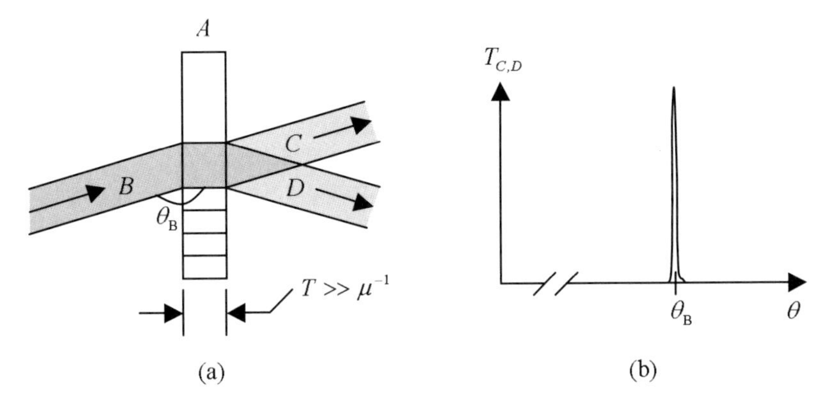

Fig. 3.17. (a) A thick near-perfect crystalline slab A is illuminated by a monochromatic plane wave B, whose propagation direction makes a Bragg angle $\theta_{\rm B}$ with respect to Bragg planes which are perpendicular to the surface of the crystal. The thickness T of the crystal is such that $\mu T \gg 1$, where μ is the linear attenuation coefficient of the crystal material. (b) The Borrmann effect, also known as anomalous transmission, refers to the non-negligible transmission coefficients $T_{C,D}$, for the beams C and D, respectively. These transmission coefficients are sharply peaked when the angle of plane-wave incidence θ coincides with the Bragg angle. Otherwise, the transmission coefficients are negligible.

their use as beam splitters, in the context of the famous Bonse–Hart X-ray interferometer (Bonse and Hart 1965).

We begin our account with the so-called 'Borrmann effect', also known as 'anomalous transmission'. Consider the symmetric-reflection Laue diffraction geometry in Fig. 3.17(a), which shows an X-ray plane wave incident upon a thick crystal at a Bragg angle $\theta_{\rm B}$. The corresponding Bragg planes are perpendicular to the face of the crystalline slab, as suggested by the series of parallel lines near the bottom of the crystal. Further, when we speak of the crystal slab as 'thick', we refer to the condition:

$$\mu T \gg 1. \tag{3.61}$$

Here, μ is the linear attenuation coefficient of the material from which the slab is composed, and T is the thickness of the slab. According to Beer's law of absorption (see eqn (2.43)), the crystal slab should allow no incident radiation to be transmitted through it, since:

$$\exp(-\mu T) \approx 0, \quad \mu T \gg 1. \tag{3.62}$$

However, experiment teaches that this is not the case. Rather, Borrmann observed that the thick crystalline slab's otherwise negligible transmission may become non-zero at the Bragg angle corresponding to a symmetric Laue reflection (Borrmann 1941). This intriguing angle-specific transparency, now known as the 'Borrmann effect', is sketched in Fig. 3.17(b). Here, we see an indicative plot of the transmission coefficients corresponding to the beams C and D in part (a) of the diagram, as a function of the angle of incidence θ for the plane wave entering the crystal. These transmission coefficients are equal to the ratio of the squared modulus of the transmitted beam, to the squared modulus of the incident beam.

Dynamical diffraction theory is required to quantitatively model the Borrmann effect. We shall not give such a development here, referring the reader to the classic accounts of Batterman and Cole (1964), Pinsker (1978), and Authier (2001). Rather, let us spend a paragraph in highlighting some of the salient features of the theory underlying the Borrmann effect.

To begin with, we note that a scalar theory is inadequate for a description of the Borrmann effect, as the polarization of the electromagnetic field must be taken into account. Only two of the three polarization vectors are relevant, namely those orthogonal vectors that are perpendicular to the wave-vector of the incident plane wave. Within the crystal, the electric displacement $\mathbf{D}$ obeys a wave equation which may be directly obtained from the Maxwell equations. In this wave equation, the effect of the crystal is quantified by its electron density, which may be expanded as a Fourier series over the reciprocal lattice vectors (cf. eqn (3.50)). Within the crystal, the electric displacement is modelled by a 'Bloch-wave expansion',[92] in which the displacement is written as a product of the incident plane wave, and a function that has the same periodicity as the crystal lattice. This function, like the electron density, may be expressed as a Fourier series over the reciprocal lattice vectors of the crystal. Next, a key assumption is made, that only two plane waves exist within the lattice (for each polarization). One of these plane waves corresponds to the incident plane wave (whose wave-vector is corrected for the mean refractive index within the crystal), with the second corresponding to a scattered plane wave which obeys the von Laue diffraction condition (3.52). This is the dynamical-theory analogue of the kinematical-theory description in which the surface of the Ewald sphere is touched by only two points of the reciprocal lattice. Since the scattered plane wave obeys the von Laue diffraction condition, its wave-vector is equal to the sum of the wave-vector of the incident field, and a reciprocal-lattice vector $\mathbf{g}_{hkl}$ corresponding to a particular reflection in the kinematical theory. With the understanding that one must account for the mean refractive index $\overline{n}$ of the crystal, one may then calculate the optical energy density of each of the two independent polarizations as being proportional to the squared modulus of the electric displacement. The resulting expression for the energy density has the same periodicity as the Bragg crystal planes corresponding to the reciprocal-lattice vector $\mathbf{g}_{hkl}$ (cf. Fig. 3.13). This periodicity may be viewed as due to the interference of the plane waves, namely the incident and scattered plane waves, existing within

[92]This is somewhat unfairly named, since the idea is due to Ewald.

the crystal. However, unlike the case of plane waves interfering in free space, the plane waves in the crystal may not be varied independently—indeed, they are coupled to one another through the wave equation mentioned earlier in this paragraph, and should therefore be considered as one entity. Of more importance, in the present context, is the fact that the above observation contains the key to the Borrmann effect. Specifically, since the energy density $\mathscr{E}(\mathbf{x})$ has the periodicity of a certain set of Bragg planes in the crystal, the possibility arises that the minima of the energy density may be made to coincide with the planes of atoms in the crystal. When this condition is fulfilled the X-ray wave-field is able to 'avoid the atoms', thereby greatly reducing photo-electric absorption by the crystal. The associated decrease, in the attenuation coefficient of the crystal, results in the 'anomalous transmission' of the Borrmann effect.

Having considered the Borrmann effect in the context of a symmetric Laue reflection from a thick near-perfect crystalline slab, it is evident that such a slab may function as a beam-splitter (i.e. an optical element which splits an incident optical beam into two or more beams). In this context, refer to Fig. 3.17(a) once more.

Now, beam-splitters are an essential component of separated-path interferometers, the common feature of which is as follows. A given incident beam is split into two spatially separated components, known as 'arms', which are then recombined to form an interference pattern. This interference pattern is sensitive to phase shifts in either or both of the arms of the interferometer, such as may be induced by placing a sample in one arm, and/or by distorting the interferometer. Phase shifts on the order of tenths or hundredths of radians can be readily detected in this manner, together with distortions of the interferometer itself which are on the order of the wavelength of the radiation being used.

In a landmark paper, Bonse and Hart demonstrated separated-path X-ray interferometry using the device sketched in Fig. 3.18 (Bonse and Hart 1965; see also Hart and Bonse 1970, together with references therein). This comprises a series of three equi-distant equal-thickness parallel slabs A, B, C, each of which are cut from a single monolithic near-perfect crystal. Since it is cut from a single crystalline block, with each blade being connected through the base of the device, there is long-range crystalline order between the atoms in each of the blades.

An illuminated Bonse–Hart interferometer is shown in plane view, in Fig. 3.19. As indicated at the bottom of each of the blades A, B, C, the active Bragg planes are considered to be perpendicular to the faces of each of the blades (cf. Fig. 3.17(a)). Note that each of these blades is optically thick, in the sense given by eqn (3.61). The incident plane wave D strikes the active planes at a Bragg angle, with the setup being such that the Borrmann effect is operative. Thus, notwithstanding the fact that blade A is optically thick, it acts as a beam-splitter by producing strong diffracted beams E and F (cf. Fig. 3.17(a)). Symmetry considerations make it evident that, if the conditions for anomalous transmission are met for beam D striking blade A, then anomalous transmission will also be observed for both beams E and F in striking blade B. Thus, beam E will exhibit

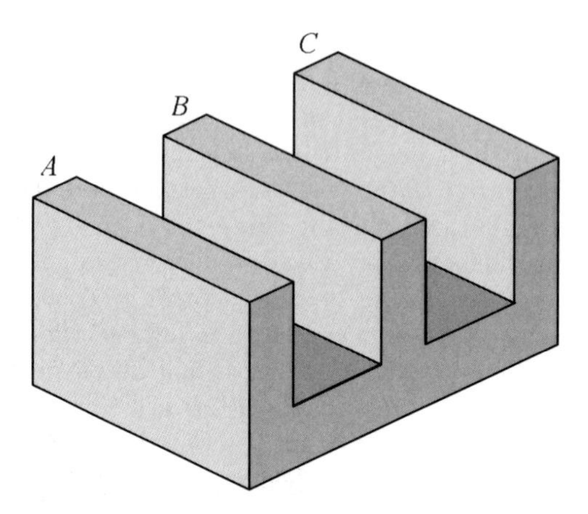

Fig. 3.18. Bonse–Hart X-ray interferometer, with three thick blades A, B, C cut from a single monolithic block of near-perfect crystal.

the Borrmann effect in yielding beams G and H, with beam F leading to beams I and J. Beams G and J are not utilized, in the setup shown here. Beams H and I overlap in the triangular region adjoining slab C. Using the same argument given in our qualitative discussion of the Borrmann effect, together with the fact that beams H and I differ by the same reciprocal lattice vector as the two beams E and F, it is evident that the interference fringe maxima, due to beams H and I in their triangular region of overlap, will lie between the atomic planes of blade C. The wave-field will thus be 'anomalously' transmitted, yielding beams L and K. An interferogram may then be recorded over the plane M, as the wave-field over this plane will be due to the coherent superposition of beams from the two arms of the interferometer.

For a perfect Bonse–Hart crystal interferometer illuminated by coherent plane waves at exactly the Bragg angle for a symmetric Laue reflection, the interferogram at M will be featureless. However, if there are any imperfections in the crystal comprising the interferometer, such as those due to strains and dislocations, then the conditions for anomalous transmission will be locally disrupted. This local disruption leads to strong absorption of the wave in the thick crystal, implying the existence of dark bands—namely interference fringes—in the pattern recorded at M. If local phase shifts between beams H and I are an odd integer multiple of π radians, interference minima will occur. On the other hand, if these phase shifts are an even integer multiple of π radians—and one may note that this includes no phase shift at all—then interference maxima will occur. Such interferograms are of crystallographic interest, insofar as they exhibit sensitivity to angstrom-scale lattice distortions in the crystal from which the Bonse–Hart interferometer is composed.

In addition to interference fringes being influenced by imperfections in the crystal, they also carry information regarding the phase shifts imparted upon

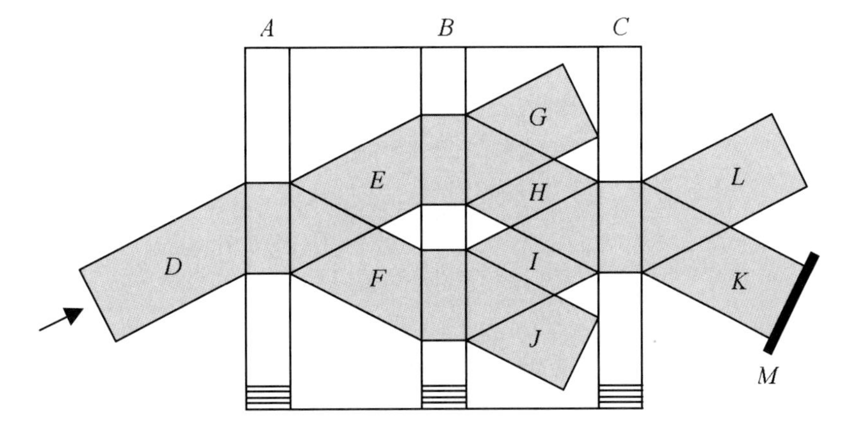

Fig. 3.19. Plane view of Bonse–Hart interferometer in action. A, B, C are three equal-thickness equi-distant crystalline slabs cut from the same crystalline block. Blade A splits the incident beam D into the transmitted beams E and F, via a symmetric Laue reflection. The second blade B further splits the beams transmitted by the first blade A. Via the action of the third blade C, beams L and K are both influenced by the interference of beams H and I. An X-ray interferogram is recorded by the detector M.

X-rays by objects placed in the separated arms of the interferometer. For example, one may place a transparent wedge ('prism') in one of the arms of the interferometer, thereby introducing a linear phase ramp across the wave-field transmitted by the wedge. This will result in a series of linear fringes in the recorded interferogram (cf. Section 3.4.1, especially the paragraph following eqn (3.82)). Such fringes were published in Bonse and Hart's original paper (1965), with these linear fringes being modulated by crystal imperfections in the interferometer. They also pointed out that small samples may be placed in one of the interferometer arms, with the wedge in the other; the resulting interferogram conveys information regarding the phase shift imparted by the sample on the beam passing through it. This application will be studied in greater detail, in Section 4.6.1.

3.2.6 *Bragg–Fresnel crystal optics*

In the past few sections, we have considered the use of atomic-scale three-dimensional gratings (i.e. crystals) as non-focussing diffractive optical elements for X-rays. Bearing in mind the preceding discussion on the use of two-dimensional gratings for X-ray focusing (i.e. zone plates), one is naturally led to the question of how to use crystals as three-dimensional diffractive focussing elements. Perhaps the most obvious answer to this question is to appropriately curve the crystal. Such a strategy, which dates back at least as far as the work of Johann, Cauchois and Johansson in the early 1930s (see, for example, Snigirev (1995)), will not be reviewed here. Rather, we consider an alternative way for crystals to

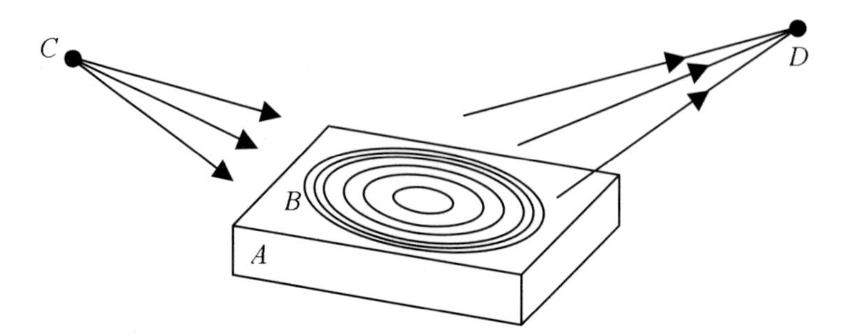

Fig. 3.20. A Bragg–Fresnel crystal optic is a near-perfect crystal slab A, one face of which has grooves B cut out in the shape of an off-axis reflective Fresnel zone plate. X-rays from a source C are then simultaneously monochromated and deviated via Bragg diffraction from the crystal, subsequently being focussed towards D by the action of the zone plate.

focus X-rays: Bragg–Fresnel crystal optics.

Consider, once again, the process of Bragg diffraction by a crystal analyser, as sketched in Figs 3.15(a) and (b). Whether the reflection be symmetric or asymmetric, we have seen that the analyser may be viewed as a form of angle-specific X-ray mirror, with reflection occurring from a given set of crystal planes (cf. Fig. 3.12). The resulting optical element is not a focussing one, as incident plane waves are transformed into diffracted plane waves. How, then, may the crystal be converted into a focussing optical element? A means of effecting this is to take one of the plane waves which are diffracted by the analyser crystal, subsequently passing it through a Fresnel zone plate so that one of its diffracted orders serves to focus the radiation to the desired point. A more sophisticated answer, which leads us directly to the concept of the Bragg–Fresnel crystal optic, is to co-locate the zone plate with the surface of the crystal. The Bragg planes of the crystal are thereby able to deviate the incident plane wave through a large angle, with the zone plate subsequently serving to focus the Bragg-diffracted beam down to a point.

This concept, due to Aristov and colleagues (Aristov *et al.* 1986a,b, 1987; see also Snigirev 1995) is sketched in Fig. 3.20. Here, we see a near-perfect crystal A into which a binary zone-plate profile B has been created. For good efficiency in a two-level zone-plate structure, the groove depth should be such that there is a phase shift of π radians, between X-rays which are Bragg-reflected from the upper surface and those reflected from the lower surface (cf. the discussion on the efficiency of phase zone plates in Section 3.2.2).

For the focussing of hard X-rays, Bragg–Fresnel crystal optics have several advantages compared to zone plates. As the energy of the radiation becomes higher, reflective zone plates become more difficult to fabricate and operate, as the angle (between the incident X-ray and the reflecting surface) required for total external reflection[93] becomes progressively smaller. Increasing radiation hardness also increases the difficulty of fabricating transmissive zone plates, as the thickness of the zones needs to be increased, while decreasing the width of the zones. This may lead to impracticably high aspect ratios in the required transmissive zone plates. In comparison, Bragg–Fresnel optics have the advantage of the high efficiency furnished by phase zone plates, without the aspect-ratio problem of transmissive optics. In addition, a higher degree of mechanical and thermal stability is provided by the fact that Bragg–Fresnel crystal optics are grooved into a monolithic crystal block, which may be cooled, if necessary, using the techniques for cryogenically cooled monochromators outlined in Section 3.2.4.

Two remarks: (i) Having considered Bragg–Fresnel crystal optics, in which crystal planes are used to produce a Bragg reflection, we note that such a reflection can also be achieved using multi-layered mirrors in place of crystal planes (cf. Section 3.3.4). The resulting Bragg–Fresnel multi-layer optics are extensively discussed by Erko *et al.* (1996). (ii) In addition to engineering Bragg–Fresnel optics which yield a point focus, whether they be based on crystals or multi-layers, one can also have so-called 'linear Bragg-Fresnel optics' that produce a line focus. Indeed, one can engineer such optics to produce an arbitrary distribution of intensity over their focal plane, for example, using the techniques discussed in remark (v) at the end of Section 3.2.2.

3.2.7 *Free space*

At the risk of being chastised for stating the obvious, and tempered with the pre-emptive retort that sometimes there is a fine line between the trivial and the profound, one may note that free space constitutes a diffractive optical element in the sense that X-rays diffract as they propagate through vacuum.

As a first example of this, recall the pattern of spots which constitutes the diffraction pattern of a small crystal, such as that shown in Fig. 2.7. In forming such a diffraction pattern, the illuminated crystal's exit-surface wave-field must propagate through a sufficiently large distance of free space, in order to evolve into a Fraunhofer diffraction pattern of a convenient size for recording by a two-dimensional imaging device. A second example is furnished by the zone plate, which requires transmitted radiation to travel through a certain distance from the plate's exit surface, in order to converge upon a focus.

As a last example, recall the Young double-slit experiment using an extended incoherent planar source, in which free space was seen to function as a diffractive optical element in at least two respects: (i) Field correlations are enhanced as partially coherent radiation propagates from a delta-correlated extended planar source to the entrance surface of the Young double slit, as quantified by our earlier discussions on the propagation of coherence functions and the van Cittert–Zernike theorem (see Section 1.9). (ii) In order for interference to be observed

[93]Note that this is the counterpart of total internal reflection for visible light, with external rather than internal reflection arising because the refractive index for X-rays is less than unity. See Section 3.3.1 for more information.

between the pair of spatially separated fields at the exit-surface of the double-slit apparatus, propagation through a sufficient distance of free space was invoked, such that the resulting pair of diffracted fields overlap with one another in the interference region (see Fig. 1.5).

The viewpoint, of free space as a diffractive optical element, will receive some prominence in the next chapter, in contexts such as the operator theory of X-ray imaging using coherent and partially coherent radiation, holography, propagation-based phase contrast, and phase retrieval. It will also figure in the final chapter's discussions on propagation-induced wave-field vortices.

3.3 Reflective optical elements

Reflective X-ray optics are the subject of this section. We begin with some general remarks concerning the reflection of X-rays from surfaces, subsequently treating capillaries, square-channel arrays, and mirrors.

3.3.1 *X-ray reflection from surfaces*

At an elementary level, the reflection of X-rays from smooth surfaces can be considered from the standpoint of geometric optics. In this context, consider Fig. 3.21(a). Here we see an X-ray A that is incident upon a flat surface BC, this being an interface between vacuum and a certain homogeneous isotropic medium. Upon striking the surface at the point O, a refracted ray D may penetrate the surface. Further, there will be a specularly reflected ray G. Quasi-monochromatic radiation of angular frequency ω will be assumed throughout. Functional dependence of refractive indices on ω will be understood as implied.

Notwithstanding the fact that we are primarily interested in reflection, let us first consider the refracted ray D. In particular, we seek a relationship between the angles θ_1 and θ_2, which the incident and refracted rays respectively make with the normal EF to the surface. While it is well known that such a relationship is given by Snell's Law, we choose to arrive at this conclusion from more fundamental considerations.

Following the discussions on geometric optics in Section 2.8, the various rays AO, OD, and OG may be viewed as streamlines of the quasi-monochromatic scalar X-ray wave-field interacting with the medium. According to Fermat's variational principle, which we now invoke, the ray streamline connecting two points—which lie within a volume of space filled with a static medium characterized by a position-dependent refractive index—is such that an extremal value for the phase is accumulated along the ray in travelling between the two points. If the extremal value for the accumulated phase is a minimum (maximum), then any fixed-endpoint infinitesimal continuous deformation of the true ray-path will lead to an increase (decrease) in the phase accumulated between the two points. Stated differently, the number of wavelengths that fit into a given ray trajectory, connecting two points which lie within a static medium of non-uniform refractive index, is either a local minimum or a local maximum with respect to fixed-endpoint infinitesimal deformations of the true ray-path. While a formula

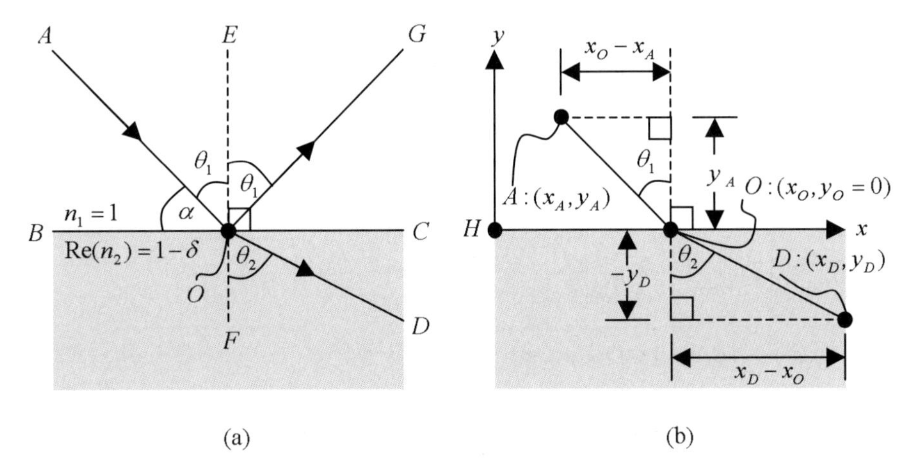

Fig. 3.21. (a) The X-ray A strikes a point O on a flat interface BC, which separates vacuum (refractive index $n_1 = 1$) from a uniform medium with refractive index n_2, whose real part is equal to $1 - \delta$. The refracted ray D, if it exists, moves away from the normal EF to the surface. The specularly reflected ray is denoted by G. (b) Diagram to aid in the derivation of Snell's Law for refraction of X-rays at interface BC of the previous figure. O, A, D, θ_1, and θ_2 are as shown in (a). An xy Cartesian coordinate system has been erected, with origin H, x-axis in the plane of the surface BC, and y-axis normal to this surface and pointing into vacuum. Cartesian coordinates of points A, O, and D are as indicated.

for the accumulated phase is given in eqn (2.119), resulting from an argument which ultimately stemmed from the Maxwell equations, we prefer to give a simple geometric argument here.

We begin with the fact that a straight line is formed by the ray connecting two points in a volume filled with an isotropic medium of constant refractive index. This conclusion may be reached via Fermat's principle, as the least phase will be accumulated along the path of shortest length which connects the two points, this path being a straight line. Next, let us consider the path taken by the ray connecting points A and D in Fig. 3.21(a), as re-sketched in part (b) of the same figure. With the end-points A and D considered fixed, and equipped with the knowledge that rays traversing through uniform-refractive-index volumes follow straight-line paths, we conclude that the actual ray-path will be a straight line from A to an as-yet undetermined point O on the surface of the medium, which is then connected to D via another straight line. By symmetry, the points A, O, and D must lie in a two-dimensional plane. Coordinatize this plane with the Cartesian xy system shown in Fig. 3.21(b), with the x-axis in the plane of the reflecting surface, the y-axis normal to the surface and pointing from the medium to the vacuum, and origin H. In this system, the coordinates of points A, O, and

D are respectively given by (x_A, y_A), $(x_O, y_O = 0)$, and (x_D, y_D). The phase ϕ_{AOD}, accumulated in traversing the indicated path from A to O to D, is given by 2π multiplied by the number of wavelengths that fit into the path. Since the wavelength of the radiation in a uniform medium is equal to $\lambda/\mathrm{Re}(n)$, where λ is the wavelength *in vacuo* and n is the complex refractive index, we have:

$$\begin{aligned}\phi_{AOD} &= \left[2\pi \times \frac{\sqrt{(x_A - x_O)^2 + y_A^2}}{\lambda/\mathrm{Re}(n_1)}\right] + \left[2\pi \times \frac{\sqrt{(x_D - x_O)^2 + y_D^2}}{\lambda/\mathrm{Re}(n_2)}\right] \\ &= \frac{2\pi}{\lambda}\left[\mathrm{Re}(n_1)\sqrt{(x_A - x_O)^2 + y_A^2} + \mathrm{Re}(n_2)\sqrt{(x_D - x_O)^2 + y_D^2}\right]. \end{aligned} \tag{3.63}$$

The only variable, in the above equation, is the x coordinate x_O of the point O. According to Fermat's principle the accumulated phase ϕ_{AOD} is an extremal value ('extremum'), so that:

$$\frac{d\phi_{AOD}}{dx_O} = 0, \tag{3.64}$$

which may be applied to eqn (3.63) to yield:

$$\frac{\mathrm{Re}(n_1)(x_A - x_O)}{\sqrt{(x_A - x_O)^2 + y_A^2}} + \frac{\mathrm{Re}(n_2)(x_D - x_O)}{\sqrt{(x_D - x_O)^2 + y_D^2}} = 0. \tag{3.65}$$

Noting from Fig. 3.21(b) that:

$$\sin\theta_1 = \frac{x_O - x_A}{\sqrt{(x_A - x_O)^2 + y_A^2}}, \qquad \sin\theta_2 = \frac{x_D - x_O}{\sqrt{(x_D - x_O)^2 + y_D^2}}, \tag{3.66}$$

we see that eqn (3.65) becomes Snell's Law:

$$\mathrm{Re}(n_1)\sin\theta_1 = \mathrm{Re}(n_2)\sin\theta_2. \tag{3.67}$$

Five remarks: (i) We leave it as an exercise to show that Snell's Law may also be obtained by taking, as a starting point, the eikonal equation for the wave-fronts of geometric optics (see eqn (2.117)). (ii) As covered in many electrodynamics texts, Snell's Law may also be derived by using Maxwell's equations to consider electric and magnetic plane waves incident upon flat boundaries between adjoining dielectric media (see, for example, Jackson (1999) or Attwood (1999)). (iii) If we set $\mathrm{Re}(n_1) = \mathrm{Re}(n_2)$ in Snell's Law, it reduces to $\theta_1 = \theta_2$, yielding straight-line propagation once more. (iv) Via an evident modification to the argument used to obtain Snell's Law from Fermat's principle, the law of specular reflection may also be so obtained. (v) Adopting language more often encountered in studies on general relativity than in X-ray optics, we note that Fermat's principle implies the ray paths to be geodesics with respect to a metric derived from the real part of the position-dependent refractive index for a static medium.

Returning to the main thread of the argument, let us see what Snell's Law has to teach regarding X-ray refraction and reflection. Applying this equation to the scenario sketched in Fig. 3.21(a), with $n_1 = 1$ and $n_2 = 1 - \delta + i\beta$ (see eqn (2.37)), we see that:

$$\sin\theta_1 = (1 - \delta)\sin\theta_2. \tag{3.68}$$

Note that δ is positive and small for X-ray energies of a few keV or higher, implying that:

$$|\theta_2| \geq |\theta_1|. \tag{3.69}$$

Therefore, as indicated in the figure, the refracted X-ray bends away from the normal EF, in passing from vacuum into the medium. Next, with θ_1 such that $\theta_2 < \frac{1}{2}\pi$, suppose that θ_1 is increased to a critical value denoted by θ_1^{crit}, which by definition corresponds to $\theta_2 = \frac{1}{2}\pi$. At this critical incident angle, for which the refracted ray D skirts the underside of the surface BC, we have:

$$\sin(\theta_1^{\text{crit}}) = (1 - \delta)\sin\left(\frac{\pi}{2}\right) = 1 - \delta. \tag{3.70}$$

Rather than being measured with respect to the normal, the angle

$$\alpha^{\text{crit}} \equiv \frac{\pi}{2} - \theta_1^{\text{crit}} \tag{3.71}$$

is more often used, this being the angle that the incident ray makes with the surface it strikes. With the above change of variables, eqn (3.70) becomes:

$$\cos(\alpha^{\text{crit}}) = 1 - \delta, \tag{3.72}$$

where use has been made of the fact that:

$$\sin\left(\frac{\pi}{2} - \alpha^{\text{crit}}\right) = \cos(\alpha^{\text{crit}}). \tag{3.73}$$

Now, the right side of eqn (3.72) is only very slightly less than unity, implying that the critical angle α^{crit} is small compared to unity. Thus, we may employ a second-order Taylor expansion, for the cosine function about zero angle, so that eqn (3.72) becomes:

$$\cos(\alpha^{\text{crit}}) \approx 1 - \frac{1}{2}(\alpha^{\text{crit}})^2 = 1 - \delta. \tag{3.74}$$

Solving for the critical angle, we obtain:

$$\alpha^{\text{crit}} = \sqrt{2\delta}. \tag{3.75}$$

If the angle α (see Fig. 3.21(a)) between the incident ray and the surface is less than this critical value, then no refracted ray can be supported and the ray will be reflected in its entirety. This X-ray phenomenon is known as 'total external reflection', which may be compared to the visible-light phenomenon of 'total internal reflection' which is covered in many elementary texts on light optics. Importantly, if the angle α is too large for total external reflection to occur, then the Fresnel reflectivity formula—obtained using a wave-optical treatment based on the Maxwell equations—implies the reflectivity to be very small (see, for example, Parratt (1954) or Attwood (1999)). Noting that the refractive index decrement δ is typically on the order of 10^{-5} radians or smaller, for hard x-rays, we see from eqn (3.75) that the critical angle is on the order of several milliradians or less. Given that total external reflection is the principle which underlies most single-surface reflective X-ray optical elements, we conclude that these optics must be operated at very glancing angles of incidence.

In order to explore this point in a little more detail, recall the discussions of Section 2.9. These culminated in the following approximate expression for the refractive index decrement of a uniform isotropic material (eqn (2.134)):

$$\delta = \frac{e^2 \rho}{2\varepsilon_0 m_{\rm e} c^2 k^2}. \tag{3.76}$$

Here, e is the charge of an electron, ρ is the number of electrons per unit volume in the material, ε_0 is the electrical permittivity of free space, $m_{\rm e}$ is the rest mass of the electron, c is the speed of light in vacuum, and k is the wave-number of the X-ray radiation. Note that this expression is often a good approximation if one is far from absorption edges, as assumed here. If the above formula is substituted into eqn (3.75), and the wave-number k written in terms of the radiation wavelength λ as $k = 2\pi/\lambda$, we obtain:

$$\alpha^{\rm crit} = \frac{e\lambda}{2\pi c}\sqrt{\frac{\rho}{\varepsilon_0 m_{\rm e}}}. \tag{3.77}$$

Thus, the critical angle $\alpha^{\rm crit}$ is proportional to the incident radiation wavelength λ, implying that single-surface X-ray reflective optics must work at progressively more glancing angles of incidence, as the radiation wavelength decreases. Also, we see that the critical angle increases as the square root of the number density ρ of electrons in the material, implying that denser materials of higher atomic number will have a higher critical angle, in comparison to less dense materials of lower atomic number. Thus, single-surface reflective X-ray mirrors are often coated with a material of high atomic number. For the same reason, multi-layer reflecting structures (see Section 3.3.4) are often made with layers of two different materials, one of which has a high atomic number and the other of which has a low atomic number. Lastly we note that, since the photon energy E is given by:

$$E = \frac{hc}{\lambda}, \tag{3.78}$$

eqn (3.77) implies the simple rule of thumb that the product of the photon energy E, and the critical angle α^{crit}, is a material-dependent constant (see, for example, Bilderback (2003)):

$$\alpha^{\text{crit}} E = \frac{eh}{2\pi}\sqrt{\frac{\rho}{\varepsilon_0 m_{\text{e}}}}. \tag{3.79}$$

3.3.2 *Capillary optics*

Single tapered capillaries, also known as 'mono-capillaries', constitute our first example of an X-ray focusing optic employing total external reflection (Stern *et al.* 1988; Yamamoto *et al.* 1988; Engström *et al.* 1989; Rindby *et al.* 1989; Thiel *et al.* 1989). The tapered mono-capillary is typically either a parabola or ellipse of revolution (i.e. a paraboloid or an ellipsoid), although conical and even tube geometries have been employed in earlier investigations. Figure 3.22(a) is indicative of the second of these four cases, with the lines A and B representing the intersection of the reflecting surface of revolution with a plane containing the axis of the optic. Such a surface of revolution is typically formed by the inner surface of a hollow tapered tube, often made of glass. Sometimes, the inner surface of the optic may be coated with a layer of material with high atomic number, such as gold, so as to increase the critical angle for total external reflection of X-rays. Two such reflected rays are shown in the diagram, including the ray C which is reflected from the point D on the inner surface of the optic. A focus is produced at the point E, which lies at a distance Δ from the exit hole of the device. This length is known as the 'working distance' of the capillary. If the device is to employ a single reflection, as sketched in Fig. 3.22(a), an ellipsoidal profile is required for point-to-point focusing. If the incident radiation is parallel, then a point focus may be achieved using a single-reflection paraboloidal optic (see, for example, Hoffman *et al.* (1994)). The working distance for such single-reflection optics can be on the order of millimetres or centimetres, with typical focal-spot sizes on the order of several microns (Balaic *et al.* 1995).

Multiple-reflection mono-capillary optics, namely tapered capillaries in which a given incident X-ray may suffer total external reflection on more than one occasion prior to exiting the device, are also possible. Unlike single-bounce devices, a true focus is not produced by multiple-reflection capillaries—as such, they are often referred to as 'concentrators' rather than 'focusers'. Multi-bounce mono-capillaries are usually constrained to have a shorter working distance than single-bounce focusing optics, with the exiting beam having its smallest diameter at the exit-hole of the device (see, for example, Hoffman *et al.* (1994)). In some sense, these multi-bounce optics may be viewed as an X-ray equivalent of an optical fibre (Engström *et al.* 1991). These devices can achieve foci as small as 50 nm (Bilderback *et al.* 1994), although foci of several microns are more typical (Bilderback 2003).

As one particular mode in which a multi-bounce capillary may operate, consider Fig. 3.22(b). Here, only one part of the capillary's inner surface is shown,

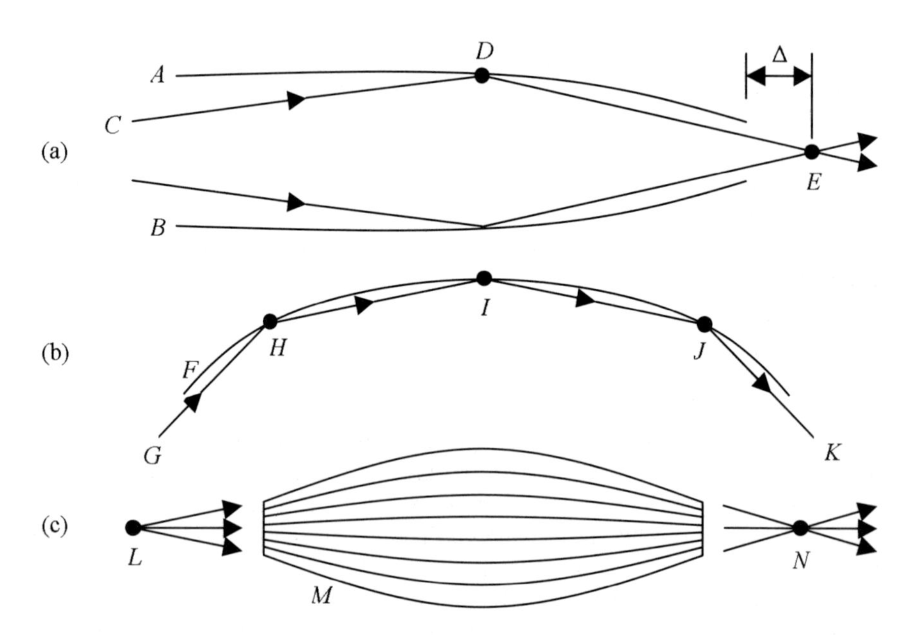

Fig. 3.22. (a) Single-bounce mono-capillary optic; (b) multiple-bounce mono-capillary optic; (c) poly-capillary optic ('Kumakhov lens').

labelled F. The incident ray G is then reflected three times from this surface, at points H, I, and J; reflection could also occur from adjacent surfaces, a possibility not indicated in the diagram. In either case, the reflection points need not all lie in a plane, notwithstanding the two-dimensional nature of our sketch. Nevertheless, the salient point is clear: if an incident ray is reflected multiple times, then the exit ray K may make an angle, with respect to the incident ray G, that is many times larger than the critical angle for total external reflection. Such 'beam steering' immediately leads to the following idea: to produce an optic which accepts a much larger range of incident X-ray angles than the mono-capillaries described earlier, one may bundle together many bent capillaries, as indicated by structure M in Fig. 3.22(c). This poly-capillary optic is known as a 'Kumakhov lens'. The Kumakhov lens may be used to concentrate or focus X-ray light from a source L, to a point N. For more information, and references to the seminal papers, see the review articles by Kumakhov and Komarov (1990) and Kumakhov (2000).

3.3.3 *Square-channel arrays*

The 'square-channel array' is an X-ray focusing device that may be regarded as a very thin Kumakhov lens employing square-shaped capillaries. However, there is the important qualitative difference that, rather than working on the 'beam steering' principle employed in the Kumakhov lens, square-channel arrays are typically sufficiently thin for most transmitted X-rays to be reflected no more than twice by each of the capillaries ('channels'). They are therefore true focusing elements, rather than radiation concentrators. For focusing applications, which include X-ray astronomy, the arrays may be curved by 'slumping' sufficiently hot elements over a spherical surface before allowing them to cool. Interestingly, such curved square-channel arrays work on the same principle as the eye of certain lobsters, with such elements often described as 'lobster-eye optics' in the literature (Angel 1979). Other channel shapes, such as tubes, may also be employed. For more information see, for example, Angel (1979), Wilkins *et al.* (1989), Chapman *et al.* (1991), Fraser *et al.* (1993), Brunton *et al.* (1995), and Peele *et al.* (1996), together with references therein.

3.3.4 *X-ray mirrors*

According to the simple ray picture given in Section 3.3.1, the total external reflection of X-rays from a flat surface requires glancing angles of incidence that are less than the critical angle given by eqn (3.75). This same analysis may be used to treat the reflection of X-rays from smooth curved surfaces, as such (idealized) surfaces may always be considered to be locally flat. The use of curved reflecting surfaces opens the possibility for using reflective X-ray optics for beam shaping, an idea that has already been explored in discussing focusing capillary optics. Here, we consider a more mainstream application of this idea, in the context of X-ray focusing mirrors.

As noted in Section 3.3.2, point-to-point focusing can be achieved using a single elliptical reflecting surface. Such devices are often difficult to fabricate with sufficient accuracy and smoothness. To remedy this, one often encounters X-ray focusers known as Kirkpatrick–Baez mirrors, which comprise a crossed pair of glancing-incidence mirrors that focus in orthogonal directions. Other designs, such as those associated with the names of Wolter or Schwarzschild, are also commonplace. In addition to machining curved mirror surfaces, one may also use mechanical bending of mirrors. As an extreme case of this, which parallels the so-called 'deformable mirrors' used in astronomical adaptive optics, X-ray mirrors are currently under development whose shape is able to be dynamically altered in real-time to adopt a given profile.

To the extent to which X-ray reflection from smooth curved surfaces can be treated under the geometric-optics approximation, single-surface reflective X-ray optics are non-dispersive/achromatic (i.e. their beam-shaping properties are independent of X-ray wavelength).[94] This advantage, which allows such optics to work with radiation of a broader spectral bandwidth than associated crystal optics, may be contrasted with the disadvantage of requiring glancing angles of incidence for their operation. For example, in the hard X-ray regime, critical angles on the order of several milliradians or less imply that single-surface re-

[94]More precisely, they are achromatic only for those energies that lie below the high-energy cutoff for total external reflection. For a fixed angle of incidence θ, this cutoff energy E_{cut} may be obtained from eqn (3.79), by making the replacements $\alpha^{\text{crit}} \rightarrow \theta$ and $E \rightarrow E_{\text{cut}}$, before solving for the cutoff energy.

flective optics should be a few hundred times longer than the width of the X-ray beam which they are required to shape. Thus a 1 mm wide beam will require a mirror of at least a few tens of centimetres in length, in order to be shaped by single-surface reflection. This imposes exacting requirements on the precision to which the surface ('figure') of the mirror must be machined. In addition to these stringent requirements on the figure of the mirror, the roughness of the reflecting surface must be kept sufficiently small, if the performance of the mirror is not to be overly compromised.

According to eqn (3.77), one may increase the critical angle for total external reflection, by increasing the number density of electrons in the reflecting material. Thus, one often encounters single-surface hard-X-ray reflective mirrors which are coated with heavy-element materials such as gold, nickel, platinum, iridium, or even depleted uranium oxide. Notwithstanding such a strategy, one is still confined to relatively small critical angles on the order of several milliradians, beyond which the X-ray reflectivity falls dramatically.

How can one transcend this limitation? Diffraction from near-perfect crystals, described earlier in this chapter, immediately springs to mind. While such crystal elements are commonly used in X-ray optics, let us here focus on what they have to teach regarding X-ray mirrors. To this end we recall the notion of Bragg planes in a crystal, as sketched in Fig. 3.14. In the accompanying description we saw that the relatively weak wave-field amplitude reflected/diffracted from a single atomic plane, may constructively interfere with and therefore be augmented by those waves which are reflected/diffracted from parallel atomic planes, provided that the Bragg condition is fulfilled (see eqn (3.57)). This same principle—and the same diagram—can be applied to reflecting surfaces in a multi-layer mirror, by replacing Fig. 3.14's single-surface reflecting planes DD', EE', etc. with multiple-layer reflecting surfaces of a given period. Rather than having reflection from parallel atomic planes in a crystal, one can have reflection from a series of amorphous layers in a so-called 'multi-layer mirror'. Under the kinematic approximation to X-ray diffraction from such amorphous surfaces, satisfaction of the Bragg Law may be used to greatly increase the X-ray reflectivity in comparison to that obtained from a single-layer reflecting surface. These Bragg angles may be somewhat larger than the corresponding angles for total external reflection. However, this increase in reflectivity comes at the cost of reducing the energy bandwidth of the accepted radiation when compared to the corresponding single-surface reflector. In this context, we recall the essentially achromatic nature of single-reflection surfaces under the geometrical-optics approximation, at glancing angles of incidence that are sufficiently small for total external reflection to be operative.

In multi-layer mirrors one often encounters alternating layers of a given heavy and light element. In accord with the finding of eqn (3.77), this will work to increase the overall reflectivity of the device. Further considerations, in the choice of materials for the layers of the mirror, include: (a) the demand that the multi-layer be chemically stable over timescales of several months or years; (b) thermal properties and stability of the multi-layer; (c) the ability to manufacture interfacial regions with roughness that is sufficiently small for the performance of the device to not be overly compromised.

In introducing the notion of the multi-layered mirror, we drew a strong parallel with the notion of Bragg planes in the context of the kinematical theory of X-ray diffraction. Such a simple kinematical analysis is often useful from both the qualitative and quantitative points of view, as a first approximation, when modelling the performance of multi-layer mirrors. However, the effect of multiple reflections is often significant, and so a dynamical theory is often required in order to properly model these optical elements. Indeed, dynamical diffraction theory is very often employed in the design of multi-layer structures, in sophisticated computer models which are used in order to optimize mirror performance. In particular, such models may be used to determine optimal thicknesses for the various layers in the mirror, and appropriate choices of layer materials, so as to maximize reflectivity in a given regime. If one allows the period of the multi-layer to vary with depth, computer modelling may be used to yield mirror designs that possess a pre-specified reflectivity versus angle (rocking curve) or pre-specified reflectivity versus energy for a given angle.

3.4 Refractive optical elements

Refractive optical elements such as prisms and lenses, which are two of the workhorses of visible-light optics, have X-ray-optical counterparts which will be respectively described in the following sub-sections.

3.4.1 *Prisms*

Consider the sketch in Fig. 3.23(a). Here, we see a prism OAB which is illuminated from the left by normally-incident monochromatic X-ray plane waves. While the X-ray deflection by the prism may be studied from the viewpoint of rays and Snell's Law, we here treat the problem from the viewpoint of waves and the projection approximation. In particular, under this approximation we wish to determine the phase of the coherent radiation over the nominal exit surface CD of the prism, thereby determining the deflection angle β shown in the figure.

As shown in Fig. 3.23(a), let x denote the distance between the ray striking the point $E(x)$ on the entrance face of the prism, and the optic axis z. The x-dependent projected thickness $T(x)$ of material in the prism, in the z direction, is given by:

$$T(x) = x \tan \alpha, \quad x \geq 0, \tag{3.80}$$

where α is the angle between the faces OA and OB of the prism. In what follows, we shall assume this angle to be much smaller than unity.

By symmetry, it is clear that the incident ray is undeviated upon traversing the point E on the entrance face of the prism. Rather, the ray is deviated at the point F on the exit face of the prism, to give the refracted ray G. The

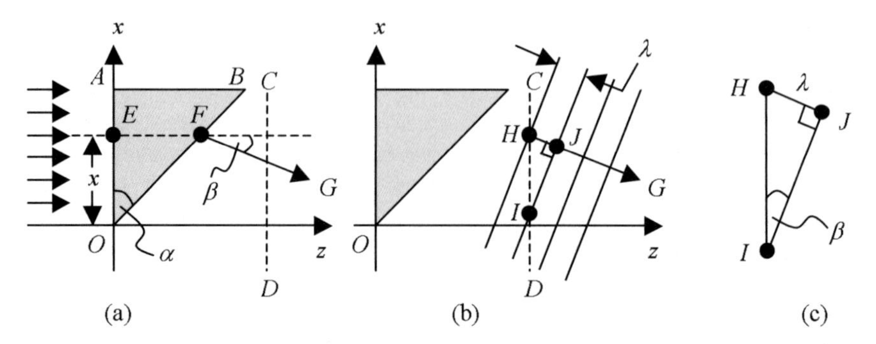

Fig. 3.23. (a) An X-ray prism OAB is illuminated from the left by normally incident monochromatic X-ray plane waves. With a view to determining the phase shift imparted by the prism over the exit surface CD, we consider a given incident ray which enters the prism at the point E, at a distance x from the optic axis z. The refracted ray G is bent through an angle β. (b) Passing from the ray to the wave picture, four parallel surfaces of constant phase are shown, corresponding to the refracted planar wave-field in the vicinity of the exit surface CD of the prism. These wavefronts have a spacing equal to the wavelength λ of the radiation. G now gives the direction of the wave-vector corresponding to the refracted plane wave. (c) Detail from (b), with deflection angle β as marked.

corresponding deflection angle is denoted by β, as marked. Since the real part $\text{Re}(n) = 1 - \delta$ of the refractive index of the prism material is very close to unity, this deflection angle will be very small.

With a view to determining this deflection angle from a wave-optical viewpoint, let us determine the position-dependent phase $\phi(x)$ over the nominal exit surface CD of the prism. Under the projection approximation, which we now assume, transverse streamline/ray deviations within the prism may be ignored (see Section 2.2). For normally incident plane-wave illumination, we may then invoke eqns (2.41) and (3.80) to see that the desired exit-surface phase $\phi(x)$ is:

$$\phi(x) = -k\delta T(x) = -k\delta x \tan\alpha \approx -k\delta x\alpha, \quad |\alpha| \ll 1, \;\; x \geq 0. \tag{3.81}$$

Before proceeding to determine the associated deflection angle β, let us do a quick back-of-the-envelope calculation to get a feel for the order of magnitude involved for X-ray phase shifts by a simple wedge. For angstrom-scale wavelengths, k will be on the order of 10^{10} m^{-1}; for wedge angles α on the order of several degrees, $\alpha \approx 0.1$ radians; lastly, the refractive index decrement may be estimated as $\delta \approx 10^{-6}$. Equation (3.81) then gives:

$$\left|\frac{\phi(x)}{x}\right| \approx k\alpha\delta \approx 10^{10}\,\text{m}^{-1} \times 10^{-1}\,\text{rad} \times 10^{-6} = 1000\,\frac{\text{rad}}{\text{m}}, \tag{3.82}$$

corresponding to phase shifts on the order of a radian per millimetre.

This estimate harmonizes with our earlier discussions on the use of transparent wedges in Bonse–Hart X-ray interferometry, as described in Section 3.2.5. In this context, let us recall Fig. 3.19, where we see an incident X-ray beam being split into two 'arms' before being recombined to form an interferogram. In the absence of a sample, and neglecting any strains or imperfections present in the monolithic crystal from which the interferometer is cut, we argued that the interferogram will be constant and featureless. If one then puts an X-ray prism in the lower arm of the interferometer, say in the region marked I in the diagram, then a transverse linear phase shift on the order of radians per millimetre will be imparted to the beam in the lower arm. Given that a typical Bonse–Hart interferometer is on the order of centimetres in size, the prism will evidently lead to several tens of linear fringes in the resulting interferogram. These linear features may be viewed as convenient 'carrier fringes', which may be distorted by introducing a sample into the upper arm of the interferometer.

We pass onto a calculation of the deflection angle β imparted on the plane wave traversing the prism in Fig. 3.23(a), from a wave-optical viewpoint. To this end, consider the sketch in Fig. 3.23(b). Here we see adjacent planar wave crests of the transmitted radiation, cutting the nominal exit surface CD of the prism. These adjacent wave-crests are separated by the wavelength λ of the radiation, as indicated in the diagram. Rather than representing the refracted ray, the vector G now indicates the direction of propagation of the transmitted plane wave. This vector, which is perpendicular to the planar wave-fronts, pierces a certain pair of adjacent wave-crests at the points H and J; the same pair of wave-crests cuts CD at the points H and I.

Form the right-angled triangle HIJ, an expanded view of which is given in Fig. 3.23(c). It is easy to see that the vertex at I subtends an angle equal to the desired deflection angle β, as indicated. Now, the phase of the transmitted plane wave changes by 2π radians as one passes from points H to J, as these are points on adjacent wave-crests. Similarly, there is a phase difference of 2π between the points H and I. Since the distance d_{HI} from H to I is given from Fig. 3.23(c) as:

$$d_{HI} = \frac{\lambda}{\sin\beta}, \tag{3.83}$$

we conclude that the magnitude of the phase gradient $d\phi(x)/dx$ in the plane CD is given by:

$$\left|\frac{d\phi(x)}{dx}\right| = \frac{2\pi}{d_{HI}} = \frac{2\pi}{\lambda/\sin\beta}. \tag{3.84}$$

Equate the right side of this expression to the magnitude of the x-derivative of eqn (3.81), before solving the resulting equation for the deflection angle β. One thereby arrives at:

$$\beta = \sin^{-1}(\alpha\delta) \approx \alpha\delta, \tag{3.85}$$

where use has been made of the fact that $|\alpha\delta| \ll 1$. Typical values for this deflection angle will be on the order of micro-radians.

3.4.2 *Compound refractive lenses*

It is only comparatively recently that refractive X-ray lenses have begun to find a place in the toolkit available to practitioners of coherent X-ray optics. Such a delay may be attributed to the closeness of X-ray refractive indices to unity, which imposes certain challenges on the fabrication of refractive X-ray focusing elements, together with a certain degree of historical inertia whereby the possible utility of refractive X-ray lenses was overlooked after having been dismissed as unfeasible.

As one of the simplest incarnations of a refractive X-ray lens, consider the sketch in Fig. 3.24(a). Here we see a cylindrical hole of radius R, which is drilled in an amorphous slab of constant thickness D. This may be considered as the X-ray equivalent of a cylindrical focusing lens for visible light, albeit an aberrated one, with the important difference that a refractive index of less than unity implies the need for a concave rather than a convex design at X-ray wavelengths.

Suppose our simple cylindrical lens to be illuminated from the left by a normally incident monochromatic scalar X-ray wave-field, as indicated by the series of parallel arrows in the diagram. Invoking the projection approximation once more, the phase $\phi(x)$ of the wave-field at the exit-surface of the cylindrical lens is proportional to the projected thickness $T(x)$ of lens material along the direction of the optic axis z, as a function of the transverse position coordinate x (see eqn (2.41) in Section 2.2):

$$\phi(x) = -k\delta T(x). \tag{3.86}$$

Here, $k = 2\pi/\lambda$ is the usual wave-number corresponding to the wavelength λ of the illuminating radiation, and the real part of the refractive index n of the material is given by $\text{Re}(n) = 1 - \delta$.

As suggested by Fig. 3.24(a), the axis of the cylindrical hole is assumed to be perpendicular to both the x- and z-axes. Further, we assume that the cylinder's axis crosses the z-axis. Thus, the projected thickness $T(x)$ of material in the lens, which is equal to the difference between the slab thickness D and the x-dependent projected thickness $2\sqrt{R^2 - x^2}$ of the cylindrical hole, is given by:

$$T(x) = D - 2\sqrt{R^2 - x^2}, \quad |x| \leq R. \tag{3.87}$$

Substitute eqn (3.87) into eqn (3.86). In the square root which appears in the resulting expression, make the binomial approximation:

$$\sqrt{R^2 - x^2} = R\left(1 - \frac{x^2}{R^2}\right)^{1/2} \approx R\left(1 - \frac{x^2}{2R^2}\right) = R - \frac{x^2}{2R}, \quad |x| \ll R, \tag{3.88}$$

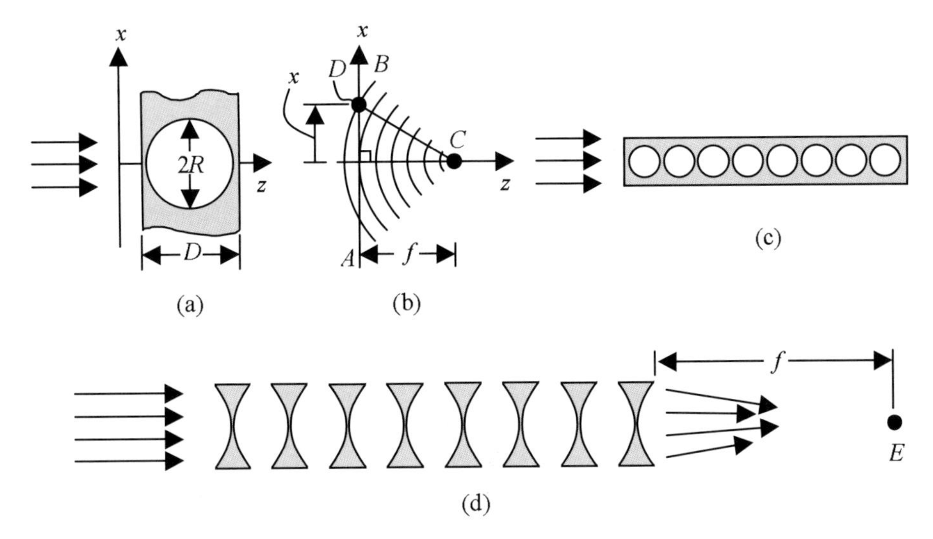

Fig. 3.24. (a) A one-dimensional refractive focusing optic comprised of a cylindrical hole of radius R, drilled in an amorphous slab of thickness D. (b) Sketch of a collapsing monochromatic cylindrical wave, the position-dependent phase variation of which is to be determined over the plane AB. The cylindrical waves converge on the line focus C. (c) Simple compound refractive lens for one-dimensional X-ray focusing, consisting of a cascaded series of cylindrical holes in an amorphous slab. (d) Compound refractive lens for two-dimensional focusing, consisting of a cascaded set of identical concave optical elements, which possess rotational symmetry about the optic axis.

and discard additive constants so as to arrive at the following expression for the phase of the wave-field over the exit surface of the refractive lens:

$$\phi(x) = -\frac{k\delta}{R}x^2. \tag{3.89}$$

Let us put this expression to one side for the moment, shifting our attention to the collapsing monochromatic scalar X-ray wave depicted in Fig. 3.24(b). In particular, we wish to determine the phase variation $\phi_{\text{cyl}}(x)$ over the plane AB, which corresponds to a cylindrical wave converging upon the line focus C with a focal length of f. If we take the line focus to correspond to zero phase, then the phase at any point $D(x)$ in the plane AB will be equal to -2π, multiplied by the number of radiation wavelengths λ that fit into a straight line whose length is equal to the distance d_{DC} between D and C. Note that the negative sign is present because the phase at $D(x)$ must be smaller than that at C, while the factor of 2π gives the phase change in radians between successive wave crests. Thus:

$$\phi_{\rm cyl}(x) = -2\pi \frac{d_{DC}}{\lambda} = -k\sqrt{f^2 + x^2} \approx -kf - \frac{k}{2f}x^2, \quad |x| \ll f, \tag{3.90}$$

where use has been made of the binomial approximation.

With the interlude of the previous paragraph under our belts, we are ready to determine the focal length f of the refractive cylindrical lens shown in Fig. 3.24(a). This may be obtained by first discarding the irrelevant additive constant $-kf$ on the right side of eqn (3.90), before equating the right-hand sides of eqns (3.89) and (3.90), and then solving for the desired focal length f. We conclude that the focal length of our cylindrical lens is proportional to the radius of curvature R of the cylindrical hole drilled into the amorphous slab, and inversely proportional to the refractive index decrement δ for the slab:

$$f = \frac{R}{2\delta}. \tag{3.91}$$

To proceed further, we seek an estimate for the magnitude of the focal length f associated with the above result. For hard X-rays, refractive index decrements typically obey the inequality:

$$\delta \leq 10^{-5}, \tag{3.92}$$

so that eqn (3.91) becomes:

$$f \geq 5 \times 10^4 R, \quad \delta \leq 10^{-5}. \tag{3.93}$$

Thus, for example, a lens with a radius of curvature of 1 mm will have a focal length of 50 m, which is unacceptably large for most practical purposes. To achieve a focal length on the order of a metre or so, the lens radius required is on the order of 10 μm, which is too small for most purposes.[95]

Historically, considerations such as those just outlined led to the conclusion that refractive lenses were not particularly useful as X-ray optical elements (Kirkpatrick and Baez 1948; Suehiro *et al.* 1991; Michette 1991; Yang 1993). This conclusion has been overruled in recent years, by the so-called 'compound refractive lens' sketched in Fig. 3.24(c) (Tomie 1994; Snigirev *et al.* 1996). Here, we see the wonderfully simple idea of having a cascaded chain of X-ray refractive lenses, which may then be illuminated from the left so as to produce a focal spot at a specified distance downstream of the optic.[96] With a total of M refractive

[95] For exceptions to this statement, see, for example, Zhang *et al.* (2001), Schroer *et al.* (2003), and Mayo and Sexton (2004). In these papers, we see that modern micro-fabrication techniques allow one to construct refractive lenses with radii of curvature on the order of microns. While we shall not pursue this point further in the text, we emphasize that micron-scale hard-X-ray refractive optical elements are a subject of continuing research. See the just-cited papers for an entry point into the literature.

[96] Note that one may reverse the ray paths in the diagram, to conclude that compound refractive lenses may also be used to collimate quasi-monochromatic radiation emerging from a small source.

lenses in series, we leave it as a simple exercise to show that the focal length f is M times smaller than the single-lens result given in eqn (3.93):

$$f = \frac{R}{2M\delta}. \tag{3.94}$$

By having a few tens of refractive lenses in series, with radii on the order of millimetres—which, importantly, is nicely matched to typical X-ray beam sizes emerging from third-generation storage-ring sources—it becomes evident that focal lengths f on the order of a metre can be achieved. Using this idea, Snigirev and co-workers demonstrated focusing of 14 keV X-rays with an array of 30 cylindrical holes drilled into an aluminium slab (Snigirev *et al.* 1996). Given that the radius of the holes was 0.3 mm and that aluminium's refractive index decrement for 14 keV X-rays is $\delta = 2.8 \times 10^{-6}$, eqn (3.94) gives a focal length of 1.8 m for the 30-element compound refractive lens. Using this device, one-dimensional focusing was demonstrated into a focal line of width approximately 8 μm.

To achieve two-dimensional focusing, one can align a pair of cylindrical compound refractive lenses at right angles to one another. However, this simple scenario will often lead to an unacceptable degree of distortion in the collapsing wavefront emerging from the exit surface of the device. Accordingly, rather than using crossed cylindrical compound refractive lenses to effect two-dimensional focusing, one more often encounters devices such as that sketched in Fig. 3.24(d). Here, we see a series of identical concave refractive X-ray lenses arranged in series, with each lens being rotationally symmetric with respect to the optic axis. This series of M optical elements is designed for two-dimensional focusing to the point marked E in the diagram.

Next, let us determine the projected thickness of each of the elements in the just-described compound refractive lens for two-dimensional focusing. Let r denote radial distance from the optic axis z. Using a trivial variant of the argument that led to expression (3.90) for a collapsing cylindrical wave, one can write down the following equation for the exit-surface phase $\phi_{\rm sph}(r)$ corresponding to a collapsing spherical wave, which one desires at the exit-surface of the device under normal-incidence illumination by a plane monochromatic scalar X-ray wave-field:

$$\phi_{\rm sph}(r) = -k\sqrt{f^2 + r^2} + kf. \tag{3.95}$$

Note that we have added the arbitrary constant kf to the above expression, so that $\phi_{\rm sph}(r = 0) = 0$. (As we shall see in a moment, this choice of constant fixes the on-axis projected thickness of the optic to be equal to its smallest possible value, namely zero, thereby minimizing the X-ray attenuation of the device.) Let $t(r)$ denote the projected thickness of any one of the M identical concave lenses sketched in Fig. 3.24(d). Adopting the projection approximation, the phase $\phi(r)$ at the exit-surface of the normally illuminated optic is given by:

$$\phi(r) = -k\delta M t(r). \tag{3.96}$$

Equating the right sides of eqns (3.95) and (3.96), and then solving for $t(r)$, we arrive at the following expression for the projected thickness of each concave lens in the aberration-free M-element rotationally symmetric compound refractive lens for two-dimensional focusing (cf. eqn (3.48)):

$$t(r) = \frac{\sqrt{f^2 + r^2} - f}{M\delta}. \tag{3.97}$$

This is a key result, which we now show to be very well-approximated by a paraboloid. To this end, insert the following binomial-series expansion:

$$\sqrt{f^2 + r^2} \approx f + \frac{r^2}{2f} - \frac{r^4}{8f^3} + \frac{r^6}{16f^5} - \cdots \tag{3.98}$$

into eqn (3.97), to see that:

$$t(r) \approx \frac{r^2}{2fM\delta} \left(1 - \frac{r^2}{4f^2} + \frac{r^4}{8f^4} - \cdots \right). \tag{3.99}$$

To first order in r/f, which is evidently an excellent approximation on account of the fact that $r \ll f$, the bracketed series may be replaced by unity to give the following paraboloidal profile for each element in the compound two-dimensional X-ray refractive lens:

$$t(r) = \frac{r^2}{2fM\delta}, \quad r \ll f. \tag{3.100}$$

For a survey of the development and application of paraboloidal compound refractive lenses, see Lengeler *et al.* (2001), and references therein. A thorough account, of the theory of image formation using such lenses, is given by Kohn (2003).

Regarding the choice of material from which compound refractive lenses may be fabricated, we note that such optics are often made from light elements such as aluminium, beryllium, boron, and lithium, or organic materials such as acrylic, polyethylene, polymethylmethacrylate, and mylar. Such light materials are chosen so as to minimize the effects of X-ray absorption, a factor neglected in the above simple analysis. Indeed, photoelectric absorption (see Section 2.10.2) will lead to an apodization of the beam, namely an increasing attenuation with distance from the optic axis. Such absorption limits the performance of compound refractive lenses at the low-energy end of the X-ray spectrum, with Compton scattering (see Section 2.10.1) limiting their performance at the high-energy end.

In addition to considerations pertaining to absorption, other factors governing the choice of refractive-lens material may include one or more of the following: resilience to radiation damage, high thermal conductivity, low thermal coefficient of expansion, chemical stability in air, ease of manufacture under the requisite techniques (e.g. drilling, stamping, lithography followed by reactive ion etching, etc.), ability to be polished to sufficient smoothness, monetary cost, and non-toxicity.

3.5 Virtual optical elements

In the previous chapter, we saw that kinematical X-ray diffraction by small crystalline samples yields far-field intensity distributions consisting of a series of spots (see Fig. 2.7). The science of X-ray crystallography aims to take one or more such diffraction patterns and thereby infer the structure of the crystal. Such a reconstruction may be properly spoken of as an atomic-scale 'image' of the crystal, notwithstanding the fact that the image is indirectly obtained via the analysis of diffraction spots that bear no direct resemblance to the crystal structure.

X-ray optics, coherent or otherwise, offers many instances of such indirect imaging methodologies. For example, the aim of X-ray tomography is to reconstruct a three-dimensional representation of a sample, given a series of two-dimensional projection images of the same. As another example, we note that interferometry seeks to reconstruct both the amplitude and phase of a given two-dimensional coherent scalar wavefront, given one or more interferograms, which are formed when the wavefront is made to interfere with a known reference wave that is often plane. As a last example we note that, like interferometry, in-line holography also seeks to reconstruct a given two-dimensional wavefront in both amplitude and phase, given one or more diffraction patterns in either the Fresnel or Fraunhofer regime.

What all of the above have in common is that imaging proceeds via a two-step process—data recording followed by image reconstruction (cf. Gabor (1948)). In the first step, data is obtained by using suitable detectors to record the intensity distributions output from physical imaging systems composed of hardware optical elements. Such optical elements are drawn from the previously described X-ray toolkit which includes interferometers, mirrors, beam-splitters, analyser crystals, and so forth. The recorded intensity data may comprise a single image, or a series of images obtained using different states of the hardware optical system (e.g. variable focus settings of an X-ray imaging system, varying sample-to-detector propagation distances for in-line holography, varying orientations of the object for tomography[97]). In the second step, namely image reconstruction, one takes the output data from the first step, together with any relevant *a priori* knowledge and pertinent assumptions, so as to computationally reconstruct certain information regarding the sample under examination by the imaging device.[98] It is this second stage, in which the computer forms an intrinsic part

[97]One obtains the same image if the object is rotated and the imaging system is left unchanged, or if the object remains static with the imaging system being rotated. In many contexts, such as laboratory-source or synchrotron-source X-ray tomography, one usually rotates the sample. Nevertheless, in view of the equivalence mentioned above, we will consider such scenarios to correspond to different states of a given imaging system, as assumed in the main text.

[98]Characterization of the optical system itself is considered to be a special case of this statement, in which sample and optical system are one and the same. Examples of such characterization may be a measurement of the aberrations of a coherent imaging system, determination of the spatial distribution and/or coherence properties of the source which illuminates a par-

of the imaging system (Lichte *et al.* 1992, 1993), which may be referred to as 'virtual optics' (Yaroslavsky and Eden 1996; Paganin *et al.* 2004).

Let us illustrate these notions, with reference to the four examples encountered in the opening paragraphs to this section. (i) In the object-reconstruction problem of crystallography, one has the software equivalent of an X-ray lens, which allows one to obtain the three-dimensional crystal structure that results in a recorded series of diffraction patterns. Such a 'software lens' is employed on account of the lack of suitable X-ray hardware that can perform the same function, at the desired resolution. Note that the reconstruction is intimately related, whether implicitly or explicitly, to the question of determining the phase of the registered Fraunhofer diffraction patterns of the sample, given that such information is lost in recording the squared modulus (i.e. the intensity) of these patterns. (ii) In the simplest incarnation of X-ray tomography, one seeks to reconstruct the three-dimensional distribution of the linear attenuation coefficient of a sample given projection radiographs of the same, which are obtained for a number of different sample orientations. Here, the virtual optics are provided by the computer embodiment of the so-called inverse Radon transformation, which allows one to transform the projection images into the desired three-dimensional reconstruction of the sample. Unlike the other cases to be discussed in this paragraph, the question of wave-field phase is irrelevant. (iii) In two-dimensional interferometry using coherent X-ray radiation, the notion of wave-field phase is paramount. Indeed, one typically first seeks to determine the two-dimensional phase of the wave-field at the exit-surface of a thin object illuminated by X-ray radiation of sufficient coherence for an interferogram to be formed. Having determined the phase of the exit-surface wave-field of the sample, one can then address the inverse-scattering problem of determining the structure of the object itself. If the projection approximation may be assumed, then the inverse scattering problem may be easily solved to yield projected information regarding the sample. Specifically, the exit-surface phase and intensity maps serve as respective inputs into eqns (2.40) and (2.42), each of which may be easily solved to respectively yield the projected real and imaginary parts of the three-dimensional complex refractive index of the sample. If three-dimensional information is required, and the projection approximation is valid, then the previously mentioned methods of tomography may subsequently be employed to reconstruct the three-dimensional complex refractive index of the sample, given as input a number of interferograms obtained for different sample orientations. (iv) Our remarks on holography are the same as those for interferometry, with the exception that the means of phase reconstruction are different.

For instances of virtual optics that involve reconstruction of both the intensity and the phase of a coherent scalar disturbance at the entrance surface of an imaging system, as is the case for interferometry and holography, one can go a

tially coherent optical system, measurement of the surface roughness of mirrors, determination of the spectral distribution of the radiation striking the entrance surface of an optical system, etc.

step further. This two-dimensional coherent wave-field contains total knowledge of the information imprinted upon the X-rays upon their passage through the sample. Therefore, if one has determined both the intensity and phase of this field, one may use software to emulate the subsequent action of an arbitrary coherent imaging system which may take such a field as input.[99] Such hybrid physical-virtual optics may be dubbed 'omni optics' on account of their great flexibility (Paganin *et al.* 2004). The virtual optics may be the computational embodiment of optical elements which correspond to hardware optical elements, although this need not necessarily be the case. Indeed, such virtual optics are of particular utility in contexts where it is impracticable or impossible to employ the corresponding hardware optics.

The main thrust of the preceding discussions is summarized in Fig. 3.25. Here we see a source A illuminating a sample B, with the resulting field over the sample's exit surface C subsequently propagating to the entrance surface D of a physical imaging system. One or more intensity distributions may be recorded by this physical optical system, corresponding to different states of the system. If one has both sufficient intensity data and the requisite algorithms, then one can reconstruct both the intensity and phase of the radiation over the entrance surface of the imaging system. One can then emulate, computationally, the subsequent action of an arbitrary imaging system in a given state parameterized by the set of real numbers $\tau \equiv \{\tau_1, \tau_2, \cdots\}$. Such a virtual imaging system may then yield an image over the virtual imaging plane E. Conversely, if one does not reconstruct the intensity and phase of the entrance-surface wave-field, one can nevertheless use virtual optics to computationally obtain a useful image, or series of images, from the measured input data.

3.6 X-ray detectors

Having discussed X-ray sources and optical elements, we are ready to move on to the third and final component which is key to all X-ray optical systems: detectors. We begin our treatment with a listing of the critical detector parameters for experiments in coherent X-ray optics. We then introduce several of the more important types of X-ray detector, briefly examining some of their relative strengths and weaknesses. We close with a discussion of the relation between detectors and coherence.

3.6.1 *Critical detector parameters for coherent X-ray optics*

Below we enumerate certain key detector parameters for coherent X-ray optics:

1. Spatial resolution: If one wants to know positional information regarding X-ray photons, either in one or two dimensions, then the spatial resolution of the detector should be sufficiently fine that the intensity of the

[99] As an obvious caveat, the exit-surface wave-field will only be reconstructed up to a given resolution limit. When emulating the subsequent action of a coherent imaging system, one will, in general, not be able to access information at length scales finer than this resolution limit.

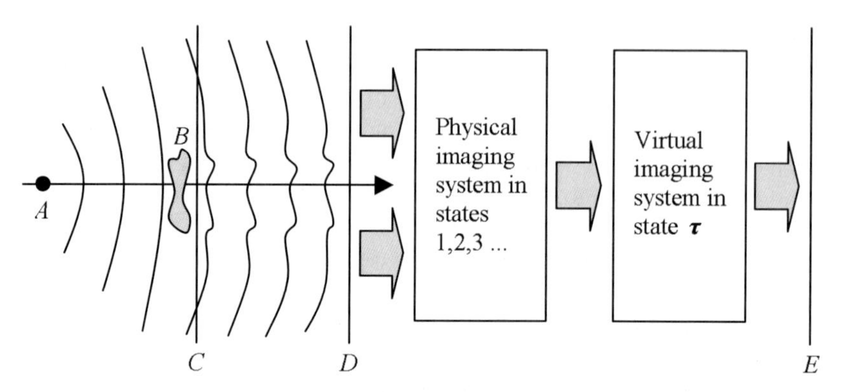

Fig. 3.25. A source A of monochromatic radiation is elastically scattered by a sample B to yield a given complex disturbance ψ_C over the plane C. This exit-surface disturbance, into which is imprinted partial information regarding the sample, then propagates to the entrance surface D of a given imaging system. Several images may then be registered, corresponding to different states of the imaging system. If these images allow one to reconstruct both the amplitude and phase of the complex disturbance ψ_C over the plane C, then one has total knowledge of the information encoded in the exit-surface wave-field by the object. 'Virtual optics' refers to the use of software, that emulates the subsequent action of any coherent imaging system, which takes the reconstructed wave-function ψ_C as input so as to yield the output image over the virtual imaging plane E. The state of this virtual imaging system is parameterized by the vector $\tau \equiv (\tau_1, \tau_2, \tau_3, \cdots)$ of real numbers $\tau_1, \tau_2, \tau_3, \cdots$. After Paganin *et al.* (2004).

X-ray field is adequately sampled. For one-dimensional position-sensitive detectors, arrays on the order of 1000 pixels or more are desirable; for two-dimensional devices, arrays on the order of 1200×1200 pixels or more are desirable. Useful pixel size can range from microns to tens of microns or more, depending on the application. Note that, in principle at least, a partially coherent X-ray wave-field carries information over length scales that are as small as the shortest wavelength for which the X-ray spectrum is non-negligible—thus there is 'plenty of room at the bottom' regarding the finest desirable pixel size quoted above.[100]

2. Temporal resolution: In order to adequately sample an X-ray beam in the context of time-resolved studies, the temporal resolution of the detector must be sufficiently fine. As an extreme example, recall that free-electron lasers are anticipated to yield hard-X-ray pulses with durations on the or-

[100]Having said this, we note that there appear to be fundamental physical limitations regarding the finest pixel size attainable. Such limitations include those due to ejected photoelectrons and Compton scattering, which are associated with the photon detection process.

der of hundreds of femtoseconds or less (see Section 3.1.4).[101] Rather less demanding is the question of performing time-resolved experiments using third-generation facilities, with storage-ring electron bunches emitting pulsed X-rays with characteristic timescales on the order of several tens of picoseconds or more. Much more modest time resolutions are often useful: for example, temporal resolutions on the order of several hundredths of a second are often adequate for the real-time imaging of biological systems such as breathing lungs and beating hearts, or in obtaining time-resolved far-field diffraction patterns of contracting muscle.

3. Spectral resolution: Spectral information, whether or not it be coupled with position information, is often useful in the context of coherent X-ray optics. Note that energy sensitivity in a detector may be viewed as a form of monochromatization of the radiation, with the energy filtering being performed by the detector rather than by an optical element such as a crystal or a multi-layer mirror. Ideally, the spectral resolution should be at least one or two orders of magnitude finer than the bandwidth of the spectra under study, although useful information can nevertheless often be gained if the spectral resolution does not meet this criterion.
4. Quantum efficiency: Whether or not one seeks to discriminate individual X-ray photons, it is often desirable that: (i) a single photon should produce a measurable signal in a given detector; and (ii) the measured signal should be proportional to the number of photons that impinge upon the detector. If both of these criteria are fulfilled, then the detector is said to have unit quantum efficiency. Many common X-ray detectors have near-unit quantum efficiency, or some appreciable fraction thereof, when operated under appropriate conditions.
5. Detective quantum efficiency (DQE): The DQE of a detector may be defined as the squared ratio, of the output signal-to-noise ratio (SNR) to the input SNR, produced by a given detector under specified conditions. Typical detector DQEs, expressed as a percentage, lie in the range from 1% to 10%. The DQE will, in general, depend on a number of factors, including the frequency (or mean frequency, as appropriate) of the X-ray photons incident on the detector, together with the intensity of the field incident on the detector.
6. Noise: Detector noise, which it is evidently desirable to minimize, is a random perturbation in the detected X-ray signal which originates from a variety of mechanisms. For detectors that require a signal to be read out, there is an associated 'read-out noise'. In the absence of any X-ray

[101]Note, however, that in view of the energy–time uncertainty principle—which implies a reciprocal relation between pulse duration and the width of the associated spectrum—that useful information regarding ultra-short X-ray pulses may be obtained from low-resolution spectra of the same. Indeed, such a strategy is employed by experimental particle physicists, in using invariant-mass histograms to measure lifetimes on the order of 10^{-23} seconds for short-lived hadrons.

illumination, detectors will always produce a signal, known as 'dark noise'. Both read-out noise and dark noise may be reduced by cooling the detector, where applicable. Lastly, there is a detector-independent source of noise known as 'photon shot noise', which is a manifestation of the quantized nature of the electromagnetic field.

7. Dynamic range: The dynamic range of a detector, which may be defined as the ratio of the largest detectable signal to the smallest detectable non-zero signal, should be sufficiently large. Imaging experiments typically have rather modest requirements on dynamic range, with 12-bit information often being adequate. Conversely, diffraction experiments typically demand many decades of dynamic range from a detector.
8. Size: Detector size should be appropriately matched to the application.
9. Acceptable photon count-rate: For detectors which count individual X-ray photons, the rate, at which photons pass through the detector, cannot be too high. Sometimes it is necessary to attenuate an X-ray beam, if the photon-counting detector is not able to cope with the rate at which incident photons are incident upon it.
10. Stability: When two given X-ray inputs are close to identical, differing only in the commencement time of the temporal interval during which they are detected, and with all other relevant variables being kept fixed (e.g. temperature and pressure), the resulting signal outputs should also be close to identical.
11. Longevity: Some detectors have a limited lifetime on account of their susceptibility to radiation damage. If this is the case, the useful lifetime of the detector must be acceptably long.
12. Cost: Monetary cost, for both detector development and for the purchase of individual detectors, may be listed among the key detector parameters. Note that costs may be reduced by adapting and/or co-developing detector technology, for example, with the high-energy-physics or astrophysics communities.

3.6.2 *Types of X-ray detector*

X-ray detectors may be broadly classified into two groups, respectively known as 'counting detectors' and 'integrating detectors'. Counting detectors, as their name implies, are able to detect and count individual X-ray photons. Integrating detectors, on the other hand, do not seek to count individual photons but rather accumulate a photon-induced signal for a given period of time, prior to the integrated signal being read out.

Examples of both classes of detectors will be examined in the following subsections. Section 3.6.2.1 treats three common types of integrating detector: film, image plates and charge-coupled devices. Section 3.6.2.2 discusses several forms of counting detector: proportional counters, multi-wire proportional counters, strip detectors, scintillators, and pixel detectors.[102] Lastly, Section 3.6.2.3 offers a brief comparison of the relative merits of the two classes of detector.

3.6.2.1 *Integrating detectors* Film is our first example of an integrating detector. This medium occupies a special place in the history of X-ray optics, for at least two reasons: (i) It was key in the discovery of X-rays, it having being recorded in Röntgen's first X-ray paper that 'Films can receive the impression' of this radiation (Röntgen 1896); (ii) Until rather recently, X-ray film was the detector of choice for two-dimensional imaging. Advantages of film include high dynamic range, high sensitivity, large area, high resolution, low cost, and the ability to be bent, together with an independence of the vicissitudes of evolving digital image formats and data-reading hardware. In addition, film performs the dual functions of data storage and display, these functions being separated in most other forms of detector. Disadvantages of film include a non-linear response to the energy per unit area which is deposited by X-rays on the film, the need to digitize the signal and correct for non-linear response if quantitative analyses are to be undertaken, degradation of developed film with age, the time delay between recording and development, and the inability to provide spectral information.

Image plates, developed in the early 1980s, are an evolution of the film concept. These plates, which may be several tens of centimetres or even metres in size, record two-dimensional images via X-ray photons exciting atoms in grains of their phosphor layer. Atoms in the phosphor are thereby rendered into a long-lived excited state. Rather than being developed, as is the case with film, image plates are subsequently read out by being digitized using a suitable scanner. This scanner passes a small visible light beam over the metastable excited atoms in the image plate, which subsequently release photons to provide a measure of the energy per unit area imparted to the plate. Having been read out, the plate may be optically erased prior to re-use. Image plates have most of the advantages listed above for film, together with the added advantages of linear response over a wide dynamic range, the ability to provide digitized image data, and re-usability. Disadvantages include the inability to work in real time[103] or close to real time, together with the inability to provide spectra at each pixel.[104]

Charge coupled devices (CCDs) are our third example of an integrating detector. The X-ray equivalent of the imaging chip in a visible-light digital camera,

[102]Notable omissions, from the above list, include micro-channel plates, streak cameras, drift chambers, TFT ('thin-film transistor') arrays, and CMOS ('complementary metal-oxide semiconductor') imagers.

[103]Note, however, that image plates may be used to achieve time-resolved (but not real-time) imaging of signals that vary in one spatial dimension. This may be done by suitably scanning the image plate during the detection process.

[104]Regarding this latter point, note that crude spectral information may be obtained by sandwiching an absorbing sheet between two image plates, prior to simultaneously exposing them. On account of the filtration of the X-ray spectrum in passing through the sheet from the first plate to the second, a degree of position-sensitive spectral information may be obtained.

the CCD is a two-dimensional array of square or rectangular pixels which is typically on the order of 1000×1000 elements in size. CCDs are semi-conductor based detectors, where each pixel accumulates electron charge in proportion to the amount of photon energy passing through the pixel. The 'buckets of charge' in each pixel may then be read out, in a 'bucket brigade' whereby consecutive pixels in a given row are read out one after the other. Advantages of CCDs include high spatial resolution, the ability to provide intrinsically digital data with excellent linearity over a moderate dynamic range, the ability to provide close to real-time imaging and re-usability. Disadvantages include a lower dynamic range and smaller image area when compared to both film and image plates, slow readout time and susceptibility to radiation damage.[105]

3.6.2.2 *Counting detectors* Gas-based detectors are the first class of counting detector to be considered here. (i) In a proportional counter, one has a gas-filled chamber permeated by a non-uniform electric field. This field exists between positive and negative electrodes, respectively consisting of a wire and a surrounding sheath, which are held at a constant potential difference with respect to one another. When an X-ray photon traverses the gas-filled volume of this detector, one or more of the gas atoms may be ionized via the photo-electric effect (see Section 2.10.2). In the vicinity of the wire, towards which ejected photo-electrons will be accelerated, the magnitude of the electric field increases as the reciprocal of the radial distance from the wire. Due to the sharp acceleration which photo-electrons experience in the vicinity of the wire, the resulting avalanche of knock-on electrons (i.e. electrons which are ejected from atoms as a result of being struck by accelerating electrons in the detector) is sufficiently large for individual X-ray photons to be detected. Importantly, the magnitude of the resulting electric pulse often permits one to determine the energy of the incident X-ray. Provided that the duration of the pulse is sufficiently small compared to the mean time between photons, X-ray-photon energies may then be counted one photon at a time. The energy resolution is typically rather poor, at around the 10% level (i.e. the ratio of the uncertainty ΔE in measured energy, to the energy E itself, is on the order of 0.1). If the photon count-rate is too high, the overlapping of various pulses will lead to a loss in the ability to determine the energy and arrival time of each photon. (ii) The proportional counter gives no positional information regarding the detected X-ray photons. This state of affairs can be remedied using multi-wire proportional counters or strip detectors,

[105]Note that use can be made of so-called 'image intensifiers', to improve the sensitivity of CCD cameras in the presence of low X-ray light levels. These amplifiers are analogous to the phosphor screens one encounters in electron microscopes. X-ray image intensifiers emit photons in the visible-light regime, in response to illumination by X-ray photons. The visible light emitted by the screen may then be recorded in a number of ways, such as (i) use of a visible-light optical system to produce a magnified or de-magnified image of the phosphorescent screen which is matched to the chip size of a visible-light CCD camera; (ii) use of optical fibres to direct the light from the phosphorescent screen to each of the pixels of a visible-light CCD camera.

in which positional information is obtained by respectively replacing the single higher-potential wire by a series of parallel wires or strips. Like the proportional counters, multi-wire proportional counters and strip detectors can provide crude spectral information with $\Delta E/E \approx 0.1$.

Scintillators are another common class of counting detector. Scintillator materials, such as sodium iodide or caesium iodide, yield visible-light photons when illuminated with X-ray photons. These visible-light photons may then be detected via a photo-multiplier tube, using a 'light pipe' if necessary in order to transfer the visible-light photons from the scintillator to the photomultiplier. Energy resolution is typically very poor, with $\Delta E/E$ on the order of 0.2.

Notwithstanding the fact that it has already been described as an integrating detector, the CCD may also function as a counting detector. Indeed, consider a CCD bathed in an X-ray beam for which the exposure time can be made sufficiently short that during this period the CCD captures a number of photons N, which is much less than the number of pixels in the detector. Ignoring all but those events which illuminate a single pixel, and making the often rather good assumption that the signal recorded by the pixel is proportional to the energy of the photon that excited that pixel, one can then make an estimate for the energy of the photon. If both the statistical properties of the beam and the illuminated sample change little over timescales much longer than the exposure time for a single frame, then many such exposures may be sequentially made so as to build up a spectrum at each pixel of the CCD array. The energy resolution, achieved with such a strategy, can approach $\Delta E/E \approx 0.01$. Such spectral imaging, in which one obtains both spectral and positional information, is evidently rather desirable.

This desideratum leads to our final example, the 'pixel detectors', which are currently under active development. Here, the basic idea is massive parallelism: by having an energy-sensitive counting detector associated with each pixel of an imaging device, which may be either one-dimensional or two-dimensional, one can have both spectral and spatial sensitivity. While the idea is simple enough in principle, in practice one must deal with the engineering complexity that it entails. The requisite challenges are currently being addressed, promising the advent of powerful new pixel detectors in the near future.

3.6.2.3 *Comparison of counting and integrating detectors* The following broad remarks may be made in comparing counting detectors to integrating detectors:

1. Counting detectors typically have a greater dynamic range than integrating detectors, often at the cost of reduced spatial resolution.
2. Integrating detectors typically have better spatial resolution than counting detectors, often at the cost of reduced dynamic range.
3. Readout noise and dark noise are almost always higher for integrating detectors than for counting detectors.
4. Counting detectors can only tolerate photon count rates that lie below a certain threshold, as they require sufficient time between photon-induced

signals in order for individual photons to be discriminated. Integrating detectors are not limited by photon count-rate, and can therefore operate in regimes where the count-rate is too high for counting detectors. However, the exposure time must be able to be made sufficiently short that the integrating detector is not saturated.

5. Energy resolution only exists for counting detectors. Energy resolution of scintillation counters is rather poor ($\Delta E/E \approx 0.2$). This energy resolution may be improved by a factor of approximately 2 to 3, for proportional counters, multi-wire proportional counters, and strip detectors. Semi-conductor counting detectors, including the photon-counting mode of the CCD, can have energy resolutions on the order of $\Delta E/E \approx 0.01$.

3.6.3 *Detectors and coherence*

At X-ray frequencies, the oscillation rates of electromagnetic fields are too fast to be directly resolved by existing detectors, a limitation which appears unlikely to change in the near future. Accordingly, the signals output by X-ray detectors necessarily represent an average over many cycles of wave-field oscillation. This observation, in the more general context of partially coherent electromagnetic disturbances at visible and higher frequencies, was considered to be an appealing factor in the development of the modern coherence theory based on correlation functions (Wolf 1954). In this context, recall the first chapter's discussions on the mutual coherence function, which was seen to be a measurable function of intensities that may be obtained via the Young's double-slit experiment.

This is indicative of the intimate relationship that exists between coherence and detectors, aspects of which it is the purpose of this section to elucidate. Before commencing, it serves us to summarize certain salient aspects of the theory of partially coherent fields, some of which were examined in the second half of the Chapter 1: (i) A given realization of a partially coherent field may be represented as a generalized Fourier integral, which decomposes the field as a sum of monochromatic plane waves, of all possible directions, energies, and polarizations; (ii) By virtue of the linearity of the free-space Maxwell equations, the various plane-wave components of a particular realization of a partially coherent field will obey the superposition principle *in vacuo*—however, this superposition principle will, in general, break down in the presence of a material medium such as a detector; (iii) The spectral width of a partially coherent field is inversely proportional to a characteristic time known as its 'coherence time'[106]; (iv) While the time-averaged intensity of a given partially coherent field may contain either no diffraction fringes or fringes of negligibly low visibility, the instantaneous in-

[106]More precisely, one has the energy–time uncertainty principle $\tau\Delta E \approx \hbar$, where ΔE is the energy width of the spectrum, τ is the coherence time, and $\hbar$ is the reduced Planck constant. Dividing through by $\hbar$, and making use of the fact that $\Delta E/\hbar = \Delta\omega$, where $\Delta\omega$ is the spread in angular frequencies of the radiation, we see that $\tau \approx 1/\Delta\omega$. Thus, roughly speaking, the coherence time is equal to the reciprocal of the spread in angular frequencies of the partially coherent field.

tensity of a given realization of the field will typically be much more highly structured; (v) While an infinite hierarchy of correlation functions is required to describe an otherwise arbitrary statistically stationary partially coherent scalar electromagnetic field, the second-order mutual coherence function is sufficient to describe the statistical aspects of the field if its statistics are Gaussian. However, the mutual coherence function does not contain enough information to describe the instantaneous intensity of a particular realization of the field, nor does it contain enough information to describe experiments in which optical intensities are averaged over times that are shorter than the coherence time.

As a first example of the relation between detectors and coherence, let us return to the quasi-monochromatic Young experiment as sketched in Fig. 1.5. In this diagram, suppose that the distance $d_{S\Sigma}$ between the quasi-monochromatic incoherent line source S and the pinhole-pierced screen Σ is reduced to zero, with the source being made sufficiently wide to cover both pinholes. One would then have a Young's experiment in which light from two independent radiators was superposed to form a pattern over the plane Π of observation. Typically, one would record this pattern by using a position-sensitive integrating detector to build up an image over timescales much longer than the coherence time of the radiation. In this case, no interference fringes are observed. However, if one were to employ a position-sensitive detector, with sufficient temporal resolution to record an image with an exposure time somewhat less than the coherence time of the field, then a non-zero fringe visibility will be observed. How does this arise, given that it runs counter to the oft-repeated statement that 'each photon ... only interferes with itself' (Dirac 1958)? With a view to answering this question, recall that the quasi-monochromatic disturbances from each of the two independent radiators will be approximately sinusoidal during an exposure whose duration is much smaller than the coherence time of the field (see eqns (1.77) and (1.78)). The relative phase of these two disturbances will be random but approximately fixed during the short exposure, so that fringes will be formed upon superposing the fields. These fringes will randomly jitter over timescales larger than the coherence time, which is why they are washed out if the exposure time is much longer than the coherence time. The formation of interference fringes, by juxtaposing light from two independent quasi-monochromatic sources and recording the resulting pattern over a timescale sufficiently shorter than the coherence time of the radiation, was demonstrated in the visible region by Magyar and Mandel (1963). The same principle holds good for X-rays, notwithstanding the greater difficulties of experimental implementation.

The above example demonstrates that sufficient temporal resolution in a position-sensitive X-ray detector may open up qualitatively new areas of coherent X-ray optics, which are not accessible when optical information is gathered by averaging over timescales much longer than the coherence time of a partially coherent X-ray field. Note, in this context, that time-averaged or ensemble-averaged correlation functions are inadequate for an analysis of optical experiments involving exposure times shorter than the coherence time of the radiation (see, for example, Magyar and Mandel (1963)). Such analyses are perhaps most directly obtained via a spectral decomposition of a particular realization of the partially coherent field.

As a second example of the relationship between detectors and coherence let us consider a counting detector, not necessarily position-sensitive, which has a sufficiently fine temporal resolution to detect the arrival time of individual photons. In being sensitive to the quanta of the electromagnetic field, namely photons, new areas of coherent X-ray optics may be opened up. For a proper analysis of such experiments, a quantum theory of coherence may be required (for text-book accounts, with references to the primary literature, see, for example, Mandel and Wolf (1995) and Loudon (2000)). Quantum coherence theory includes as a limiting case the classical coherence theory outlined in Chapter 1. Importantly, quantum coherence theory encompasses phenomena that cannot be described using the classical theory.

In discussing the relationship between detectors and coherence, let us pass from considerations of temporal resolution to those of spatial resolution, noise, and dynamic range. As a simple albeit idealized example about which these discussions shall revolve, let us consider the interference of two planar monochromatic scalar electromagnetic waves ψ_1 and ψ_2, which are inclined with respect to one another so as to produce a series of linear sinusoidal fringes over the surface of a pixellated two-dimensional position-sensitive detector. Under these approximations, the intensity $I(x, y)$ over the imaging plane will be independent of time, being given by the following expression:

$$I(x,y) = |\psi_1 + \psi_2|^2 = |\psi_1|^2 + |\psi_2|^2 + A\cos(Bx + C). \tag{3.101}$$

Here (x, y) are Cartesian coordinates in the image plane which are such that the linear fringes are parallel to the y-axis, $A \equiv 2|\psi_1\psi_2|$, B is a positive real number which is a function of the angle between the interfering plane waves, and C is a real constant that is a function of the relative phase between the interfering plane waves. The first two terms, on the right side of the above equation, represent the uniform intensity predicted by the ray theory of light. The third term, which gives linear sinusoidal fringes, is indicative of the interference between the two coherent plane waves. Let us now bring a position-sensitive detector into the picture, respectively considering the effects of spatial resolution, noise, and dynamic range in the following three different scenarios. (i) Suppose that the period $2\pi/B$ of the fringes is very much smaller than the pixel size of the position-sensitive detector. Then, neglecting any subtleties due to aliasing, the detected intensity will be insensitive to the interference term present in eqn (3.101). The detected intensity will therefore be well approximated by the geometric theory, which neglects any interference between the two plane waves. From a wave perspective, the coarse spatial resolution of the detector has not allowed it to sample the field over a sufficiently fine length scale, to be sensitive to the interference fringes present in the field. From a coherence-theory perspective, the effects of field correlations are invisible to the detector in this instance. Regardless of the perspective adopted, in considering this idealized example, the point is clear: spatial resolution of one's detector may be a key factor in determining whether or not an experiment is sensitive to the coherence properties of a given X-ray wave-field. (ii) Suppose, instead, that the period of the fringes is on the order of a few pixel widths, so that the interference pattern may be adequately sampled by the detector. However, further suppose that the readout noise of the detector is sufficiently high, adopting the idealization that the noise between adjacent pixels is uncorrelated. The readout noise may then be described as 'white', meaning that its two-dimensional Fourier transform has an approximately constant modulus for each of the spatial frequencies to which the detector is sensitive. If this modulus is sufficiently large, that is the detector is too noisy, the signal due to the sinusoidal fringes will not be resolvable. (iii) Suppose that $|\psi_1| \gg |\psi_2|$, so that one of the interfering plane waves is much weaker than the other. This implies that the amplitude $A = 2|\psi_1\psi_2|$ of the sinusoidal interference fringes will be much less than the mean intensity $|\psi_1|^2 + |\psi_2|^2 \approx |\psi_1|^2$ of the interference pattern. One therefore has the problem of detecting a rather weak signal in the presence of a strong one. If the dynamic range of the detector is not sufficiently high, then these small-amplitude interference fringes will not be resolved against the large uniform background intensity.

Having separately considered the roles of temporal and spatial resolution in detectors for coherent X-ray optics,[107] together with noise and dynamic range, let us turn to the question of energy resolution. As an example, let us consider the interference pattern produced by two small spatially separated independent localized sources of polychromatic scalar X-ray radiation. For fields which do not possess any particular symmetries the complex time-dependent beating, between the different monochromatic components of a given realization of this partially coherent field, will typically yield a speckled distribution in both space and time (cf. Chapter 5).[108] We assume, as is often but not always the case, that this speckled distribution evolves over spatial and temporal scales that are much shorter than the detector's spatial and temporal resolutions, respectively. This speckled structure is therefore washed out in forming the time-and-space-averaged intensity measured by a position-sensitive integrating detector such as a CCD camera or an image plate. Suppose, however, that one was in possession of a pixel detector which was able to yield both high-resolution spectral and spatial information. Rather than producing a single two-dimensional pixellated image, such a detector would produce a series of such images, each of which would cor-

[107]Note that, in general, there does not exist a strict separation between the properties of spatial and temporal resolution.

[108]Regarding spatially distributed speckle, one can easily observe this phenomenon by bouncing light from a laser pointer off a rough wall. Those with corrective lenses should remove them before looking at the resulting patch of light on the wall; those without corrective lenses should sufficiently defocus their eyes in looking at the patch of light on the wall. In both instances, one will observe that the reflected light has a grainy appearance known as speckle. Note that this phenomenon is closely related to the notion of phase vortices, which shall be discussed in Chapter 5.

respond to a narrow spectral range in the illuminating radiation. In this sense, the detector would function as a series of monochromators, each of which have a specified narrow bandpass. Interestingly, this serves to blur the distinction, between optical elements and detectors, which has been made for the purposes of segmenting the present chapter. Of more importance in the present context, however, is the observation that if the spectral resolution of our hypothetical pixel detector is sufficiently fine, then each monochromated image registered by the detector would be a Young's two-pinhole interferogram with non-zero visibility. These patterns will pass over to speckled distributions, if the size of each of the independent radiators is not sufficiently small.

3.7 Summary

The present chapter was devoted to a study of the three key components of any experiment in coherent X-ray optics: sources, optical elements, and detectors.

Regarding sources, we began our discussions with the notions of brightness and emittance, these being simple measures of the quality of a source in the context of coherent optics. Equipped with these measures, we then treated a series of X-ray sources: fixed-anode and rotating-anode sources, the three generations of synchrotron-light sources, free-electron lasers, energy-recovering linear accelerators, and soft X-ray lasers. The last three sources, listed above, have yet to be realized at hard-X-ray energies. Both free-electron lasers and energy-recovering linear accelerators are likely to become operational in this regime, by the end of the first decade of the 21st century.

Our treatment of X-ray optical elements had four parts, which respectively considered diffractive, reflective, refractive, and virtual optics. (i) Our examination of diffractive optical elements included diffraction gratings, Fresnel zone plates, analyser crystals, crystal monochromators, crystal beam-splitters, the Bonse–Hart X-ray interferometer, Bragg–Fresnel crystal optics, and free space. Each of these was seen to have the common feature of using diffraction to transform a given input X-ray wave-field into one or more diffracted output fields. (ii) X-ray reflection from flat surfaces was examined in the ray picture, leading to the conclusion that single-surface reflective X-ray optics must operate at progressively more glancing angles of incidence as the X-ray energy increases. Such single-surface reflective optics include mono-capillaries, poly-capillaries (Kumakhov lenses), and square-channel arrays. If one wishes to go beyond the glancing angles of incidence required by single-surface reflective optics, one may employ multiple parallel reflecting layers in a so-called multi-layer mirror. (iii) Refractive optical elements, which form the mainstay of many experiments in the visible-light regime, were seen to be rather less common in coherent X-ray optics. This is due in part to the weak refracting power of most materials in the X-ray regime, as embodied by the real part of their refractive index being very close to unity. This feebleness of refraction may be addressed by having many refractive lenses in series, leading to the idea of the compound refractive lens. Alternatively, one may work with single refractive lenses that have very small radii of curvature, on the order of tens of microns. This latter possibility has recently become viable, on account of advances in microfabrication techniques. In addition to lenses, we also gave a brief treatment of prisms. These have long been a part of the toolkit of X-ray optical elements, a state of affairs which may be contrasted with the relatively late development of X-ray refractive lenses. (iv) Finally, we treated virtual optics, in which a computer forms an intrinsic part of an optical imaging system.

Detectors constitute the third and final part of any experiment in coherent X-ray optics. Depending on the context, key criteria for assessing X-ray detectors were seen to include a suitable subset of the following list: spatial resolution, temporal resolution, spectral resolution, quantum efficiency, detective quantum efficiency, noise, dynamic range, size, acceptable photon count rate, stability, monetary cost, and radiation resilience. We saw that X-ray detectors may be classified as either counting or integrating devices. Counting devices, as their name implies, are detectors which are able to count individual photons. Integrating detectors output a signal that is accumulated over a given period, during which many photons may be absorbed by the detector. Integrating detectors discussed include film, image plates, and CCDs. Counting detectors were seen to include proportional counters, multi-wire proportional counters, strip detectors, scintillators, and pixel detectors. Having introduced these various incarnations of counting and integrating detectors, a brief comparison was given of the relative merits of each, measured against some of the key criteria mentioned above. Lastly, we offered a discussion of the relationship between detectors and coherence, with some emphasis on the question of the new experiments in coherent X-ray optics which will become possible on account of advances in detector technology.

Further reading

For a broad and elementary introduction covering many of the topics in this chapter, see Margaritondo (2002). More detailed general information can be found in Attwood (1999) and the volume edited by Mills (2002b). Regarding the history of X-ray sources, good starting points are the collection edited by Michette and Pfauntsch (1996), together with the history of synchrotron radiation by Lea (1978). Textbook accounts, of the theory of synchrotron radiation, are given in Jackson (1999), Duke (2000), Wiedemann (2003), Hofmann (2004), and Clarke (2004). Conventional laser theory is developed in Loudon (2000), with a wealth of historical references to the primary literature in Mandel and Wolf (1995). For more detail on free-electron lasers, see the reviews by Murphy and Pellegrini (1998), Pellegrini (2001), O'Shea and Freund (2001), Krinsky (2002), Pellegrini and Stöhr (2003), Krishnagopal and Kumar (2004), and references therein. Textbooks treating the same subject include Marshall (1985), Brau (1990), Freund and Antonsen (1996), Saldin *et al.* (1999), and Shiozawa (2004). For a selection of key early papers on ultraviolet, extreme ultraviolet, and X-ray lasers, see the volume edited by Waynant and Ediger (1993). Attwood (1999) gives an excellent text-book account of soft X-ray lasers, with recent review articles including Daido (2002) and Tallents (2003). Diffractive X-ray optics are treated in Erko *et al.* (1996) and Attwood (1999). General treatments of diffraction gratings include the rather mathematical analysis of Wilcox (1984), together with the comprehensive text of Loewen and Popov (1997). Maystre (1993) collects many primary references on diffraction gratings. Entry points, to the fascinating early primary literature on X-ray diffraction gratings, include Compton and Doan (1925), Osgood (1927), Hunt (1927), Weatherby (1928), and Bearden (1935). For the role of gratings in the context of various methods for soft X-ray spectroscopy, see the review by Henke (1980). The volume edited by Ojeda-Castañeda and Gómez-Reino (1996) collects much primary literature on zone plates. Seminal papers, regarding the use of zone plates for focussing X-rays, include Myers (1951), Baez (1952, 1961), Kirz (1974), and Yun *et al.* (1999). See also the review article by Michette (1988), together with the textbook accounts of zone plates in Erko *et al.* (1996) and Attwood (1999). Regarding dynamical theory for the diffraction of plane waves from perfect crystal slabs, classic entry points to the literature include Zachariasen (1945), Batterman and Cole (1964), Pinsker (1978), and Authier (2001). For a review of the various designs for monochromator crystals, see Caciuffo *et al.* (1987), Authier (2001), and Mills (2002a). An interesting personal account of the history of cryogenically cooled monochromators is given by Bilderback *et al.* (2000). Regarding X-ray interferometers, see Hart (1980) and Bonse (1988), together with references therein. Bragg–Fresnel optics are reviewed in Snigirev (1995) and Erko *et al.* (1996). Regarding mono-capillaries, see the review by Bilderback (2003), while for poly-capillaries optics we refer the reader to Kumakhov and Komarov (1990) and Kumakhov (2000). For textbook accounts of X-ray mirrors, see Attwood (1999) and Als-Nielsen and McMorrow (2001). For micro-channel plates, see Angel (1979), Wilkins *et al.* (1989), Chapman *et al.* (1991), Fraser *et al.* (1993), Brunton *et al.* (1995), and Peele *et al.* (1996). Key articles on compound refractive lenses include Tomie (1994), Snigirev *et al.* (1996), and Lengeler *et al.* (2001). Regarding detectors, see the overviews by Thompson (2001), Lewis (2003), and Brügemann and Gerndt (2004). For an in-depth treatment of the fundamentals of radiation detection, see Knoll (2000). The state of the art in detector development is perhaps best appreciated by studying the proceedings of a recent conference (e.g. the series of international conferences on synchrotron radiation instrumentation, of which the latest installment is Warwick *et al.* (eds) (2004)).

References

J. Als-Nielsen and D. McMorrow, *Elements of modern X-ray physics*, John Wiley & Sons, New York (2001).

J. Andruszkow, B. Aune, V. Ayvazyan, N. Baboi, R. Bakker, *et al.*, *First observation of self-amplified spontaneous emission in a free-electron laser at 109 nm wavelength*, Phys. Rev. Lett. **85**, 3825–3829 (2000).

J.R.P. Angel, *The lobster-eye telescope*, Astrophys. J. **233**, 364–373 (1979).

M. von Ardenne, *Zur Leistungsfähigkeit des Elektronen-Schattenmikroskopes und über ein Röntgenstrahlen-Schattenmikroskop*, Naturwiss. **27**, 485–486 (1939).

M. von Ardenne, *Elektronen-Übermikroskopie*, Springer, Berlin (1940).

V.V. Aristov, A.A. Snigirev, Yu.A. Basov, and A.Yu. Nikulin, *X-ray Bragg optics*, in D.T. Attwood and J. Bokor (eds), *Short wavelength coherent radiation: generation and application*, American Institute of Physics Conference Proceedings, American Institute of Physics, New York, **147**, pp. 253–259 (1986a).

V.V. Aristov, A.I. Erko, A.Yu. Nikulin, and A.A. Snigirev, *Observation of X-ray Bragg diffraction on the periodic surface relief of a perfect silicon crystal*, Opt. Commun. **58**, 300–302 (1986b).

V.V. Aristov, Yu.A. Basov, S.V. Redkin, A.A. Snigirev, and V.A. Yunkin, *Bragg zone plates for hard X-ray focusing*, Nucl. Instr. Meth. Phys. Res. A **261**, 72–74 (1987).

L. Arzimovich and I. Pomeranchuk, *The radiation of fast electrons in a magnetic field* (in Russian), J. Phys. (USSR) **9**, 267 (1945).

D. Attwood, *Soft X-rays and extreme ultraviolet radiation: principles and applications*, Cambridge University Press, Cambridge (1999).

A. Authier, *Dynamical theory of X-ray diffraction*, Oxford University Press, Oxford (2001).

A.V. Baez, *A study in diffraction microscopy with special reference to X-rays*, J. Opt. Soc. Am. **42**, 756–762 (1952).

A.V. Baez, *Fresnel zone plate for optical image formation using extreme ultraviolet and soft x radiation*, J. Opt. Soc. Am. **51**, 405–409 (1961).

D.X. Balaic, K.A. Nugent, Z. Barnea, R.F. Garrett, and S.W. Wilkins, *Focusing of X-rays by total external reflection from a paraboloidally tapered glass capillary*, J. Synchrotron Rad. **2**, 296–299 (1995).

Yu.A. Basov, D.V. Roshchupkin, I.A. Schelokov, and A.E. Yakshin, *Two-dimensional X-ray focusing by a phase Fresnel zone plate at grazing incidence*, Opt. Commun. **114**, 9–12 (1995).

B.W. Batterman and H. Cole, *Dynamical diffraction of X rays by perfect crystals*, Rev. Mod. Phys. **36**, 681–716 (1964).

J.A. Bearden, *The measurement of X-ray wavelengths by large ruled gratings*, Phys. Rev. **48**, 385–390 (1935).

D.H. Bilderback, *The potential of cryogenic silicon and germanium X-ray monochromators for use with large synchrotron heat loads*, Nucl. Instr. Meth. Phys. Res. A **246**, 434–436 (1986).

D.H. Bilderback, *Review of capillary X-ray optics from the 2nd International Capillary Optics Meeting*, X-Ray Spectrom. **32**, 195–207 (2003).

D.H. Bilderback, S.A. Hoffman, and D.J. Thiel, *Nanometer spatial resolution achieved in hard X-ray imaging and Laue diffraction*, Science **263**, 201–203 (1994).

D.H. Bilderback, A.K. Freund, G.S. Knapp, and D.M. Mills, *The historical development of cryogenically cooled monochromators for third-generation synchrotron radiation sources*, J. Synchrotron Rad. **7**, 53–60 (2000).

D.H. Bilderback, I.V. Bazarov, K. Finkelstein, S.M. Gruner, H.S. Padamsee, C.K. Sinclair, *et al.*, *Energy-recovery linac project at Cornell University*, J. Synchrotron Rad. **10**, 346–348 (2003).

J.P. Blewett, *Radiation losses in the induction electron accelerator*, Phys. Rev. **69**, 87–95 (1946).

J.P. Blewett, *Synchrotron radiation—early history*, J. Synchrotron Rad. **5**, 135–139 (1998).

R. Bonifacio, C. Pellegrini, and L. M. Narducci, *Collective instabilities and high-gain regime in a free electron laser*, Opt. Commun. **50**, 373–378 (1984).

U. Bonse and M. Hart, *An X-ray interferometer*, Appl. Phys. Lett. **6**, 155–156 (1965).

U. Bonse, *Recent advances in X-ray and neutron interferometry*, Physica B **151**, 7–21 (1988).

G. Borrmann, *Über Extinktionsdiagramme von Quartz*, Phys. Z. **42**, 157–162 (1941).

C.A. Brau, *Free-electron lasers*, Academic Press, San Diego, CA (1990).

L. Brügemann and E.K.E. Gerndt, *Detectors for X-ray diffraction and scattering: a user's overview*, Nucl. Instr. Meth. Phys. Res. A **531**, 292–301 (2004).

A.N. Brunton, G.W. Fraser, J.E. Lees, W.B. Feller, and P.L. White, *X-ray focusing with 11-μm square-pore microchannel plates*, Proc. SPIE **2519**, 40–49 (1995).

R. Caciuffo, S. Melone, F. Rustichelli, and A. Boeuf, *Monochromators for X-ray synchrotron radiation*, Phys. Rep. **152**, 1–71 (1987).

H.N. Chapman, K.A. Nugent, and S.W. Wilkins, *X-ray focusing using square channel capillary arrays*, Rev. Sci. Instrum. **62**, 1542–1561 (1991).

J.A. Clarke, *The science and technology of undulators and wigglers*, Oxford University Press, Oxford (2004).

A.H. Compton and R.L. Doan, *X-ray spectra from a ruled diffraction grating*, Proc. Natl. Acad. Sci. **11**, 598–601 (1925).

M. Cornacchia, J. Arthur, K. Bane, P. Bolton, R. Carr, F.J. Decker, *et al.*, *Future possibilities of the Linac Coherent Light Source*, J. Synchrotron Rad. **11**, 227–238 (2004).

V.E. Cosslett and W.C. Nixon, *X-ray shadow microscope*, Nature **168**, 24–25 (1951).

V.E. Cosslett and W.C. Nixon, *X-ray shadow microscopy*, Nature **170**, 436–438 (1952).

V.E. Cosslett and W.C. Nixon, *The X-ray shadow microscope*, J. Appl. Phys. **24**, 616–623 (1953).

V.E. Cosslett and W.C. Nixon, *X-ray microscopy*, Cambridge University Press, Cambridge (1960).

H. Daido, *Review of soft X-ray laser researches and developments*, Rep. Prog. Phys. **65**, 1513–1576 (2002).

C.G. Darwin, *The theory of X-ray reflexion*, Phil. Mag. **27**, 315–333 (1914a).

C.G. Darwin, *The theory of X-ray reflexion. Part II*, Phil. Mag. **27**, 675–690 (1914b).

P. Desauté, H. Merdji, V. Greiner, T. Missalla, C. Chenais-Popovics, and P. Troussel, *Characterization of a high resolution transmission grating*, Opt. Commun. **173**, 37–43 (2000).

E. Di Fabrizio, F. Romanato, M. Gentili, S. Cabrini, B. Kaulich, J. Susini, *et al.*, *High-efficiency multilevel zone plates for keV X-rays*, Nature **401**, 895–898 (1999).

E. Di Fabrizio, S. Cabrini, D. Cojoc, F. Romanato, L. Businaro, M. Altissimo, *et al.*, *Shaping X-rays by diffractive coded nano-optics*, Microelectr. Eng. **67–68**, 87–95 (2003).

P.A.M. Dirac, *Quantum mechanics*, Oxford University Press, Oxford (1958).

P.J. Duke, *Synchrotron radiation: production and properties*, Oxford University Press, Oxford (2000).

A. Einstein, *Zur Quantentheorie der Strahlung*, Phys. Z. **18**, 121–128 (1917). Translated as *On the quantum theory of radiation*, in D. ter Haar (ed.), *The old quantum theory*, Pergamon, Oxford (1967).

F.R. Elder, A.M. Gurewitsch, R.V. Langmuir, and H.C. Pollock, *Radiation from electrons in a synchrotron*, Phys. Rev. **71**, 829–830 (1947a).

F.R. Elder, A.M. Gurewitsch, R.V. Langmuir, and H.C. Pollock, *A 70-MeV synchrotron*, J. Appl. Phys. **18**, 810–818 (1947b).

F.R. Elder, R.V. Langmuir, and H.C. Pollock, *Radiation from electrons accelerated in a synchrotron*, Phys. Rev. **74**, 52–56 (1948).

L.R. Elias, W.M. Fairbank, J.M.J. Madey, H.A. Schwettman, and T.I. Smith, *Observation of stimulated emission of radiation by relativistic electrons in a spatially periodic transverse magnetic field*, Phys. Rev. Lett. **36**, 717–720 (1976).

P. Engström, S. Larsson, A. Rindby, and B. Stocklassa, *A 200 μm X-ray microbeam spectrometer*, Nucl. Instr. Meth. Phys. Res. B **36**, 222–226 (1989).

P. Engström, S. Larsson, A. Rindby, A. Buttkewitz, S. Garbe, G. Gaul, *et al.*, *A submicron synchrotron X-ray beam generated by capillary optics*, Nucl. Instr. Meth. Phys. Res. A **302**, 547–552 (1991).

A.I. Erko, V.V. Aristov, and B. Vidal, *Diffraction X-ray optics*, Institute of Physics Publishing, Bristol (1996).

P.P. Ewald, *Zur Begründung der Kristalloptik. Teil I. Theorie der Dispersion*, Ann. Physik **49**, 1–38 (1916a).

P.P. Ewald, *Zur Begründung der Kristalloptik. Teil II. Theorie der Reflexion und Brechung*, Ann. Physik **49**, 117–143 (1916b).

P.P. Ewald, *Zur Begründung der Kristalloptik. Teil III. Die Kristalloptik der Röntgenstrahlen*, Ann. Physik **54**, 519–597 (1917).

G.W. Fraser, A.N. Brunton, J.E. Lees, J.F. Pearson, and W.B. Feller, *X-ray focusing using square-pore microchannel plates: first observation of cruxiform image structure*, Nucl. Instr. Meth. Phys. Res. A **324**, 404–407 (1993).

H.P. Freund and T.M. Antonsen, *Principles of free-electron lasers*, second edition, Chapman and Hall, London (1996).

D. Gabor, *A new microscopic principle*, Nature **161**, 777–778 (1948).

S.M. Gruner, *Concepts and applications of energy recovery linacs (ERLs)*, in T. Warwick, J. Arthur, H.A. Padmore, and J. Stöhr (eds), *Synchrotron radiation instrumentation: eighth international conference on synchrotron radiation instrumentation, San Francisco, California, 25–29 August 2003 (AIP conference proceedings 705)*, Springer-Verlag, New York, pp. 153–156 (2004).

M. Hart, *The application of synchrotron radiation to X-ray interferometry*, Nucl. Instr. Meth. Phys. Res. **172**, 209–214 (1980).

M. Hart and U. Bonse, *Interferometry with x rays*, Phys. Today, 26–31 (August 1970).

B.L. Henke, *X-ray spectroscopy in the 100–1000 eV region*, Nucl. Instr. Meth. Phys. Res. **177**, 161–171 (1980).

S.A. Hoffman, D.J. Thiel, and D.H. Bilderback, *Developments in tapered monocapillary and polycapillary glass X-ray concentrators*, Nucl. Instr. Meth. Phys. Res. A **347**, 384–389 (1994).

A. Hofmann, *The physics of synchrotron radiation*, Cambridge University Press, Cambridge (2004).

F.L. Hunt, *X-rays of long wave-length from a ruled grating*, Phys. Rev. **30**, 227–231 (1927).

D. Iwanenko and I. Pomeranchuk, *On the maximal energy attainable in a betatron*, Phys. Rev. **65**, 343 (1944).

J.D. Jackson, *Classical electrodynamics*, third edition, John Wiley & Sons, New York (1999).

K.-J. Kim, *Brightness, coherence and propagation characteristics of synchrotron radiation*, Nucl. Instr. Meth. Phys. Res. A **246**, 71–76 (1986).

K.-J. Kim, *Characteristics of synchrotron radiation*, in A. Thompson, D. Attwood, E. Gullikson, M. Howells, K.-J. Kim, J. Kirz, J. Kortright, I. Lindau, P. Pianetta, A. Robinson, J. Scofield, J. Underwood, D. Vaughan, G. Williams, and H. Winick, *X-ray data booklet*, second edition, Lawrence Berkeley National Laboratory, Berkeley, CA, pp. 2-1–2-16 (2001).

L. Kipp, M. Skibowski, R.L. Johnson, R. Berndt, R. Adelung, S. Harm, *et al.*, *Sharper images by focusing soft X-rays with photon sieves*, Nature **414**, 184–188 (2001).

P. Kirkpatrick and A.V. Baez, *Formation of optical images by X-rays*, J. Opt. Soc. Am. **38**, 766–774 (1948).

J. Kirz, *Phase zone plates for X-rays and the extreme UV*, J. Opt. Soc. Am. **64**, 301–309 (1974).

G.F. Knoll, *Radiation detection and measurement*, third edition, John Wiley & Sons, New York (2000).

V. Kohn, *An exact theory of imaging with a parabolic continuously refractive X-ray lens*, J. Exp. Theor. Phys. (USSR) **97**, 204–215 (2003).

S. Krinsky, *Fundamentals of hard X-ray synchrotron radiation sources*, in D.M. Mills (ed.), *Third generation hard X-ray synchrotron radiation sources*, John Wiley & Sons, New York, pp. 1–40 (2002).

S. Krishnagopal and V. Kumar, *Free-electron lasers*, Rad. Phys. Chem. **70**, 559–569 (2004).

G.N. Kulipanov, A.N. Skrinsky, and N.A. Vinokurov, *Synchrotron light sources and recent developments of accelerator technology*, J. Synchrotron Rad. **5**, 176–178 (1998).

M.A. Kumakhov, *Capillary optics and their use in X-ray analysis*, X-Ray Spectrom. **29**, 343–348 (2000).

M.A. Kumakhov and F.F. Komarov, *Multiple reflection from surface X-ray optics*, Phys. Rep. **191**, 289–350 (1990).

S.S. Kurennoy, D.C. Nguyen, and L.M. Young, *Waveguide-coupled cavities for energy recovery linacs*, Nucl. Instr. Meth. Phys. Res. A **528**, 220–224 (2004).

A.N. Kurokhtin and A.V. Popov, *Simulation of high-resolution X-ray zone plates*, J. Opt. Soc. Am. A. **19**, 315–324 (2002).

M. von Laue, *Die dynamische theorie der Röntgenstrahlinterferenzen in neuer form*, Ergeb. Exakt. Naturwiss. **10**, 133–158 (1931).

K.R. Lea, *Highlights of synchrotron radiation*, Phys. Rep. **43**, 337–375 (1978).

B. Lengeler, C.G. Schroer, B. Bennera, T.F. Günzler, M. Kuhlmann, J. Tümmler, *et al.*, *Parabolic refractive X-ray lenses: a breakthrough in X-ray optics*, Nucl. Instr. Meth. Phys. Res. A **467–468**, 944–950 (2001).

R.A. Lewis, *Medical applications of synchrotron radiation X-rays*, Phys. Med. Biol. **42**, 1213–1243 (1997).

R.A. Lewis, *Position sensitive detectors for synchrotron radiation studies: the tortoise and the hare?*, Nucl. Instr. Meth. Phys. Res. A **513**, 172–177 (2003).

H. Lichte, E. Völkl, and K. Scheerschmidt, *First steps of high resolution electron holography into materials science*, Ultramicroscopy **47**, 231–240 (1992).

H. Lichte, P. Kessler, D. Lenz, and W.-D. Rau, *0.1 nm information limit with the CM30FEG-Special Tübingen*, Ultramicroscopy **52**, 575–580 (1993).

A. Liénard, *Champ électrique et magnétique, produit par une charge électrique concentrée en un point et animée d'un mouvement quelconque*, L'Éclairage Élec. **16**, 5–14; 53–59; 106–112 (1898).

E.G. Loewen and E. Popov, *Diffraction gratings and applications*, Marcel Dekker Inc., New York (1997).

R. Loudon, *The quantum theory of light*, third edition, Oxford University Press, Oxford (2000).

J.M.J. Madey, *Spontaneous emission of bremsstrahlung in a periodic magnetic field*, J. Appl. Phys. **42**, 1906–1913 (1971).

G. Magyar and L. Mandel, *Interference fringes produced by superposition of two independent maser light beams*, Nature **198**, 255–256 (1963).

L. Mandel and E. Wolf, *Optical coherence and quantum optics*, Cambridge University Press, Cambridge (1995).

G. Margaritondo, *Elements of synchrotron light*, Oxford University Press, Oxford (2002).

T.C. Marshall, *Free-electron lasers*, Macmillan, New York (1985).

L. Marton, Internal Report, RCA laboratories, Princeton, New Jersey, NJ (1939).

S.C. Mayo, P.R. Miller, S.W. Wilkins, T.J. Davis, D. Gao, T.E. Gureyev, *et al.*, *Quantitative X-ray projection microscopy: phase-contrast and multi-spectral imaging*, J. Microsc. **207**, 79–96 (2002).

S.C. Mayo and B. Sexton, *Refractive microlens array for wave-front analysis in the medium to hard X-ray range*, Opt. Lett. **29**, 866–868 (2004).

D. Maystre (ed.), *Selected papers on diffraction gratings*, SPIE Milestone Series volume MS83, SPIE Optical Engineering Press, Bellingham (1993).

A.G. Michette, *X-ray microscopy*, Rep. Prog. Phys. **51**, 1525–1606 (1988).

A.G. Michette, *No X-ray lens*, Nature **353**, 510–510 (1991).

A. Michette and S. Pfauntsch (eds), *X-rays: the first hundred years*, John Wiley & Sons, Chichester (1996).

D.M. Mills, *X-ray optics for third-generation synchrotron radiation sources*, in D.M. Mills (ed.), *Third generation hard X-ray synchrotron radiation sources*, John Wiley & Sons, New York, pp. 41–99 (2002a).

D.M. Mills (ed.), *Third generation hard X-ray synchrotron radiation sources*, John Wiley & Sons, New York (2002b).

K. Miyamoto, *The phase Fresnel lens*, J. Opt. Soc. Am. **51**, 17–20 (1961).

H. Motz, *Applications of the radiation from fast electron beams*, J. Appl. Phys. **22**, 527–535 (1951).

J.B. Murphy and C. Pellegrini, *Introduction to the physics of the free electron laser*, in M. Month, S. Turner, H. Araki *et al.* (eds), *Frontiers of particle beams, Lecture notes in physics* **296**, pp. 163–212, Springer, Berlin (1988).

O.E. Myers, *Studies of transmission zone plates*, Am. J. Phys. **19**, 359–365 (1951).

V. Nazmov, L. Shabel'nikov, F.-J. Pantenburg, J. Mohr, E. Reznikova, A. Snigirev, *et al.*, *Kinoform X-ray lens creation in polymer materials by deep X-ray lithography*, Nucl. Instr. Meth. Phys. Res. B **217**, 409–416 (2004).

G.R. Neil, C.L. Bohn, S.V. Benson, G. Biallas, D. Douglas, H.F. Dylla, *et al.*, *Sustained kilowatt lasing in a free-electron laser with same-cell energy recovery*, Phys. Rev. Lett. **84**, 662–665 (2000).

J. Ojeda-Castañeda and C. Gómez-Reino (eds), *Selected papers on zone plates*, SPIE Milestone Series volume M128, SPIE Optical Engineering Press, Bellingham (1996).

T.H. Osgood, *X-ray spectra of long wave-length*, Phys. Rev. **30**, 567–573 (1927).

P.G. O'Shea and H.P. Freund, *Free-electron lasers: status and applications*, Science **292**, 1853–1858 (2001).

D. Paganin, T.E. Gureyev, S.C. Mayo, A.W. Stevenson, Ya.I. Nesterets, and S.W. Wilkins, *X-ray omni-microscopy*, J. Microsc. **214**, 315–327 (2004).

L.G. Parratt, *Surface studies of solids by total reflection of X-rays*, Phys. Rev. **95**, 359–369 (1954).

A.G. Peele, K.A. Nugent, A.V. Rode, K. Gabel, M.C. Richardson, R. Strack, *et al.*, *X-ray focusing with lobster-eye optics: a comparison of theory with experiment*, Appl. Opt. **35**, 4420–4425 (1996).

C. Pellegrini, *Design considerations for a SASE X-ray FEL*, Nucl. Instr. Meth. Phys. Res. A **475**, 1–12 (2001).

C. Pellegrini and J. Stöhr, *X-ray free-electron lasers – principles, properties and applications*, Nucl. Instr. Meth. Phys. Res. A **500**, 33–40 (2003).

Z.G. Pinsker, *Dynamical theory of X-ray scattering in ideal crystals*, Springer-Verlag, Berlin (1978).

I. Pomeranchuk, *On the maximal energy which the primary electrons of cosmic rays can have on the earth's surface due to radiation in the earth's magnetic field* (in Russian), J. Phys. (USSR) **2**, 65–69 (1940).

F.M. Quinn, E.A. Seddon, W.F. Flavell, P. Weightman, M.W. Poole, B. Todd, *et al.*, *The 4GLS Project: update and technological challenges*, in T. Warwick, J. Arthur, H.A. Padmore, and J. Stöhr (eds), *Synchrotron radiation instrumentation: eighth international conference on synchrotron radiation instrumentation, San Francisco, California, 25–29 August 2003 (AIP conference proceedings 705)*, Springer-Verlag, New York, pp. 93-96 (2004).

A. Rindby, P. Engström, S. Larsson, and B. Stocklassa, *Microbeam technique for energy-dispersive X-ray fluorescence*, X-Ray Spectrom. **18**, 109–112 (1989).

G.L. Rogers, *Gabor diffraction microscopy: the hologram as a generalized zone-plate*, Nature **166**, 237–237 (1950).

W.C. Röntgen, *On a new kind of rays*, Nature **53**, 274–276 (1896).

J.J. Sakurai, *Advanced quantum mechanics*, Addison-Wesley, Reading, MA (1967).

E.L. Saldin, E.A. Schneidmiller, and M.V. Yurkov, *The physics of free electron lasers*, Springer, Berlin (1999).

G.A. Schott, *Electromagnetic radiation, and the mechanical reactions arising from it*, Cambridge University Press, Cambridge (1912).

C.G. Schroer, M. Kuhlmann, U.T. Hunger, T.F. Günzler, O. Kurapova, S. Feste, *et al.*, *Nanofocusing parabolic refractive X-ray lenses*, Appl. Phys. Lett. **82**, 1485–1487 (2003).

J. Schwinger, *On the classical radiation of accelerated electrons*, Phys. Rev. **75**, 1912–1925 (1949).

T. Shintake, H. Kitamura, and T. Ishikawa, *X-ray FEL project at SPring-8 Japan*, in T. Warwick, J. Arthur, H.A. Padmore, and J. Stöhr (eds), *Synchrotron radiation instrumentation: eighth international conference on synchrotron radiation instrumentation, San Francisco, California, 25–29 August 2003 (AIP conference proceedings 705)*, Springer-Verlag, New York, pp. 117–120 (2004).

T. Shiozawa, *Classical relativistic electrodynamics: theory of light emission and application to free electron lasers*, Springer-Verlag, Berlin (2004).

T.I. Smith, H.A. Schwettman, R. Rohatgi, Y. Lapierre, and J. Edighoffer, *Development of the SCA/FEL for use in biomedical and materials science experiments*, Nucl. Instr. Meth. Phys. Res. A **259**, 1–7 (1987).

A. Snigirev, *The recent development of Bragg–Fresnel crystal optics. Experiments and applications at the ESRF (invited)*, Rev. Sci. Instrum. **66**, 2053–2058 (1995).

A. Snigirev, V. Kohn, I. Snigireva, and B. Lengeler, *A compound refractive lens for focusing high-energy X-rays*, Nature **384**, 49–51 (1996).

H.H. Solak, C. David, and J. Gobrecht, *Fabrication of high-resolution zone plates with wideband extreme-ultraviolet holography*, Appl. Phys. Lett. **85**, 2700–2702 (2004).

E.A. Stern, Z. Kalman, A. Lewis, and K. Lieberman, *Simple method for focusing x rays using tapered capillaries*, Appl. Opt. **27**, 5135–5139 (1988).

S. Suehiro, H. Miyaji, and H. Hayashi, *Refractive lens for X-ray focus*, Nature **352**, 385–386 (1991).

G.J. Tallents, *The physics of soft X-ray lasers pumped by electron collisions in laser plasmas*, J. Phys. D: Appl. Phys. **36**, R259–R276 (2003).

R. Tatchyn, J. Arthur, M. Baltay, K. Bane, R. Boyce, M. Cornacchia, *et al.*, *Research and development towards a 4.5–1.5 Å linac coherent light source (LCLS) at SLAC*, Nucl. Instr. Meth. Phys. Res. A **375**, 274–283 (1996).

D.J. Thiel, E.A. Stern, D.H. Bilderback, and A. Lewis, *The focusing of synchrotron radiation using tapered glass capillaries*, Physica B **158**, 314–316 (1989).

A. Thompson, *X-ray detectors*, in A. Thompson, D. Attwood, E. Gullikson, M. Howells, K.-J. Kim, J. Kirz, J. Kortright, I. Lindau, P. Pianetta, A. Robinson, J. Scofield, J. Underwood, D. Vaughan, G. Williams, and H. Winick, *X-ray data booklet*, second edition, Lawrence Berkeley National Laboratory, Berkeley, CA, pp. 4-33–4-39 (2001).

M. Tigner, *A possible apparatus for electron clashing-beam experiments*, Nuovo Cimento **37**, 1228–1231 (1965).

D.H. Tomboulian and P.L. Hartman, *Spectral and angular distribution of ultraviolet radiation from the 300–MeV Cornell synchrotron*, Phys. Rev. **102**, 1423–1447 (1956).

T. Tomie, *X-ray lens*, Japanese Patent No. 6-045288 (1994).

Y. Wang, W. Yun, and C. Jacobsen, *Achromatic Fresnel optics for wideband extreme-ultraviolet and X-ray imaging*, Nature **424**, 50–53 (2003).

B.E. Warren, *X-ray diffraction*, Addison-Wesley, Reading, MA (1969).

T. Warwick, J. Arthur, H.A. Padmore, and J. Stöhr (eds), *Synchrotron radiation instrumentation: eighth international conference on synchrotron radiation instrumentation, San Francisco, California, 25–29 August 2003 (AIP conference proceedings 705)*, Springer-Verlag, New York (2004).

R.W. Waynant and M.N. Ediger (eds), *Selected papers on UV, VUV and X-ray lasers*, SPIE Milestone Series volume M71, SPIE Optical Engineering Press, Bellingham (1993).

B.B. Weatherby, *A determination of the wave-length of the $K\alpha$ line of carbon*, Phys. Rev. **32**, 707–711 (1928).

H. Wiedemann, *Synchrotron radiation*, Springer-Verlag, Berlin (2003).

C.H. Wilcox, *Scattering theory for diffraction gratings*, Springer-Verlag, New York (1984).

S.W. Wilkins, A.W. Stevenson, K.A. Nugent, H.N. Chapman, and S. Steenstrup, *On the concentration, focusing and collimation of X-rays and neutrons using microchannel plates and configurations of holes*, Rev. Sci. Instrum. **60**, 1026–1036 (1989).

E. Wolf, *Optics in terms of observable quantities*, Nuovo Cimento **12**, 884–888 (1954).

E. Wolf, *Coherence and radiometry*, J. Opt. Soc. Am. **68**, 6–17 (1978).

N. Yamamoto and Y. Hosokawa, *Development of an innovative 5 $\mu m\phi$ focused X-ray beam energy-dispersive spectrometer and its applications*, Jpn. J. Appl. Phys. **27**, L2203–L2206 (1988).

B.X. Yang, *Fresnel and refractive lenses for X-rays*, Nucl. Instr. Meth. Phys. Res. A **328**, 578–587 (1993).

L. Yaroslavsky and M. Eden, *Fundamentals of digital optics*, Birkhauser, Boston, MA (1996).

W. Yun, B. Lai, Z. Cai, J. Maser, D. Legnini, E. Gluskin, *et al.*, *Nanometer focusing of hard x rays by phase zone plates*, Rev. Sci. Instrum. **70**, 2238–2241 (1999).

W.H. Zachariasen, *Theory of X-ray diffraction in crystals*, John Wiley & Sons, New York (1945). Reprinted by Dover Publications, New York (1967).

Y. Zhang, T. Katoh, Y. Kagoshima, J. Matui, and Y. Tsusaka, *Focusing hard X-ray with a single lens*, Jpn. J. Appl. Phys. **40**, L75–L77 (2001).

4

Coherent X-ray imaging

The present chapter treats the subject of two-dimensional imaging in coherent X-ray optics. The emphasis will be on the imaging of coherent fields, although polychromatic fields and partially coherent fields will also be considered.

We open with a discussion of the operator theory of imaging. Here, one can view a given two-dimensional X-ray imaging system in 'input–output' terms. The input wave-field is often taken to be that which exists over the nominal planar exit surface of an illuminated sample, although this is not necessarily the case. One can then view the imaging system as an operator which transforms the two-dimensional input field into a given two-dimensional output field, the time-averaged intensity of which may then be registered using a position-sensitive detector. We shall outline the operator theory of imaging for three important classes of imaging system, in order of decreasing generality. (i) The first such class is a generic two-dimensional 'coherent imaging system', all optical elements of which are taken to elastically scatter the X-ray photons coursing through them. Thus, if the input field is monochromatic, the exit field will also be monochromatic. If the incident field is polychromatic, then each monochromatic component of the input beam will be assumed to be transmitted by the system independently. (ii) The second class of imaging system, which is a subset of the first, is known as a 'coherent linear imaging system'. These systems have the property that, at the level of fields rather than that of intensities, a given linear superposition of inputs yields the corresponding linear superposition of outputs. This assumption of linearity is often an excellent approximation for coherent X-ray imaging, provided that the fields are not so intense that nonlinear optical effects become significant. (iii) As a subset of the second class, one has the so-called 'shift invariant coherent linear imaging systems'. These are coherent linear imaging systems which have the additional restriction that a transverse shift in the input leads to a corresponding transverse shift in the output. As we shall see, the action of such systems may be described using a two-dimensional convolution integral. The Fourier representation of this integral leads to the extremely useful notion of a transfer function, which will be used in many subsequent sections of the chapter.

We then pass onto a discussion of self-imaging fields, which have the unusual property that the act of free-space propagation—through certain suitable distances—allows an image of the input field to be registered over a given output plane. Two particular classes of such self-imaging fields will be treated, respectively known as 'Talbot fields' and 'Montgomery fields'. Both monochromatic and polychromatic variants of these fields will be examined.

Next, we consider three different forms of holographic imaging: in-line holography, off-axis holography, and Fourier holography. Each of these means of imaging have the common feature that the input field is free-space propagated through a certain distance, prior to reaching the output plane. Unlike the case of self-imaging fields, however, the recorded intensity distribution does not bear a simple relationship to the input field. Rather, the image must be regarded as an encrypted form of the input field. Holography seeks to take one or more such images, known as holograms, which are then decoded into order to yield both the amplitude and phase of the input field.

Phase-contrast imaging systems are tackled next. All such systems seek to form images of transparent or near-transparent features in a sample, which have a greater effect on the phase than the intensity of the radiation passing through them. The goal, therefore, is to visualize the transverse phase shifts in the input wave-field, in the intensity of the corresponding output wave-field. Hence the term 'phase contrast'. Several means of achieving X-ray phase contrast will be considered: Zernike phase contrast, differential interference contrast, analyser-based imaging and propagation-based phase contrast. We will also make some remarks on the notion of hybrid phase-contrast imaging systems, which represent a conceptual amalgam of two or more means of phase-contrast imaging.

Phase-contrast imaging is largely qualitative. Rather than seeking to merely visualize the transverse phase shifts in the input beam, one may seek to take one or more phase contrast images and hence infer both the amplitude and the phase of the input wave-field. We outline three methods for such quantitative phase-contrast imaging, also known as phase retrieval: iterative methodologies based on the so-called Gerchberg–Saxton algorithm and its variants, non-iterative phase retrieval using the transport-of-intensity equation, and a certain means for phase retrieval using far-field diffraction patterns of one-dimensional coherent fields.

Interferometric means for coherent X-ray imaging will also be considered, in the respective contexts of the Bonse–Hart X-ray interferometer, the Young interferometer, and the intensity interferometer. In the first of these, transverse phase variations in an input coherent scalar field are visualized by interfering it with a plane wave. This yields a two-dimensional interferogram, comprising a modulated series of linear fringes which may then be processed to yield the desired phase distribution. In the second method the Young interferometer is used to study the spatial coherence of an X-ray source. Lastly, intensity interferometry—which considers interference at the level of time-dependent intensities rather than at the level of fields—is discussed in the context of the X-ray analogue of the famous experiment of Hanbury Brown and Twiss. Some other means of interferometric coherence measurement are also briefly discussed.

Virtual X-ray optics is the subject of the final section of this chapter. As mentioned in Chapter 3, one may view the computer as an intrinsic part of a coherent X-ray imaging system. Part of the optics comprises optical hardware, which yields one or more images which may then be decoded using appropriate software, to yield the complex disturbance incident upon the coherent imaging system. This process may be interpreted as using a software lens to bring the incident wave-field into 'focus'. In principle, this wave-field may be reconstructed using any of the methods for quantitative phase imaging outlined above, including the various means for phase retrieval, together with X-ray interferometry. In thereby reconstructing the wave-field incident upon the coherent imaging system, one has total knowledge of the information encoded in a given wave-field upon its passage through a given sample. One can then emulate, in software, the subsequent action of an arbitrary coherent imaging system. A demonstration of this idea will be given, making using of phase retrieval to reconstruct the incident complex wave-field.

4.1 Operator theory of imaging

The operator theory of imaging is a convenient and powerful formalism which shall be employed in most of the remaining sections of this chapter. We consider the operator theory for imaging two-dimensional coherent and partially coherent fields, respectively, in Sections 4.1.1 and 4.1.2. Finally, cascaded systems are treated in Section 4.1.3.

4.1.1 *Operator theory of imaging using coherent fields*

Two-dimensional coherent imaging systems take a given two-dimensional forward-propagating coherent scalar wave-field as input, and yield as output a different coherent wave-field with the same wavelength as the input. In the operator theory of such imaging systems, one can consider them to be characterized by an operator which acts on the input to yield the corresponding output. The intensity of the output field may then be registered using a position-sensitive detector. The phase of the output cannot be measured directly, on account of the extreme rapidity of electromagnetic wave-field oscillations in the X-ray regime.

We separately consider three classes of coherent imaging system: arbitrary two-dimensional coherent imaging systems (Section 4.1.1.1), linear coherent imaging systems (Section 4.1.1.2), and linear shift-invariant coherent imaging systems (Section 4.1.1.3). Each of these, which are listed in order of decreasing generality, are completely characterized by a given mapping from the input to the output two-dimensional coherent fields. Such a mapping will be considered to depend on a set of real 'control parameters' $\tau \equiv (\tau_1, \tau_2, \cdots)$, which specify the state of the optical system.

4.1.1.1 *Coherent imaging systems* Let us consider an X-ray imaging system, which takes as input a given coherent two-dimensional complex scalar electromagnetic field $\psi_{\rm IN}(x, y)$ in order to produce a field $\psi_{\rm OUT}(x, y, \tau_1, \tau_2, \cdots)$ as output. Both input and output fields are assumed to be monochromatic and forward-propagating, with the angular frequencies ω of input and output being equal. Harmonic time evolution $\exp(-i\omega t)$ is suppressed throughout.

The action of our coherent imaging system may be represented in operational terms as:

$$\psi_{\text{OUT}}(x,y,\tau_1,\tau_2,\cdots) = \mathcal{W}(\tau_1,\tau_2,\cdots)\psi_{\text{IN}}(x,y). \tag{4.1}$$

Here, $\mathcal{W}(\tau_1,\tau_2,\cdots)$ is an operator which completely characterizes the coherent imaging system, whose state is specified by the values of a series of real parameters $\tau_1,\tau_2,\cdots$. In words, the above equations states that the operator $\mathcal{W}(\tau_1,\tau_2,\cdots)$ acts on the input field $\psi_{\text{IN}}(x,y)$ so as to yield the corresponding output field $\psi_{\text{OUT}}(x,y,\tau_1,\tau_2,\cdots)$.

4.1.1.2 *Linear coherent imaging systems* Suppose the operator $\mathcal{W}(\tau_1,\tau_2,\cdots)$ maps the input $\psi^{(1)}_{\text{IN}}(x,y)$ to the output $\psi^{(1)}_{\text{OUT}}(x,y,\tau_1,\tau_2,\cdots)$, with the same operator mapping the input $\psi^{(2)}_{\text{IN}}(x,y)$ to the output $\psi^{(2)}_{\text{OUT}}(x,y,\tau_1,\tau_2,\cdots)$:

$$\begin{aligned}\psi^{(1)}_{\text{OUT}}(x,y,\tau_1,\tau_2,\cdots) &= \mathcal{W}(\tau_1,\tau_2,\cdots)\psi^{(1)}_{\text{IN}}(x,y),\\ \psi^{(2)}_{\text{OUT}}(x,y,\tau_1,\tau_2,\cdots) &= \mathcal{W}(\tau_1,\tau_2,\cdots)\psi^{(2)}_{\text{IN}}(x,y).\end{aligned} \tag{4.2}$$

If the following holds for any pair of complex numbers a and b, and any pair of inputs $\psi^{(1)}_{\text{IN}}(x,y)$ and $\psi^{(2)}_{\text{IN}}(x,y)$, then the coherent imaging system is said to be 'linear':

$$\begin{aligned}&\mathcal{W}(\tau_1,\tau_2,\cdots)\left[a\psi^{(1)}_{\text{IN}}(x,y) + b\psi^{(2)}_{\text{IN}}(x,y)\right]\\ &\quad = a\psi^{(1)}_{\text{OUT}}(x,y,\tau_1,\tau_2,\cdots) + b\psi^{(2)}_{\text{OUT}}(x,y,\tau_1,\tau_2,\cdots).\end{aligned} \tag{4.3}$$

Thus, for a two-dimensional coherent linear imaging system, a given linear combination of inputs yields the same linear combination of corresponding outputs.

If the imaging system is linear, and we can safely assume that a zero input field always yields a zero output field, then the imaging system will be completely characterized by its Green function $G(x,y,x',y',\tau_1,\tau_2,\cdots)$. In this case, input and output wave-fields are related by the following linear integral transform:

$$\psi_{\text{OUT}}(x,y,\tau_1,\tau_2,\cdots) = \iint dx'dy' G(x,y,x',y',\tau_1,\tau_2,\cdots)\psi_{\text{IN}}(x',y'). \tag{4.4}$$

We shall provide a physical motivation for writing down such an expression, near the end of this sub-section. For the moment, we merely note—from a mathematical rather than from a physical point of view—that the above transform satisfies the required property of linearity given in eqn (4.3).

With a view to obtaining a physical interpretation for the Green function appearing in the above expression, let us consider the case of a unit-strength impulse-like input which is zero everywhere except for a given point (x_0,y_0) on the entrance surface of the imaging system. Such an input may be written as:

$$\psi_{\rm IN}(x,y) = \delta(x - x_0, y - y_0), \tag{4.5}$$

where $\delta(x,y)$ is the two-dimensional Dirac delta. Insert this into eqn (4.4), and then invoke the sifting property of the Dirac delta (see Appendix A) to obtain the following expression for the output corresponding to our pointlike input:

$$\begin{aligned}\psi_{\rm OUT}(x,y,\tau_1,\tau_2,\cdots) &= \iint dx'dy' G(x,y,x',y',\tau_1,\tau_2,\cdots)\delta(x'-x_0,y'-y_0) \\ &= G(x,y,x_0,y_0,\tau_1,\tau_2,\cdots).\end{aligned} \tag{4.6}$$

This result gives us the promised physical interpretation for the Green function: $G(x,y,x_0,y_0,\tau_1,\tau_2,\cdots)$ represents the wave-field output by the coherent linear imaging system which is in a state specified by $\tau_1,\tau_2,\cdots$, as a function of the coordinates (x,y) over the output plane, in response to a unit-strength pointlike input field located at (x_0,y_0). As such, we may also speak of the Green function as a 'position dependent complex point spread function' or as an 'impulse response'.

Indeed, one may take this notion as a starting point from which to obtain eqn (4.4). Let us close this sub-section by showing how this may be done. Via the sifting property of the Dirac delta, a given input field can be written as:

$$\psi_{\rm IN}(x,y) = \iint dx'dy' \psi_{\rm IN}(x',y')\delta(x-x',y-y'). \tag{4.7}$$

This amounts to decomposing the input field as a sum of pointlike inputs $\delta(x - x', y - y')$, each of which is weighted by the factor $\psi_{\rm IN}(x',y')$ prior to being summed over all x' and y'. Since the imaging system is linear, by assumption, each of these pointlike inputs can be propagated through the system independently, before being summed via the double integral to produce the output $\psi_{\rm OUT}(x,y)$ corresponding to $\psi_{\rm IN}(x,y)$.

By definition a given pointlike input $\delta(x-x',y-y')$, which is non-zero only at the point $(x,y)=(x',y')$ in the input plane, yields the corresponding output $G(x,y,x',y',\tau_1,\tau_2,\cdots)$. Therefore the pointlike input $\psi_{\rm IN}(x',y')\delta(x-x',y-y')$ yields the corresponding output $\psi_{\rm IN}(x',y')G(x,y,x',y',\tau_1,\tau_2,\cdots)$. As stated earlier, each of these pointlike inputs propagates independently through the imaging system, by virtue of the linearity of the system. Therefore if we make the following replacement on the right side of eqn (4.7):

$$\psi_{\rm IN}(x',y')\delta(x-x',y-y') \longrightarrow \psi_{\rm IN}(x',y')G(x,y,x',y',\tau_1,\tau_2,\cdots), \tag{4.8}$$

then the left side will become the desired output $\psi_{\rm OUT}(x,y)$. This replacement yields eqn (4.4), as required.

4.1.1.3 *Linear shift-invariant coherent imaging systems* The notion of a coherent linear imaging system, treated in Section 4.1.1.2, can be specialized still further if the imaging system is assumed to be 'shift invariant'. The property of shift invariance refers to the demand that a transverse shift in the input should lead to the same shift in the corresponding output. Thus, if input and output are related via eqn (4.1), the following should hold for any real numbers a and b[109]:

$$\psi_{\mathrm{OUT}}(x-a, y-b, \tau_1, \tau_2, \cdots) = \mathcal{W}(\tau_1, \tau_2, \cdots)\psi_{\mathrm{IN}}(x-a, y-b). \tag{4.9}$$

Since the system is assumed to be linear as well as shift invariant, its action may be characterized by the Green function appearing in eqn (4.4). Further, the property of shift invariance implies that this Green function must depend only on coordinate differences, so that:

$$G(x, y, x', y', \tau_1, \tau_2, \cdots) = G(x-x', y-y', \tau_1, \tau_2, \cdots). \tag{4.10}$$

This transforms eqn (4.4) into a convolution integral (see eqn (A.8) of Appendix A):

$$\psi_{\mathrm{OUT}}(x, y, \tau_1, \tau_2, \cdots) = \iint dx'dy' G(x-x', y-y', \tau_1, \tau_2, \cdots)\psi_{\mathrm{IN}}(x', y'). \tag{4.11}$$

The physical interpretation of this shift-invariant Green function is very similar to that given in the previous sub-section: $G(x-x_0, y-y_0, \tau_1, \tau_2, \cdots)$ represents the wave-field output by the shift-invariant coherent linear imaging system which is in a state specified by $\tau_1, \tau_2, \cdots$, as a function of the coordinates (x, y) over the output plane, in response to a unit-strength pointlike input field $\delta(x-x_0, y-y_0)$ located at (x_0, y_0). As such, we may again speak of the Green function as a 'complex point spread function' or an 'impulse response'. Each point in the input field yields the corresponding complex point-spread function in the output plane, centred at the source point in the input plane. Equation (4.11) says that one must add up the field produced by each point in the input plane, to yield the total disturbance in the output plane.

It is often very convenient, from both a practical and from a conceptual point of view, to transform the convolution integral in eqn (4.11) into the corresponding Fourier-space representation. Accordingly, let us Fourier transform eqn (4.11) with respect to x and y, using the convention for the Fourier integral which is given in eqns (A.4) and (A.5) of Appendix A. Making use of the convolution theorem in eqn (A.9), one arrives at:

[109] In practice, no imaging system—except perhaps for free space propagation from plane to plane—will be able to meet this requirement for very large a and b. In practice, the range of acceptable transverse shifts must be kept suitably small.

$$\mathcal{F}[\psi_{\rm OUT}(x,y,\tau_1,\tau_2,\cdots)] = \mathcal{F}[\psi_{\rm IN}(x,y)] \times T(k_x,k_y,\tau_1,\tau_2,\cdots). \quad (4.12)$$

Here, $\mathcal{F}$ denotes Fourier transformation with respect to x and y,

$$T(k_x,k_y,\tau_1,\tau_2,\cdots) \equiv 2\pi\mathcal{F}[G(x,y,\tau_1,\tau_2,\cdots)] \quad (4.13)$$

is the 'transfer function' associated with a shift-invariant coherent linear imaging system and (k_x, k_y) are the Fourier coordinates dual to (x, y).

By taking the inverse Fourier transformation of eqn (4.12) with respect to k_x and k_y, and then comparing the resulting expression to eqn (4.1), we conclude that $\mathcal{W}(\tau_1,\tau_2,\cdots)$ takes a particularly simple form for coherent linear shift-invariant imaging systems:

$$\mathcal{W}(\tau_1,\tau_2,\cdots) = \mathcal{F}^{-1}T(k_x,k_y,\tau_1,\tau_2,\cdots)\mathcal{F}. \quad (4.14)$$

More explicitly, we have what shall be the most-invoked result of this chapter:

$$\psi_{\rm OUT}(x,y,\tau_1,\tau_2,\cdots) = \mathcal{F}^{-1}T(k_x,k_y,\tau_1,\tau_2,\cdots)\mathcal{F}\psi_{\rm IN}(x,y). \quad (4.15)$$

In words, eqn (4.15) states that, to calculate the two-dimensional wave-field output by a given coherent shift-invariant linear imaging system, one should perform the following three steps: (i) Fourier transform the input two-dimensional coherent wave-field, with respect to x and y; (ii) filter the Fourier transform by multiplying by the transfer function $T(k_x,k_y,\tau_1,\tau_2,\cdots)$, which depends on the state of the imaging system completely characterized by the values of the real parameters $\tau_1,\tau_2,\cdots$; (iii) inverse Fourier transform the resulting expression, with respect to k_x and k_y, in order to yield the desired output wave-field.

We seek a physical interpretation for the transfer function $T(k_x,k_y,\tau_1,\tau_2,\cdots)$. To this end, consider an input field:

$$\psi_{\rm IN}(x,y) = \exp\left[i\left(k_x^{(0)}x + k_y^{(0)}y\right)\right] \quad (4.16)$$

comprising a single Fourier harmonic with wave-number $(k_x, k_y) = (k_x^{(0)}, k_y^{(0)})$. Note that, since the Fourier transform of the above input yields a Dirac delta in Fourier space:

$$\mathcal{F}[\psi_{\rm IN}(x,y)] = \mathcal{F}\left\{\exp\left[i\left(k_x^{(0)}x + k_y^{(0)}y\right)\right]\right\} = 2\pi\delta\left(k_x - k_x^{(0)}, k_y - k_y^{(0)}\right), \quad (4.17)$$

it may be viewed as the Fourier analogue of the pointlike input used in obtaining a physical interpretation for the Green function (see eqn (4.5)). Next, insert eqn

(4.16) into eqn (4.15), and then make use of eqn (4.17). Thus our single-harmonic input is transformed to the following output[110]:

$$\begin{aligned}\psi_{\rm OUT}(x,y,\tau_1,\tau_2,\cdots) &= \mathcal{F}^{-1}T(k_x,k_y,\tau_1,\tau_2,\cdots)2\pi\delta\left(k_x-k_x^{(0)},k_y-k_y^{(0)}\right)\\ &= \frac{1}{2\pi}\iint\left[T(k_x,k_y,\tau_1,\tau_2,\cdots)2\pi\delta\left(k_x-k_x^{(0)},k_y-k_y^{(0)}\right)\right]\\ &\qquad\times\exp[i(k_x x+k_y y)]dk_x dk_y\\ &= T(k_x=k_x^{(0)},k_y=k_y^{(0)},\tau_1,\tau_2,\cdots)\exp[i(k_x^{(0)}x+k_y^{(0)}y)]\\ &= T(k_x=k_x^{(0)},k_y=k_y^{(0)},\tau_1,\tau_2,\cdots)\psi_{\rm IN}(x,y). \end{aligned} \quad (4.18)$$

This yields the promised physical interpretation for the transfer function: if a single Fourier harmonic $\exp[i(k_x^{(0)}x+k_y^{(0)}y)]$ is input into a shift-invariant coherent linear imaging system whose state is characterized by $\tau_1,\tau_2,\cdots$, then the resulting output will be equal to the input field multiplied by the number $T(k_x=k_x^{(0)},k_y=k_y^{(0)},\tau_1,\tau_2,\cdots)$. This number is a function of the two-dimensional wave-vector $(k_x^{(0)},k_y^{(0)})$ characterizing the single Fourier harmonic which is input into the system.

Pushing this idea further, let us return consideration to eqn (4.15). This may now be viewed in very physical terms as taking the input wave-field $\psi_{\rm IN}(x,y)$ and using the two-dimensional Fourier transform to decompose it into a continuous sum of its constituent Fourier harmonics $\exp[i(k_x x+k_y y)]$, before multiplying each harmonic by a wave-number-dependent weighting factor and then re-summing the resulting filtered harmonics to yield the output disturbance $\psi_{\rm OUT}(x,y)$.

In light of the filtration mentioned above, the meaning of the modulus and phase of the transfer function is now clear. The modulus of the transfer function tells us how the amplitude of a given input Fourier harmonic is altered in passing through the shift-invariant coherent linear imaging system, with the phase of the transfer function representing a constant phase bias imparted on the said harmonic. With this point is mind, we note that it is often convenient to explicitly write the transfer function in terms of its modulus $|T(k_x,k_y,\tau_1,\tau_2,\cdots)|$ and phase $\varphi(k_x,k_y,\tau_1,\tau_2,\cdots)$:

$$T(k_x,k_y,\tau_1,\tau_2,\cdots)=|T(k_x,k_y,\tau_1,\tau_2,\cdots)|\exp\left[i\varphi(k_x,k_y,\tau_1,\tau_2,\cdots)\right]. \quad (4.19)$$

Any deviation, of $|T(k_x,k_y,\tau_1,\tau_2,\cdots)|$ from unity or $\varphi(k_x,k_y,\tau_1,\tau_2,\cdots)$ from a constant, will result in the output intensity differing from the input intensity. Accordingly, one may consider the modulus and phase of the transfer function

[110]Note: In the second line of the next equation, we have made use of the explicit form for the inverse two-dimensional Fourier transform, given in eqn (A.4). In the fourth line, we have made use of the sifting property of the Dirac delta.

as quantifying the 'aberrations' of a coherent shift-invariant linear imaging system. Adopting the commonly used albeit misleading terminology, any deviation of $|T(k_x, k_y, \tau_1, \tau_2, \cdots)|$ from unity is spoken of as representing the 'incoherent aberrations' of the system. Any deviation of $\varphi(k_x, k_y, \tau_1, \tau_2, \cdots)$ from a constant is spoken of as embodying the 'coherent aberrations' of the system.

Prior to discussing the notion of coherent aberrations in more detail, let us make an apparent digression in recalling the operator form for the Fresnel diffraction integral which was given in eqn (1.28):

$$\psi_\omega(x, y, z = \Delta) = \exp(ik\Delta)\mathcal{F}^{-1} \exp\left[\frac{-i\Delta(k_x^2 + k_y^2)}{2k}\right] \mathcal{F}\psi_\omega(x, y, z = 0). \quad (4.20)$$

Here, all symbols are as defined in Chapter 1. Compare eqns (4.15) and (4.20), respectively associating input and output fields with unpropagated and propagated fields. We conclude that free-space propagation from plane to parallel plane, under the Fresnel approximation, is an example of a shift-invariant coherent linear imaging system described by the following 'free space' transfer function:

$$T(k_x, k_y, k, \Delta) = \exp(ik\Delta) \exp\left[\frac{-i\Delta(k_x^2 + k_y^2)}{2k}\right]. \quad (4.21)$$

Considered as an imaging system, such free-space propagation has no incoherent aberrations, with coherent aberrations given by the following second-order Taylor series in k_x and k_y:

$$\varphi(k_x, k_y, k, \Delta) = k\Delta - \frac{\Delta(k_x^2 + k_y^2)}{2k}. \quad (4.22)$$

Returning to an arbitrary shift-invariant coherent linear imaging system, it is often the case that $\varphi(k_x, k_y, \tau_1, \tau_2, \cdots)$ is a sufficiently slowly varying function of k_x and k_y to be written as a Taylor-series expansion in these variables. If this is the case, then:

$$\varphi(k_x, k_y, \tau_1, \tau_2, \cdots) = \sum_{m=0}^{\infty}\sum_{n=0}^{\infty} \varphi_{mn}(\tau_1, \tau_2, \cdots)(k_x)^m (k_y)^n. \quad (4.23)$$

The real numbers φ_{mn}, which depend on the state of the imaging system, are termed 'coherent aberration coefficients'. Dropping their explicit functional dependence on $\tau_1, \tau_2, \cdots$ for the sake of notational simplicity, let us write down the first six terms of our Taylor series for the coherent aberrations:

$$\varphi(k_x, k_y, \tau_1, \tau_2, \cdots) = \varphi_{00} + \varphi_{10}k_x + \varphi_{01}k_y + \varphi_{20}k_x^2 + \varphi_{02}k_y^2 + \varphi_{11}k_x k_y + \cdots \quad (4.24)$$

We may discard φ_{00} since it only serves to contribute an irrelevant constant phase to the output. Further, φ_{10} and φ_{01} are irrelevant as far as imaging is concerned, since, by the Fourier shift theorem, they serve only to laterally displace the output image. The remaining coefficients $\varphi_{20}, \varphi_{02}, \varphi_{11}, \cdots$ are closely related to commonly encountered coherent aberrations which have particular names such as 'defocus', 'spherical aberration', 'coma', and 'astigmatism'. Regarding the first-named coherent aberration, namely defocus, we note that (4.24) reduces to (4.22) if:

$$\varphi_{20} = \varphi_{02} = -\frac{\Delta}{2k}, \qquad \varphi_{mn} = 0 \text{ otherwise.} \tag{4.25}$$

Even if the remaining aberration coefficients are non-zero, we can meaningfully speak of $-2k\varphi_{20}$ and $-2k\varphi_{02}$ as being equal to the defocus in the x and y directions, respectively. If $\varphi_{20} = \varphi_{02} \equiv d/(-2k)$, then one would simply speak of 'the defocus d'. Similarly, the other named aberration coefficients are closely related to the set of numbers φ_{mn}. For example, the coefficient of $(k_x^2 + k_y^2)^2$ is proportional to the 'spherical aberration' of the imaging system. Defocus and spherical aberration correspond to rotationally symmetric coherent aberrations; however, the other listed coherent aberrations do not possess rotational symmetry.

4.1.2 *Operator theory of imaging using partially coherent fields*

Here we briefly consider the generalization of the operator theory of imaging, to the case of partially coherent scalar fields. The treatment is cursory in comparison to that of coherent imaging, as partially coherent imaging lies largely outside the scope of this text (for further information, see, for example, Goodman (1985)). Our presentation is broken into two parts, which separately consider imaging of a polychromatic field, and imaging of a statistically stationary partially coherent field.

4.1.2.1 *Polychromatic imaging systems* Suppose that a particular realization $\Psi_{\text{IN}}(x, y, z, t)$ of a forward-propagating partially coherent scalar electromagnetic wave-field is incident upon a given two-dimensional linear imaging system. The resulting two-dimensional output is denoted by $\Psi_{\text{OUT}}(x, y, z, t)$, with a position-sensitive detector subsequently measuring a time-averaged intensity in forming a given image.

Assume the field to be pulsed—that is, assume the field to be non-zero only over a certain finite time interval. In practice this will always be the case, since no real source exists for an infinite time. The field may therefore be represented as a Fourier series over finitely many distinct frequencies, this being a sum of the monochromatic components of the field. Assume the imaging system is elastically scattering, so that there is no mixing between the various monochromatic components of the polychromatic field. The linearity of the imaging system, by assumption, requires that each monochromatic component propagate independently through the system. If this is a good approximation, then one may use the formalism of Section 4.1.1.1 in order to separately propagate each monochromatic component of the field through the imaging system. Each monochromatic output may then be summed, before taking the squared modulus and then averaging over the detection time in forming the registered time-averaged intensity.[111] If this detection time is long compared to the reciprocal of the difference in radiation frequency between any two monochromatic components, then the detector will register the summed intensities of each monochromatic component.

4.1.2.2 *Partially coherent imaging systems* More generally, one may consider two-dimensional linear imaging systems operating with statistically stationary forward-propagating scalar electromagnetic fields. Note that such fields cannot be pulsed—indeed, in principle if not in practice, they must exist for an infinite time. Suppose that the statistical properties of the input field are well described by the mutual coherence function. If this is the case, then the input into the partially coherent linear imaging system is given by the input mutual coherence function $\Gamma_{\rm IN}(x_1, y_1, x_2, y_2, \tau)$, as a function of any pair of points (x_1, y_1) and (x_2, y_2) over the input plane, together with all time lags τ. The linear imaging system then maps the input mutual coherence into the corresponding output $\Gamma_{\rm OUT}(x_1, y_1, x_2, y_2, \tau)$:

$$\Gamma_{\rm IN}(x_1, y_1, x_2, y_2, \tau) \longrightarrow \Gamma_{\rm OUT}(x_1, y_1, x_2, y_2, \tau). \tag{4.26}$$

Given the output mutual coherence function, one can compute the time-averaged intensity $\overline{I}_{\rm OUT}(x, y)$ of the resulting image, using[112]:

$$\overline{I}_{\rm OUT}(x, y) = \Gamma_{\rm OUT}(x_1 = x, y_1 = y, x_2 = x, y_2 = y, \tau = 0). \tag{4.27}$$

Since the partially coherent imaging system is assumed to be linear, its action may be characterized by the appropriate generalization of eqn (4.4). In an obvious notation we then have:

$$\begin{aligned} &\Gamma_{\rm OUT}(x_1, y_1, x_2, y_2, \tau, \tau_1, \tau_2, \cdots) \\ &\quad = \int \cdots \int dx_1' dy_1' dx_2' dy_2' d\tau' G(x_1, y_1, x_2, y_2, \tau, x_1', y_1', x_2', y_2', \tau', \tau_1, \tau_2, \cdots) \\ &\qquad \times \Gamma_{\rm IN}(x_1', y_1', x_2', y_2', \tau'). \end{aligned} \tag{4.28}$$

An analogue of the transfer function also exists, for shift-invariant systems, although it will not be written here. See, for example, Goodman (1985) for further information.

4.1.3 *Cascaded systems*

The operator theory of imaging may be of utility when studying 'cascaded' two-dimensional optical imaging systems. In a cascaded optical system, the passage

[111]This implicitly assumes an ideal detector whose efficiency is equal to unity for all radiation frequencies present in the field. If this is not a good approximation then the argument may be readily modified to account for a frequency-dependent efficiency.

[112]To obtain eqn (4.27), take eqn (1.88) and set the spatial arguments equal to one another, with the time lag τ being taken to zero.

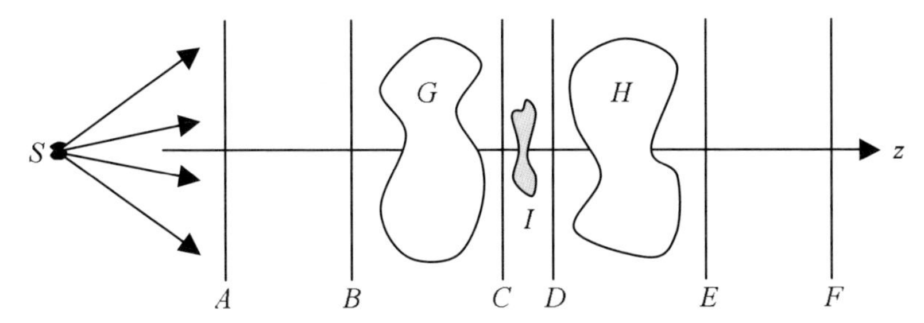

Fig. 4.1. Example of a cascaded optical system. A source S yields radiation which traverses the planes A through F of a two-dimensional imaging system. Two sets of optical elements, G and H, are as indicated. A sample I lies between the planes C and D. Free space is considered to lie between the planes A and B, together with the planes E and F. An image is registered over the plane F. Note that the planes A through F are not necessarily parallel to one another.

of a given field through the system may be broken down into a number of consecutive stages. The two-dimensional output of a given stage is taken to be the input for the subsequent stage, with an appropriate operator being used to map the input for a given stage into the corresponding output.

This notion may be clarified using the example of a cascaded optical system in Fig. 4.1. Suppose, for the moment, that the source S may be well approximated as monochromatic. The field over the entrance surface A of the cascaded system is then assumed to be a forward-propagating scalar monochromatic field. If there are any apertures present in this plane A, then they will be assumed to block out any radiation incident upon it. Since free space is assumed to lie between the planes A and B, one may use an appropriate diffraction operator (see, for example, eqn (1.25)) in mapping the field over A to the field over B. The field over B is then considered to be the input into a linear optical imaging system G. One can make use of the formalism of Section 4.1.1 in calculating the output over the plane C, as a function of the state of the optical system G. Having propagated the field to the plane C, it subsequently interacts with an elastically scattering sample I in reaching the plane D. To model the passage from planes C to D, one must use an appropriate scattering theory. Such scattering theories may include the projection approximation (Section 2.2), the first Born approximation (Section 2.5), the multi-slice approximation (Section 2.7), and so forth. Irrespective of the particular scattering theory employed, one may use it to evolve the wave-field from the plane C to the plane D. Passage through the remaining planes, including the linear optical system H lying between the planes D and E, together with the slab of free space between the planes E and F, may then be computed. This yields the complex disturbance over the plane F, the intensity of which may be registered using a position-sensitive detector.

Suppose, now, that the monochromatic source in the above example is replaced by a polychromatic source whose emitted radiation is non-zero only during a specified finite time interval. The resulting forward-propagating polychromatic field can then be decomposed into a sum of forward-propagating monochromatic fields. Both the sample and the optical elements are assumed to be elastically scattering, for each of the monochromatic components present in the polychromatic beam. Further, the sample is assumed to be static, so that its structure is unchanged during the period of illumination. Under the above assumptions each monochromatic component can be independently evolved through the cascaded optical system, and then summed over the plane F to yield the resulting complex disturbance over this plane. The intensity of this field can then be averaged over the detection time, accounting if appropriate for frequency-dependent detector efficiency, to yield the registered image over plane F (cf. Section 4.1.2.1).

Lastly, consider the source in Fig. 4.1 to emit forward-propagating statistically stationary partially coherent scalar electromagnetic radiation. Assume the statistics of this field to be such that it is well described by its mutual coherence function, as a function of all pairs of points over the plane A, for all time lags τ. The mutual coherence function can then be propagated through the slab of free space from planes A to B, using either eqns (1.127) or (1.137). If the Green function for the linear optical element G is known, then the mutual coherence over plane B can be evolved to that over plane C, using eqn (4.28). Passage from planes C to D, which involves scattering from the sample I, requires an appropriate scattering theory for the mutual coherence function. Such a theory will not be outlined in this text (see, for example, Goodman (1985)). One can continue in a similar fashion, to compute the mutual coherence function for all pairs of points in the plane F of the position-sensitive detector. The time-averaged intensity over this output plane may then be calculated using eqn (4.27).

We close this section by noting that use of the operator theory of imaging, for the study of cascaded optical systems, often yields rather unwieldy expressions if all operators are written out in full. While judicious approximations may render meaningful analytic expressions, it is often the case that numerical analysis is required for the study of realistic cascaded systems in coherent X-ray optics.

4.2 Self imaging

Consider the sketch of a two-dimensional imaging system given in Fig. 4.2(a). Here we see a sample A, illuminated from the left by scalar X-ray radiation. This radiation may be coherent or partially coherent. Upon passing through the sample, a field is formed over the nominal exit surface $z = 0$, this being parallel to the specified optic axis z. If the incident field is monochromatic, then the exit-surface field may be described by a complex wave-function $\psi_\omega(x, y, z = 0)\exp(-i\omega t)$ whose modulus is independent of time; if the incident field is polychromatic, then the exit-surface field may be described by a wave-function which is a superposition of monochromatic wave-functions of various angular frequencies ω. Upon passing through the imaging system denoted by B, an image of the field

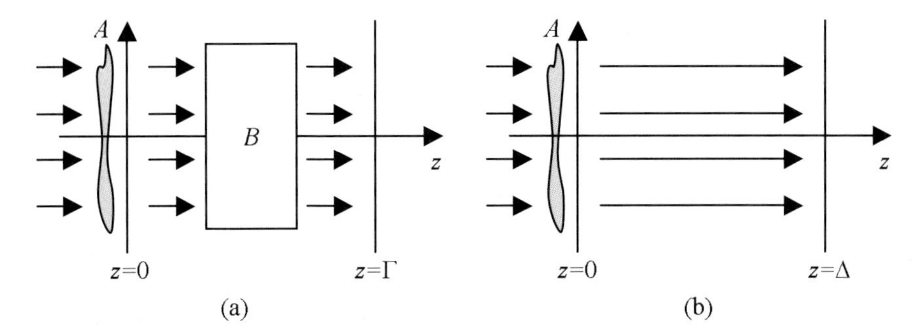

Fig. 4.2. (a) A sample A is illuminated by X-rays from the left, leading to a given field over the exit-surface plane $z = 0$ which is perpendicular to the nominal optic axis z. This field may be coherent or partially coherent. Using a suitable imaging system B, one may form an image over the plane $z = \Gamma$. (b) Suppose the optical system B to be removed, with all else remaining unchanged. 'Self imaging' refers to the phenomenon whereby an image of the field at $z = 0$ is formed over the plane $z = \Delta$, downstream of the exit surface of the object, in the absence of any imaging system.

at $z = 0$ is formed over the plane $z = \Gamma$. If the imaging system is perfect, then the time-averaged intensities over the planes $z = 0$ and $z = \Gamma$ will be equal, up to transverse and multiplicative scale factors.

As shown in Fig. 4.2(b), let us now suppose that the imaging system B is removed. In general, the time-averaged intensity distribution formed over the plane $z = \Delta$ will not be a direct representation of the intensity over the exit surface $z = 0$ of the sample. However, under certain circumstances the field over the plane $z = \Delta$ may have the same time-averaged intensity as the field over the plane $z = 0$, notwithstanding the fact that there is only free space between the two planes. One would then speak of the field as being 'self imaging'. Such self-imaging fields are remarkable insofar as they are able to image themselves through the act of free-space propagation from plane to parallel plane, without the need for an intervening imaging system.

In this section we shall consider two broad categories of self-imaging fields. The first such class are monochromatic and polychromatic fields which are periodic over the plane $z = 0$. The self-imaging of such fields, at certain propagation distances $z = \Delta$, is known as the 'Talbot effect'. We shall separately consider the Talbot effect for monochromatic and polychromatic scalar fields, in Sections 4.2.1 and 4.2.2, respectively. The second class of self-imaging fields, to be considered here, are those whose two-dimensional Fourier transforms are non-zero only over one or more circles in two-dimensional reciprocal space. Such self-imaging fields are said to exhibit the 'Montgomery effect'. This effect, for monochromatic and polychromatic scalar fields, will be sequentially treated in Sections 4.2.3 and 4.2.4.

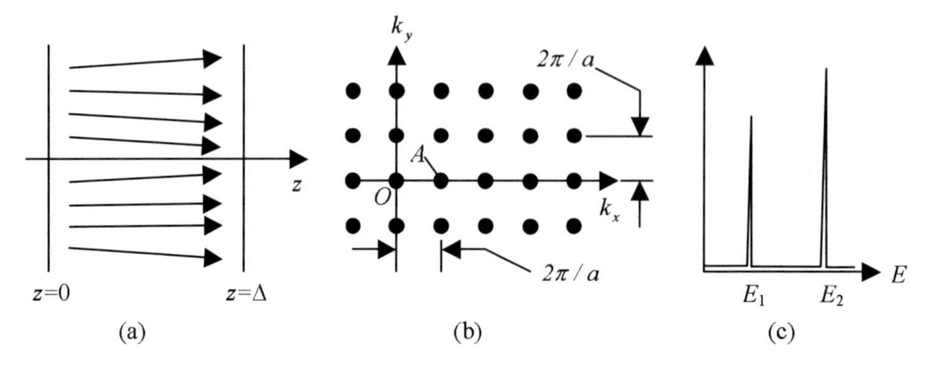

Fig. 4.3. (a) Two planes $z = 0$ and $z = \Delta \geq 0$ are shown, these being perpendicular to a given optic axis z. According to the Talbot effect, a forward-propagating coherent scalar field in the plane $z = 0$ will be re-imaged over certain planes downstream of it, if: (i) the field is periodic over the plane $z = 0$, with the same period in the x and y directions, and (ii) the field over the plane $z = 0$ is sufficiently slowly varying for the Fresnel diffraction theory to be applicable. In our discussions, we shall assume the field over $z = 0$ to be periodic in the Cartesian coordinates x and y, with period a in both of these directions. (b) If one takes the Fourier transform of the unpropagated disturbance with respect to x and y, the result will be non-zero only over the specified grid of points. Here, k_x and k_y are the Fourier-space coordinates dual to the real-space coordinates x and y. (c) Spectrum of a certain polychromatic field which contains two monochromatic components, corresponding to energies E of E_1 and E_2.

4.2.1 *Talbot effect for monochromatic fields*

With reference to Fig. 4.3(a), suppose that one has a strictly monochromatic forward-propagating scalar disturbance over the plane $z = 0$, this plane being perpendicular to a specified optic axis z. As is customary, the spatial part of this disturbance may be quantified by the wave-function $\psi_\omega(x, y, z = 0)$, where (x, y) are Cartesian coordinates in the plane perpendicular to the optic axis, and ω denotes the angular frequency of the radiation. If $\psi_\omega(x, y, z = 0)$ is periodic in x and y, with the same period in both directions, and $\psi_\omega(x, y, z = 0)$ is sufficiently slowly varying in x and y for free-space propagation to be well described by the Fresnel diffraction theory, the intensity of the field over $z = 0$ will be re-imaged over certain planes $z = \Delta$ downstream of the exit-surface plane $z = 0$. As mentioned earlier, this phenomenon is known as the 'Talbot effect', after its discoverer (Talbot 1836). Later researches include those of Lord Rayleigh (1881), together with Cowley and Moodie (1957a,b,c). Note that this last-mentioned series of papers speaks of 'Fourier images' rather than the 'Talbot effect', with both terms being common in the current literature.

Under the stated assumptions, one may employ the operator form (1.28) of the Fresnel diffraction integral, to propagate the monochromatic disturbance from the plane $z = 0$ to the plane $z = \Delta \geq 0$, assuming that free space lies between these two planes. Thus we have:

$$\psi_\omega(x, y, z = \Delta) = \exp(ik\Delta) \times \mathcal{F}^{-1} \exp\left[\frac{-i\Delta(k_x^2 + k_y^2)}{2k}\right] \mathcal{F}\psi_\omega(x, y, z = 0), \quad \Delta \geq 0. \tag{4.29}$$

Here, $k = 2\pi/\lambda$ is the wave-number corresponding to the wavelength λ of the field, $\mathcal{F}$ denotes Fourier transformation with respect to x and y using the convention given in eqn (A.5), $\mathcal{F}^{-1}$ denotes the corresponding inverse Fourier transformation and (k_x, k_y) are the Fourier-space coordinates dual to the real-space coordinates (x, y).

As stated previously the Talbot effect will be operative if the unpropagated field $\psi_\omega(x, y, z = 0)$ is periodic in x and y, with the same period a in both directions.[113] We may therefore decompose the unpropagated field as the following Fourier series:

$$\psi_\omega(x, y, z = 0) = \sum_p \sum_q \psi_\omega^{(p,q)} \exp\left[\frac{2\pi i}{a}(xp + yq)\right], \tag{4.30}$$

with the double sum being taken over all integers p and q.

In order to propagate this field through the distance $z = \Delta$, we insert the above Fourier decomposition into the operator form of the Fresnel diffraction integral, as given in eqn (4.29). In the resulting expression, the order of the summations and integrations may be interchanged, leading to:

$$\psi_\omega(x, y, z = \Delta) = \exp(ik\Delta) \times \sum_p \sum_q \psi_\omega^{(p,q)} \mathcal{F}^{-1} \exp\left[\frac{-i\Delta(k_x^2 + k_y^2)}{2k}\right] \left\{\mathcal{F} \exp\left[\frac{2\pi i}{a}(xp + yq)\right]\right\}. \tag{4.31}$$

To proceed further we need to evaluate the term in braces above. To do so, we make use of the form for the two-dimensional Fourier integral given in eqn (A.5), and then make use of the Fourier representation for the Dirac delta given in eqn (A.2). This yields the following expression:

$$\mathcal{F} \exp\left[\frac{2\pi i}{a}(xp + yq)\right] = 2\pi\delta\left(\frac{2\pi p}{a} - k_x\right) \delta\left(\frac{2\pi q}{a} - k_y\right), \tag{4.32}$$

which may be substituted into eqn (4.31) to give:

[113]Note that these periods need not be equal in order for the Talbot effect to be operative. However, we will not discuss such a generalization in this text.

$$\psi_\omega(x,y,z=\Delta) = 2\pi \exp(ik\Delta)$$
$$\times \sum_p \sum_q \psi_\omega^{(p,q)} \mathcal{F}^{-1} \exp\left[\frac{-i\Delta(k_x^2+k_y^2)}{2k}\right] \delta\left(\frac{2\pi p}{a} - k_x\right) \delta\left(\frac{2\pi q}{a} - k_y\right). \tag{4.33}$$

As is clear from the above equation, the two-dimensional Fourier transform of the unpropagated disturbance is only non-zero at points in the following cubic lattice of coordinates in Fourier space:

$$(k_x, k_y) = \left(\frac{2\pi p}{a}, \frac{2\pi q}{a}\right). \tag{4.34}$$

This set of points, which may be identified with the two-dimensional reciprocal lattice of the unpropagated periodic disturbance, is represented in Fig. 4.3(b).

Let us put physics to one side for the moment, to make a simple mathematical observation which contains the key to the Talbot effect. Consider the following quadratic function $v(k_x, k_y)$, which is defined at each point (k_x, k_y) in the two-dimensional Fourier space sketched in Fig. 4.3(b):

$$v(k_x, k_y) = \frac{\Delta}{2k}(k_x^2 + k_y^2). \tag{4.35}$$

We now prove the following theorem: if $v(k_x, k_y)$ is equal to an integer multiple of 2π at the point $A \equiv (k_x = 2\pi/a, k_y = 0)$ in the sketch, then $v(k_x, k_y)$ is equal to an integer multiple of 2π at every lattice point $(k_x, k_y) = (2\pi p/a, 2\pi q/a)$. The proof is as follows. By assumption, at the point A the function $v(k_x, k_y)$ takes the value:

$$v\left(k_x = \frac{2\pi}{a}, k_y = 0\right) = \frac{\Delta}{2k} \times \left(\frac{2\pi}{a}\right)^2 = 2\pi m, \tag{4.36}$$

where m is an integer. From the second equality above, we have:

$$\frac{\pi\Delta}{ka^2} = m. \tag{4.37}$$

To complete the proof, we need to show that $v(k_x, k_y)$ is equal to an integer multiple of 2π at every lattice point $(k_x, k_y) = (2\pi p/a, 2\pi q/a)$. At such lattice points, $v(k_x, k_y)$ is equal to:

$$\begin{aligned} v\left(k_x = \frac{2\pi p}{a}, k_y = \frac{2\pi q}{a}\right) &= \frac{\Delta}{2k}\left[\left(\frac{2\pi p}{a}\right)^2 + \left(\frac{2\pi q}{a}\right)^2\right] \\ &= 2\pi \times \left\{\frac{\pi\Delta}{ka^2}\right\} \times (p^2+q^2) \\ &= 2\pi m(p^2+q^2). \end{aligned} \tag{4.38}$$

Note that we have made use of eqn (4.37) in the final line of the above expression. This completes our proof, since $2\pi m(p^2+q^2)$ will be an integer multiple of 2π if m, p, and q are all integers.

In light of the above finding, let us return consideration to result (4.33). If the propagation distance Δ is such that eqn (4.37) is obeyed, then the phase

$$-v(k_x,k_y) = -\frac{\Delta(k_x^2+k_y^2)}{2k}, \tag{4.39}$$

of the Fresnel propagator, will be an integer multiple of 2π at each of the points at which the Fourier transform of the unpropagated disturbance is non-zero. The Fresnel propagator

$$\exp\left[\frac{-i\Delta(k_x^2+k_y^2)}{2k}\right] \tag{4.40}$$

is therefore equal to unity at each of these lattice points, implying that the propagated field

$$\psi_\omega\left(x,y,z=\Delta=\frac{mka^2}{\pi}\right) \tag{4.41}$$

is the same in both modulus and phase as the unpropagated field $\psi_\omega(x,y,z=0)$, up to a trivial constant phase factor $\exp(ik\Delta)$.

This completes our derivation of the Talbot effect. In summary, the analysis shows that a periodic paraxial coherent scalar field, with period a in both the x and y directions, is re-imaged in both modulus and phase over planes at the following 'Talbot distances' Δ_m downstream of the initial field:

$$\Delta_m = \frac{mka^2}{\pi} = \frac{2ma^2}{\lambda}, \quad m=1,2,\cdots \tag{4.42}$$

Thus the field, which is transversely periodic in x and y, with period a, is also longitudinally periodic, with period $ka^2/\pi = 2a^2/\lambda$.

We now turn to an experimental demonstration of the Talbot effect using hard X-rays, due to Cloetens and colleagues (Cloetens *et al.* 1999). This was performed at the European Synchrotron Radiation Facility (ESRF) in Grenoble, France. Wiggler radiation was monochromated using a silicon 111 double-bounce monochromator, yielding near-plane quasi-monochromatic X-ray radiation with a mean wavelength of $\lambda = 0.85$ Å. This was then used to illuminate a gold grid, periodic in both x and y with period $a = 12.5$ μm. According to the $m=1$ case of eqn (4.42), the exit-surface wave-field of the gold mesh will be re-imaged at a distance of:

$$\Delta_{m=1} = \frac{2a^2}{\lambda} = \frac{2\times(12.5\times10^{-6})^2}{0.85\times10^{-10}} = 3.7\,\text{m}. \tag{4.43}$$

Figure 4.4(a) shows the intensity which Cloetens and colleagues recorded 4 mm downstream of the exit surface of the gold grid. The absorptive effects of the grid bars are clearly evident in this image. Images (b) through (e) respectively show the intensity distributions recorded at propagation distances $z = \frac{1}{4}\Delta_{m=1}$, $z = \frac{1}{2}\Delta_{m=1}$, $z = \frac{3}{4}\Delta_{m=1}$, and $z = \Delta_{m=1}$. We see that the exit-surface intensity is re-imaged at the Talbot distance, as shown in part (e) of the figure. Some blurring is apparent in the Talbot image, in comparison to the exit-surface intensity in (a). This is due to the partially coherent nature of the illuminating beam, a factor not accounted for in the above analysis (cf. Section 4.6.4). Further, we see from part (c) of the figure that re-imaging is observed at half the Talbot distance, albeit with a transverse displacement (not indicated in the figure). It is left as an exercise to show that self-imaging is also observed midway between the Talbot planes defined by eqn (4.42), up to a transverse displacement (see, for example, Cowley (1995)).

We close by noting that the analysis of this section may be generalized to the case of point-source illumination, by making use of the Fresnel scaling theorem given in Appendix B. In this case, the Talbot distances no longer correspond to equally spaced planes. We refer the reader to the previously cited papers by Cowley and Moodie, together with the text by Cowley (1995), for more information on this point.

4.2.2 *Talbot effect for polychromatic fields*

Consider a two-component polychromatic beam that has a spectrum as shown in Fig. 4.3(c). As indicated there, the polychromatic beam is a sum of two monochromatic beams, with energies $E_1 = \hbar c k_1$ and $E_2 = \hbar c k_2$. The corresponding wave-numbers are respectively equal to k_1 and k_2, with angular frequencies denoted ω_1 and ω_2. The wave-function $\Psi(x, y, z, t)$ of this beam may be written as a sum of the time-dependent wave-functions of two monochromatic beams:

$$\Psi(x, y, z, t) = \psi_{\omega_1}(x, y, z) \exp(-i\omega_1 t) + \psi_{\omega_2}(x, y, z) \exp(-i\omega_2 t). \tag{4.44}$$

Here, (x, y, z) are the usual Cartesian coordinates with nominal optic axis z, and t denotes time.

By taking the squared modulus of the above expression, we may write the time-dependent intensity $I(x, y, z, t) = |\Psi(x, y, z, t)|^2$ of the polychromatic beam in terms of the time-independent intensities $I_{\omega_1}(x, y, z) = |\psi_{\omega_1}(x, y, z)|^2$ and $I_{\omega_2}(x, y, z) = |\psi_{\omega_2}(x, y, z)|^2$ of each of its monochromatic components:

$$\begin{aligned} I(x, y, z, t) =& I_{\omega_1}(x, y, z) + I_{\omega_2}(x, y, z) \\ &+ 2\,\mathrm{Re}\left\{\psi^*_{\omega_1}(x, y, z)\psi_{\omega_2}(x, y, z) \exp[i(\omega_1 - \omega_2)t]\right\}. \end{aligned} \tag{4.45}$$

We see that the instantaneous intensity of the polychromatic beam consists of a sum of three terms. The first two terms respectively represent the time-independent intensities of each monochromatic component of the beam. The

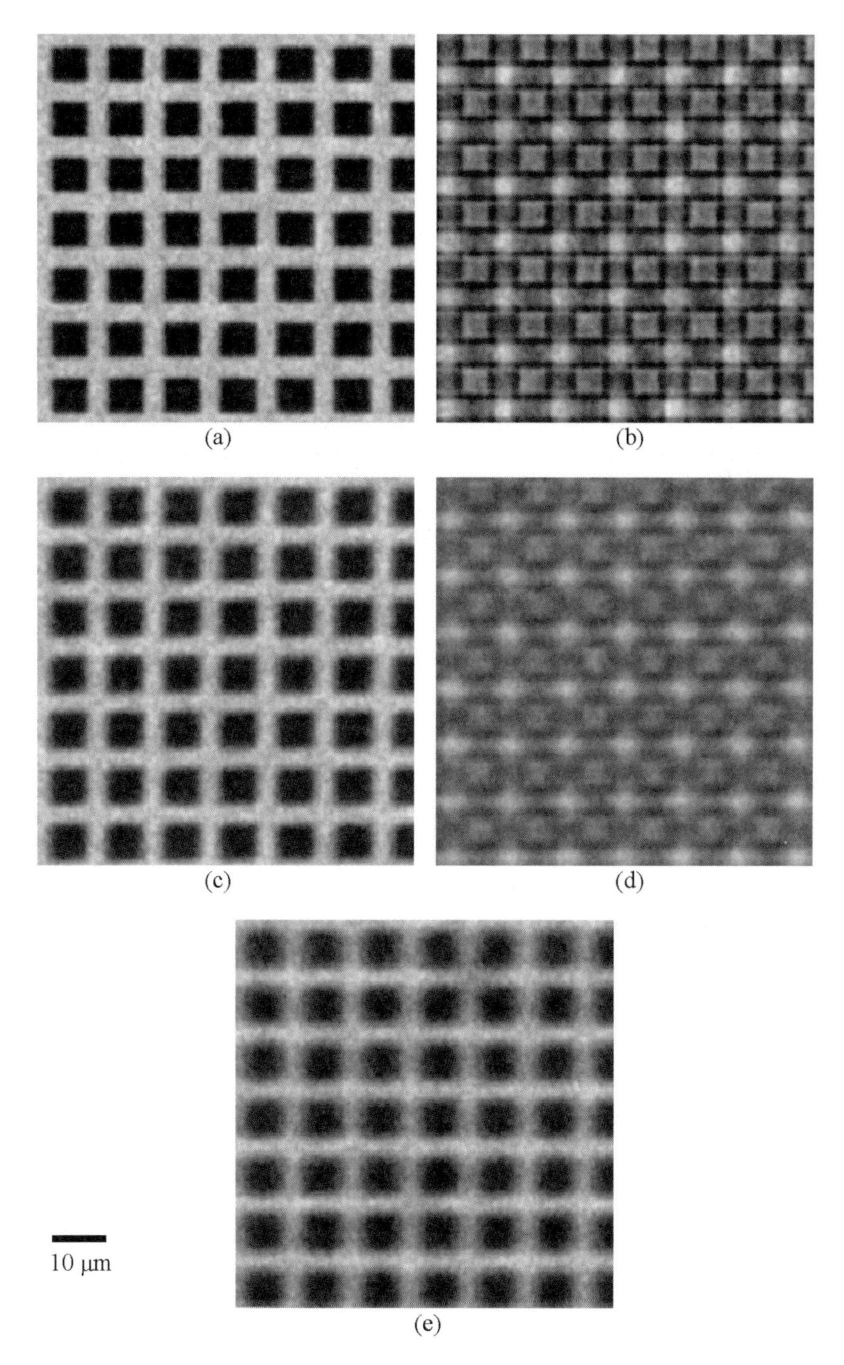

Fig. 4.4. Experimental demonstration of the Talbot effect, for hard-X-ray imaging of a gold mesh, obtained at the ESRF synchrotron. Details in main text. Images taken from Cloetens *et al.* (1999). Used with permission.

third term represents a time-dependent 'beating' between the two monochromatic components. At each point (x, y, z) in space, the magnitude of this beating term has a modulus of $2|\psi^*_{\omega_1}(x,y,z)\psi_{\omega_2}(x,y,z)| = 2\sqrt{I_{\omega_1}(x,y,z)I_{\omega_2}(x,y,z)}$, with a sinusoidal time evolution of angular frequency equal to $\omega_2 - \omega_1$.

Assume that the period $2\pi/|\omega_2 - \omega_1|$, of this time-dependent beating, is much shorter than the time over which an ideal position-sensitive integrating detector records the time-averaged intensity $\overline{I}(x, y, z)$ of the beam. The beating term may then be neglected in this time average, so that the intensity registered by the detector is simply the sum of the intensities due to each of the monochromatic components[114]:

$$\overline{I}(x, y, z) = I_{\omega_1}(x, y, z) + I_{\omega_2}(x, y, z). \tag{4.46}$$

To proceed further we assume the boundary value $\Psi(x, y, z = 0, t)$ of our polychromatic beam to coincide with the exit-surface wave-field which results when a thin periodic sample is illuminated with a z-directed two-component polychromatic plane wave with component X-ray energies equal to E_1 and E_2. The sample is considered to be periodic in both x and y, with period a. Further, the sample is assumed to be elastically scattering, so that the exit-surface beam contains only the energies E_1 and E_2. Invoking the translational symmetry of this scenario, we conclude that each of the two monochromatic components in the exit-surface plane $z = 0$ will have a periodicity of a.

According to eqn (4.46), the time-averaged intensity recorded by a two-dimensional position-sensitive detector in a plane perpendicular to the optic axis at some distance downstream of the sample, will be equal to the sum of the intensities of each monochromatic component in the beam. By the analysis of Section 4.2.1, the intensity of the first monochromatic component will be self-imaged over planes at a distance:

$$\Delta_m^{(1)} = \frac{mk_1a^2}{\pi} = \frac{ma^2E_1}{\hbar c\pi}, \quad m = 1, 2, 3, \cdots \tag{4.47}$$

downstream of the exit surface of the sample. Similarly, the intensity of the second monochromatic component will be self-imaged over planes at the following distances downstream of the exit surface of the object:

$$\Delta_n^{(2)} = \frac{nk_2a^2}{\pi} = \frac{na^2E_2}{\hbar c\pi}, \quad n = 1, 2, 3, \cdots \tag{4.48}$$

[114]We have assumed that the efficiency of the ideal detector is the same for each of the two X-ray energies. This is often a bad assumption, which may be accounted for by multiplying each of the terms on the right side of eqn (4.46) by constants which are indicative of the efficiency for the two X-ray energies. The resulting conclusions are unchanged by this complication, as the reader may readily verify.

Since the time-averaged intensities add, a polychromatic version of the Talbot effect will be observed if we can find integers m and n such that:

$$\Delta_m^{(1)} = \Delta_n^{(2)}. \tag{4.49}$$

Making use of the eqns (4.47) and (4.48), the above condition becomes:

$$\frac{E_1}{E_2} = \frac{n}{m}. \tag{4.50}$$

Thus, the ratio of the X-ray energies should be expressible as a ratio of integers.

Suppose, then, that the ratio of E_1 to E_2 may be expressed as a ratio of integers. We may assume, without loss of generality, that these integers m and n have no common divisor. The period Δ_p of the resulting polychromatic Talbot effect will then be given by:

$$\Delta_p = \frac{ma^2 E_1}{\hbar c \pi} = \frac{na^2 E_2}{\hbar c \pi}. \tag{4.51}$$

This polychromatic Talbot distance is evidently equal to a certain integer number of Talbot distances for one of the monochromatic components of the polychromatic field, and a different integer number of Talbot distances for the other monochromatic component.

With the above ideas in mind, we recall the discussion on harmonic contamination from crystal monochromators which was given towards the end of Section 3.2.4. There we saw that a crystal monochromator may transmit radiation frequencies which are an integer multiple of a lowest-order harmonic. If both the fundamental wavelength λ and a single higher-order harmonic λ/j is passed by a monochromator, where j is an integer greater than 1, then condition (4.50) will be fulfilled for the two frequencies present in the beam. Each of these frequencies will have an associated Talbot distance respectively given by $2a^2/\lambda \equiv \Delta$ and $2a^2/(\lambda/j) = j\Delta$. For this case, the associated polychromatic Talbot distance Δ_p will coincide with the Talbot distance $2ja^2/\lambda$ of the harmonic which contaminates the fundamental. More generally, one may have the case that a number of harmonics are present. The associated polychromatic Talbot distance Δ_p will then be the shortest distance which may be divided by any Talbot distance associated with a given frequency component of the beam, such that one always obtains an integer. For example, if the fundamental wavelength λ is contaminated by both third-order and fifth-order harmonics, with all higher-order harmonics being negligible, then the polychromatic Talbot distance will be equal to $15 \times 2a^2/\lambda$.

4.2.3 *Montgomery effect for monochromatic fields*

In the previous pair of sub-sections, we examined the Talbot effect for the self-imaging of transversely periodic two-dimensional scalar fields. This analysis was predicated on the Fresnel approximation, valid for paraxial fields. Here, we examine the so-called 'Montgomery effect' for self-imaging forward-propagating coherent scalar fields, in an analysis which does not assume paraxiality (Montgomery 1967, 1968; Durnin *et al.* 1987; Durnin 1987).

Montgomery self imaging may be subdivided into two classes. The second such class is analogous to Talbot imaging, insofar as the intensity of the propagated wave-field is periodic in the propagation direction. In the first class, which is still more remarkable, the image is unchanged under free-space propagation. Wave-fields belonging to this first class are known as 'diffraction-free beams'. Below, we separately treat the two classes of Montgomery self imaging for monochromatic scalar fields.

4.2.3.1 *Diffraction-free beams* Consider Fig. 4.3(a) once more. Suppose that we have a forward-propagating, but not necessarily paraxial, monochromatic scalar X-ray field over the plane $z = 0$. Denote the spatial part of the wave-function of this field by $\psi_\omega(x, y, z = 0)$. Suppose, further, that the two-dimensional Fourier transform of the field is only non-zero over a circle (i.e. an infinitely-thin annulus) of radius κ in Fourier space, which is centred at the origin, with the radius of the circle being less than or equal to the wave-number k of the radiation. Thus, by assumption, we have:

$$\mathcal{F}[\psi_\omega(x, y, z = 0)] = \delta\left(\sqrt{k_x^2 + k_y^2} - \kappa\right) f(k_\theta), \quad 0 \leq \kappa \leq k. \tag{4.52}$$

Here δ denotes the Dirac delta, $k_\theta = \tan^{-1}(k_y/k_x)$ denotes the angle made by the point (k_x, k_y) in plane polar coordinates about the origin $k_x = k_y = 0$ of the two-dimensional Fourier space, and $f(k_\theta)$ is an arbitrary function of this polar angle.

The above field, known as a 'diffraction-free beam', has the remarkable property that its intensity is the same for all propagation distances z (Montgomery 1967, 1968):

$$|\psi_\omega(x, y, z)|^2 = |\psi_\omega(x, y, z = 0)|^2. \tag{4.53}$$

This property can be derived by substituting the above expression into the operator form (1.24) and (1.25) of the angular-spectrum representation for diffracting a forward-propagating coherent scalar field between two parallel planes separated by vacuum. We shall do this in a moment. However, prior to writing any more formulae, let us give a verbal argument to explain why the field $\psi_\omega(x, y, z)$ is diffraction free.

According to eqn (1.25), the Fourier representation $\exp[i\Delta(k^2 - k_x^2 - k_y^2)^{1/2}]$ of the free-space propagator has the following two properties: (i) It is rotationally symmetric in the two-dimensional Fourier plane (k_x, k_y), this being a consequence of the rotational symmetry of an expanding spherical wave (outgoing free-space Green function) about any axis passing through its point of origin. (ii) In the (k_x, k_y) plane, within a disc of radius equal to the wave-number of the radiation, the Fourier representation of the propagator has unit modulus. This is a consequence of the fact that the modulus of a non-evanescent plane wave is unchanged upon propagation from plane to parallel plane.

According to the second property listed above, the Fourier-transformed field in eqn (4.52) is multiplied by a complex function of k_x and k_y, of modulus unity, in propagating from plane to plane. According to the first property, this complex function is the same for each point on the Fourier-space circle where $\mathcal{F}[\psi_\omega(x, y, z = 0)]$ is non-vanishing. Therefore free-space propagation through any distance only serves to multiply the Fourier transform, of the unpropagated disturbance, by a complex constant of modulus unity. This amounts to multiplying the real-space representation of the unpropagated disturbance by a complex constant of modulus unity. Thus the intensity is invariant under propagation, so that the field is diffraction free in the sense specified by eqn (4.53).

The mathematics, corresponding to the verbal description of the above paragraph, is as follows. Substitute eqn (1.25) into eqn (1.24), so that:

$$\psi_\omega(x, y, z = \Delta) = \mathcal{F}^{-1} \exp\left[i\Delta\sqrt{k^2 - k_x^2 - k_y^2}\right] \mathcal{F}\psi_\omega(x, y, z = 0), \quad \Delta \geq 0. \tag{4.54}$$

Making us of eqn (4.52), this becomes the following expression for the propagated field:

$$\psi_\omega(x, y, z = \Delta) = \mathcal{F}^{-1} \exp\left[i\Delta\sqrt{k^2 - \{k_x^2 + k_y^2\}}\right] \delta\left(\sqrt{k_x^2 + k_y^2} - \kappa\right) f(k_\theta),$$
$$\Delta \geq 0, \quad 0 \leq \kappa \leq k. \tag{4.55}$$

By the sifting property of the Dirac delta (see Appendix A), together with the rotational symmetry of the free-space propagator, we may replace the term in braces by κ^2. Thus:

$$\psi_\omega(x, y, z = \Delta) = \exp\left(i\Delta\sqrt{k^2 - \kappa^2}\right) \mathcal{F}^{-1}\delta\left(\sqrt{k_x^2 + k_y^2} - \kappa\right) f(k_\theta),$$
$$\Delta \geq 0, \quad 0 \leq \kappa \leq k. \tag{4.56}$$

The squared modulus of this expression is independent of the propagation distance Δ, implying the beam to obey the diffraction-free property in eqn (4.53).

Nine remarks: (i) Diffraction-free beams maintain this property when passed through a shift-invariant coherent linear imaging system with no incoherent aberrations, provided that all coherent aberrations of the system are rotationally symmetric. (ii) Suppose that a diffraction-free beam is normally incident, from vacuum, upon a slab of transparent stratified medium. The refractive index of this medium is real, by assumption, and varies only with z between its entrance and exit surfaces. If backscattering can be neglected then the emerging beam will be diffraction free, with a transverse intensity distribution equal to that of the incident beam. This may be demonstrated, for example, using the multislice concept outlined in Section 2.7. (iii) If $\kappa > k$ then one has what may be termed an 'evanescent diffraction-free beam'. Its intensity will decay with propagation distance Δ, with the propagated intensity being equal to the unpropagated intensity multiplied by $|\exp(i\Delta\sqrt{k^2-\kappa^2})|^2 = \exp(-2\Delta\sqrt{\kappa^2-k^2})$. (iv) Monochromatic forward-propagating non-evanescent scalar plane waves are a special case of diffraction free beam, with $f(k_\theta) = a\delta(k_\theta - b)$. Here, a is a complex constant and b is a real constant. Like all diffraction-free beams, together with the Talbot beams considered earlier, the plane wave has infinite energy over any transverse plane. This divergent energy is a consequence of the singular nature of the two-dimensional spatial Fourier spectra of these beams. (v) Since diffraction-free beams have infinite energy, they can only ever be approximated in practice. Thus, rather than being termed 'non-diffracting', physical approximations to such beams should perhaps be more properly termed 'slowly diffracting'. For example, one might consider the singular Fourier spectrum in eqn (4.52) to be replaced by a non-singular spectrum which is only non-zero over a thin annulus $\kappa_1 \leq (k_x^2 + k_y^2)^{1/2} \leq \kappa_2$ in Fourier space, where $\kappa_1 < \kappa_2 \leq k$. Such a beam would be approximately non-diffracting over propagation distances Δ sufficiently small that the wave-field propagator $\exp[i\Delta(k^2 - k_x^2 - k_y^2)^{1/2}]$ could be considered approximately constant over the Fourier-space annulus $\kappa_1 \leq (k_x^2 + k_y^2)^{1/2} \leq \kappa_2$. (vi) In line with the previous remark, a diffraction-free beam could be approximately realized by normally illuminating a thin annular aperture with z-directed plane waves, with the aperture lying at the back focal plane of a Fresnel zone plate. The zone plate and the aperture should be co-axial with the optic axis. The visible-light version of this experiment was published by Durnin *et al.* (1987), using refractive lenses rather than zone plates. The resulting beam was seen to have an on-axis intensity which was approximately constant up to a certain rather long propagation distance, beyond which the on-axis intensity was seen to rapidly decay. (vii) Diffraction-free beams may also be generated by illuminating suitable diffraction gratings. See, for example, Vasara *et al.* (1989), and references therein, regarding the visible-light demonstration of this. Suitably scaled forms of these gratings may be used to create diffraction-free beams using X-rays. (viii) For the case where $f(k_\theta) = 1$, one can readily show that the diffraction-free beam takes the form of a Bessel function whose argument is proportional to the radial distance from the optic axis. Such beams are known as 'Bessel beams'. For a text-book account, see Nieto-Vesperinas (1991) and references therein. (ix) Phase vortices (see Chapter 5) may be embedded in diffraction-free beams (see, for example, Vasara *et al.* (1989)).

4.2.3.2 *Longitudinally periodic Montgomery beams* Rather than having a two-dimensional Fourier transform which is only non-vanishing along a certain circle in Fourier space, as quantified by eqn (4.52), consider the unpropagated field to be only non-vanishing over a pair of concentric Fourier-space circles. Both circles should have respective radii κ_1 and $\kappa_2 > \kappa_1$ which are less than or equal to the wave-number of the radiation, with their centres located at the origin $(k_x, k_y) = (0, 0)$. The two-dimensional Fourier transform of the unpropagated field can then be written as the following generalization of eqn (4.52):

$$\mathcal{F}[\psi_\omega(x,y,z=0)] = \delta\left(\sqrt{k_x^2+k_y^2}-\kappa_1\right) f_1(k_\theta) + \delta\left(\sqrt{k_x^2+k_y^2}-\kappa_2\right) f_2(k_\theta),$$
$$0 \le \kappa_1 < \kappa_2 \le k, \quad (4.57)$$

where $f_1(k_\theta)$ and $f_2(k_\theta)$ are arbitrary functions.

The self-imaging property of these beams can be obtained using a variation on the argument of the previous sub-section. Specifically, the form for the free-space propagator in eqn (4.54) implies that the field in eqn (4.57) will be self-imaged over propagation distances Δ_m which are such that the propagator has the same phase (modulo 2π) over each of the Fourier-space circles with respective radii of κ_1 and κ_2 (Montgomery 1967, 1968). Thus the required self-imaging propagation distances obeys the equation:

$$\Delta_m\sqrt{k^2-\kappa_1^2} = \Delta_m\sqrt{k^2-\kappa_2^2} + 2m\pi, \qquad (4.58)$$

where m is an integer. Solving for the self-imaging distance Δ_m, we arrive at:

$$\Delta_m = \frac{2m\pi}{\sqrt{k^2-\kappa_1^2}-\sqrt{k^2-\kappa_2^2}}. \qquad (4.59)$$

The propagated field therefore has longitudinal periodicity, in the same sense as was discussed earlier in the context of the Talbot effect for paraxial monochromatic scalar fields (see Section 4.2.1).

The generalization of Montgomery's idea, to the case of forward-propagating scalar fields which have two-dimensional Fourier transforms over a specified plane which are non-vanishing over more than two concentric Fourier-space circles, centred at the origin, is left as an exercise.

4.2.4 *Montgomery effect for polychromatic fields*

Montgomery self imaging may be generalized from the case of monochromatic radiation to that of polychromatic radiation, using similar logic to that employed in passing from Sections 4.2.1 to 4.2.2. This generalization is left as an exercise. In particular, the reader may wish to prove the following statement: The superposition, of finitely many z-directed scalar diffraction-free beams, each of which is monochromatic and all of which have distinct frequencies which do not differ infinitesimally from one another, will have a time-averaged intensity that is also diffraction free. This time average should be taken over an interval which is much longer than the reciprocal of the smallest frequency difference between any two of the monochromatic components.

4.3 Holography

In the 1940s, electron microscopists were well aware of the fact that strong aberrations in then-available magnetic lenses placed strong constraints on the resolution attainable with their instruments. It was therefore sensible to seek to improve this resolution by building better magnetic lenses for electron microscopes, with much effort being expended towards achieving this end. Notwithstanding such an 'obvious' attack on the problem, which continues to this day, Dennis Gabor took a quite different approach in a beautiful paper which would ultimately earn him the 1971 Nobel prize (Gabor 1948). Rather than seeking to build better magnetic lenses, Gabor's lateral thought was to 'dispense altogether with electron objectives' (i.e. electron objective lenses) and instead illuminate the sample with a point source, as shown in Fig. 4.5 (a). The resulting intensity distribution, recorded using a planar imaging device B, is not a true image of the object. Indeed the aberrations of the imaging system have been exacerbated, rather than improved, by discarding all lenses from the imaging system. Gabor viewed the intensity distribution ('in-line hologram'), produced by such a system, as an encrypted representation of the sample which could subsequently be decoded to create the image which would have been obtained if one possessed an electron microscope which was relatively free from aberrations. This ingenious 'two-step process', of recording followed by reconstruction, will form the subject of this sub-section.

Before proceeding, two remarks are in order: (i) We are, of course, more concerned with the optics of X-rays than that of electrons, but in view of the fact that the time-independent free-space Schrödinger equation for monochromatic electrons is identical in form to the free-space Helmholtz equation for monochromatic scalar electromagnetic waves, there are evidently direct parallels between electron and X-ray holography. (ii) Notwithstanding Gabor's invention of holography in the context of aberration correction for imperfect imaging systems, the emphasis of most mainstream work on holography has shifted from aberration correction towards sample reconstruction and wave-function reconstruction.

4.3.1 *In-line holography*

We begin with a brief description of the basic concepts of in-line holography, as originally formulated by Gabor. Figure 4.5(a) shows the concentric spherical surfaces of constant phase in the radiation emitted by an idealized point source of monochromatic scalar X-ray waves, such as might be produced by using an X-ray zone plate to focus monochromated plane synchrotron radiation onto a small pinhole. If one immerses an object A located in the plane $z = 0$ into the disturbance created by the point source, the spatial part of the wave-field $\psi(x, y, z > 0)$ downstream of the object can always be decomposed into a sum of two disturbances: the unscattered wave $\psi_O(x, y, z > 0)$ plus the wave $\psi_S(x, y, z > 0)$ scattered from the object. Here, as usual, (x, y, z) denotes a right-handed Cartesian coordinate system, with z being identified with the nominal optic axis. Note, also, that in all our discussions on holography we shall suppress explicit

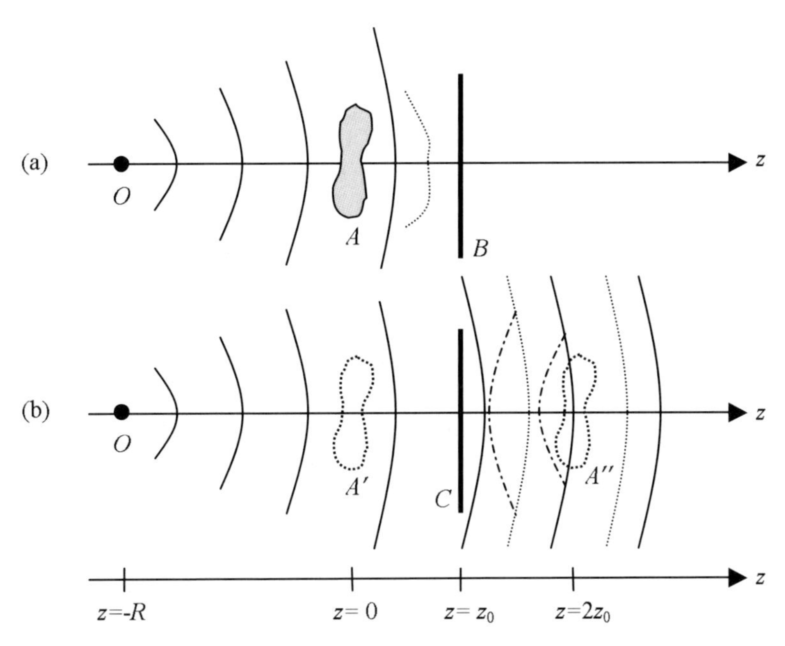

Fig. 4.5. Setup for in-line holography. (a) Step 1: Recording. A point source O of coherent scalar radiation, which lies on the negative optic axis z, emanates spherical waves which impinge upon a sample A. The radiation downstream of A can be taken as a coherent superposition of the unscattered spherical wave (solid line) and the wave scattered by the sample (dotted line). The in-line hologram, which is registered over plane B, records the interference between these waves. (b) Step 2: Reconstruction. In the absence of the sample, the in-line hologram C is illuminated by the same coherent radiation with which it was recorded. The wave-field downstream of the hologram can be decomposed into three components, corresponding to the unscattered reference wave (solid line), a virtual image A' of the sample (dotted line), and a real image A'' of the sample (dash-dot line). Note that the process of holographic reconstruction can either be undertaken using a physical optical system, or, as is now commonly the case, computationally.

dependence of field quantities on angular frequency ω.

Let an imaging device B, such as a photographic plate or a CCD camera, be located in the plane $z = z_0$ (see Fig. 4.5(a)). The intensity distribution $I_{\rm H}(x, y)$ over this plane, which is termed an 'in-line hologram', is the interference pattern formed between the unscattered and scattered wave-fields:

$$I_{\rm H}(x,y) = |\psi_{\rm O}(x,y,z_0) + \psi_{\rm S}(x,y,z_0)|^2. \tag{4.60}$$

For the sake of simplicity, take $R \to \infty$ in Fig. 4.5; this assumption may be relaxed using the scaling theorem outlined in Appendix B, which shows that the Fresnel approximation allows one to map the $R \to \infty$ results onto the finite-R result. For now, note that our assumption implies that the unscattered ('reference') wave $\psi_0(x, y, z)$ is a plane wave which is normally incident on the detector, allowing us to write $\psi_0(x, y, z_0) \approx A$, where A is a given complex constant. Equation (4.60) therefore becomes:

$$I_{\mathrm{H}}(x, y) \approx |A|^2 + A^* \psi_{\mathrm{S}}(x, y, z_0) + A\psi_{\mathrm{S}}^*(x, y, z_0) + |\psi_{\mathrm{S}}(x, y, z_0)|^2, \quad (4.61)$$

where we have assumed that the scattered wave has the same frequency as the unscattered wave (elastic scattering approximation). If the scattered wave is assumed to be much weaker than the reference wave, so that $|\psi_{\mathrm{S}}| \ll |A|$, we can discard the last term on the right side of eqn (4.61). We shall assume this to be the case for the remainder of this sub-section.

We have now completed the 'recording' stage of in-line holography. Suppose that one now takes a digitized image of $I_{\mathrm{H}}(x, y)$, as given in eqn (4.61), and then computationally applies the free-space wave-field propagator $\mathcal{D}_{z-z_0}$ for diffracting this wave-field through a distance $z - z_0$, where $z \geq z_0$ (see eqn (1.25)). Physically, this amounts to performing the experiment shown in Fig. 4.5(b), whereby an absorbing screen C is placed in front of the original reference wave, in the same location that the in-line hologram was recorded; the screen C has a transmission function that is equal to $I_{\mathrm{H}}(x, y)$, thereby implying that the complex disturbance at the exit surface of C is equal to $AI_{\mathrm{H}}(x, y)$. Bearing eqn (4.61) in mind, and making use of the previously mentioned approximation that $|\psi_{\mathrm{S}}| \ll |A|$, we see that the reconstructed wave-field $\psi_{\mathrm{recon}}(x, y, z), z > z_0$, is given by a sum of three terms:

$$\Psi_{\mathrm{recon}}(x, y, z) \approx \mathcal{D}_{z-z_0}[A|A|^2 + |A|^2 \psi_{\mathrm{S}}(x, y, z_0) + A^2 \psi_{\mathrm{S}}^*(x, y, z_0)], \quad z \geq z_0. \quad (4.62)$$

Evidently, at the exit surface of the in-line hologram, the reconstructed wave-field $\psi_{\mathrm{recon}}(x, y, z_0)$ consists of a superposition of three terms: (i) a term $A|A|^2$ which, up to a multiplicative constant $|A|^2$, is equal to the unscattered ('reference') wave—this propagates into the half space $z \geq z_0$ as indicated by the solid lines in Fig. 4.5(b); (ii) a term $|A|^2 \psi_{\mathrm{S}}(x, y, z_0)$ which, up to a multiplicative constant $|A|^2$, is equal to the wave-field which has been scattered by the sample—this propagates into the half space $z \geq z_0$ as indicated by the dotted lines in Fig. 4.5(b), forming a virtual image A' of the object; (iii) a term $A^2 \psi_{\mathrm{S}}^*(x, y, z_0)$ which, up to a multiplicative constant, is equal to the complex conjugate of the wave-field scattered by the sample—this propagates into the half space $z \geq z_0$ as indicated by the dash-dot lines in Fig. 4.5(b), and, as shall be demonstrated in a moment, forms a real image A'' of the sample.

From the point of view of a computational reconstruction of the sample from a digitized hologram, it is convenient to choose $z = 2z_0$ in eqn (4.62), giving:

$$\psi_{\text{recon}}(x, y, 2z_0) \approx A|A|^2 \exp(ikz_0) + |A|^2\psi_{\text{S}}(x, y, 2z_0) + A^2\mathcal{D}_{z_0}[\psi_{\text{S}}^*(x, y, z_0)]. \tag{4.63}$$

To evaluate the last term on the right side of eqn (4.63), we make use of the reciprocity theorem outlined in Appendix C, which implies that:

$$\mathcal{D}_{z_0}[\psi_{\text{S}}^*(x, y, z_0)] = [\psi_{\text{S}}(x, y, z = 0)]^*. \tag{4.64}$$

In words, this theorem states that the forward free-space propagation of a given two-dimensional coherent field which is free of evanescent components, through a distance d, gives the complex conjugate of the (inverse) free-space propagation of the complex conjugate of that coherent field through a distance $-d$. We therefore arrive at the following expression for the reconstructed wave-field ψ_{recon} over the plane $z = 2z_0$:

$$\begin{aligned}
&\psi_{\text{recon}}(x, y, z = 2z_0) \\
&\quad \equiv \mathcal{D}_{z-z_0}[AI_{\text{H}}(x, y)] \\
&\quad = A|A|^2 \exp(ikz_0) + |A|^2\psi_{\text{S}}(x, y, z = 2z_0) + A^2[\psi_{\text{S}}(x, y, z = 0)]^* \\
&\quad = A^2[A \exp(-ikz_0) + \psi_{\text{S}}(x, y, z = 0)]^* + |A|^2\psi_{\text{S}}(x, y, z = 2z_0).
\end{aligned} \tag{4.65}$$

The left side of eqn (4.65) is readily evaluated numerically and, as stated earlier, amounts to computationally applying a free-space propagator to the digitized hologram. One thereby obtains the complex conjugate of the reconstructed exit-surface wave-field, given by the first term in square brackets on the last line of eqn (4.65), superposed over a 'twin image' term given by the second term in the last line of eqn (4.65).

Herein lies a fundamental limitation of in-line holography in the form outlined here: the wave-fields on the right side of eqn (4.65) are overlapping in space. In particular, the overlap of the real and virtual image is known as the 'twin image problem' of in-line holography. This problem serves to reduce the quality of reconstructions obtained using Gabor's method.

The application of Gabor's ideas to X-ray holography was outlined by Baez (1952), with early experimental studies including those of Aoki and Kikuta (1974). As a more contemporary example of the experimental application of Gabor's ideas to coherent X-ray imaging, see, for example, Watanabe *et al.* (1997).

We close this section by pointing out that there have been many studies concerned with means of eliminating the twin image problem from Gabor holography. This remains an active field of research, which shall not be reviewed here. Note, however, that the question of twin-image removal from in-line holograms is quite closely related to the problem of phase retrieval, which is treated later in this chapter (Section 4.5). Note, further, that the twin-image problem may be circumvented using the methods of off-axis holography and Fourier holography, which are respectively treated in the following pair of sub-sections.

4.3.2 *Off-axis holography*

As we have just seen, in-line holography suffers from the limitations of the twin-image problem, which arises because the real and virtual images are overlapping in space. This limitation may be overcome using off-axis holography, as developed by Leith and Upatneiks (1962, 1963, 1964).

Suppose one could incline the 'reference wave' of the previous section, with respect to the imaging device which records the hologram. With an inclined propagating reference wave, one replaces A in the previous section's equations with $A\exp(i\alpha x)$, where α is a non-zero real number which specifies the tilt of the reference wave with respect to the hologram plane. Equation (4.61) becomes:

$$I_{\mathrm{H}}(x,y) \approx |A|^2 + A^* \exp(-i\alpha x)\psi_{\mathrm{S}}(x,y,z_0) + A\exp(i\alpha x)\psi_{\mathrm{S}}^*(x,y,z_0) + |\psi_{\mathrm{S}}(x,y,z_0)|^2. \tag{4.66}$$

Note that, contrary to the case for in-line holography, we will not need to assume that $|\psi_{\mathrm{S}}| \ll |A|$ so as to eliminate the last term on the right side of this equation. Evidently, this allows us to work with samples which are more strongly scattering than those required for in-line holography.

The second and third terms on the right side of eqn (4.66), which respectively lead to the real and virtual images of the object, are 'carried' on the oppositely-inclined plane waves $\exp(-i\alpha x)$ and $\exp(i\alpha x)$, implying that these terms will become spatially separated upon free-space propagation. The first and fourth terms have no such carrier wave. We conclude that, provided α is sufficiently large, the real and virtual images are spatially separated both from one another, and from the wave-fields corresponding to the remaining terms. It is this ability to isolate the reconstruction of the wave-field scattered by the sample, which allows off-axis holography to overcome the twin-image problem of in-line holography.

4.3.3 *Fourier holography*

Consider the schematic for so-called Fourier holography which is sketched in Fig. 4.6. Here, a point source A radiates spherical reference waves ψ_0 whose surfaces of constant phase are indicated by the solid concentric circles. In the presence of a sample B, a scattered wave-field ψ_{S} is produced, as indicated by the dashed lines. The intensity distribution of the resulting far-field diffraction pattern, which is formed by the interference between the scattered and unscattered wave-fields, comprises the so-called 'Fourier hologram' which is measured over the plane C.

As indicated in the figure, let the optic axis z be perpendicular to the plane C, which is taken to lie in the plane $z = z_0$. Both the point source and the sample are taken to lie in the vicinity of the plane $z = 0$. Let us approximate the disturbance due to the point source by a two-dimensional Dirac delta centred at $(x,y,z) = (x_0,y_0,0)$, so that the total complex disturbance in the plane $z = 0$ can be written as:

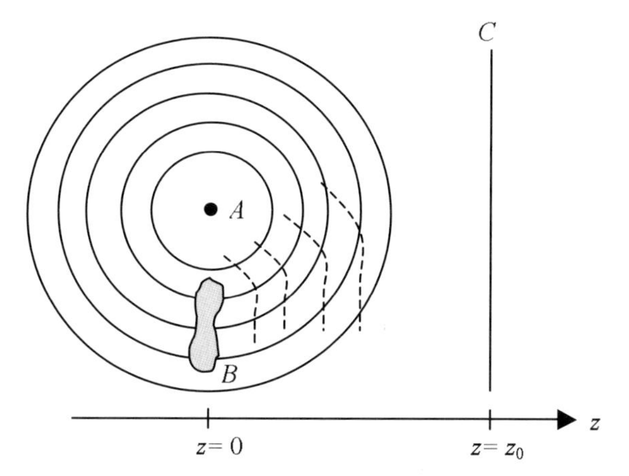

Fig. 4.6. Schematic for Fourier holography. A point source A emanates spherical reference waves, as indicated by the concentric circles. This reference wave, when scattered from a sample B, leads to the scattered wave indicated by the dashed lines. The far-field diffraction pattern, resulting from the coherent superposition of reference and scattered wave, is recorded over the plane C. Note that z_0 must be sufficiently large for the recorded pattern to lie in the far zone for a given experimental configuration. After Faigel and Tegze (2003).

$$\psi(x, y, z = 0) = \delta(x - x_0, y - y_0) + \psi_{\rm S}(x, y, z = 0). \tag{4.67}$$

If the plane C is sufficiently far downstream of the sample for it to be located in the far field, then the intensity $I_{\rm H}$ recorded over this plane will be a Fraunhofer diffraction pattern. Up to constant transverse and multiplicative scaling factors, which will be suppressed here, this pattern is given by the squared modulus of the Fourier transform of $\psi(x, y, z = 0)$ with respect to x and y, so that:

$$\begin{aligned} I_{\rm H}(k_x, x_y) &= |\mathcal{F}[\psi(x, y, z = 0)]|^2 \\ &= \frac{1}{4\pi^2} + |\mathcal{F}[\psi_{\rm S}(x, y, z = 0)]|^2 \\ &\quad + \frac{1}{2\pi} \exp[i(k_x x_0 + k_y y_0)]\mathcal{F}[\psi_{\rm S}(x, y, z = 0)] \\ &\quad + \frac{1}{2\pi} \exp[-i(k_x x_0 + k_y y_0)]\{\mathcal{F}[\psi_{\rm S}(x, y, z = 0)]\}^*. \end{aligned} \tag{4.68}$$

Here, (k_x, k_y) are the usual Fourier variables dual to x and y.

Suppose that one takes the inverse Fourier transform of the intensity distribution of the Fourier hologram, using numerical methods such as the inverse fast Fourier transform of a digitized hologram, or using optical methods. Making use of the shift theorem, the Fourier representation of the Dirac delta and the convolution theorem (see Appendix A), we arrive at:

$$2\pi\mathcal{F}^{-1}[I_{\mathrm{H}}(k_x,k_y)] = \delta(x,y) + [\psi_{\mathrm{S}}(x,y,z=0) \star \psi_{\mathrm{S}}^*(-x,-y,z=0)] + \psi_{\mathrm{S}}(x+x_0,y+y_0,z=0) + \psi_{\mathrm{S}}^*(-x+x_0,-y+y_0,z=0). \quad (4.69)$$

We examine, in turn, each term appearing on the right side of eqn (4.69). The first term will lead to a sharp peak at the origin of coordinates in the reconstructed wave-field. This may be thought of as the zeroth diffracted order of the holographic 'grating', this being the unscattered component of the illuminating beam. The second term will also typically be localized about the origin of coordinates in the reconstructed wave-field. The third term represents a reconstruction of the scattered wave-field $\psi_{\mathrm{S}}(x,y,z=0)$, which has been laterally displaced by $-x_0$ and $-y_0$ in the x and y directions, respectively. Finally, the fourth term gives an inverted and displaced form of the complex conjugate $\psi_{\mathrm{S}}^*(x,y,z=0)$ of the scattered wave-field, with the lateral displacement being by x_0 and y_0 in the x and y directions, respectively.

As was the case with off-axis holography, the reconstructed wave-field, as given by the third term in eqn (4.69), is spatially separated from the remaining terms present in the reconstruction. One thereby avoids the twin-image problem of Gabor holography. Note that Fourier holography may be thought of as a form of off-axis holography, with an off-axis reference wave being given by a spherical wave emanating from the plane of the sample, rather than by an off-axis plane wave.

For an experimental demonstration of the principles of Fourier holography, we consider the experiment by Eisebitt *et al.* (2004). In this work, the sample was a magnetic nanostructure comprising the domains of a 50-period cobalt/platinum multilayer fabricated on one side of a Si_3N_4 membrane. The thickness of the alternating cobalt and platinum layers was 4 and 7 Å, respectively. On the other side of the Si_3N_4 membrane a 600 nm thick gold film was deposited, this being opaque to the soft X-rays used in the experiment. To provide a 'window' for the sample, a circular hole of 1.5 μm diameter was drilled into the gold layer, from the surface of the gold layer down to the surface of the Si_3N_4 membrane. The windowed part of the sample then served as the object B in Fig. 4.6—note that this differs from the situation shown in Fig. 4.6, since the radiation emerging from the object is a transmitted rather than a scattered beam. Rather than having a point source A as shown in the same diagram, Eisebitt and co-workers drilled a conical hole right through the magnetic-multilayer/Si_3N_4/gold mask. The axis of this hole was at a distance of 3 μm from that of the window, with a diameter of 350 nm on the gold side of the structure and a diameter of 100 nm on the opposite side. This hole, when illuminated by soft X-rays, yielded the 'reference beam' emanating from the point A in Fig. 4.6. The sample-plus-aperture structure was illuminated by 778 eV right circularly polarized soft X-rays. These X-rays were produced by taking the output from a circular undulator at the BESSY-II storage ring in Berlin, Germany, spectrally filtering it through a spherical grating monochromator, and then spatially filtering the resulting radiation by passing it through a 20 μm pinhole at a distance of 6 m downstream of the focus of the monochromator. On emerging from the spatial filter, the radiation propagated through a distance of 723 mm before illuminating the sample-plus-aperture structure described above. The resulting Fourier hologram was recorded over a plane at a distance of 315 mm downstream of the structure. To register this image, 50 frames were summed, each being obtained with an exposure time of approximately 10 s using a 2048×2048 pixel CCD with a pixel size of 13.5 μm. The resulting Fourier hologram, shown in Fig. 4.7(a), has a number of salient features including a series of diagonal lines due to the interference of the object and reference beams[115], a series of rotationally symmetric fringes due to the circular window around the object, and a speckled structure due to the scattering from the sample. Numerically taking the modulus of the Fourier transform of this CCD image[116], Eisebitt and colleagues obtained the reconstruction shown in Fig. 4.7(b). The large central disc corresponds to the first pair of terms on the right side of eqn (4.69). In this context, we note that the self-correlation of a disc with the same disc yields another disc, with twice the radius. The two lobes of the reconstruction, evident in the figure, correspond to the last pair of terms in eqn (4.69). Each lobe is an independent image of the modulus of the wave-field at the exit surface of the sample window, with one lobe being an inverted form (through both x and y) of the other. The rippled structures in each lobe correspond to the magnetic domain structure of the multilayer. An effect known as 'magnetic circular dichroism' is responsible for one class of domain being more strongly absorbing than the other, for incident right circularly polarized X-rays. Note that this contrast is expected to be reversed upon replacing right circularly polarized X-rays with left circularly polarized X-rays. This was observed to be the case, by reconstructing the corresponding in-line hologram obtained with the helicity of the incident radiation reversed (not shown). The spatial resolution of the reconstruction, in both cases, was 50 nm.

4.4 Phase contrast

Suppose, for the sake of argument, that one was able to construct an essentially aberration-free X-ray imaging system. Suppose, further, that one was interested in the study of perfectly transparent samples such as soft biological tissue illuminated by hard X-rays. These materials, if they are sufficiently thin for the projection approximation (Section 2.2) to hold, change only the phase of the radiation transmitted through them. The intensity of the exit-surface wave-function therefore encodes no information about the structure of the sample, with all such information being present in the phase of the exit wave-function. Our putative perfect X-ray imaging system will therefore yield an image which tells us nothing

[115]Some of these are forked, this being indicative of the presence of phase vortices in the wave-field scattered by the sample. See Chapter 5 for more on phase vortices.

[116]In writing eqn (4.69) the inverse Fourier transform was taken, rather than the forward Fourier transform. The reconstruction is the same in both cases.

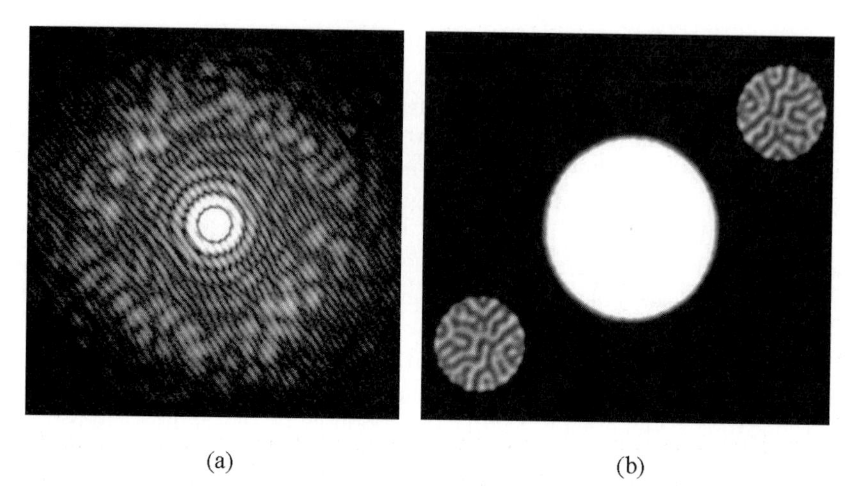

(a) (b)

Fig. 4.7. Demonstration of Fourier holography for the imaging of a magnetic nanostructure comprising a 50-period cobalt/platinum multilayer (details in main text). (a) Fourier hologram; (b) reconstruction. Diameter of each side-lobe in the reconstruction, in which the domain structure of the magnetic multilayer is evident, is 1.5 μm. Image taken from Eisebitt *et al.* (2004). Used with permission.

about the sample. If one defines a 'perfect' imaging system as one which faithfully reproduces the square modulus of the sample's exit surface wave-function, up to transverse and multiplicative scale factors, then we are led to the slightly paradoxical conclusion that perfect imaging systems are useless for the study of non-absorbing samples which change only the phase of X-ray radiation passing through them. Well known examples of this fact include: (i) the poor contrast evident in an in-focus image of a transparent biological sample which is taken on a good visible-light microscope; (ii) the poor contrast of untreated biological tissue in conventional medical X-ray radiographs.

Since perfect imaging systems are useless for the study of non-absorbing thin samples, we are led to deliberately introduce 'aberrations' in order to render visible the phase shifts which have been imparted to the probe radiation upon its passage through the sample of interest. Such systems are known as 'phase contrast imaging' systems, since they render phase shifts visible as intensity variations.

The most famous method of rendering phase shifts visible as intensity variations, namely interferometry, is treated in Section 4.6.1. Here, we restrict ourselves to the study of non-interferometric phase-contrast imaging systems which are coherent, shift-invariant linear imaging systems (Section 4.1.1.3). Recall that the action of such an imaging system is to take a given forward-propagating monochromatic scalar wave-field $\psi_{\mathrm{IN}}(x, y)$ as input and to yield the wave-function $\psi_{\mathrm{OUT}}(x, y) = \mathcal{F}^{-1}T(k_x, k_y)\mathcal{F}\psi_{\mathrm{IN}}(x, y)$ as output (see eqn (4.15)). Here, x, y are

Cartesian coordinates in the plane of the object, $\mathcal{F}$ denotes the Fourier transform with respect to x and y using the convention given in Appendix A, and k_x, k_y are the Fourier-space variables dual to x and y. As discussed in Section 4.1.1.3, a coherent shift-invariant linear imaging system is completely characterized by the transfer function $T(k_x, k_y)$, which equals unity for all k_x, k_y in a 'perfect' imaging system.[117]

Perhaps surprisingly, almost any deviation of $T(k_x, k_y)$, from the 'perfect' value of unity for all spatial frequencies k_x, k_y, will yield phase contrast in the sense that phase variations in the sample's exit surface wave-function will lead to intensity variations in the image produced by the imaging system. At once trivial and profound, this observation can be understood in terms of the interference between different Fourier components of the exit-wave-function being altered by the reweighting of these components which is brought about by the non-ideal transfer function. This is easily seen by reference to Fig. 4.8. Consider a given point (x, y) in the uniform-intensity exit-surface wave-function $\exp[i\phi(x, y)]$, which serves as the input into the imaging system. Via a Fourier integral, the complex disturbance at this point (x, y) is given by a coherent superposition of infinitely many plane-wave terms; the sum of all of these components will yield a complex number lying on a circle of unit radius in the complex plane, as suggested in the figure. As one moves around the (x, y) plane, the complex numbers which are superposed to give $\exp[i\phi(x, y)]$ will of course change, but their sum will still be constrained to lie on the circle of complex numbers with unit modulus. Now, if this exit surface wave-function is 'filtered' by an imaging system which has a non-constant transfer function $T(k_x, k_y)$, then each of the complex numbers will be re-weighted differently, so that, in general, the modulus of their coherent superposition will in general vary with (x, y). Thus we come to the remarkable conclusion that almost any non-uniform choice for $T(k_x, k_y)$ will result in a phase-contrast imaging system. This assertion is borne out in the following, where we consider four different forms of X-ray phase-contrast imaging which correspond to four different choices for $T(k_x, k_y)$: Zernike phase contrast (Section 4.4.1), differential interference contrast (Section 4.4.2), analyser-based imaging (Section 4.4.3), and propagation-based phase contrast (Section 4.4.4). We shall then offer some remarks on hybrid phase-contrast imaging systems (Section 4.4.5).

4.4.1 *Zernike phase contrast*

Suppose that a thin non-absorbing sample, when illuminated by coherent normally incident plane X-rays of uniform unit intensity, leads to an exit-surface wave-function $\exp[i\phi(x, y)]$, where the real function $\phi(x, y)$ describes the position-dependent phase shift imprinted by the sample on the X-ray beam. As discussed earlier, an imaging system which perfectly reproduces this exit-surface wave-function will give an image containing no information about the sample.

Let us decompose the exit-surface wave-function $\exp[i\phi(x, y)]$ as:

[117] In making this statement, we ignore both multiplicative constants in the transfer function, and transverse scaling factors due to magnification.

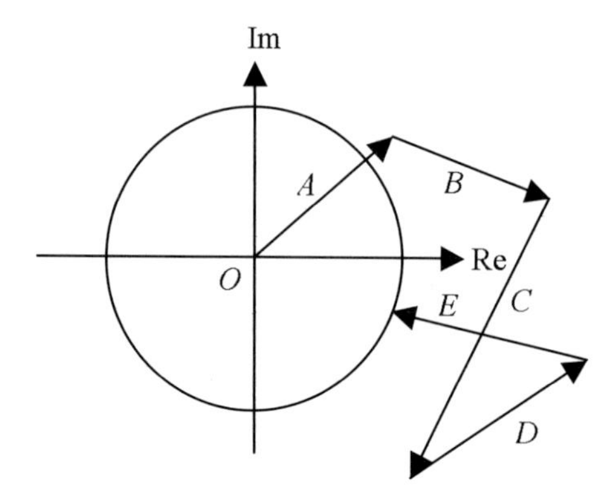

Fig. 4.8. Complex plane, where real and imaginary axes are respectively denoted by 'Re' and 'Im'. Suppose that five different complex numbers, each parameterized by some set of numbers τ, are represented by the vectors ('phasors') $A(\tau), B(\tau), C(\tau), D(\tau), E(\tau)$, and are constrained such that their sum must have modulus unity. Their sum is therefore represented by a vector which stretches from the origin of the complex plane, to a point on the circle with radius unity, which is centred at the origin of the complex plane. If these phasors, which are considered to represent different Fourier components of a given uniform-intensity disturbance $\exp[i\phi(x,y)]$, are reweighted by different complex coefficients (say, $\alpha, \beta, \gamma, \delta, \varepsilon$), then the sum $\alpha A+\beta B+\gamma C+\delta D+\varepsilon E$ will in general have a modulus that is no longer constrained to lie on a circle of unit radius in the complex plane. In general, this modulus varies with τ.

$$\exp[i\phi(x,y)] = \langle \exp[i\phi(x,y)] \rangle + f(x,y), \tag{4.70}$$

where the constant $\langle \exp[i\phi(x,y)] \rangle$ is by definition equal to the mean value of $\exp[i\phi(x,y)]$ over the entrance surface of a given coherent imaging system. Thus, by construction, the spatially averaged value of the complex function $f(x,y)$ is equal to zero. The Fourier transform of eqn (4.70), with respect to x and y, is:

$$\mathcal{F}\{\exp[i\phi(x,y)]\} = 2\pi \langle \exp[i\phi(x,y)] \rangle \delta(k_x,k_y) + \mathcal{F}f(x,y), \tag{4.71}$$

where $\delta(k_x,k_y)$ is the Dirac delta. Since the average value of $f(x,y)$ is zero, its two-dimensional Fourier transform $\mathcal{F}f(x,y)$ vanishes at the origin $k_x = k_y = 0$ of Fourier space. Conversely, $2\pi\langle \exp[i\phi(x,y)] \rangle \delta(k_x,k_y)$ is non-zero only at the origin of Fourier space. We conclude that, when one physically forms the Fourier transform of the exit-wave-function $\exp[i\phi(x,y)]$—as may be done, for example, by making the said wave-field coincide with the entrance-surface of a zone plate and then allowing the exit-surface radiation to propagate to the focal plane downstream of the plate—the contributions due to $\langle \exp[i\phi(x,y)] \rangle$ and $f(x,y)$ are spatially separated.

Herein lies the germ of an idea which would ultimately earn Zernike his 1953 Nobel prize. In its simplest incarnation, the method of Zernike phase contrast supposes the Fourier-transformed exit-surface wave-function in eqn (4.71) to impinge upon a thin non-absorbing screen of uniform thickness, which has a phase-shifting non-absorbing dot located at $k_x = k_y = 0$. Suppose this dot to impart a phase bias of $\phi_{\rm B}$ to the beam passing through it, and that the imaging system subsequently forms the inverse Fourier transform of the resulting wave-function. One thereby obtains the complex disturbance $\langle \exp[i\phi(x,y)]\rangle \exp(i\phi_{\rm B}) + f(x,y)$, so that the action of the Zernike phase contrast system is to alter the entrance wave-field as follows:

$$\begin{aligned} \exp[i\phi(x,y)] &= \langle \exp[i\phi(x,y)]\rangle + f(x,y) \\ &\longrightarrow \exp(i\phi_{\rm B})\langle \exp[i\phi(x,y)]\rangle + f(x,y) \\ &= [1+\exp(i\phi_{\rm B})-1]\langle \exp[i\phi(x,y)]\rangle + f(x,y) \\ &= \langle \exp[i\phi(x,y)]\rangle + f(x,y) + [\exp(i\phi_{\rm B})-1]\langle \exp[i\phi(x,y)]\rangle \\ &= \exp[i\phi(x,y)] + [\exp(i\phi_{\rm B})-1]\langle \exp[i\phi(x,y)]\rangle. \end{aligned} \qquad (4.72)$$

If we introduce the following complex constant ϖ, which depends on the phase bias $\phi_{\rm B}$ introduced by the phase-shifting dot:

$$\varpi \equiv [\exp(i\phi_{\rm B})-1]\langle \exp[i\phi(x,y)]\rangle \qquad (4.73)$$

then the action of the Zernike phase-contrast system can be written as:

$$\exp[i\phi(x,y)] \longrightarrow \exp[i\phi(x,y)] + \varpi. \qquad (4.74)$$

Thus, the Zernike system serves to add the bias-dependent complex constant ϖ to each point in the uniform-intensity input disturbance $\exp[i\phi(x,y)]$, to yield the output disturbance $\exp[i\phi(x,y)] + \varpi$. This may be viewed in interferometric terms as yielding an interference pattern between the incident disturbance and a certain z-directed plane wave, with the plane wave taking the value ϖ at each point in the output plane of the imaging system. The phase variations over the input plane are thereby visualized as intensity variations over the output plane, so that we can indeed speak of this system as exhibiting phase contrast.

This result can also be considered from a more geometric point of view, as sketched in Fig. 4.9. This shows the complex plane, with a circle of unit radius centred at the origin O. At any given point (x,y) in the input field, the associated complex disturbance $\exp[i\phi(x,y)]$ will correspond to a point $P(x,y)$ in the complex plane. As (x,y) is varied, the point $P(x,y)$ is constrained to move about the circle, implying that the intensity (modulus squared) of the field is always unity. Now, as we saw earlier, the action of a Zernike phase-contrast system is to add the complex constant ϖ to the complex disturbance at each point in the input field. In the complex plane this may be viewed as displacing the circle, originally centred at O, to a new circle of unit radius which is centred

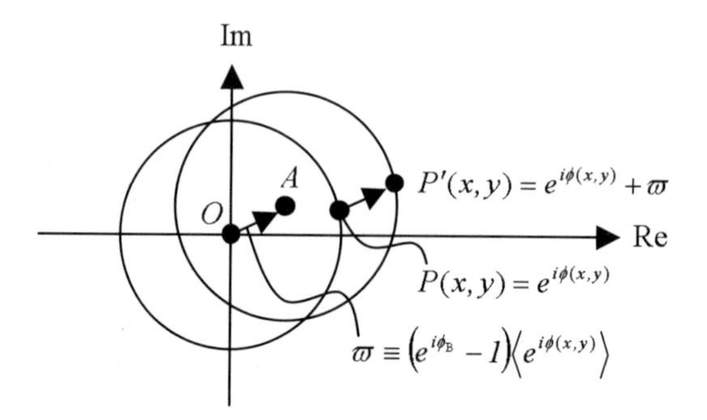

Fig. 4.9. Geometric construction in the complex plane, to aid interpretation of Zernike phase contrast. The complex number $P(x,y) \equiv \exp[i\phi(x,y)]$ lies on a circle of unit radius centred at the origin of the complex plane, with real and imaginary axes respectively denoted by 'Re' and 'Im'. The complex number $P'(x,y) \equiv \exp[i\phi(x,y)] + \varpi$ lies on a circle of unit radius centred at $A \equiv \varpi$.

at A. Thus each point $P(x,y)$ is mapped to the corresponding point $P'(x,y)$, with the output field being constrained to lie on the circle centred at A. As one moves from point to point in the output field, the corresponding disturbance will move about the circle centred at A. The resulting intensity will now depend on the phase of the input field, since the distance of $P'(x,y)$ from the origin now depends on $\phi(x,y)$. Again, we see that this system exhibits phase contrast.

For an early experimental implementation of X-ray Zernike phase contrast using soft X-rays, see, for example, Schmahl *et al.* (1994), and references therein. Here, we give a brief discussion of a more contemporary implementation of X-ray Zernike phase contrast microscopy due to Neuhäusler *et al.* (2003). In the experiment sketched in Fig. 4.10, 4 keV X-rays were used to image buried copper interconnects in a microelectronics circuit, using Fresnel zone plates for both the condenser and objective lenses. These gold zone plates had outermost zone widths of 60 and 50 nm respectively, resulting in an X-ray micrograph with 256-times magnification and 60 nm resolution. Rather than using a spot to phase shift the unscattered radiation, use was made of a phase shifting nickel ring, in the Fourier plane[118] of the objective lens. Bearing in mind the beam stop in Fig. 4.10, it is evident that the unscattered radiation occupies an annular region in the previously mentioned Fourier plane. Hence the use of a nickel phase ring, of thickness 0.7 μm, in order to shift the unscattered radiation by $\pi/2$

[118] A lens may be viewed as a Fourier transformer, since it converts incident plane waves into spots over the focal plane (more precisely, these are smeared Dirac deltas). The disturbance incident on the lens may be Fourier-decomposed as a continuous sum of such plane waves, implying the disturbance in the focal plane to be the Fourier transform of the incident disturbance. Hence we speak of the focal plane of the lens, over which each incident plane wave is brought to a point focus, as the 'Fourier plane'. Note that these remarks are closely related to the Abbe theory of imaging, which will not be covered here (see, for example, Born and Wolf (1999)).

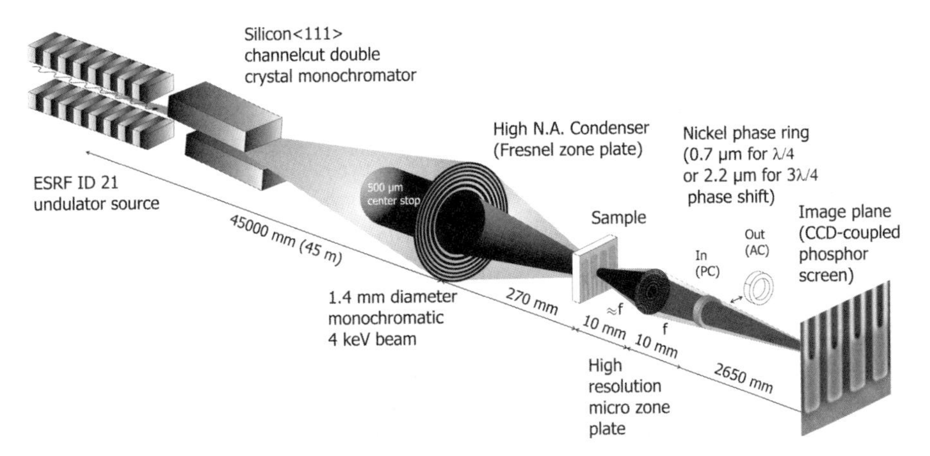

Fig. 4.10. Experimental setup for Zernike phase contrast imaging using 4 keV X-rays, by Neuhäusler and colleagues. Fresnel zone plates are used for the condenser and objective lenses, with the unscattered radiation being phase shifted by $\pi/2$ radians using a nickel phase ring in the Fourier-transform plane of the objective lens. Image taken from Neuhäusler *et al.* (2003). Used with permission.

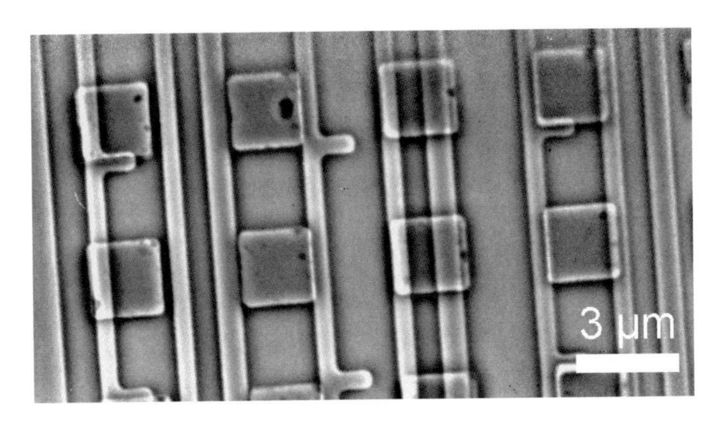

Fig. 4.11. Zernike phase contrast micrograph, obtained using the X-ray microscope drawn in Fig. 4.10. This clearly shows buried copper interconnects in an integrated circuit. Image taken from Neuhäusler *et al.* (2003). Used with permission.

radians and thereby achieve Zernike phase contrast. An example of the resulting image, of copper interconnects buried within the silicon dioxide dielectric of a microelectronics device, is given in Fig. 4.11.

4.4.2 *Differential interference contrast*

Consider, once more, a thin transparent object which is normally illuminated by unit-modulus monochromatic z-directed scalar plane waves, such that the complex wave-field at the exit surface of the object is $\exp[i\phi(x,y)]$. Suppose, further, that one has an imaging system which is able to create a replica $\exp[i\phi(x+\varepsilon,y)]$ of this wave-field, which has been transversely displaced by a distance ε in the negative x direction. If these two wave-fields may be overlaid, then the action of the optical system will be to transform input to output as follows:

$$\exp[i\phi(x,y)] \longrightarrow \exp[i\phi(x,y)] + \exp[i\phi(x+\varepsilon,y)]. \tag{4.75}$$

The intensity $I(x,y)$ of the resulting disturbance, which will be an interferogram formed between the field and the slightly displaced version of itself, is obtained by taking the squared modulus of the right side of the above expression:

$$I(x,y) = |\exp[i\phi(x,y)] + \exp[i\phi(x+\varepsilon,y)]|^2 . \tag{4.76}$$

To proceed further, assume ε to be sufficiently small that one may make the following first-order Taylor approximation to the phase $\phi(x+\varepsilon,y)$:

$$\phi(x+\varepsilon,y) \approx \phi(x,y) + \varepsilon\frac{\partial\phi(x,y)}{\partial x}. \tag{4.77}$$

Inserting approximation (4.77) into eqn (4.76), we see that:

$$\begin{aligned} I(x,y) &\approx \left|\exp[i\phi(x,y)] + \exp\left\{i\left[\phi(x,y) + \varepsilon\frac{\partial\phi(x,y)}{\partial x}\right]\right\}\right|^2 \\ &= \left|1 + \exp\left[i\varepsilon\frac{\partial\phi(x,y)}{\partial x}\right]\right|^2 \\ &= 2\left\{1 + \cos\left[\varepsilon\frac{\partial\phi(x,y)}{\partial x}\right]\right\}. \end{aligned} \tag{4.78}$$

The above result is good insofar as our imaging system exhibits phase contrast, with the intensity of the resulting output depending on the x-derivative of phase of the input. Since the contrast depends on the derivative of the phase, we may speak of this as exhibiting 'differential phase contrast' or 'differential interference contrast' (DIC). However, bearing in mind the Taylor-series expansion of the cosine term:

$$\cos\left[\varepsilon\frac{\partial\phi(x,y)}{\partial x}\right] = 1 - \frac{1}{2}\left[\varepsilon\frac{\partial\phi(x,y)}{\partial x}\right]^2 + \cdots, \tag{4.79}$$

it is evident that this differential phase contrast is only of second order in the small quantity ε.

This situation may be improved to give differential phase contrast which is of first order in ε, by introducing a phase bias ϕ_{B} into the displaced field

$\exp[i\phi(x+\varepsilon, y)]$, prior to interfering it with the non-displaced copy of itself. The intensity of the resulting disturbance, denoted $I_{\rm DIC}(x, y)$, is then:

$$I_{\rm DIC}(x, y) = |\exp[i\phi(x, y)] + \exp(i\phi_{\rm B}) \exp[i\phi(x+\varepsilon, y)]|^2 . \tag{4.80}$$

Making use of the Taylor series approximation in eqn (4.77), valid for small transverse displacements ε, we arrive at:

$$I_{\rm DIC}(x, y) \approx 2\left\{1 + \cos\left[\phi_{\rm B} + \varepsilon\frac{\partial\phi(x, y)}{\partial x}\right]\right\}. \tag{4.81}$$

Next, we make use of the following double-angle formula:

$$\cos(A + B) = \cos A \cos B - \sin A \sin B \tag{4.82}$$

for real A and B, to re-write eqn (4.81) as:

$$I_{\rm DIC}(x, y) \approx 2\left\{1 + \cos(\phi_{\rm B}) \cos\left[\varepsilon\frac{\partial\phi(x, y)}{\partial x}\right] - \sin(\phi_{\rm B}) \sin\left[\varepsilon\frac{\partial\phi(x, y)}{\partial x}\right]\right\}. \tag{4.83}$$

Assume that:

$$\left|\varepsilon\frac{\partial\phi(x, y)}{\partial x}\right| \ll 1, \tag{4.84}$$

so that we can make the following first-order Taylor approximations:

$$\cos\left[\varepsilon\frac{\partial\phi(x, y)}{\partial x}\right] \approx 1, \qquad \sin\left[\varepsilon\frac{\partial\phi(x, y)}{\partial x}\right] \approx \varepsilon\frac{\partial\phi(x, y)}{\partial x}. \tag{4.85}$$

Thus, to first order in ε, eqn (4.83) becomes:

$$I_{\rm DIC}(x, y) \approx 2\left\{1 + \cos(\phi_{\rm B}) - \varepsilon \sin(\phi_{\rm B})\frac{\partial\phi(x, y)}{\partial x}\right\}. \tag{4.86}$$

Introduction of a phase bias $\phi_{\rm B}$ thereby leads to differential interference contrast of first order in the transverse displacement ε of the two superposed fields. The resulting signal is linear in the x-derivative of the phase, under the assumptions stated above. This phase-contrast signal is maximized by choosing $\phi_{\rm B}$ such that $\sin(\phi_{\rm B}) = \pm 1$, that is, with the phase bias being an odd integer multiple of $\pi/2$.

Differential interference contrast, as described above, is an example of a coherent shift-invariant linear imaging system, in the sense outlined in Section 4.1.1.3. What is the associated transfer function $T(k_x, k_y)$ which completely characterizes such a system? Denoting input and output wave-fields by $\psi_{\rm IN}(x, y)$ and $\psi_{\rm OUT}(x, y)$ respectively, we have (cf. eqn (4.80)):

$$\psi_{\mathrm{OUT}}(x,y) = \psi_{\mathrm{IN}}(x,y) + \exp(i\phi_{\mathrm{B}})\psi_{\mathrm{IN}}(x+\varepsilon,y)$$
$$= \psi_{\mathrm{IN}}(x,y) \star [\delta(x,y) + \exp(i\phi_{\mathrm{B}})\delta(x+\varepsilon,y)], \qquad (4.87)$$

where '$\star$' denotes two-dimensional convolution. Fourier transform both sides of this expression with respect to x and y, and then make use of the convolution theorem (eqn (A.9)) to see that:

$$\mathcal{F}\psi_{\mathrm{OUT}}(x,y) = \mathcal{F}\left\{\psi_{\mathrm{IN}}(x,y) \star [\delta(x,y) + \exp(i\phi_{\mathrm{B}})\delta(x+\varepsilon,y)]\right\}$$
$$= 2\pi\left\{\mathcal{F}\psi_{\mathrm{IN}}(x,y)\right\} \times \left\{\mathcal{F}[\delta(x,y) + \exp(i\phi_{\mathrm{B}})\delta(x+\varepsilon,y)]\right\}$$
$$= 2\pi\left\{\mathcal{F}\psi_{\mathrm{IN}}(x,y)\right\} \times \left\{\frac{1}{2\pi}\left[1 + \exp(i\phi_{\mathrm{B}})\exp(i\varepsilon k_x)\right]\right\}. \qquad (4.88)$$

Take the inverse Fourier transform with respect to k_x and k_y to give

$$\psi_{\mathrm{OUT}}(x,y) = \mathcal{F}^{-1}\left[1 + \exp(i\phi_{\mathrm{B}})\exp(i\varepsilon k_x)\right]\mathcal{F}\psi_{\mathrm{IN}}(x,y), \qquad (4.89)$$

which may be compared to eqn (4.15) to obtain the following expression for the differential interference contrast transfer function:

$$T(k_x, k_y, \phi_{\mathrm{B}}, \varepsilon) = 1 + \exp(i\phi_{\mathrm{B}})\exp(i\varepsilon k_x). \qquad (4.90)$$

This differs from unity—that is, from the transfer function of a 'perfect' imaging system—by the single Fourier harmonic $\exp(i\phi_{\mathrm{B}})\exp(i\varepsilon k_x)$.

For an experimental demonstration of X-ray differential interference contrast microscopy using zone-plate optics, see, for example, Kaulich *et al.* (2002a,b) and references therein. The former paper contains a particularly useful list of key references to the earlier literature.

4.4.3 *Analyser-based phase contrast*

In all of the examples of phase-contrast imaging considered so far, use of a suitable 'aberrated' transfer function—that is, a transfer function differing from unity—was seen to yield phase contrast in a coherent shift-invariant linear imaging system (cf. Section 4.1.1.3). Thus Zernike phase contrast employs a transfer function that is unity everywhere with the exception of a phase offset at the origin (Section 4.4.1), while differential interference contrast has a transfer function that differs from unity by the addition of a single Fourier harmonic (Section 4.4.2). In this context, we reiterate the fact that almost any deviation from unity, of the transfer function associated with a two-dimensional coherent shift-invariant linear imaging system, will yield phase contrast in the sense that phase variations in the input field will contribute to intensity variations in the resulting output.

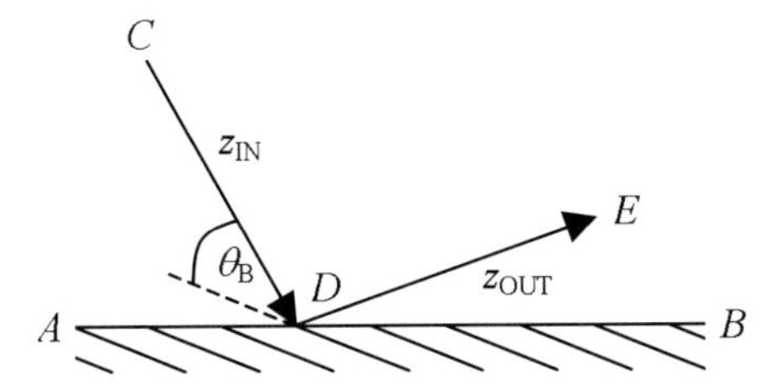

Fig. 4.12. Geometry for analyser-based phase contrast imaging.

The present section, on the use of analyser crystals (cf. Section 3.2.3) to yield phase contrast, is no exception. Stated more precisely, we shall see that the technique of analyser-based imaging makes use of reflection of a given two-dimensional forward-propagating coherent scalar wave from a near-perfect crystalline surface, in order to filter the Fourier spectrum and thereby achieve phase contrast (Förster *et al.* 1980; Somenkov *et al.* 1991; Ingal and Beliaevskaya 1995; Davis *et al.* 1995a,b; Davis 1996; Davis and Stevenson 1996; Chapman *et al.* 1997).

Recall the fact that Bragg's Law is often formulated as the statement that the planar surface, of a perfect semi-infinite crystal, will act as a perfect mirror for an incident coherent plane wave impinging at a so-called Bragg angle $\theta_{\rm B}$. Otherwise, the crystal acts as an absorber (see eqn (3.57)). This was seen to be a consequence of the kinematical theory of X-ray diffraction. The situation is more complex when a dynamical theory is employed, with the crystal now strongly reflecting over a range of plane-wave angles of incidence that are sufficiently close to the Bragg angle (cf. Fig. 3.15(c)).

To proceed further, let us set up two optic axes associated with a given nominal Bragg angle $\theta_{\rm B}$, as sketched in Fig. 4.12.[119] Here, $z_{\rm IN}$ is an optic axis giving the propagation direction CD of a coherent plane wave incident on the crystal surface AB at a given Bragg angle $\theta_{\rm B}$ to the active crystal planes, while $z_{\rm OUT}$ denotes the optic axis giving the propagation direction DE of the corresponding asymmetrically-Bragg-diffracted plane wave. Denote by $(x_{\rm IN}, y_{\rm IN})$ the Cartesian ordered pair which coordinatizes the plane perpendicular to $z_{\rm IN}$; the Fourier coordinates dual to $(x_{\rm IN}, y_{\rm IN})$ will be denoted by $(k_x^{\rm IN}, k_y^{\rm IN})$. Both $(x_{\rm OUT}, y_{\rm OUT})$ and $(k_x^{\rm OUT}, k_y^{\rm OUT})$ are similarly defined.

A plane wave $\exp[i(k_x^{\rm IN} x_{\rm IN} + k_y^{\rm IN} y_{\rm IN})]$, whose angle of incidence θ is sufficiently close to the Bragg angle, will be reflected by the crystal, this reflection being associated with a given phase-amplitude shift $T(k_x^{\rm IN}, k_y^{\rm IN})$. Thus the passage from incident to reflected plane wave may be denoted as follows:

[119] This shows a so-called 'Bragg geometry', in which incident and diffracted waves lie on the same side of the crystalline slab. 'Laue geometry' may also be considered, in which incident and diffracted waves lie on opposite sides of the crystalline slab. This case will not be treated in the present text.

$$\exp[i(k_x^{\mathrm{IN}} x_{\mathrm{IN}} + k_y^{\mathrm{IN}} y_{\mathrm{IN}})] \to T(k_x^{\mathrm{IN}}, k_y^{\mathrm{IN}}) \exp[i(k_x^{\mathrm{OUT}} x_{\mathrm{OUT}} + k_y^{\mathrm{OUT}} y_{\mathrm{OUT}})]. \quad (4.91)$$

In writing down this expression, we have implicitly assumed that the input–output relation, due to reflection from the crystal, defines a linear shift-invariant imaging system as described in Section 4.1.1.3 (cf. Davis 1996). This is an excellent approximation for a sufficiently high quality crystal.[120] Indeed, $T(k_x^{\mathrm{IN}}, k_y^{\mathrm{IN}})$ evidently plays the role of a transfer function for the crystal.

For notational simplicity, we henceforth assume the active Bragg planes of the semi-infinite crystal to be parallel to its surface, so that the superscripts 'IN' and 'OUT' can be dropped from both real-space and Fourier-space coordinate systems. Note that this restriction is easily dropped at the cost of increased complexity in the resulting formulae, if necessary. Returning to the main thread of the argument, let us again consider an incident plane wave $\exp[i(k_x x + k_y y)]$, whose angle of incidence θ is sufficiently close to the Bragg angle such that it is either reflected or 'close' to being reflected. As shown in Fig. 4.13(a) let us interpose a sample of interest between the plane wave and the crystal surface, with the sample leading to a disturbance

$$\psi_{\mathrm{IN}}(x, y) = \sqrt{I_{\mathrm{IN}}(x, y)} \exp[i\phi_{\mathrm{IN}}(x, y)] \quad (4.92)$$

over some plane z = constant (the nominal exit surface of the sample). This plane is sufficiently close to the entrance surface of the crystal for the effects of free-space propagation to be negligible (cf. Sections 4.4.4 and 4.4.5, together with Fig. 4.13(b)). Via a Fourier integral, the disturbance over the exit surface of the sample can be represented as a continuous sum of plane waves nominally centred about the unscattered plane wave:

$$\psi_{\mathrm{IN}}(x, y) = \frac{1}{2\pi} \iint \breve{\psi}_{\mathrm{IN}}(k_x, k_y) \exp[i(k_x x + k_y y)] dk_x dk_y. \quad (4.93)$$

Here, $\breve{\psi}_{\mathrm{IN}}(k_x, k_y)$ denotes the Fourier transform of $\psi_{\mathrm{IN}}(x, y)$ with respect to x and y, with (k_x, k_y) being Fourier coordinates dual to the (x, y) coordinates in the plane perpendicular to the incident direction CD:

$$\breve{\psi}_{\mathrm{IN}}(k_x, k_y) \equiv \mathcal{F}\psi_{\mathrm{IN}}(x, y). \quad (4.94)$$

If the sample is sufficiently thin, as we shall assume to be the case, then we can take the incident Fourier spectrum $\breve{\psi}_{\mathrm{IN}}(k_x, k_y)$ to be sufficiently narrow that the influence of all other Bragg peaks, apart from that which we are here considering, can be safely ignored. Bearing in mind the fact that our near-perfect analyser crystal is a linear imaging system, each incoming plane-wave component

[120] Our assumption will break down in the presence of sufficiently strong defect networks in the crystal, or in the presence of significant strain fields.

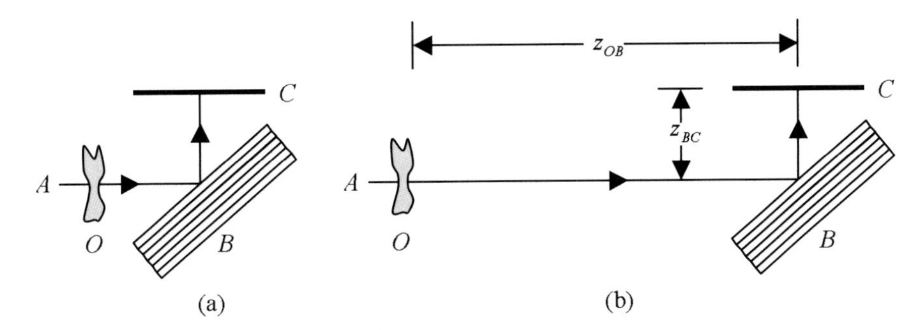

Fig. 4.13. (a) Schematic for analyser-based phase contrast imaging, with an incident plane wave from A passing through an object O before being reflected by an analyser crystal B. A phase-contrast image is subsequently recorded over a two-dimensional position-sensitive detector C. Distances from the object to the crystal, and from the crystal to the detector, are chosen to be sufficiently small for the effects of Fresnel diffraction to be negligible. Active Bragg planes are assumed to be parallel to the surface of the crystal. (b) As for A, but with the object-to-crystal distance z_{OB} and/or the crystal-to-detector distance z_{BC} chosen to be sufficiently long for the effects of Fresnel diffraction to be non-negligible. Treatment of this latter case is deferred to Section 4.4.5.

$\exp[i(k_x x + k_y y)]$ may be mapped to the corresponding outgoing component using eqn (4.91), leading to the following expression for the wave-function reflected by the crystal:

$$\psi_{\mathrm{OUT}}(x,y) = \frac{1}{2\pi}\iint dk_x dk_y \breve{\psi}_{\mathrm{IN}}(k_x,k_y) T(k_x,k_y) \exp[i(k_x x + k_y y)]. \quad (4.95)$$

Here, the (x, y) coordinates on the left side are in the plane perpendicular to the outgoing direction DE. The above equation, which may be written in the form:

$$\breve{\psi}_{\mathrm{OUT}}(k_x,k_y) = T(k_x,k_y)\breve{\psi}_{\mathrm{IN}}(k_x,k_y) \quad (4.96)$$

is simply a statement that the crystal possesses the transfer function $T(k_x, k_y)$. In this context, note that the inverse Fourier transform of eqn (4.96) is identical to eqn (4.15).

The functional form of the transfer function $T(k_x, k_y)$, for a given crystal, may be calculated from first principles using dynamical diffraction theory. However, for the purposes of a discussion focussing on the fundamentals of coherent X-ray imaging, we will continue to work with a generic expression for this transfer function.

Introduce the following expression:

$$R(k_x, k_y) \equiv |T(k_x, k_y)|^2, \tag{4.97}$$

which is known as the 'rocking curve' of the crystal (cf. Fig. 3.15(c)). This function is typically strongly peaked about a given Bragg angle for a particular crystal, rapidly reducing to zero (sometimes in a non-monotonic manner) as one moves away from that Bragg angle. Typical widths for rocking curves are on the order of micro-radians. Herein lies the 'analysing' power of the analyser crystal, which evidently possesses an exquisite sensitivity in its reflectivity as a function of the angle with which plane waves are incident upon it (cf. Section 3.2.3). If one decomposes the exit-surface complex amplitude emerging from a sample, into a continuous sum (Fourier integral) of plane waves of different directions, the crystal's filtration of this Fourier spectrum will lead to great sensitivity to the phase and amplitude modulations imposed by the sample upon the wave-field.

Following Gureyev and Wilkins (1997), one may consider two opposite extremes of analyser-based X-ray phase contrast imaging. The first extreme, which they termed the 'refractometric mode', corresponds to the case where the Fourier spectrum of the incident wave-field is much broader than the transfer function of the crystal. The second mode, termed the 'differential' mode, corresponds to the case where the Fourier spectrum of the incident wave-field is much narrower than the transfer function of the crystal. We consider each mode in turn.

In all further discussions, we assume the crystal transfer function $T(k_x, k_y)$ to be a function of one Fourier-space direction only, which may be taken as k_x upon a suitable rotation of coordinate system. This is an excellent approximation for many practical applications.

In the refractometric mode of analyser-based phase contrast imaging, the Fourier spectrum of the incident wave-field (i.e. the wave-field transmitted through the sample, and incident upon the analyser crystal) is much broader than the transfer function of the crystal. Therefore, the analyser crystal serves to filter the incident Fourier spectrum, to be non-zero only along a small strip centred along a line:

$$k_x = k_x^0 \tag{4.98}$$

in Fourier space. Let us take this strip ('reciprocal-space slit'), within which the transfer function is non-negligible in modulus, to be the Fourier-space region

$$k_x^0 - \Delta < k_x < k_x^0 + \Delta. \tag{4.99}$$

This is somewhat analogous to Schlieren imaging[121] as it corresponds to a pair of knife edges, one of which blocks all Fourier frequencies in the region:

[121] In Schlieren phase contrast, not considered in detail in this text, the corresponding transfer function $T(k_x, k_y)$ is equal to zero if $k_x < 0$, being equal to unity otherwise. This may be achieved by placing a so-called 'knife edge' in the back focal plane of a lens, thereby blocking out half of the Fourier spectrum of the incident disturbance.

$$k_x < k_x^0 - \Delta, \tag{4.100}$$

and the other of which blocks all Fourier frequencies in the region:

$$k_x > k_x^0 + \Delta. \tag{4.101}$$

A crucial difference between knife edges and analyser crystals, considered as Fourier space filters, lies in the fact that, for the former, the transfer function has uniform modulus over the transmitted regions of Fourier space, whereas the latter is a complex function of Fourier-space location for the transmitted regions of Fourier space. In the refractometric mode, the analyser crystal passes only a narrow prism of propagation vectors in the plane-wave decomposition of the wave-field which is incident upon it, corresponding to the Fourier-space strip in eqn (4.99) (Davis 1996). Each spatial frequency, in this 'pass band' of the imaging system, will in general be weighted by a different complex factor $T(k_x)$, corresponding to the transfer function of the analyser crystal. Further, as one alters the angle of incidence between the unscattered plane wave and the surface of the analyser crystal, the pass band of the system will sweep through various regions in the Fourier spectrum of the incident field. Viewed somewhat crudely in terms of geometric optics, one may state that the crystal only reflects incident 'rays' that lie within the previously mentioned prism of angles, corresponding to the Fourier-space pass-band of the analyser crystal. This picture, however, is only valid in the geometric-optics limit of the theory, which will not be discussed here.

In the second mode of analyser-based phase contrast, namely the differential mode, the Fourier spectrum of the input field is much narrower than the transfer function of the crystal. Provided that the analyser is oriented such that the transfer function has no zeros in regions where the spectrum of the incident wave is non-negligible, we conclude that no spatial frequencies are blocked by the imaging system. Rather than acting as a dual Schlieren-type knife edge, as was the case in the refractometric mode described earlier, the analyser serves to re-weight each Fourier component of the wave-field incident upon it. Once again, this leads to phase contrast.

We are ready to put some mathematics to the preceding statements regarding the differential mode of analyser-based phase contrast imaging. Make the admittedly restrictive assumption that the Fourier spectrum $\breve{\psi}_{\rm IN}(k_x, k_y)$ of the input wave-field, appearing in eqn (4.96), is sufficiently narrow for the transfer function $T(k_x, k_y) = T(k_x)$ to be safely approximated by its first order Taylor-series expansion about k_x^0, so that:

$$T(k_x) \approx T(k_x^0) + (k_x - k_x^0)A, \tag{4.102}$$

where:

$$A \equiv \left[\frac{dT(k_x)}{dk_x} \right]_{k_x = k_x^0}. \tag{4.103}$$

Hence eqn (4.96) becomes:

$$\breve{\psi}_{\mathrm{OUT}}(k_x, k_y) = \left[T(k_x^0) + (k_x - k_x^0) A \right] \breve{\psi}_{\mathrm{IN}}(k_x, k_y). \tag{4.104}$$

Note that, by assumption, $\breve{\psi}_{\mathrm{IN}}(k_x, k_y)$ is negligible at Fourier space coordinates (k_x, k_y) for which the approximation in square brackets ceases to be valid, so that the product on the right-hand side of eqn (4.104) is a good approximation for all Fourier-space coordinates (k_x, k_y).

Next, inverse Fourier transform eqn (4.104) with respect to k_x and k_y, bearing in mind the Fourier derivative theorem, to give:

$$\psi_{\mathrm{OUT}}(x, y) = B\psi_{\mathrm{IN}}(x, y) - iA \frac{\partial}{\partial x} \psi_{\mathrm{IN}}(x, y), \tag{4.105}$$

where:

$$B \equiv T(k_x^0) - Ak_x^0. \tag{4.106}$$

Write $\psi_{\mathrm{OUT}}(x, y)$ in terms of its intensity $I_{\mathrm{OUT}}(x, y)$ and phase $\phi_{\mathrm{OUT}}(x, y)$ as:

$$\psi_{\mathrm{OUT}}(x, y) = \sqrt{I_{\mathrm{OUT}}(x, y)} \exp[i\phi_{\mathrm{OUT}}(x, y)], \tag{4.107}$$

with:

$$\psi_{\mathrm{IN}}(x, y) = \sqrt{I_{\mathrm{IN}}(x, y)} \exp[i\phi_{\mathrm{IN}}(x, y)] \tag{4.108}$$

similarly defined. Substitute these expressions into eqn (4.105), and then take the squared modulus of the result, to give (Paganin *et al.* 2004a):

$$\begin{aligned} \frac{I_{\mathrm{OUT}}(x, y)}{I_{\mathrm{IN}}(x, y)} =& |B|^2 + 2\,\mathrm{Re}(A^* B) \frac{\partial \phi_{\mathrm{IN}}(x, y)}{\partial x} - \mathrm{Im}(A^* B) \frac{\partial}{\partial x} \log_e I_{\mathrm{IN}}(x, y) \\ &+ |A|^2 \left[\frac{\partial \phi_{\mathrm{IN}}(x, y)}{\partial x} \right]^2 + \frac{|A|^2}{4} \left[\frac{\partial}{\partial x} \log_e I_{\mathrm{IN}}(x, y) \right]^2. \end{aligned} \tag{4.109}$$

The registered phase contrast image $I_{\mathrm{OUT}}(x, y)$ is the product of the incident intensity $I_{\mathrm{IN}}(x, y)$ with the five terms on the right-hand side of eqn (4.109). Let us examine each term in turn. (i) In the absence of the analyser crystal, $A = 0$ and so eqn (4.109) reduces to:

$$I_{\mathrm{OUT}}(x, y) = |B|^2 I_{\mathrm{IN}}(x, y), \tag{4.110}$$

which exhibits no phase contrast. Therefore, up to a multiplicative constant, the first term represents the absorption-contrast image which would have been registered in the absence of an analyser crystal. Note that, in many practical contexts, this term is not the dominant term in image formation; indeed, if this term is dominant, then there is often less need to perform a phase contrast imaging experiment, since absorption contrast gives an adequate signal in the image. (ii) The second term is proportional to the derivative, in the x direction, of the phase $\phi_{\rm IN}(x, y)$ of the incident wave-field. Hence one may call $I_{\rm OUT}(x, y)$ a phase-contrast image, with this image being sensitive to phase gradients in one direction only (cf. eqn (4.86)). Indeed all terms but the first, on the right side of eqn (4.109), introduce a strong directional bias in the resulting analyser-based phase contrast image. (iii) The third term is proportional to the derivative of the incident intensity, in the x direction. Note that the image is sensitive to both the incident intensity, and its derivative with respect to x. This may be compared to the fact that the image is sensitive to phase gradients in the x direction, but not to the phase itself. (iv) The fourth term is quadratic in the derivative of the incident phase in the x direction, while (v) the fifth term is quadratic in the derivative of the incident intensity in the x direction.

To illustrate the theory developed above we now turn to an example of analyser-based phase contrast imaging, as shown in Fig. 4.14. This is an image of four identical nylon inclusions, each of length 10 mm, which are embedded in a 4.3 cm thick TOR(MAM)© mammographic test phantom (Department of Medical Physics, Leeds University, United Kingdom). This is intended to emulate a full thickness of female human breast tissue, forming a common standard which may be used to compare the relative performance of various mammographic imaging systems. In the present experiment, a silicon 111 monochromator was used to pass 20 keV synchrotron X-rays, emanating from the ELETTRA storage ring in Trieste, Italy. The analyser crystal was detuned from the Bragg condition for the unscattered X-ray radiation, such that the reflectivity for the unscattered radiation was approximately 50% of its maximum value. For a more detailed explanation of the experimental configuration, the reader is referred to Lewis *et al.* (2002).

Figure 4.14 clearly demonstrates a number of salient features of analyser-based phase contrast imaging. One immediately notices that, despite the fact that all four nylon fibres are identical, they have very different appearances in the analyser-based image. The vertical fibre has the greatest contrast, since this fibre will produce the greatest phase and intensity gradients in the x direction, which is identified with the horizontal direction in the displayed image (cf. eqn (4.109)). The reversal in contrast, from black on the left side of the vertical fibre, through to white on the right side, is a direct manifestation of the second term on the right-hand side of eqn (4.109). The two diagonal fibres display an intermediate degree of contrast, due to the x-derivative of the phase of the wave-field transmitted by the sample. The horizontal fibre displays no 'derivative' contrast, since, with the exception of its ends, it introduces no modulations in the x direction. One can see the weak absorptive effects of the horizontal fibre, which are represented by the first term on the right side of eqn (4.109). These weak

Fig. 4.14. Analyser-based phase contrast image of four identical nylon inclusions, each of length 10 mm, embedded in a mammographic test phantom designed to emulate a 4.3 cm thickness of female human breast tissue. Image taken from Paganin *et al.* (2004a). Used with permission.

absorptive effects are present for all four fibres; however, for the non-horizontal fibres, the 'derivative' contrast is the dominant contrast mechanism. Also, note the white dot on the right side of the horizontal fibre, which exhibits derivative phase contrast. The effects of the final three terms, in eqn (4.109), are not apparent to the eye in the present image. Indeed in many treatments these three terms may be safely ignored (see, for example, Chapman *et al.* (1997)). Lastly, we note that the analyser crystal is able to reject the vast majority of the radiation which is inelastically or incoherently scattered by the sample (Section 2.10)—this is a considerable advantage in contexts which generate a significant degree of such scatter, including the imaging of thick biological samples.

4.4.4 *Propagation-based phase contrast*

Consider such everyday phenomena as the heat shimmer over a hot road, the twinkling of stars on a clear night, the focusing of sunlight by a good glass lens, and the network of bright lines which dance about the floor of a swimming pool on a cloudless day. Each of these are examples of phase contrast imaging systems, in the general sense of the term which was defined earlier in this chapter. What each of these four examples have in common is the use of propagation to

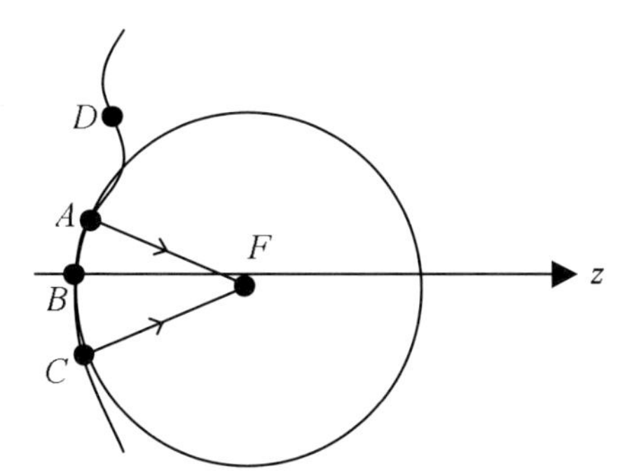

Fig. 4.15. Propagation-based phase contrast, from the point of view of geometric optics.

render phase variations, which are introduced over a given exit-surface plane downstream of a non-absorbing structure, visible as intensity fluctuations over a plane suitably far downstream of the initial plane.

Bearing this in mind, consider Fig. 4.15. The wavy line schematically denotes the wave-front (surface of constant phase) associated with a paraxial wave-field:

$$\psi_\omega(x, y, z = 0) = \sqrt{I_\omega(x, y, z = 0)} \exp[i\phi_\omega(x, y, z = 0)] \tag{4.111}$$

which exists over an exit-surface plane, taken as $z = 0$, of a semi-transparent structure (not shown) illuminated from the left with paraxial monochromatic scalar electromagnetic radiation. The quantities $I_\omega(x, y, z = 0)$ and $\phi_\omega(x, y, z = 0)$ respectively denote the intensity and phase of the radiation over the plane $z = 0$, with x and y being Cartesian coordinates over planes parallel to the optic axis z, and ω denoting the angular frequency of the radiation. At a point B over the wave-front, draw a tangent sphere ABC centred at F, with F lying near the optic axis on account of the fact that the radiation is assumed to be paraxial. At point B, our exit-surface wave-field looks like a collapsing spherical wave, on account of the fact that this tangent sphere is on the right side of the wave-front. Therefore, within the geometric-optics approximation, the intensity will increase as one moves from B towards F. For the same reason, the intensity will increase as one moves in the positive z direction through a small distance, from the point B to a point to the right of B. Using similar reasoning, one can consider tangent spheres to points such as D on the wave-front. Here, the tangent sphere is on the left side of the wave-front, rather than on the right side; locally, the wave-front then looks like an expanding spherical wave, with the intensity therefore decreasing as one moves through a small distance in the positive z direction.

This line of reasoning gives a simple geometric picture for what may be termed 'propagation-based phase contrast', namely a form of phase contrast imaging which uses free-space propagation in order to convert phase variations over a plane, into intensity variations in a registered image. Importantly, no optical elements are required in order to create such a phase-contrast image. We now turn to a wave-optical formulation of this concept of propagation-based phase contrast.

Let us begin with the operator form for the Fresnel diffraction integral, given in eqn (1.28):

$$\psi_\omega(x,y,z=\Delta) = \exp(ik\Delta) \times \mathcal{F}^{-1}\exp\left[\frac{-i\Delta(k_x^2+k_y^2)}{2k}\right]\mathcal{F}\psi_\omega(x,y,z=0), \quad \Delta \geq 0. \tag{4.112}$$

Here, $\psi_\omega(x,y,z=\Delta)$ denotes the wave-field that is obtained by propagating the disturbance $\psi_\omega(x,y,z=0)$ through a distance $\Delta \geq 0$, $\mathcal{F}$ denotes Fourier transformation with respect to x and y using the convention outlined in Appendix A, $\mathcal{F}^{-1}$ denotes the corresponding inverse Fourier transformation, k is the wave-number of the radiation and (k_x, k_y) are the Fourier coordinates dual to (x, y).

Assume the propagation distance Δ to be sufficiently small that one may make the following approximation to the second exponent (Fresnel propagator) in eqn (4.112):

$$\exp\left[\frac{-i\Delta(k_x^2+k_y^2)}{2k}\right] \approx 1 - \frac{i\Delta(k_x^2+k_y^2)}{2k}. \tag{4.113}$$

Making subsequent use of the Fourier derivative theorem, and writing the un-propagated wave-field in the form given by eqn (4.111), we see that eqn (4.112) becomes:

$$\psi_\omega(x,y,z=\Delta) = \exp(ik\Delta) \times \left[1+\frac{i\Delta\nabla_\perp^2}{2k}\right]\sqrt{I_\omega(x,y,z=0)}\exp[i\phi_\omega(x,y,z=0)], \quad \Delta \geq 0. \tag{4.114}$$

Here, $\nabla_\perp^2 \equiv \partial^2/\partial x^2 + \partial^2/\partial y^2$ denotes the Laplacian in the xy plane.

Since we are interested in calculating the intensity:

$$I_\omega(x,y,z=\Delta) = |\psi_\omega(x,y,z=\Delta)|^2 \tag{4.115}$$

of the propagation-based phase contrast image, take the squared modulus of eqn (4.114). Discarding terms that are quadratic in the propagation distance Δ, which will be permissible if the propagation distance is sufficiently small, we arrive at:

$$I_\omega(x,y,z=\Delta) = I_\omega(x,y,z=0) + 2\,\mathrm{Re}\{\sqrt{I_\omega(x,y,z=0)}\exp[-i\phi_\omega(x,y,z=0)]\frac{i\Delta}{2k} \times \nabla_\perp^2\sqrt{I_\omega(x,y,z=0)}\exp[i\phi_\omega(x,y,z=0)]\}. \quad (4.116)$$

This expression may be manipulated into the following form[122]:

$$I_\omega(x,y,z=\Delta) = I_\omega(x,y,z=0) - \frac{\Delta}{k}\nabla_\perp \cdot [I_\omega(x,y,z=0)\nabla_\perp\phi_\omega(x,y,z=0)]. \quad (4.117)$$

This result shows how the propagation-based phase contrast image, recorded over the plane $z = \Delta \geq 0$, is related to both the phase and intensity variations of the field over the plane $z = 0$. The more strongly curved the wave-front over a given point in the plane $z = 0$, the greater will be the increase or decrease in the intensity at the corresponding point in the propagation-based phase contrast image. This observation allows us to make contact with the simple geometric picture, for propagation-based phase contrast, with which we opened the present sub-section.

If the intensity of the unpropagated field is sufficiently slowly varying in x and y, then its x and y derivative may be neglected. Equation (4.117) then becomes (Bremmer 1952):

$$I_\omega(x,y,z=\Delta) \approx I_\omega(x,y,z=0)\left[1 - \frac{z\lambda}{2\pi}\nabla_\perp^2\phi_\omega(x,y,z=0)\right]. \quad (4.118)$$

Here, we have made use of the fact that $k = 2\pi/\lambda$, where λ is the radiation wavelength.

In the regime of small defocus and negligible wave-front tilt, eqn (4.118) describes phase contrast by defocus, with contrast proportional to the Laplacian of the phase of the wave-function being visualized. This contrast increases linearly with z, until the value of z is sufficiently large for our approximations, used in obtaining this expression, to break down. Note that if one makes use of an imaging system to replicate, possibly in magnified or demagnified form, the wave-field in the plane $z = 0$, then z can evidently be made negative as well as positive. Indeed, eqn (4.118) models the well-known contrast reversal which is observed when one passes from under-focus to over-focus, when using a bright-field imaging system[123] to form an image of a thin transparent sample (see, for example,

[122]Note that this is a form of the 'transport-of-intensity equation' (Teague 1983), which will be discussed at greater length in Section 4.5.2.

[123]By 'bright-field imaging system' we refer to a coherent two-dimensional imaging system which is not specifically engineered to yield phase contrast, namely one which is designed to form as faithful as possible a reproduction of the intensity of the incident wave-field, in forming the output wave-field corresponding to the registered image.

Zernike (1942)). Note in this context, that one does not need an imaging system to achieve such 'propagation based' phase contrast, since free-space propagation will do the job. In the absence of an imaging system, one must choose z to be non-negative, corresponding to free-space propagation, from the exit surface of the sample to the imaging plane, through such a distance. Accordingly, while this section will speak of the case of achieving phase contrast via propagation through free space, it should be understood that the same ideas hold true for defocus (which, as already stated, may be either positive or negative) in a bright-field X-ray imaging system.

An important feature, of propagation-based phase contrast, is its relative insensitivity to polychromaticity (i.e. lack of temporal coherence) in the radiation source. However, the method is rather sensitive to lack of spatial coherence in the source: if the size of an incoherent source is too large, propagation-based phase contrast is destroyed. Thus propagation-based X-ray phase contrast can be achieved using fully polychromatic X-ray radiation from a small source (Wilkins *et al.* 1996).

To explore this idea in more detail, let us obtain a polychromatic generalization of eqn (4.118). This equation holds for each monochromatic component, which has given wavelength λ corresponding to angular frequency ω, of an incident polychromatic field. Indicate this dependence of the relevant quantities on angular frequency by writing eqn (4.118) as:

$$I_\omega(x,y,z) \approx I_\omega(x,y,z=0)\left[1 - \frac{zc}{\omega}\nabla_\perp^2 \phi_\omega(x,y,z=0)\right]. \qquad (4.119)$$

Here, we have made use of the fact that:

$$\lambda = \frac{2\pi c}{\omega}, \qquad (4.120)$$

where c is the speed of light in vacuum.

To keep our formulae simple, assume that polychromatic plane waves are normally incident upon a non-absorbing structure which is sufficiently thin for the projection approximation (Section 2.2) to be valid.[124] Assume further that the structure is composed of a certain number M of different materials, each of which have a given ω-dependent refractive index denoted by:

$$n_{\omega,j} \equiv 1 - \delta_{\omega,j}, \quad j = 1,2,\cdots,M. \qquad (4.121)$$

The projected thickness of each material is written as $T_j(x,y)$. Under these approximations the phase shift $\Delta\phi_\omega(x,y,z=0)$, suffered by a given normally

[124]Note, however, that the assumptions of both planar incident radiation, and a transparent sample, are easily lifted. At the price of working with a non-trivial scattering theory, one may also lift the assumption of the projection approximation by using, for example, one of the scattering theories outlined in Chapter 2 (e.g. multislice or the first Born approximation).

incident monochromatic plane wave of angular frequency ω which passes through the structure, is given by the following generalized form of eqn (2.41):

$$\Delta\phi_\omega(x,y,z=0) = -\frac{2\pi}{\lambda}\sum_{j=1}^{M}\delta_{\omega,j}T_j(x,y) = -\frac{\omega}{c}\sum_{j=1}^{M}\delta_{\omega,j}T_j(x,y). \qquad (4.122)$$

Hence the phase of a given monochromatic component in the plane $z = 0$, that is, at the exit surface of the structure of interest, is given by:

$$\phi_\omega(x,y,z=0) \approx \phi_{\rm IN}^{(\omega)} - \frac{\omega}{c}\sum_{j=1}^{M}\delta_{\omega,j}T_j(x,y). \qquad (4.123)$$

Here, $\phi_{\rm IN}^{(\omega)}$ is a given constant phase offset of a particular monochromatic component of the polychromatic plane wave. Substitute eqn (4.123) into eqn (4.119), letting $I_\omega(x,y,z = 0) \equiv f(\omega)$ indicate the different spectral weights of the monochromatic components of the polychromatic plane wave. Note that the independence of $f(\omega)$ on x and y amounts to the previously stated assumption that the object is non-absorbing at all wavelengths for which the incident spectrum is non-negligible. One may then integrate the resulting formula over all angular frequencies, to obtain (Wilkins *et al.* 1996; Pogany *et al.* 1997)[125]:

$$\overline{I}(x,y,z) \approx \overline{I}(x,y,z=0) + z\sum_{j=1}^{M}\left[\int \delta_{\omega,j}f(\omega)d\omega\right]\nabla_\perp^2 T_j(x,y). \qquad (4.124)$$

Here, we have defined:

$$\overline{I}(x,y,z) \equiv \int I_\omega(x,y,z)d\omega, \qquad (4.125)$$

this being the time-averaged intensity of the polychromatic field, which is obtained by summing the intensities of each of its monochromatic components (cf. Section 4.1.2.1). Again, we see the now-familiar Laplacian term, indicating that the polychromatic phase contrast image is sensitive to the curvature (i.e. the Laplacian) of the projected thickness of each material in the sample.

Regarding the experimental demonstration of propagation-based phase contrast—albeit in a non-imaging context—it is worth noting that Röntgen himself investigated this problem, within the context of the geometric-optics concept of refraction. Specifically, he enquired whether his newly discovered X-rays were

[125]We recall that the time-averaged intensity, when averaged over a time sufficiently long with respect to the coherence time of a finite-duration polychromatic field, is the sum of the intensities of each monochromatic component.

'susceptible of refraction', concluding that 'I have obtained no evidence of refraction' (Röntgen 1896). In retrospect, we now know that X-rays are a form of electromagnetic radiation, a hypothesis considered and then rejected in Röntgen's pioneering paper. This then leads us onto a consideration of experimental work in propagation-based phase contrast using electromagnetic radiation, of which visible light was the earliest known example. In its simplest incarnation, such propagation-based phase contrast is at least as old as the lens itself, the origins of which will not be traced here. In the context of visible light microscopy, propagation-based phase contrast (via over- or under-focus of a bright-field image of a transparent sample) has long been known. In this context, note that Zernike opened one of his earliest phase-contrast papers with the lines, 'Every microscopist knows that transparent objects show light or dark contours under the microscope in different ways varying with defocus...' (Zernike 1942).

Regarding experimental demonstrations of propagation-based X-ray phase contrast imaging using hard X-rays, the key papers are by Snigirev *et al.* (1995, 1996), Wilkins (1996), Wilkins *et al.* (1996), Nugent *et al.* (1996), and Cloetens *et al.* (1996). Of these, the papers by Snigirev *et al.*, Nugent *et al.*, and Cloetens *et al.* all used monochromatic synchrotron radiation, while the work cited of Wilkins and colleagues involved using polychromatic radiation from a microfocus lab-based source and has laid the foundations for phase-contrast imaging using conventional X-ray sources. A propagation-based phase-contrast image, from Snigirev *et al.* (1995), is shown in Fig. 4.16 while one from Wilkins *et al.* (1996) is shown in Fig. 4.17. In both sets of images we see the effects of propagation-based phase contrast, due to the transverse redistribution of optical energy arising from transverse phase modulations imposed on the X-ray beam upon its passage through the sample.[126]

4.4.5 *Hybrid phase contrast*

We have repeatedly seen that almost any deviation from unity, of the transfer function characterizing a two-dimensional coherent shift-invariant linear imaging system, yields phase contrast in the sense that phase variations in the input field lead to intensity variations in the corresponding output. Examples include the use of a Zernike system to phase shift the central spot of the Fourier transform of the input (Section 4.4.1), the transfer function of differential interference contrast which differs from unity by the addition of single Fourier harmonic (Section 4.4.2), the use of a crystal transfer function in analyser-based phase contrast (Section 4.4.3) and the use of the free-space transfer function in propagation-based phase contrast (Section 4.4.4).

The second and third systems, listed above, suffer from the deficiency of directional bias in the resulting phase-contrast image, being sensitive to transverse phase gradients in a given direction while being insensitive to phase gradients in the orthogonal transverse direction. This is a direct consequence of the fact that

[126]Note, also, the example of propagation-based X-ray phase contrast—in the non-imaging context of X-ray lithography—by White and Cerrina (1992).

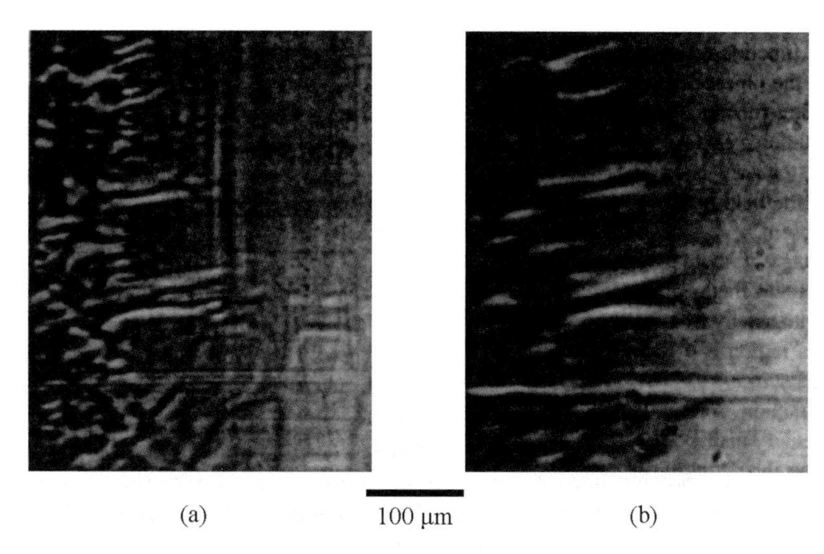

Fig. 4.16. Propagation-based phase contrast images of partially melted aluminium foil, obtained using 16 KeV monochromated synchrotron radiation. (a) Image obtained at 25 cm propagation distance from foil; (b) image obtained at 150 cm from the foil. Note that, since aluminium is essentially non-absorbing at the X-ray energy used to produce the above data, no contrast would be observed in an absorption-contrast image (i.e. one taken at close to zero propagation distance, between the exit surface of the sample and the surface of the detector). Image taken from Snigirev *et al.* (1995). Used with permission.

the associated transfer functions vary with one Fourier-space coordinate only, taken to be k_x in our preceding analyses. Most non-trivial means of removing this symmetry, of the transfer function, will overcome the deficiency mentioned above. This is easily obtained by cascading (Section 4.1.3), hybridizing either of these phase-contrast systems (whose transfer function depends only on k_x) with another phase-contrast system whose transfer function depends on k_y (Pavlov *et al.* 2004). One simple means of doing so is to defocus one's differential interference phase contrast images, or one's analyser-based phase contrast images, thereby incorporating propagation-based phase contrast into each system. These hybrid phase-contrast scenarios are respectively described in the following pair of sub-sections.

4.4.5.1 *Combined differential interference contrast and propagation-based phase contrast* Inserting the transfer function for differential interference contrast (eqn (4.90)) into the operator form for the action of an arbitrary shift-invariant coherent linear imaging system (eqn (4.15)), we see that the relation between input and output fields is given by:

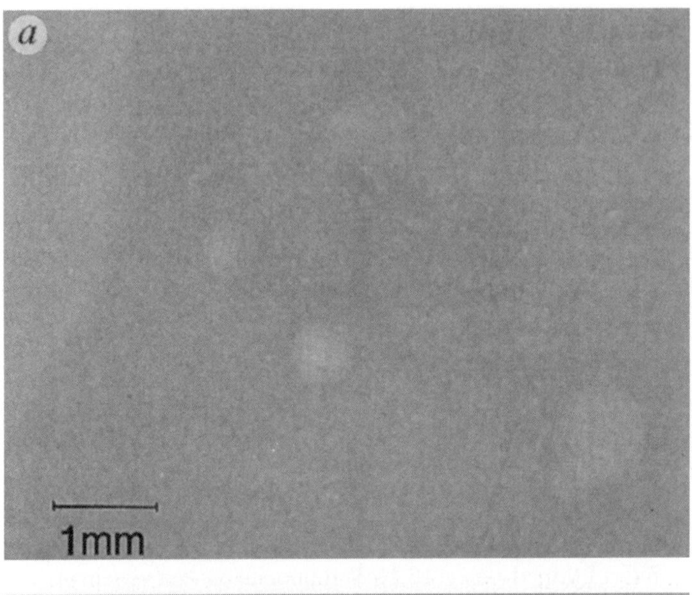

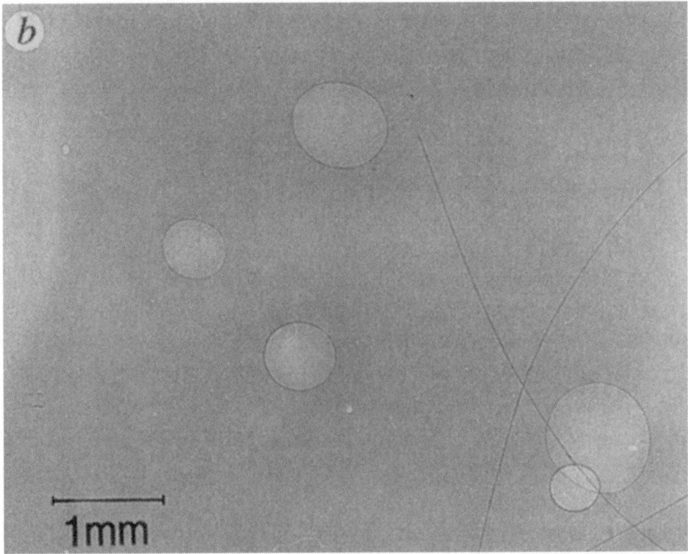

Fig. 4.17. Propagation-based phase contrast images of air bubbles and glass fibres embedded in glue, obtained using broad-spectrum polychromatic radiation from a micro-focus laboratory X-ray source at 60 kV tube voltage. Anode was copper, with size of polychromatic X-ray source at approximately 20 μm. For both images, source-to-sample distance was 200 mm. (a) Absorption image, obtained using a sample-to-detector distance of 1 mm, shows very little contrast as the sample is almost perfectly transparent; (b) Polychromatic propagation-based phase contrast image, showing strong contrast, obtained using a sample-to-detector distance of 1200 mm. Image taken from Wilkins *et al.* (1996). Used with permission.

$$\psi_{\mathrm{OUT}}(x, y, \phi_{\mathrm{B}}, \varepsilon) = \mathcal{F}^{-1}\left[1 + \exp(i\phi_{\mathrm{B}}) \exp(i\varepsilon k_x)\right] \mathcal{F}\psi_{\mathrm{IN}}(x, y). \tag{4.126}$$

Recall that ε is the transverse shift between the two copies of the incident coherent scalar wave-field which are created in a differential interference contrast system, and ϕ_{B} is the phase bias induced on the displaced field (see Section 4.4.2). All other symbols are as defined previously.

Next, suppose that this output wave-field is propagated through a distance $\Delta \geq 0$. Assuming the field to be paraxial, we may apply the operator form of the Fresnel diffraction operator to the right side of eqn (4.126), as given in eqn (1.28), so that the output field is now:

$$\psi_{\mathrm{OUT}}(x, y, \phi_{\mathrm{B}}, \varepsilon, \Delta, k) = \exp(ik\Delta) \\ \times \mathcal{F}^{-1} \exp\left[\frac{-i\Delta(k_x^2 + k_y^2)}{2k}\right] \left[1 + \exp(i\phi_{\mathrm{B}}) \exp(i\varepsilon k_x)\right] \mathcal{F}\psi_{\mathrm{IN}}(x, y). \tag{4.127}$$

In the lower line of this equation, assume that both Δ and ε are sufficiently small that one may replace all exponential functions involving these quantities, by their first-order Taylor expansions. Making subsequent use of the Fourier derivative theorem, we arrive at:

$$\psi_{\mathrm{OUT}}(x, y, \phi_{\mathrm{B}}, \varepsilon, \Delta, k) \approx \exp(ik\Delta) \\ \times \left(1 + \frac{i\Delta\nabla_{\perp}^2}{2k}\right) \left[1 + \exp(i\phi_{\mathrm{B}}) + \varepsilon \exp(i\phi_{\mathrm{B}}) \frac{\partial}{\partial x}\right] \psi_{\mathrm{IN}}(x, y). \tag{4.128}$$

To avoid undue complexity in our formulae we assume that the incident field has no intensity modulations, so that we may write:

$$\psi_{\mathrm{IN}}(x, y) = \exp[i\phi_{\mathrm{IN}}(x, y)]. \tag{4.129}$$

Insert this into eqn (4.128), expand the resulting expression and then discard the cross term proportional to the product $\varepsilon\Delta$, leaving:

$$\psi_{\mathrm{OUT}}(x, y, \phi_{\mathrm{B}}, \varepsilon, \Delta, k) \exp\{-i[\phi_{\mathrm{IN}}(x, y) + k\Delta]\} \\ \approx 1 + \exp(i\phi_{\mathrm{B}}) + i\varepsilon \exp(i\phi_{\mathrm{B}}) \frac{\partial \phi_{\mathrm{IN}}(x, y)}{\partial x} \\ - \frac{\Delta}{2k}[1 + \exp(i\phi_{\mathrm{B}})] \left[\nabla_{\perp}^2 \phi_{\mathrm{IN}}(x, y) + i|\nabla_{\perp}\phi_{\mathrm{IN}}(x, y)|^2\right]. \tag{4.130}$$

Take the squared modulus of the above expression, keeping only those terms which are linear in either Δ or ε, discarding higher-order terms in either variable together with any cross terms proportional to $\varepsilon\Delta$. This yields the following approximate formula for the intensity $I_{\mathrm{OUT}}(x, y, \varepsilon, \phi_{\mathrm{B}}, \Delta, k)$ output by our hybrid imaging system, which incorporates differential interference contrast with propagation-based phase contrast:

$$\begin{aligned} &I_{\mathrm{OUT}}(x, y, \phi_{\mathrm{B}}, \varepsilon, \Delta, k) \\ &\quad \approx 2\left[1 + \cos\phi_{\mathrm{B}} - \varepsilon\sin(\phi_{\mathrm{B}})\frac{\partial\phi_{\mathrm{IN}}(x, y)}{\partial x} - \frac{\Delta}{k}(\cos\phi_{\mathrm{B}} + 1)\nabla_{\perp}^{2}\phi_{\mathrm{IN}}(x, y)\right]. \end{aligned} \tag{4.131}$$

As the above result demonstrates, the hybrid imaging system is sensitive to phase variations in both the x and y directions. Note that the above formula reduces to eqn (4.86) when $\Delta = 0$, corresponding to 'pure' differential interference contrast. If we set $\varepsilon = 0$ and $\phi_{\mathrm{B}} = 0$, then the right side of the above equation is equal to four times the right side of eqn (4.118), for the case of uniform unpropagated intensity. This factor of four arises because two identical fields are considered to be superposed over one another, prior to propagation, in the above calculation. This doubles the amplitude and quadruples the intensity.

4.4.5.2 *Combined analyser-based contrast and propagation-based phase contrast*

The essential feature of the general methodology for studying hybrid phase contrast in shift-invariant coherent linear imaging systems, a special case of which is described in the previous sub-section, is as follows: (i) When cascading phase-contrast imaging systems, the corresponding transfer functions are multiplied by one another in the relation mapping input to output fields (cf. eqn (4.127)); (ii) Assume that the transfer functions are sufficiently slowly varying, over the region of Fourier space where the two-dimensional Fourier transform of the incident disturbance is non-negligible, for them to be well approximated by low-order Taylor expansions; (iii) Invoke the Fourier derivative theorem, to convert terms in the Taylor series into differential operators in real space, in the relation which maps the input to the output fields (cf. eqn (4.128)); (iv) Evaluate the intensity of the output produced by the cascaded phase-contrast system (Pavlov *et al.* 2004).

This methodology may also be used to study the hybrid phase contrast imaging system sketched in Fig. 4.13(b), which unites analyser-based phase contrast with propagation-based phase contrast (Pavlov *et al.* 2004; Coan *et al.* 2005). Here analyser-based phase contrast is augmented by propagation-based phase contrast via the simple expedient of introducing sufficient distance, between the object and the crystal and/or between the crystal and the detector, in order for propagation-based phase contrast to be active. This serves to redress the direction specificity of analyser-based phase contrast imaging, as epitomized by Fig. 4.14. Pavlov *et al.* (2004) have developed a theoretical formalism, based on the methodology described above, for studying this hybrid of analyser-based and propagation-based phase contrast imaging. For an experimental study, see Coan *et al.* (2005).

4.5 Phase retrieval

Consider, once again, the fine network of bright lines which dance about the bottom of an outdoor swimming pool's floor on a cloudless day. This intensity distribution is a phase-contrast image, in the sense described at the start of Section 4.4: phase variations in a given wave-field are rendered visible as intensity variations, in the image produced by a phase-contrast imaging system.

The notion of 'phase retrieval' goes a step beyond phase contrast, in the sense that one now wants to measure, or 'retrieve', a given phase distribution from one or more phase contrast images of that distribution, which are obtained as output from a given phase-contrast imaging system. Adopting this viewpoint, such phase-contrast images are viewed as encrypted or coded images of the desired phase distribution $\phi(x, y)$. The problem of phase retrieval can then be phrased as the problem of decoding such images to obtain the phase $\phi(x, y)$ of the input radiation field which led to the said images.

The problem of phase retrieval, often termed the 'phase problem' for brevity, arises on account of the extremely rapid nature of wave-field oscillations at X-ray frequencies. The existing state of the art, in X-ray detector technology, is still many orders of magnitudes too slow to directly detect the position-dependent phase of a coherent X-ray wave-field, or the position-and-time-dependent phase of a partially coherent X-ray wave-field. Rather, imaging experiments are only sensitive to the time-averaged intensity, with this time average being taken over very many cycles of wave-field oscillation. If one measures the time-averaged intensity image (or a set of such intensity images) produced by a phase-contrast imaging system, such as those which have previously been described, then the intensity of this image (or set of images) will in general be a function of both the intensity and phase of the radiation wave-field which was input into the system. How can one retrieve the phase of the input wave-field, given one or more phase-contrast images of the wave-field?

This phase problem is an example of an 'inverse problem' and may be profitably studied within the context of the theory of inverse problems, which has generated a huge and profound literature. Before turning to a detailed study of certain methods of phase retrieval in coherent X-ray optics, it is therefore fitting that we say a few words concerning the theory of inverse problems.

The laws of physics may be abstractly viewed as algorithms governing the flow of information (Landauer 1996), a notion central to the considerations of inverse-problem theory. Loosely, a large class of 'direct problems' consist of determining the effect which arises from a given cause, while the associated 'inverse problem' attempts to infer the cause from a given measured effect. In the context of imaging, the forward problem is associated with a model for a particular imaging scenario, with the goal being to determine the imaging data which results from the model of the object, sources, and detector, together with the interactions between them. The associated inverse imaging problem is then to determine something about the object and/or the radiation, given measured image data. Solution of the direct problem is always possible by using the constructive methods of mathematics (see, for example, Anger (1990)) to write down a map such as an integral transform between the known parameters of a given model and the desired data. This is not so for the inverse problem, because the inverse map, from data to the model, does not necessarily exist. Apart from having a detailed understanding of the related direct problem, solution of an inverse problem requires one to consider the existence and uniqueness of the solution, and, if these criteria are met, the stability of the retrieved information with respect to perturbations in the input data (see, for example, Baltes (1978)). Each of these questions must be considered before constructing an appropriate algorithm for the solution of a given inverse problem.

One may identify two broadly distinct schools of thought regarding the methodology for approaching inverse problems such as that of phase retrieval. These different approaches shall respectively be considered in the following two paragraphs.

An inverse problem is said to be 'well-posed in the sense of Hadamard' if there exists a unique solution depending continuously on the data (Hadamard 1923; see also Kress 1989). Otherwise, the inverse problem is 'ill-posed'. Let us deal with ill-posed problems first. If the data are inconsistent with the mapping to be inverted, then no solution exists. One may proceed by projecting the measured data onto a function space such that it is now consistent with the existence of an inverse mapping from the data back to the model of the object which led to such data. Alternatively one may have the opposite situation, where an inverse problem is ill-posed because there are a large set of model parameters consistent with the measured data. One means of coping with this is to choose the solution of minimum norm (the so-called generalized solution), thus suppressing 'invisible' or 'ghost' degrees of freedom which lie in the nullspace of the direct mapping (see, for example, Bertero *et al.* (1990)). The role of *a priori* knowledge is crucial in attempting such analyses of ill-posed problems, and one must be very careful that the results so obtained are meaningful. A much more satisfactory alternative, where attainable, is to increase the scope or nature of the measurements and use *a priori* knowledge to render the inverse problem well-posed.

An alternative viewpoint on the inverse problem, to that discussed above, is provided by the so-called 'Bayesian' methods which attack inverse problems from the standpoint of statistical inference. A non-Bayesian may state that a given inverse problem cannot be uniquely solved, as either (i) there are an infinite multiplicity of non-unique solutions which may not be distinguished from one another or (ii) no solution exists. The non-Bayesian may then conclude that extra data and/or *a priori* knowledge is required in order to proceed further. In contrast to this impasse reached by the non-Bayesian, the Bayesian might associate a statistical probability with each of the possible solutions, taking account of uncertainty and working within a given model, and then choose that which has the highest probability of being correct. This is an essential feature of the 'maximum entropy' principle of inference pioneered by Jaynes. This maximum-entropy principle, as a broad approach to the inverse problem, has generated a vast literature which, like its non-Bayesian counterpart, spans an impressive variety of fields. For further information on the maximum-entropy method, the reader is referred to a collection of Jaynes' papers (Jaynes 1983), together with his posthumously published book (Jaynes 2003) and the text by Wu (1997).

In the following sub-sections, three different approaches to the inverse problem of phase retrieval are outlined. The first of these (Section 4.5.1) is an iterative methodology of phase retrieval based on the Gerchberg–Saxton algorithm and its extensions. Modified forms of this algorithm have received significant attention, in recent years, as a means of attacking the phase problem of non-crystalline sample reconstruction given far-field X-ray diffraction patterns of the same. This goal goes under the twin names of the 'non-crystallographic phase problem' or 'coherent diffractive imaging'. The second phase-retrieval algorithm to be considered is based on the transport-of-intensity equation (Section 4.5.2), this being the continuity equation associated with the paraxial equation (eqn (2.34)). This may be used to non-iteratively retrieve the phase of a given coherent paraxial two-dimensional disturbance from a series of two defocused images of the same. Extensions to the case of polychromatic and partially coherent radiation are also possible. Lastly, we consider the problem of phase retrieval given the modulus of the Fourier transform of a one-dimensional coherent scalar disturbance, in Section 4.5.3.

4.5.1 *The Gerchberg–Saxton algorithm and its extensions*

Recall the discussion of Section 1.5 regarding the Fraunhofer (far-field) diffraction pattern produced when a given two-dimensional coherent wave-field $\psi(x, y)$ is propagated through a sufficiently large distance. We saw that, up to known multiplicative and transverse scale factors, the two-dimensional far-field disturbance is proportional to the Fourier transform, with respect to x and y, of the initial disturbance $\psi(x, y)$. Ignoring the just-mentioned scale factors, to avoid clutter in our formulae, we may therefore state that the resulting Fraunhofer diffraction pattern is given by $|\mathcal{F}\psi(x, y,)|^2$. This Fraunhofer diffraction pattern, which cannot be regarded as an 'image' as such, exhibits both phase and intensity contrast in the sense that it depends on both the phase and intensity of the unpropagated field $\psi(x, y)$.

One is naturally led to the phase-retrieval problem of reconstructing an image of both the intensity and phase of a coherent disturbance $\psi(x, y)$, given a Fraunhofer diffraction pattern of the same. This sub-section will discuss a number of attempts to address this question, together with closely related problems. Before proceeding we note that, given the vast literature on the phase problem for X-ray crystallography, we will henceforth focus our attention on the phase problem for aperiodic functions $\psi(x, y)$.

Suppose, for the present, that one considers a less ambitious phase-retrieval problem, which may be stated as follows: given both $|\psi(x, y)|^2$ and $|\mathcal{F}\psi(x, y)|^2$, determine $\psi(x, y)$. This question is termed the 'Pauli problem', as it was first considered by Pauli in the context of a mathematically equivalent problem in quantum mechanics (Pauli 1933).[127]

We now outline a famous technique due to Gerchberg and Saxton (1972), for addressing the optical-physics form of the Pauli problem. We consider its application to the phase retrieval of two-dimensional wave-fields, leaving the problem of one-dimensional phase retrieval to Section 4.5.3. For the case of two-dimensional wave-fields, the Gerchberg–Saxton algorithm is:

$$\psi(x,y) = \lim_{N\to\infty} \left(\mathcal{P}_1\mathcal{F}^{-1}\mathcal{P}_2\mathcal{F}\right)^N \{|\psi(x,y)| \exp[i\phi_{\text{initial}}(x,y)]\}. \quad (4.132)$$

Here, the projection operator $\mathcal{P}_1$ replaces the modulus of the function on which it acts with the known quantity $|\psi(x,y)|$, the projection operator $\mathcal{P}_2$ acts to replace the modulus of the function on which it acts with the known quantity $|\mathcal{F}\psi(x,y)|$, N is the number of times that the operator $\mathcal{P}_1\mathcal{F}^{-1}\mathcal{P}_2\mathcal{F}$ is iterated, and $\phi_{\text{initial}}(x,y)$ is the initial guess for the unknown phase.[128] As usual all cascaded operators act from right to left, so that:

$$\mathcal{P}_1\mathcal{F}^{-1}\mathcal{P}_2\mathcal{F}|\psi(x,y)| \equiv \mathcal{P}_1(\mathcal{F}^{-1}(\mathcal{P}_2(\mathcal{F}(|\psi(x,y)|)))). \quad (4.133)$$

In words, eqn (4.132) comprises the following iterative procedure for reconstructing $\psi(x,y)$: (i) start with an initial crude estimate $|\psi(x,y)| \exp[i\phi_{\text{initial}}(x,y)]$ for the desired wave-field $\psi(x,y)$, which has the known modulus and either a random or constant phase; (ii) apply the Fourier transform operator, $\mathcal{F}$, to the current estimate for the reconstructed wave-field; (iii) the resulting estimate, for the far-field disturbance $\mathcal{F}\psi(x,y)$, is improved by replacing its modulus with the known modulus $|\mathcal{F}\psi(x,y)|$, which amounts to acting on the estimate with the operator $\mathcal{P}_2$; (iv) apply the inverse Fourier transform operator $\mathcal{F}^{-1}$ to the current estimate for the reconstructed far-field disturbance; (v) the resulting estimate, for the unpropagated disturbance $\psi(x,y)$, is improved by replacing its modulus with the known modulus $|\psi(x,y)|$, which amounts to acting on the estimate with the operator $\mathcal{P}_1$; (vi) if a suitable convergence criterion has been met, or if the algorithm has stagnated by running for too long without converging, stop the procedure and output the current iterate as the algorithm's estimate for $\psi(x,y)$—otherwise, iterate by going to step (ii).

One can show that the error, in the wave-function reconstructed using the Gerchberg–Saxton algorithm, is a non-increasing function of the number of iterations (Gerchberg and Saxton 1972). However, the algorithm often stagnates,

[127] As has already been mentioned on several occasions, the Helmholtz equation, governing the free-space evolution of the spatial component $\psi(x,y,z)$ of a monochromatic scalar electromagnetic field, is mathematically identical in form to the time-independent free-space Schrödinger equation for the spatial part of a monoenergetic non-relativistic wave-function for a spinless particle. For a two-dimensional quantum-mechanical wave-function, the equivalent of the far-field disturbance is termed the 'momentum-space wave-function'. Pauli was the first to pose the phase problem of reconstructing a given wave-function ψ, given both its modulus and the modulus of the corresponding momentum-space wave-function.

[128] This may be chosen as a random distribution of phases, for each point in the initial image. It may also be chosen as a constant.

with the error often remaining approximately constant for a very large number of iterations. Accordingly, a number of generalizations of the algorithm have been proposed.

The most prevalent of these generalizations are due to Fienup (1982), known as the 'error reduction algorithm' and the 'hybrid input–output algorithm'. These consider a more difficult form of the Pauli problem than that presented earlier, namely that of reconstructing a given two-dimensional complex wave-field $\psi(x, y)$ given only the modulus $|\mathcal{F}\psi(x, y)|$ of its Fourier transform (i.e. the Fraunhofer diffraction pattern). This problem is particularly important in the context of the so-called 'non-crystallographic phase problem', where one measures only the Fraunhofer diffraction pattern of a non-crystalline object, and wishes to form an image of both the intensity and phase at the exit-surface of the sample—that is, one wishes to reconstruct $\psi(x, y)$.

In the absence of any additional *a priori* knowledge the non-crystallographic problem is insoluble, as one could choose any two-dimensional function for the phase of the Fraunhofer diffraction pattern. However, as emphasized by Fienup, the *a priori* knowledge that the sample has 'finite support', namely that the sample is confined to a finite region of the xy plane, is a powerful tool in solving the non-crystallographic phase problem that has just been posed. Such knowledge imposes constraints on the class of allowed wave-functions in the xy plane. Another powerful constraint is the knowledge that, at X-ray energies, the real part of the refractive index of most materials is less than unity (Miao *et al.* 1998; see also Section 2.9). Therefore under the projection approximation one often has the additional constraint that the imaginary part of the exit-surface wave-field $\psi(x, y)$ must be non-positive, assuming the sample to impart only weak phase shifts on normally-incident coherent scalar plane waves passing through it (see eqn (2.40)).

The first of Fienup's algorithms, to be discussed here, is the 'error-reduction algorithm' (Fienup 1982):

$$\psi(x, y) = \lim_{N \to \infty} \left(\mathcal{P}_1' \mathcal{F}^{-1} \mathcal{P}_2' \mathcal{F}\right)^N \psi_{N=0}(x, y). \tag{4.134}$$

All symbols are as previously defined, with the exception of: (i) $\psi_{N=0}(x, y) \equiv \mathcal{F}^{-1}\{|\mathcal{F}\psi(x, y)| \exp[i\varphi_{\text{initial}}(k_x, k_y)]\}$, where $\varphi_{\text{initial}}(k_x, k_y)$ is a random initial guess for the phase of the wave-field's complex amplitude in the Fraunhofer plane, which has known modulus $|\mathcal{F}\psi(x, y)|$, where k_x and k_y are the usual Fourier variables dual to x and y; (ii) $\mathcal{P}_2'$, which is a projection operator imposing the known constraints on $\mathcal{F}\psi(x, y)$; and (iii) $\mathcal{P}_1'$, which imposes the known constraints on $\psi(x, y)$. For example, if the support of $\psi(x, y)$ is known to be the region Ω of the xy plane, then:

$$\mathcal{P}_1' \psi(x, y) \equiv \begin{cases} \psi(x, y), & (x, y) \in \Omega, \\ 0, & \text{otherwise}. \end{cases} \tag{4.135}$$

One can estimate the support of $\psi(x,y)$, using the fact that the inverse Fourier transform of the Fraunhofer diffraction pattern, namely $\mathcal{F}^{-1}[|\mathcal{F}\psi(x,y)|^2]$, gives the autocorrelation of $\psi(x,y)$; one can then estimate the support of $\psi(x,y)$ to be half of the diameter of this autocorrelation. Note also that, in Fienup's error-reduction algorithm, one typically takes $\mathcal{P}_2' = \mathcal{P}_2$, so that this operator replaces the modulus of the function upon which it acts by the known modulus $|\mathcal{F}\psi(x,y)|$.

As was the case with the Gerchberg–Saxton algorithm, the error-reduction algorithm often stagnates. To combat this problem, Fienup developed a second generalization of the Gerchberg–Saxton algorithm, known as the 'hybrid input–output algorithm', which generalizes the error-reduction algorithm in eqn (4.134). Let us write this expression in the following suggestive form, valid for $N > 0$:

$$\psi_{N+1}(x,y) = \begin{cases} \left(\mathcal{F}^{-1}\mathcal{P}_2'\mathcal{F}\right)\psi_N(x,y), & (x,y) \in \Omega', \\ 0, & \text{otherwise.} \end{cases} \tag{4.136}$$

where Ω' is the set of points, in the xy plane, where $\psi_N(x,y)$ satisfies the given constraints, and:

$$\psi_N(x,y) \equiv \left(\mathcal{P}_1'\mathcal{F}^{-1}\mathcal{P}_2'\mathcal{F}\right)^N \psi_{N=0}(x,y) \tag{4.137}$$

is the current iterate of the reconstructed wave-function.

We are now ready to write down Fienup's hybrid input–output algorithm, valid for $N > 0$:

$$\psi_{N+1}(x,y) = \begin{cases} \left(\mathcal{F}^{-1}\mathcal{P}_2'\mathcal{F}\right)\psi_N(x,y), & (x,y) \in \Omega', \\ \psi_N(x,y) - \beta\left(\mathcal{F}^{-1}\mathcal{P}_2'\mathcal{F}\right)\psi_N(x,y), & \text{otherwise.} \end{cases} \tag{4.138}$$

Here, $\psi_{N=0}(x,y)$ is the previously defined first guess for the reconstructed complex amplitude at the exit-surface of the sample, and β is real number between 0.5 and 1. Importantly, $\psi_N(x,y)$ is no longer viewed as the current best estimate for the desired function $\psi(x,y)$, which is updated to give the better estimate $\psi_{N+1}(x,y)$. Rather, $\psi_N(x,y)$ it is viewed as an 'input' which is the driving function for a given mathematical transformation that yields $\psi_{N+1}(x,y)$ as output. As seen in eqn (4.138), this output is a hybrid of both the input function $\psi_N(x,y)$, and that which would have been produced if one had been using the error-reduction algorithm in eqn (4.136). The control parameter β governs the degree of 'hybridization' in the hybrid input–output algorithm.

This hybrid input–output algorithm has proven pivotal in contemporary studies on the non-crystallographic phase retrieval problem, of reconstructing $\psi(x,y)$ for a finite-support sample, given only its Fraunhofer diffraction pattern. Here, we describe a pioneering experimental demonstration by Miao *et al.* (1999), which draws from and extends earlier work by Boyes-Watson *et al.* (1947), Sayre (1952), Bates (1982), and Miao *et al.* (1998).

Consider the scanning electron micrograph in Fig. 4.18(a), which shows a sample consisting of gold dots (diameter approximately 100 nm, thickness approximately 80 nm) on a silicon nitride membrane. When illuminated with planar, monochromated X-ray radiation of wavelength 1.7 nm, Miao and colleagues obtained the far-field diffraction pattern shown in Fig. 4.18(b). The central portion of this pattern was inaccessible, due to the presence of a beam stop in the X-ray experiment. To supply this lost low-frequency information, use was made of the low-resolution optical micrograph of the sample shown in Fig. 4.18(c). In order to reconstruct an image of the sample from the diffraction pattern, use was made of the hybrid input–output algorithm given by eqn (4.138). In addition to the support constraint mentioned in our earlier discussions, use was also made of the object-plane constraint that, since the X-ray refractive index of the sample is always less than unity, this implies that the imaginary part of the exit-surface wave-function will always have the same sign (for full details, see the two papers by Miao *et al.*, which were cited in the previous paragraph). These ideas were used to obtain the sample reconstruction in Fig. 4.18(d), using only the diffraction pattern in Fig. 4.18(b), together with the low-resolution micrograph and the *a priori* knowledge that the refractive index of the weak sample is always less than unity.

4.5.2 *The transport-of-intensity equation*

Consider a z-directed coherent paraxial scalar wave-field $\psi(x,y,z)\exp(-i\omega t)$. As pointed out in Chapter 2, the spatial part of such a wave-field obeys the paraxial equation (see eqn (2.34)):

$$\left(2ik\frac{\partial}{\partial z}+\nabla_{\perp}^{2}\right)\psi(x,y,z)=0. \tag{4.139}$$

We remind the reader that $\nabla_{\perp}^{2}\equiv\partial^{2}/\partial x^{2}+\partial^{2}/\partial y^{2}$ is the Laplacian operator in the xy plane, and that the wave-number k is related to the wavelength λ of the coherent radiation via $k=2\pi/\lambda$. Explicit dependence on the angular frequency ω will not be indicated in this section.

Write $\psi(x,y,z)$ in terms of its intensity $I(x,y,z)\equiv|\psi(x,y,z)|^{2}$ and phase $\phi(x,y,z)\equiv\arg\psi(x,y,z)$, so that:

$$\psi(x,y,z)=\sqrt{I(x,y,z)}\exp[i\phi(x,y,z)]. \tag{4.140}$$

If we substitute this last expression into the paraxial equation, expand the result and then cancel a common factor, the imaginary part of the resulting equation yields a continuity equation which is now commonly termed the 'transport-of-intensity equation' in the context of phase retrieval (Teague 1983):

$$\nabla_{\perp}\cdot[I(x,y,z)\nabla_{\perp}\phi(x,y,z)]=-k\frac{\partial I(x,y,z)}{\partial z}. \tag{4.141}$$

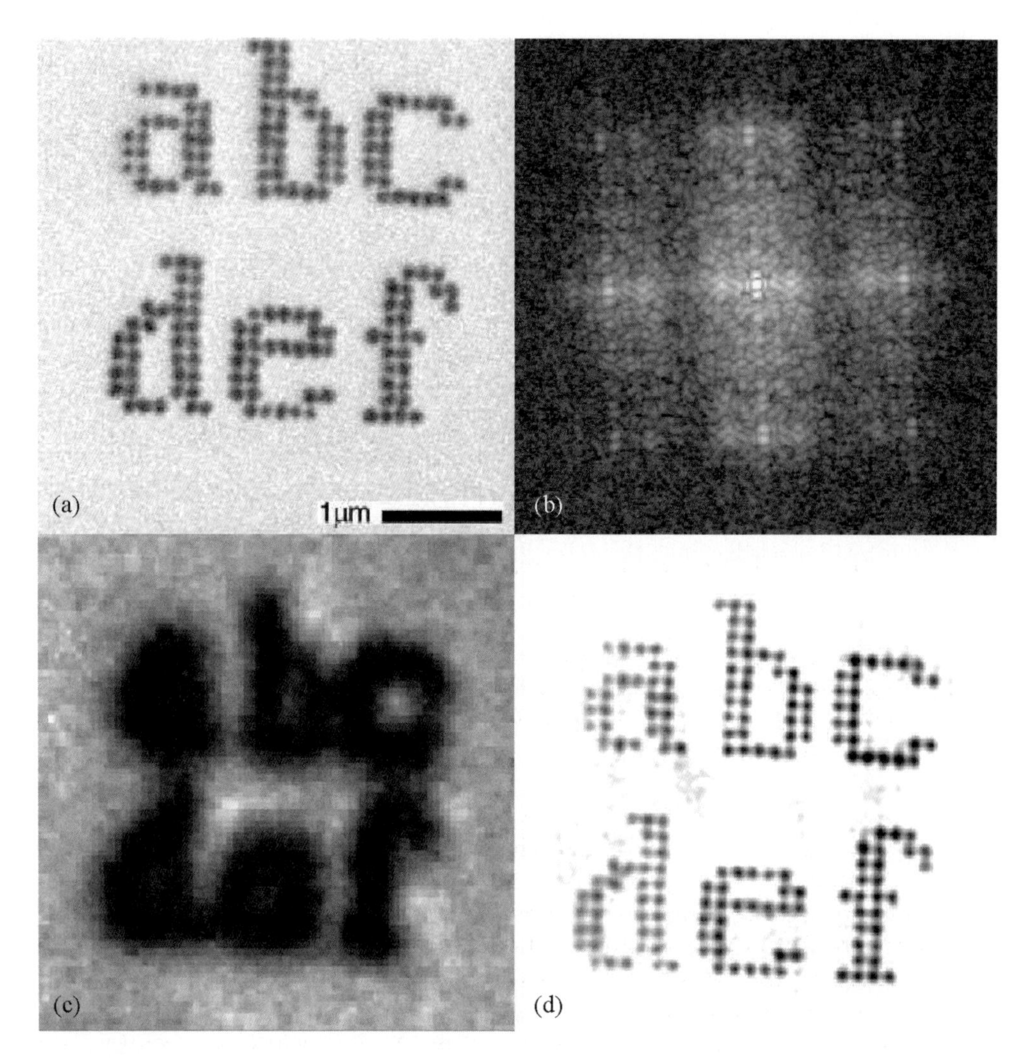

Fig. 4.18. (a) Image of test sample obtained using a scanning electron microscope. Sample consists of a series of gold dots, approximately 100 nm wide and 80 nm thick, deposited on a silicon nitride membrane. (b) X-ray far-field diffraction pattern of the sample, obtained by illuminating it with monochromated planar X-rays of wavelength 1.7 nm. The centre of this pattern was blocked by a beam stop, with the resulting missing information being supplied by the power spectrum of the low-resolution optical micrograph shown in (c). (d) Reconstructed sample, obtained from the far-field diffraction pattern. Image from Miao *et al.* (1999). Used with permission.

Being a continuity equation, this second-order elliptic partial differential equation expresses the conservation of optical energy as it evolves from the plane $z =$ constant to a parallel plane infinitesimally downstream.

To further explore this point, make the finite-difference approximation:

$$\frac{\partial I(x,y,z)}{\partial z} \approx \frac{I(x,y,z+\delta z) - I(x,y,z)}{\delta z}, \tag{4.142}$$

where the propagation distance δz is sufficiently small for the evolution of intensity to be linear in z. The above approximation can then be substituted into eqn (4.141). If one solves the resulting expression for the propagated intensity $I(x,y,z+\delta z)$, this gives:

$$I(x,y,z+\delta z) \approx I(x,y,z) - \frac{\delta z}{k} \nabla_\perp \cdot [I(x,y,z)\nabla_\perp \phi(x,y,z)]\,. \tag{4.143}$$

If we make the approximation:

$$\begin{aligned}\nabla_\perp \cdot [I(x,y,z)\nabla_\perp \phi(x,y,z)] &= I(x,y,z)\nabla_\perp^2 \phi(x,y,z) + \nabla_\perp I(x,y,z) \cdot \nabla_\perp \phi(x,y,z) \\ &\approx I(x,y,z)\nabla_\perp^2 \phi(x,y,z),\end{aligned} \tag{4.144}$$

which will be valid if the transverse intensity gradient and/or the transverse phase gradient is not too strong, then we see that eqn (4.143) is identical to eqn (4.118), which was derived using an argument based on the operator form of the Fresnel diffraction integral. The second term on the right side of this equation represents the local increase or decrease in intensity which is due to the local curvature of the phase in the plane $z =$ constant; a positive wave-front curvature implies a locally expanding wave which (on account of the negative sign which pre-multiplies the final term of eqn (4.143)) leads to a decrease in the propagated intensity, with negative wave-front curvature implying an increase in propagated intensity. The local redistribution of optical energy upon propagation is identical to the propagation-based phase contrast discussed in Section 4.4.4.

We see that the transport-of-intensity equation, being a mathematical embodiment of propagation-based phase contrast in the small-defocus regime, gives some insight into the formation of such images. However, it allows us to go further that this, to ask the question of how phase may be retrieved from one or more propagation-based phase contrast images.

As a starting point towards developing this point of view, we note that eqn (4.141) relates the intensity $I(x,y,z)$ of a coherent paraxial wave-field to both its longitudinal intensity derivative $\partial I(x,y,z)/\partial z$ and its phase $\phi(x,y,z)$. If one measures both the intensity over a given plane, and estimates the intensity derivative using such finite difference approximations as:

$$\frac{\partial I(x,y,z)}{\partial z} \approx \frac{I(x,y,z+\delta z) - I(x,y,z-\delta z)}{2\delta z}, \tag{4.145}$$

then the transport-of-intensity equation can be uniquely solved for the phase $\phi(x, y, z)$ in the plane over which the intensity $I(x, y, z)$ is measured (Gureyev *et al.* 1995). As a caveat on this claim, the intensity must be strictly positive over a simply connected region over which the phase is to be retrieved,[129] with appropriate boundary conditions being supplied for the phase. If Dirichlet boundary conditions on the phase are provided, which corresponds to specification of the phase over the one-dimensional boundary of the two-dimensional simply connected region over which the phase is to be recovered, then the retrieved phase is unique. If Neumann boundary conditions are supplied, which corresponds to specification of the normal derivative of the phase over the one-dimensional boundary of the two-dimensional simply connected region over which the phase is to be recovered, then the retrieved phase is unique up to an arbitrary and irrelevant additive constant. Interestingly, if the intensity of the field is known to vanish outside a certain simply-connected region in the plane of interest, and to be strictly positive inside, then the transport-of-intensity equation may be uniquely solved for the phase (up to an arbitrary and irrelevant additive constant), in the absence of any knowledge of the boundary conditions on the phase (Gureyev and Nugent 1996).

A number of algorithms exist for numerical solution of the transport-of-intensity equation. These include methods based on the full-multigrid algorithm (Gureyev *et al.* 1999; Allen and Oxley 2001), a Green-function method (Teague 1983), a fast-Fourier-transform-based method due to Gureyev and Nugent (1996, 1997), and another such method due to Paganin and Nugent (1998). We now give an outline of the last-mentioned algorithm.

Following a suggestion due to Teague (1983) we make the approximation:

$$I(x, y, z)\nabla_\perp \phi(x, y, z) \approx \nabla_\perp \xi(x, y, z), \tag{4.146}$$

which serves to define the scalar potential $\xi(x, y, z)$. Substitute this into eqn (4.141) to give the Poisson-type equation:

$$\nabla_\perp^2 \xi(x, y, z) \approx -k \frac{\partial I(x, y, z)}{\partial z}, \tag{4.147}$$

a symbolic solution for which can immediately be written as:

$$\xi(x, y, z) \approx -k \nabla_\perp^{-2} \frac{\partial I(x, y, z)}{\partial z}. \tag{4.148}$$

Apply $\nabla_\perp$ to both sides of this expression, then use eqn (4.146) to re-write the left side of the result, to give:

[129]Note that this excludes discontinuous phase maps, such as the phase screw dislocations and edge dislocations considered in the final chapter, from the class of admissible solutions for the phase which are obtained via the transport-of-intensity equation (cf. Gureyev *et al.* (1995)). As shall be seen in this final chapter such phase discontinuities are associated with zeros of intensity, which is why they are implicitly excluded by the demand that the intensity be strictly positive over a certain simply connected region in the plane of interest.

$$I(x,y,z)\nabla_{\perp}\phi(x,y,z) = -k\nabla_{\perp}\left[\nabla_{\perp}^{-2}\frac{\partial I(x,y,z)}{\partial z}\right]. \quad (4.149)$$

Divide both sides by $I(x,y,z)$, then apply the divergence operator $\nabla_{\perp}\cdot$ and make use of the identity $\nabla_{\perp}\cdot\nabla_{\perp} = \nabla_{\perp}^2$. This gives:

$$\nabla_{\perp}^2\phi(x,y,z) = -k\nabla_{\perp}\cdot\left\{\frac{1}{I(x,y,z)}\nabla_{\perp}\left[\nabla_{\perp}^{-2}\frac{\partial I(x,y,z)}{\partial z}\right]\right\}, \quad (4.150)$$

the inverse Laplacian of which yields the desired result (Paganin and Nugent 1998):

$$\phi(x,y,z) = -k\nabla_{\perp}^{-2}\left(\nabla_{\perp}\cdot\left\{\frac{1}{I(x,y,z)}\nabla_{\perp}\left[\nabla_{\perp}^{-2}\frac{\partial I(x,y,z)}{\partial z}\right]\right\}\right). \quad (4.151)$$

Regarding the numerical implementation of this symbolic solution to the transport-of-intensity equation, it is convenient to make use of the Fourier derivative theorem, as we now show. Consider a function $f(x,y)$ which is suitably well behaved to permit its representation as a Fourier integral, using the convention and notation outlined in Appendix A:

$$f(x,y) = \frac{1}{2\pi}\iint \breve{f}(k_x,k_y)\exp[i(k_x x + k_y y)]dk_x dk_y. \quad (4.152)$$

If we now apply the operator $(\partial^m/\partial x^m + \partial^n/\partial y^n)^p$ to both sides of this equation, where m, n, p are non-negative integers, then the operator can be brought inside the integral in the resulting expression. This leads to the following modified form of the Fourier derivative theorem:

$$\left(\frac{\partial^m}{\partial x^m} + \frac{\partial^n}{\partial y^n}\right)^p f(x,y)$$
$$= \frac{1}{2\pi}\iint [(ik_x)^m + (ik_y)^n]^p \breve{f}(k_x,k_y)\exp[i(k_x x + k_y y)]dk_x dk_y. \quad (4.153)$$

In words, this equation states that $(\partial^m/\partial x^m + \partial^n/\partial y^n)^p f(x,y)$ may be calculated by taking the Fourier transform $\breve{f}(k_x,k_y)$ of $f(x,y)$, multiplying by the function $[(ik_x)^m + (ik_y)^n]^p$ and then taking the inverse Fourier transform of the result. We immediately conclude that:

$$\left(\frac{\partial^m}{\partial x^m} + \frac{\partial^n}{\partial y^n}\right)^p = \mathcal{F}^{-1}\left[(ik_x)^m + (ik_y)^n\right]^p \mathcal{F}, \quad (4.154)$$

where we remind the reader that $\mathcal{F}$ denotes the two-dimensional Fourier transform operator with respect to x and y, with $\mathcal{F}^{-1}$ denoting the corresponding inverse transform. Importantly, we can now lift our restriction that m, n, p be non-negative integers, since the right-hand side of eqn (4.154) evidently remains well defined for certain non-positive-integer values of m, n, p. Such generalizations of the familiar integer-order differential operators are variously known as 'Fourier-integral operators' or 'pseudo-differential operators'. For our present purposes, we note the following special case of eqn (4.154):

$$\nabla_{\perp}^{-2} = -\mathcal{F}^{-1}\frac{1}{k_x^2 + k_y^2}\mathcal{F}, \quad m = n = 2, \ p = -1, \tag{4.155}$$

Using a simple extension of the ideas which have just been outlined, we can also give the following Fourier representation for the gradient operator:

$$\nabla_{\perp} = i\mathcal{F}^{-1}(k_x, k_y)\mathcal{F}. \tag{4.156}$$

By making use of eqns (4.155) and (4.156), with the fast Fourier transform being used to numerically implement $\mathcal{F}$ and $\mathcal{F}^{-1}$, eqn (4.151) may be numerically computed, provided that both $I(x, y, z)$ and $\partial I(x, y, z)/\partial z$ are sampled over Cartesian grids. The singularity at $k_x = k_y = 0$, in expression (4.155) for the inverse Laplacian, can be tamed by replacing $1/(k_x^2 + k_y^2)$ by zero at the origin $(k_x, k_y) = (0, 0)$ of Fourier space—this amounts to numerically evaluating the Cauchy principal value of the integral appearing in eqn (4.155).

This completes our derivation of the algorithm, due to Paganin and Nugent (1998), for phase retrieval using the transport-of-intensity equation. This procedure uniquely, rapidly and robustly yields the retrieved phase, with considerably reduced coherence requirements when compared to interferometry. Further, one retrieves the phase (up to an additive constant) without the modulo-2π problems associated with interferometry.[130,131] Finally, we note that the algorithm based on eqn (4.151) has been successfully applied to phase retrieval using electrons, neutrons, visible light, atoms and X-rays (see, for example, the references in the review by Paganin and Nugent (2001)).

As an example of the application of eqn (4.151) to phase retrieval using hard X-rays, we consider the experiment of McMahon *et al.* (2003) which is shown in Fig. 4.19. As indicated there, collimated 1.83 keV hard-X-ray waves were normally incident upon a 20 μm beam defining aperture. The radiation then passed through the sample, which was a tapered atomic-force microscope tip. The beam then struck a nickel zone plate, with the beam being sufficiently far from the optic axis for the zeroth and first diffracted orders to be spatially separated

[130]Note, however, that the method of phase-stepped interferometry is able to recover phase maps which are not wrapped modulo 2π (Section 4.6.1.2).

[131]The same comments apply to other means of numerically solving the transport-of-intensity equation, such as the previously mentioned methods based on the fast Fourier transform, the full multi-grid algorithm or Green-function methods.

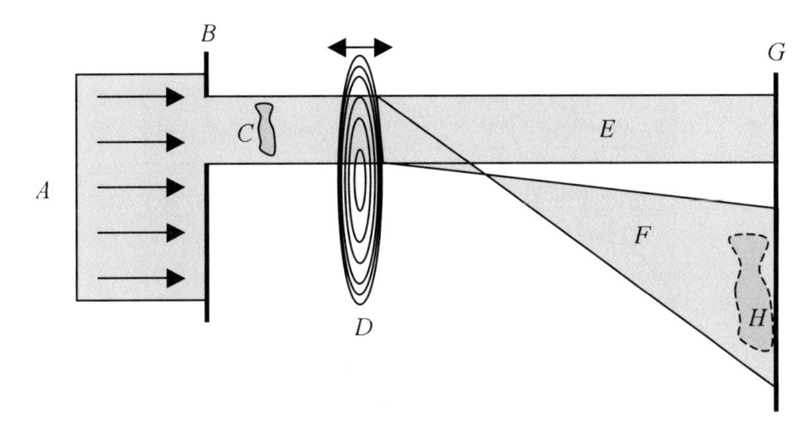

Fig. 4.19. Experiment to demonstrate X-ray phase retrieval using the transport-of-intensity equation (McMahon *et al.* 2003). Collimated 1.83 keV hard X-rays A are passed through a 20 μm diameter beam-defining aperture B, after which they traverse a sample C and then strike a zone plate D. The beam-defining aperture is sufficiently displaced, from the optic axis passing through the centre of the zone plate, for the zeroth order E and first order F diffracted beams to be spatially separated over the image plane G, which coincides with the surface of a CCD camera. This CCD records a propagation-based phase contrast image H. After McMahon *et al.* (2003).

from one another over the imaging plane of the CCD used to record the images. The geometry was such that a magnification of $M = 160$ was achieved. Examples of in-focus, over-focus and under-focus images are given in Figs. 4.20(a)–(c) respectively. Note that the variation in focus was obtained by translation of the zone plate along the optic axis. After computationally rescaling all images to unit magnification, and making use of the Fresnel scaling theorem (see Appendix B) to replace the defocus distance $z = 1$ mm with $z/M = z/160$, the symbolic solution in eqn (4.151) was evaluated using fast Fourier transforms, to yield the reconstructed phase map given in Fig. 4.20(d).

We close with two remarks: (i) Phase retrieval using the transport-of-intensity equation can also be considered for the case of polychromatic or partially coherent radiation. In this context see, for example, Paganin and Nugent (1998), Gureyev (1999), Nugent and Paganin (2000), and Gureyev *et al.* (2004). (ii) A phase reconstruction based on the transport-of-intensity equation may be used as a first guess for generalized Gerchberg–Saxton algorithms which take through-focal series as input data (Gureyev 2003).

4.5.3 *One-dimensional phase retrieval*

Consider the following problem of one-dimensional phase retrieval: Given the modulus $|\mathcal{F}[\psi(x)]|$ of the Fourier transform of a given complex function $\psi(x)$ of a real variable x, reconstruct $\psi(x)$. Here, $\mathcal{F}$ denotes the one-dimensional Fourier

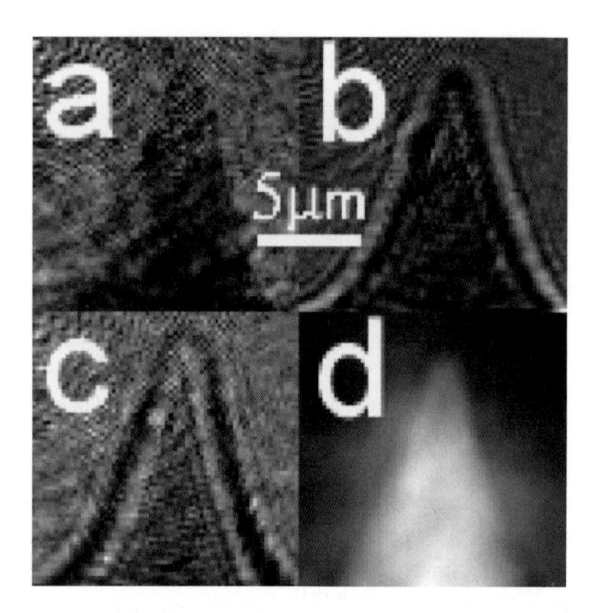

Fig. 4.20. Phase retrieval using hard-X-ray images of the tip used in an atomic force microscope. (a) In-focus image; (b) Over-focused image; (c) Under-focused image; (d) Reconstructed phase map. Further details in main text. Image taken from McMahon *et al.* (2003). Used with permission.

transform operator, which acts on a given function $\psi(x)$ to produce its Fourier transform with respect to x (see Appendix A):

$$\mathcal{F}[\psi(x)] \equiv \breve{\psi}(k_x) \equiv \frac{1}{\sqrt{2\pi}} \int_{-\infty}^{\infty} \psi(x) e^{-ik_x x} dx. \tag{4.157}$$

Further, k_x is the Fourier-space coordinate conjugate/dual/reciprocal to x.

In the rather general setting in which it has just been posed, the one-dimensional phase-retrieval problem is insoluble. To see this, choose any real function $\varphi(k_x)$, which allows one to form the wave-function $|\breve{\psi}(k_x)| \exp[i\varphi(k_x)]$. The inverse Fourier transform of this wave-function will evidently be consistent with the given data $|\breve{\psi}(k_x)|$, with the arbitrary nature of $\varphi(k_x)$ implying infinite ambiguity in the 'reconstructed' real-space wave-function $\mathcal{F}^{-1}|\breve{\psi}(k_x)| \exp[i\varphi(k_x)]$. To proceed beyond this impasse use must be made of suitable *a priori* knowledge, which must be such as to reduce the previously mentioned ambiguity to within acceptable levels. This *a priori* knowledge shall take the form of a suitable restriction of the class of allowable real-space wave-functions, as shall become clear later in this sub-section.

It is at this point that the theory of functions of a complex variable (i.e. complex analysis) becomes useful, a fact which has been pointed out by Wolf

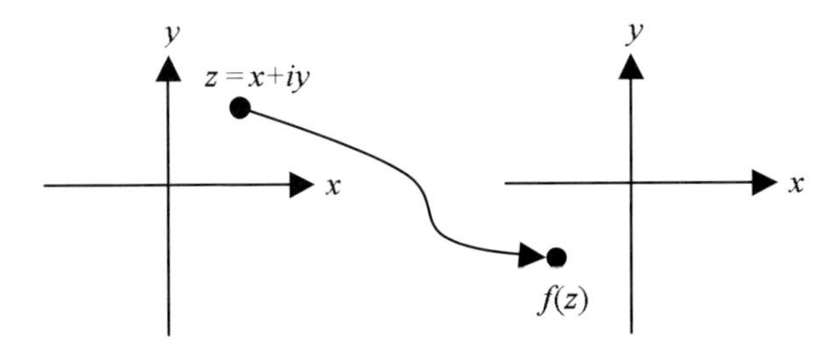

Fig. 4.21. A point z in the xy plane, with Cartesian coordinates (x, y), may be identified with the complex number $z = x+iy$. The resulting point z occupies the complex plane, as shown on the left side of the figure. A complex function of a complex variable, denoted $f(z)$, maps points z in the complex plane to points $f(z)$ in another complex plane, as indicated.

(1962) in the related context of reconstructing coherence functions. The use of complex analysis in the one-dimensional phase-retrieval problem, together with certain contemporary applications in coherent X-ray optics, will form the subject of this sub-section. Before proceeding, however, we give a brief review of the requisite formalism from the theory of functions of a single complex variable.

As shown in Fig. 4.21, set up a two-dimensional real Cartesian coordinate system, with axes labelled x and y. With each point in this plane we may associate the complex number $z = x + iy$. With this identification the xy plane is known as the 'complex plane'. A given complex function $f(z)$, of a single complex variable z, may be visualized as a mapping which takes points z on the complex plane as input, and returns points $f(z)$ on a second complex plane as output. Such a function is said to be 'analytic' at a given point z_0 in the complex plane, if its derivative with respect to z exists at that point, that is, if:

$$\lim_{\Delta z \to 0} \frac{f(z_0 + \Delta z) - f(z_0)}{\Delta z} \tag{4.158}$$

exists at the point z_0. Note that, in order for this limit to exist, it must be independent of the direction of the vector Δz, in the complex plane, which points from z_0 to $z_0 + \Delta z$. If the function is analytic for all points in a given region, then it is said to be analytic in that region. As an example, if $f(z)$ is analytic in the region $y \geq 0$ of the complex plane (the 'upper half plane'), then this function is 'analytic in the upper half plane'. Also, if a given function is analytic at every point z, of finite modulus, in the complex plane, then this function is said to be 'entire'.

We are now ready to introduce two remarkable theorems from complex analysis: the Cauchy–Goursat theorem and Cauchy's integral formula. Proofs for each of these theorems, which are not given here, can be found in most texts on complex analysis (e.g. Morse and Feshbach (1953) or Markusevich (1983)).

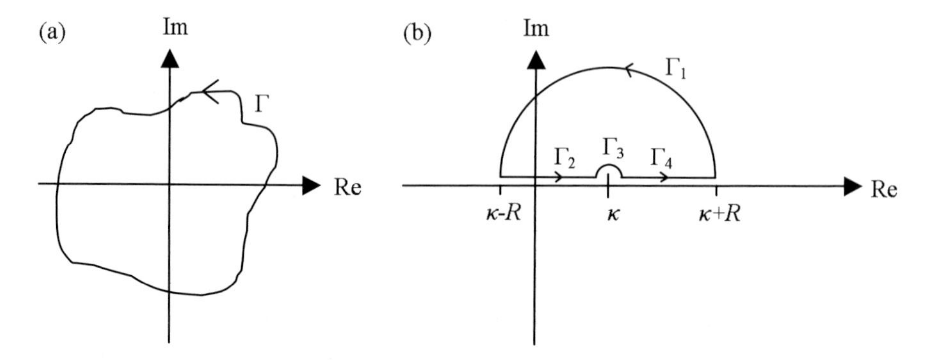

Fig. 4.22. (a) A simple closed contour Γ in the complex plane. (b) Contour for evaluation of an integral in the text.

(a) *Cauchy–Goursat theorem.* Suppose that $f(z)$ is a complex function of a complex variable, which is analytic at every point both on and inside the simple closed curve[132] Γ—see Fig. 4.22(a). Then the integral of $f(z)$ over Γ is zero, according to the Cauchy–Goursat theorem:

$$\oint_\Gamma f(z)dz = 0. \tag{4.159}$$

By convention, in all integrals over simple closed curves the curves will be understood to be traversed in the anti-clockwise sense, as indicated in Fig. 4.22(a).

(b) *Cauchy integral formula.* Suppose that $f(z)$ is a complex function of a complex variable, which is analytic at every point both on and inside the simple closed curve Γ. Suppose, further, that α is any point which lies inside Γ. Then the value of f at $z = \alpha$ can be determined from its value at each point on Γ, using the Cauchy integral formula:

$$\frac{1}{2\pi i}\oint_\Gamma \frac{f(z)}{z-\alpha}dz = \begin{cases} f(\alpha) & \text{if } \alpha \text{ is inside } \Gamma, \\ 0 & \text{if } \alpha \text{ is outside } \Gamma. \end{cases} \tag{4.160}$$

Consider the Fourier integral given in eqn (4.157). Let us replace the real variable k_x with the complex variable $\xi \equiv k_x + ik'_x$, where k'_x is real. The Fourier integral in eqn (4.157) then becomes:

$$\breve{\psi}(\xi) \equiv \breve{\psi}(k_x + ik'_x) \equiv \frac{1}{\sqrt{2\pi}}\int_{-\infty}^{\infty} \psi(x)e^{-i(k_x+ik'_x)x}dx. \tag{4.161}$$

[132]By ‘simple closed curve’, we mean a closed curve which does not intersect itself.

In this way, we can extend the complex function $\breve{\psi}(k_x)$ of a real variable k_x, to a complex function of a complex variable: $\breve{\psi}(\xi \equiv k_x + ik'_x)$. The original function $\breve{\psi}(k_x)$ then coincides with the value that $\breve{\psi}(\xi \equiv k_x + ik'_x)$ takes on the real axis. According to the Paley–Wiener theorem (see, for example, Boas (1954)) this extension of a function into the complex plane will result in an entire function whose modulus grows no faster than exponentially, if and only if $\psi(x)$ is of finite support (i.e. if there exists an interval $A < x < B$, with A and B both being finite real numbers, such that $\psi(x)$ is only non-zero within the interval).

We now have the necessary mathematical machinery to make use of complex analysis in attacking the one-dimensional phase problem posed at the beginning of this sub-section: given the modulus $|\breve{\psi}(k_x)|$ of the Fourier transform of $\psi(x)$ with respect to x, reconstruct $\psi(x)$. Note that this problem reduces to that of finding the phase of $\breve{\psi}(k_x)$, since $\psi(x)$ can then be synthesized by taking the inverse Fourier transform of $\breve{\psi}(k_x)$, which is itself constructed from its measured modulus and retrieved phase. As mentioned earlier, this one-dimensional phase problem is intractable if no restrictions are made to the class of admissible solutions for the reconstructed wave-function. In the absence of any additional data further progress can be made by suitably restricting the class of admissible solutions. Here, we restrict the class of admissible solutions for the real-space wave-functions, by demanding that admissible wave-functions be complex functions of finite support. Note for future reference that a common idiom refers to $\breve{\psi}(k_x)$ as belonging to the class of 'band-limited functions'. With this restriction of being band-limited, further progress can be made on the one-dimensional phase problem which forms the principal concern for the present section. The treatment given here follows that of Klibanov *et al.* (1995), whose review paper gives many references to the original literature.

Consider the problem of finding a relationship between the modulus and phase of a one-dimensional complex band-limited function $\breve{\psi}(k_x)$. By the Paley–Wiener theorem, we may extend this function into the space of complex arguments, giving an entire function $\breve{\psi}(\xi)$, where $\xi \in \mathbb{C} \equiv k_x + ik'_x$. Now, the natural logarithm of an analytic function is, at every point for which the function is non-zero, itself analytic. Therefore $\log_e[\breve{\psi}(\xi)]$ is analytic at all points, in the complex plane, for which $\breve{\psi}(\xi)$ is non-zero. Assume, for the moment, that no such zeros exist in the upper half plane $\mathrm{Re}(\xi) \geq 0$. With this assumption, which shall be lifted in due course, we conclude that $\log_e[\breve{\psi}(\xi)]$ is analytic in the upper-half plane $\mathrm{Re}(\xi) \geq 0$.

Since $\log_e[\breve{\psi}(\xi)]$ is analytic in the upper half plane, the Cauchy integral formula implies that:

$$\oint_\Gamma \frac{\log_e[\breve{\psi}(\xi)]}{\xi - \kappa} d\xi = 0, \tag{4.162}$$

where κ is a real number, and Γ is any closed contour in the upper-half plane which does not have $\xi = \kappa$ as an interior point. Let us choose, for Γ, the contour sketched in Fig. 4.22(b). The function $\log_e[\breve{\psi}(\xi)]/(\xi - \kappa)$ is then analytic at every point on and inside this contour. This contour consists of four parts: Γ_1, the upper half of an anticlockwise-traversed circle of radius R which is centred on κ; the directed line segment Γ_2 consisting of real numbers from $\kappa - R$ to $\kappa - \varepsilon$; the upper half Γ_3 of a clockwise-traversed circle of radius ε which is centred on κ; the directed line segment Γ_4 consisting of real numbers from $\kappa + \varepsilon$ to $\kappa + R$. With this decomposition of Γ, and taking R to be arbitrarily large while taking ε to be arbitrarily small but positive, we may write eqn (4.162) as:

$$\lim_{R\to\infty} \int_{\Gamma_1} \frac{\log_e[\breve{\psi}(\xi)]}{\xi - \kappa} d\xi + \lim_{R\to\infty,\varepsilon\to 0+} \int_{\Gamma_2} \frac{\log_e[\breve{\psi}(\xi)]}{\xi - \kappa} d\xi$$
$$+ \lim_{\varepsilon\to 0+} \int_{\Gamma_3} \frac{\log_e[\breve{\psi}(\xi)]}{\xi - \kappa} d\xi + \lim_{R\to\infty,\varepsilon\to 0+} \int_{\Gamma_4} \frac{\log_e[\breve{\psi}(\xi)]}{\xi - \kappa} d\xi = 0. \qquad (4.163)$$

We now consider, in turn, each of the four integrals appearing in this expression.

(a) *First integral in eqn (4.163).* Take this integral, which is over the semicircular region Γ_1 sketched in Fig. 4.22(b), and let ξ over this curve be parameterized by the angle θ, where:

$$\xi(\theta) = \kappa + Re^{i\theta}, \quad 0 \le \theta \le \pi. \qquad (4.164)$$

With this change of variables from ξ to θ, and using the fact that:

$$\log_e[\breve{\psi}(\xi)] = \log_e |\breve{\psi}(\xi)| + i\text{Arg}\breve{\psi}(\xi), \qquad (4.165)$$

our first integral becomes:

$$\lim_{R\to\infty} \int_{\Gamma_1} \frac{\log_e[\breve{\psi}(\xi)]}{\xi - \kappa} d\xi = i \lim_{R\to\infty} \int_0^\pi \log_e |\breve{\psi}(\kappa + Re^{i\theta})| d\theta$$
$$- \lim_{R\to\infty} \int_0^\pi \text{Arg}\breve{\psi}(\kappa + Re^{i\theta}) d\theta. \qquad (4.166)$$

Put this expression to one side, for the moment. In order to proceed further, we need to investigate the asymptotic behaviour of $\breve{\psi}(\xi)$ as $|\xi| \to \infty$ in the upper half plane. Since we have already assumed $\breve{\psi}(k_x) \equiv \breve{\psi}[\text{Re}(\xi)]$ to be band-limited, the function $\psi(x)$, which appears in eqn (4.161), is of finite support. Let this support be contained within the real interval $A < x < B$, which allows us to rewrite eqn (4.161) with the lower and upper limits of integration replaced by the real numbers A and B, respectively. Take the resulting formula, and integrate it by parts $m + 1$ times, where m is a non-negative integer. The resulting formula is:

$$\begin{aligned}\breve{\psi}(\xi) &= \frac{1}{\sqrt{2\pi}}\int_A^B \psi(x)e^{-i\xi x}dx \\ &= \frac{1}{\sqrt{2\pi}}\sum_{j=0}^{m}(-1)^j\left\{\left[\frac{d^j}{dx^j}\psi(x)\right]\left[\frac{e^{-i\xi x}}{(-i\xi)^{j+1}}\right]\right\}_A^B \\ &\quad + \frac{(-1)^{m+1}}{\sqrt{2\pi}}\int_A^B\left[\frac{d^{m+1}}{dx^{m+1}}\psi(x)\right]\frac{e^{-i\xi x}}{(-i\xi)^{m+1}}dx. \end{aligned} \tag{4.167}$$

To proceed further, assume that $\psi(x)$ can be represented by a finite-order polynomial[133] in x, for real x. Let m be the highest power of x which appears in this polynomial. This implies that the integral in the previous expression will vanish, leaving us with:

$$\breve{\psi}(\xi) = \frac{1}{\sqrt{2\pi}}\sum_{j=0}^{m}\frac{(-1)^j}{(-i\xi)^{j+1}}\left[\alpha_j^B e^{-i\xi B} - \alpha_j^A e^{-i\xi A}\right] \tag{4.168}$$

where:

$$\alpha_j^A \equiv \left[\frac{d^j\psi(x)}{dx^j}\right]_{x=A}, \qquad \alpha_j^B \equiv \left[\frac{d^j\psi(x)}{dx^j}\right]_{x=B}. \tag{4.169}$$

From eqn (4.168) we may deduce the asymptotic behaviour of $\breve{\psi}(\xi)$ as $|\xi| \to \infty$ in the upper half of the complex plane:

$$\breve{\psi}(\xi) \to A\xi^{-p}e^{-i\xi q}, \quad \mathrm{Re}(\xi) \geq 0, \ |\xi| \to \infty. \tag{4.170}$$

Here, A is a complex number, q is a real number, and p is a positive integer.

Equipped with this expression, eqn (4.166) becomes:

$$\begin{aligned}\lim_{R\to\infty}\int_{\Gamma_1}\frac{\log_e[\breve{\psi}(\xi)]}{\xi-\kappa}d\xi =& i\pi\log_e|A| - \lim_{R\to\infty} ip\pi\log_e R \\ &+ \lim_{R\to\infty} 2iRq - \pi\mathrm{Arg}A + \pi\kappa q + \frac{p\pi^2}{2}. \end{aligned} \tag{4.171}$$

(b) *Second and fourth integrals in eqn (4.163).* When summed together, these integrals are identically equal to the 'Cauchy principal value' of the integral of $\log_e[\breve{\psi}(k_x)]/(k_x - \kappa)$ with respect to k_x, with the integrand being a complex function of the real variable k_x:

[133] Those who do not wish to make this assumption may note that it may be avoided by invoking the Riemann–Lebesgue lemma, as discussed in the review by Klibanov *et al.* (1995).

$$\lim_{R\to\infty,\varepsilon\to 0^+}\int_{\Gamma_2}\frac{\log_e[\breve{\psi}(\xi)]}{\xi-\kappa}d\xi+\lim_{R\to\infty,\varepsilon\to 0^+}\int_{\Gamma_4}\frac{\log_e[\breve{\psi}(\xi)]}{\xi-\kappa}d\xi$$
$$=\lim_{R\to\infty,\varepsilon\to 0^+}\left(\int_{-R+\kappa}^{\kappa-\varepsilon}+\int_{\kappa+\varepsilon}^{R+\kappa}\right)\frac{\log_e[\breve{\psi}(k_x)]}{k_x-\kappa}dk_x$$
$$\equiv ⨍_{-\infty}^{\infty}\frac{\log_e[\breve{\psi}(k_x)]}{k_x-\kappa}dk_x. \tag{4.172}$$

The final line serves to define the Cauchy principal value, denoted by a dash through the integral sign. This expression has the form of a 'logarithmic Hilbert transform'.

(c) *Third integral in eqn (4.163).* Let Γ_5 denote an anti-clockwise traversed circular contour, of radius $\varepsilon\to 0^+$, which is centred at the point $\xi=\kappa$ in the complex plane. Then, from Cauchy's integral theorem, the third integral in eqn (4.163) is given by:

$$\lim_{\varepsilon\to 0^+}\int_{\Gamma_3}\frac{\log_e[\breve{\psi}(\xi)]}{\xi-\kappa}d\xi=-\frac{1}{2}\lim_{\varepsilon\to 0^+}\int_{\Gamma_5}\frac{\log_e[\breve{\psi}(\xi)]}{\xi-\kappa}d\xi=-\pi i\log_e[\breve{\psi}(\kappa)]. \tag{4.173}$$

Next, substitute eqns (4.171), (4.172), and (4.173) into (4.163). Take the real part of the resulting expression, to yield:

$$\mathrm{Arg}\breve{\psi}(\kappa)=-\frac{1}{\pi}⨍_{-\infty}^{\infty}\frac{\log_e|\breve{\psi}(k_x)|}{k_x-\kappa}dk_x-\kappa q+\beta, \tag{4.174}$$

where:

$$\beta\equiv\mathrm{Arg}(A)-\frac{p\pi}{2} \tag{4.175}$$

is a real constant.

Equation (4.174) shows how one may obtain the phase of $\breve{\psi}(\kappa)$ from its modulus $|\breve{\psi}(\kappa)|$. We recall that the derivation of this expression is subject to the following assumptions:

- Assumption 1: $\breve{\psi}(\kappa)$ is band-limited, that is, its inverse Fourier transform $\psi(x)$ is non-zero only over a certain finite interval $A<x<B$ of the real line;
- Assumption 2: Over the interval $A<x<B$, $\psi(x)$ may be exactly represented by a finite-order polynomial;
- Assumption 3: When the complex function $\breve{\psi}(\kappa)$ of a real variable is extended to a complex function of a complex variable using eqn (4.161), the resulting function has no zeros in the upper half plane.

We now remove the third of the assumptions listed above. Let $\{\xi_1, \xi_2, \cdots, \xi_n\}$ denote all of the zeros of $\breve{\psi}(\xi)$ which lie in the upper-half plane $\mathrm{Re}(\xi) \geq 0$. Construct the function $\breve{\Psi}(\xi)$, which is by definition given by:

$$\breve{\Psi}(\xi) \equiv \breve{\psi}(\xi) \prod_{j=1}^{n} \frac{\xi - \xi_j^*}{\xi - \xi_j}. \tag{4.176}$$

Note that, in the product above, zeros are understood to be repeated if they have a multiplicity in excess of unity. Both $\breve{\Psi}(\xi)$ and $\breve{\psi}(\xi)$ are analytic in the upper half plane. By construction, $\breve{\Psi}(\xi)$ has no zeros in the upper-half plane, as is easily verified; hence we may rewrite eqn (4.174) with $\breve{\psi}(\xi)$ replaced by $\breve{\Psi}(\xi)$. Take the resulting expression, and then make use of the fact that both $\breve{\Psi}(\xi)$ and $\breve{\psi}(\xi)$ have the same modulus along the real line (i.e. $|\breve{\Psi}(k_x)| = |\breve{\psi}(k_x)|$). Substitute (4.176) into the resulting expression, to give:

$$\mathrm{Arg}\breve{\psi}(\kappa) = -\frac{1}{\pi} \int_{-\infty}^{\infty} \frac{\log_e |\breve{\psi}(k_x)|}{k_x - \kappa} dk_x - \kappa q + \beta - \mathrm{Arg}\left(\prod_{j=1}^{n} \frac{\kappa - \xi_j^*}{\kappa - \xi_j} \right). \tag{4.177}$$

One can readily show that:

$$-\mathrm{Arg}\left(\prod_{j=1}^{n} \frac{\kappa - \xi_j^*}{\kappa - \xi_j} \right) = 2\pi s + 2 \sum_{j=1}^{n} \mathrm{Arg}(\kappa - \xi_j), \tag{4.178}$$

where s is an integer. This allows us to write our final result:

$$\mathrm{Arg}\breve{\psi}(\kappa) = -\frac{1}{\pi} \int_{-\infty}^{\infty} \frac{\log_e |\breve{\psi}(k_x)|}{k_x - \kappa} dk_x - \kappa q + \gamma + 2 \sum_{j=1}^{n} \mathrm{Arg}(\kappa - \xi_j), \tag{4.179}$$

where $\gamma \equiv \beta + 2\pi s$ is a real number. This 'logarithmic dispersion relation' is the central result of the present section, for it relates the modulus $|\breve{\psi}(k_x)|$ and phase $\mathrm{Arg}\breve{\psi}(k_x)$ of a function $\breve{\psi}(k_x)$ whose inverse Fourier transform $\psi(x)$ is a function of finite support that can be exactly represented by a finite-order polynomial over the support.

There are four ambiguities associated with this expression, which we shall now examine. Note that the first three of these ambiguities coincide with the 'trivial characteristics' identified by Bates in a seminal article (Bates 1982). *First ambiguity:* The real constant γ is, in general, both unknown and irrelevant, since it constitutes a constant additive phase factor on $\breve{\psi}(\kappa)$ which serves only to change the absolute phase of the reconstructed wave-function $\psi(x)$. We may therefore take $\gamma = 0$. *Second ambiguity:* The real constant q is also unknown. By the Fourier shift theorem, this unknown linear phase factor κq serves only to laterally displace the reconstructed real-space wave-function $\psi(x)$. The associated ambiguity is unsurprising, since one can easily show the modulus of the Fourier transform of $\psi(x)$ is the same as the modulus of the Fourier transform of $\psi(x - x_0)$, for real x_0. We cannot distinguish between $\psi(x)$ and its laterally shifted version $\psi(x - x_0)$. This is a fundamental ambiguity, yet a trivial one for most purposes. We may therefore take $q = 0$, with the understanding that this implies that $\psi(x)$ is reconstructed up to an unknown transverse shift. *Third ambiguity:* One can easily show that the Fourier transform of $\psi(x)$ has the same modulus as the Fourier transform of $\psi^*(-x)$. Therefore, one cannot distinguish between $\psi(x)$ and $\psi^*(-x)$. This third ambiguity is implicitly manifest as a symmetry of eqn (4.179), whose mathematical form is invariant under the transformation $\breve{\psi}(\kappa) \rightarrow \breve{\psi}^*(\kappa)$, with the exception of a negative sign introduced on the Hilbert-transform term. As with the second ambiguity, this third ambiguity (namely the inability to distinguish between $\psi(x)$ and $\psi^*(-x)$) is a real ambiguity. Unless one can resolve the ambiguity via invoking relevant *a priori* knowledge or by taking additional intensity measurements, it remains a fundamental ambiguity of the one-dimensional phase retrieval problem. *Fourth ambiguity:* The final ambiguity concerns the final term in eqn (4.179): this term is not known because the complex upper-half-plane zeros $\{\xi_1, \xi_2, \cdots, \xi_n\}$ are not known. Geometrically, each term in this summation corresponds to the angle, measured from the negative real axis and with positive angle corresponding to clockwise winding, made by the line extending from κ to ξ_j. Again, without any further *a priori* knowledge or additional experimental measurements, the final term of eqn (4.179) represents a fundamental ambiguity of the one-dimensional phase retrieval method outlined here. Note that this ambiguity may be resolved by taking additional data (see, for example, Nikulin (1998)), or by introducing additional *a priori* knowledge regarding the sample.

We close by noting that extensions of the methodology for one-dimensional phase retrieval, as described above, may be used for the analysis of kinematical X-ray diffraction patterns of layered structures that exhibit density variations in one dimension only. For entry points to the literature, see, for example, Petrashen' and Chukhovskii (1989), Reiss and Lipperheide (1996), Nikulin (1998), Nikulin and Zaumseil (1999), Siu *et al.* (2001), and Nikulin and Zaumseil (2004), together with references therein.

4.6 Interferometry

Interferometry, in which one superposes two or more coherent or partially coherent disturbances and then registers the time-averaged intensity of an associated signal, is the topic of this section. The subject is vast, and will we only scratch the surface in the four sub-sections which follow. Section 4.6.1 treats the phase imaging of two-dimensional coherent scalar fields using the Bonse–Hart interferometer. We consider the use of the Young interferometer to image the coherence properties of an X-ray source, in Section 4.6.2. Section 4.6.3 treats the X-ray incarnation of the intensity interferometer of Hanbury Brown and Twiss, in the context of coherence measurement. Lastly, Section 4.6.4 gives some entry points to the literature on other important means of interferometric coherence measurement, besides Young interferometry and intensity interferometry.

4.6.1 *Interferometric imaging using the Bonse–Hart interferometer*

Let us consider a forward-propagating two-dimensional coherent scalar wave-field $\exp[i\phi_{\rm IN}(x,y)]$, of unit modulus. Here, as usual, (x,y) denotes Cartesian coordinates in the plane perpendicular to a given optic axis. Our field serves at the input for a Bonse–Hart interferometer made from a single monolithic strain-free slab of perfect crystal (Bonse and Hart (1965); see also the review by Hart and Bonse (1970), and the reviews by Hart (1980) and Bonse (1988), together with references therein)—see Section 3.2.5. In forming the resulting interferogram, one may consider the input field to be overlaid with the boundary value $\exp[i(k_x^{(0)}x+k_y^{(0)}y)]$ of a coherent 'reference' plane wave, where $k_x^{(0)}$ and $k_x^{(0)}$ are related to the tilts of the reference wave in the x and y directions. Thus the action of the interferometer may be viewed as mapping input to output at follows:

$$\exp[i\phi_{\rm IN}(x,y)] \longrightarrow \exp[i\phi_{\rm IN}(x,y)] + \exp[i(k_x^{(0)}x+k_y^{(0)}y)]. \qquad (4.180)$$

The intensity $I_{\rm OUT}(x,y)$ of the resulting 'interferogram' may be obtained in the usual way, by taking the squared modulus of the above expression:

$$\begin{aligned} I_{\rm OUT}(x,y) &= \left|\exp[i\phi_{\rm IN}(x,y)] + \exp[i(k_x^{(0)}x+k_y^{(0)}y)]\right|^2 \\ &= 2 + \exp[i(k_x^{(0)}x+k_y^{(0)}y)]\exp[-i\phi_{\rm IN}(x,y)] \\ &\quad + \exp[-i(k_x^{(0)}x+k_y^{(0)}y)]\exp[i\phi_{\rm IN}(x,y)] \\ &= 2\left\{1+\cos\left[k_x^{(0)}x+k_y^{(0)}y-\phi_{\rm IN}(x,y)\right]\right\}. \end{aligned} \qquad (4.181)$$

As should be clear from the last line of the above equation, the phase $\phi_{\rm IN}(x,y)$ of the input field serves to distort the otherwise linear set of 'carrier' fringes $\cos(k_x^{(0)}x+k_y^{(0)}y)$. Thus the interferogram may be spoken of as a phase contrast image, insofar as its intensity distribution is sensitive to transverse phase variations in the input beam.

4.6.1.1 *Fourier-transform method of interferogram analysis*

The inverse problem of interferometry deals with the question of how $\phi_{\rm IN}(x,y)$—or, more generally, the input complex wave-field itself—may be determined given one or more interferograms of the same. One such method, commonly known as the 'Fourier transform method' of interferogram analysis, will now be described (Takeda *et al.* 1982). Consider the representation of the interferogram which is given in the second and third lines of eqn (4.181). Take the two-dimensional Fourier transform of this expression with respect to x and y, using the convention outlined in Appendix A. Making note of the Fourier integral representation of the Dirac delta, we arrive at:

$$\begin{aligned}\mathcal{F}\left[I_{\mathrm{OUT}}(x,y)\right] &= 4\pi\delta(k_x,k_y)\\ &+\frac{1}{2\pi}\iint \exp[-i\phi_{\mathrm{IN}}(x,y)]\exp\left\{-i\left[\left(k_x-k_x^{(0)}\right)x+\left(k_y-k_y^{(0)}\right)y\right]\right\}dxdy\\ &+\frac{1}{2\pi}\iint \exp[i\phi_{\mathrm{IN}}(x,y)]\exp\left\{-i\left[\left(k_x+k_x^{(0)}\right)x+\left(k_y+k_y^{(0)}\right)y\right]\right\}dxdy.\end{aligned} \tag{4.182}$$

The first term, on the right side of the above expression, is sharply localized to the origin of the two-dimensional Fourier space. The second term is equal to the two-dimensional Fourier transform of the complex conjugate of the input field, displaced by an amount $(k_x^{(0)}, k_y^{(0)})$. The final term is equal to the two-dimensional Fourier transform of the input field, displaced by an amount $(-k_x^{(0)}, -k_y^{(0)})$. Provided that the input wave-field is sufficiently slowly varying, the three terms mentioned above will be spatially separated in Fourier space. If this is the case, then one may place a 'window' about the region of Fourier space corresponding to the third term, setting to zero all other terms. The origin of Fourier space can then be re-located to the point $(-k_x^{(0)}, -k_y^{(0)})$, before taking the inverse Fourier transform of the resulting expression in order to obtain the input field $\exp[i\phi_{\mathrm{IN}}(x,y)]$. If one subsequently seeks to determine the phase of the recovered field, use can be made of:

$$\phi_{\mathrm{IN}}(x,y) = \tan^{-1}\left\{\frac{\mathrm{Im}\exp[i\phi_{\mathrm{IN}}(x,y)]}{\mathrm{Re}\exp[i\phi_{\mathrm{IN}}(x,y)]}\right\}. \tag{4.183}$$

The recovered phase is modulo 2π, since the phase of a complex number is only defined up to an integer multiple of 2π. The problem of 'phase unwrapping', which tackles the task of computing a two-dimensional phase map which is not modulo 2π given one or more phase maps which are, will not be considered here.

4.6.1.2 *Phase-shifting method of interferogram analysis* An alternative means of solving the inverse problem of interferometry is known as 'phase shifting' or 'phase stepping'. The essential idea here is to take several interferograms, adding a different constant phase bias to the reference beam before registering each of the corresponding interferograms. Given these images, one can then determine the phase of the input disturbance. An advantage of this method is that it can analyse more rapidly varying input phase maps than can be dealt with using the Fourier-transform method. This comes at the price of requiring more than one interferogram.

There are a great variety of methods for phase-shifting interferometry—see, for example, Hariharan (2003) and references therein. Here, we outline a variant which makes use of phase biases of 0, $\pi/2$, π, and $3\pi/2$ radians. We then show how this can be generalized to the case of M equally spaced phase biases ranging from 0 through to $2\pi(M-1)/M$ radians in steps of $2\pi/M$ radians.

Phase shifting using four interferograms. With a given phase bias B_j on the reference plane wave, eqn (4.180) becomes:

$$\exp[i\phi_{\mathrm{IN}}(x,y)] \longrightarrow \exp[i\phi_{\mathrm{IN}}(x,y)] + \exp[i(k_x^{(0)}x + k_y^{(0)}y + B_j)] \tag{4.184}$$

with eqn (4.181) being transformed to:

$$I_{\mathrm{OUT},j}(x,y) = 2\left\{1 + \cos\left[k_x^{(0)}x + k_y^{(0)}y + B_j - \phi_{\mathrm{IN}}(x,y)\right]\right\}. \tag{4.185}$$

Here, j is an integer. For the four-image variant of phase stepping being considered here, we choose:

$$B_1 = 0, \qquad B_2 = \frac{\pi}{2}, \qquad B_3 = \pi, \qquad B_4 = \frac{3\pi}{2}. \tag{4.186}$$

Letting:

$$\eta(x,y) \equiv k_x^{(0)}x + k_y^{(0)}y, \tag{4.187}$$

we have the series of four interferograms:

$$\begin{aligned}
I_{\mathrm{OUT},1}(x,y) &= 2\left\{1 + \cos\left[\eta(x,y) - \phi_{\mathrm{IN}}(x,y)\right]\right\}, \\
I_{\mathrm{OUT},2}(x,y) &= 2\left\{1 + \cos\left[\eta(x,y) + \frac{\pi}{2} - \phi_{\mathrm{IN}}(x,y)\right]\right\}, \\
I_{\mathrm{OUT},3}(x,y) &= 2\left\{1 + \cos\left[\eta(x,y) + \pi - \phi_{\mathrm{IN}}(x,y)\right]\right\}, \\
I_{\mathrm{OUT},4}(x,y) &= 2\left\{1 + \cos\left[\eta(x,y) + \frac{3\pi}{2} - \phi_{\mathrm{IN}}(x,y)\right]\right\}.
\end{aligned} \tag{4.188}$$

Using the relevant double-angle formulae, the above equations become:

$$\begin{aligned}
\frac{I_{\mathrm{OUT},1}(x,y)}{2} - 1 &= \cos\eta(x,y)\cos\phi_{\mathrm{IN}}(x,y) + \sin\eta(x,y)\sin\phi_{\mathrm{IN}}(x,y), \\
\frac{I_{\mathrm{OUT},2}(x,y)}{2} - 1 &= \cos\eta(x,y)\sin\phi_{\mathrm{IN}}(x,y) - \sin\eta(x,y)\cos\phi_{\mathrm{IN}}(x,y), \\
\frac{I_{\mathrm{OUT},3}(x,y)}{2} - 1 &= -\cos\eta(x,y)\cos\phi_{\mathrm{IN}}(x,y) - \sin\eta(x,y)\sin\phi_{\mathrm{IN}}(x,y), \\
\frac{I_{\mathrm{OUT},4}(x,y)}{2} - 1 &= -\cos\eta(x,y)\sin\phi_{\mathrm{IN}}(x,y) + \sin\eta(x,y)\cos\phi_{\mathrm{IN}}(x,y).
\end{aligned} \tag{4.189}$$

We thereby arrive at (see, for example, Hariharan (2003))[134]:

$$\phi_{\rm IN}(x,y) = k_x^{(0)}x + k_y^{(0)}y - \tan^{-1}\left[\frac{I_{{\rm OUT},4}(x,y) - I_{{\rm OUT},2}(x,y)}{I_{{\rm OUT},1}(x,y) - I_{{\rm OUT},3}(x,y)}\right]. \quad (4.190)$$

Phase shifting using M interferograms. Return to eqn (4.185), letting the phase bias B_j take on the following series of M equally spaced values:

$$B_j = \frac{2\pi j}{M}, \quad j = 0, 1, 2, \cdots, M-1. \quad (4.191)$$

Thus we measure M interferograms, with intensity maps $I_{{\rm OUT},j}(x,y)$ given by:

$$I_{{\rm OUT},j}(x,y) = 2\left\{1 + \cos\left[k_x^{(0)}x + k_y^{(0)}y + \frac{2\pi j}{M} - \phi_{\rm IN}(x,y)\right]\right\},$$
$$j = 0, 1, 2, \cdots, M-1. \quad (4.192)$$

Next, suppose that we take our series of M interferograms and then form the complex function $\aleph(x,y)$, which is defined by:

$$\aleph(x,y) \equiv \sum_{j=0}^{M-1} I_{{\rm OUT},j}(x,y)\exp\left(-\frac{2\pi i j}{M}\right). \quad (4.193)$$

Insert eqn (4.192) into (4.193), and then write the cosine function in terms of complex exponentials, to obtain:

$$\begin{aligned}\aleph(x,y) =& 2\sum_{j=0}^{M-1}\exp\left(-\frac{2\pi i j}{M}\right) \\ &+ \sum_{j=0}^{M-1}\exp\left(-\frac{2\pi i j}{M}\right)\exp\left\{i\left[k_x^{(0)}x + k_y^{(0)}y + \frac{2\pi j}{M} - \phi_{\rm IN}(x,y)\right]\right\} \\ &+ \sum_{j=0}^{M-1}\exp\left(-\frac{2\pi i j}{M}\right)\exp\left\{-i\left[k_x^{(0)}x + k_y^{(0)}y + \frac{2\pi j}{M} - \phi_{\rm IN}(x,y)\right]\right\}.\end{aligned} \quad (4.194)$$

The first term vanishes, on the right side of the above equation. This can be seen by invoking the formula for summing a geometric series, but the point can

[134]Note that there is no 'quadrant ambiguity' with the expression below, which gives the phase modulo 2π. This quadrant ambiguity refers to the fact that, for any angle θ, $\tan\theta = \tan(\theta+\pi)$, so that the inverse tangent is only defined modulo π. However, the numerator and denominator, of the fraction in square brackets, are respectively equal to the sine and cosine of $k_x^{(0)}x + k_y^{(0)}y - \phi_{\rm IN}(x,y)$. The sign of the numerator and denominator therefore allows one to resolve the quadrant ambiguity, thereby recovering the phase modulo 2π rather than merely modulo π. Note, further, that a similar argument can be applied to eqn (4.183).

be arrived at more directly using a symmetry argument in which each term in the sum is drawn as a point in the complex plane. In the second term on the right side of eqn (4.194), the two terms involving j both cancel, rendering the summation trivial. In the third term, factor all quantities out the front of the summation sign, which do not depend on j. The summation in the resulting expression vanishes, for the same reason as was given for the vanishing of the first term on the right side of eqn (4.194). Thus:

$$\aleph(x,y) = M \exp\left\{i\left[k_x^{(0)}x + k_y^{(0)}y - \phi_{\mathrm{IN}}(x,y)\right]\right\}, \tag{4.195}$$

so that (see, for example, Momose (2002)):

$$\phi_{\mathrm{IN}}(x,y) = k_x^{(0)}x + k_y^{(0)}y - \tan^{-1}\left\{\frac{\mathrm{Im}[\aleph(x,y)]}{\mathrm{Re}[\aleph(x,y)]}\right\}. \tag{4.196}$$

4.6.1.3 *Example of X-ray interferometry* X-ray interferometry, employing the Bonse–Hart interferometer, was first demonstrated in 1965 (Bonse and Hart 1965). For a more contemporary experimental realization, see the schematic in Fig. 4.23 (Momose 2002). Here we see white wiggler radiation from the Photon Factory in Tsukuba, Japan, which is incident upon a double-bounce monochromator (cf. Fig. 3.16) in the upper left of the diagram. An energy of 17.7 keV was selected. Using an asymmetric reflection from a 'collimating' crystal (see Fig. 3.15(b)), the width of the monochromated beam was expanded prior to entering the three-blade Bonse–Hart interferometer (see Figs. 3.18 and 3.19). As an ingeniously simple method to introduce the various phase biases required for the phase-shifting method of interferogram analysis, a 0.5-mm swivelling plastic plate ('phase shifter') was introduced into the reference arm of the interferometer, between the first and second blades. As the plate was swivelled, the reference beam travelled through a different constant thickness of the plate, thereby introducing a tunable phase bias B_j. A fresh mouse liver was placed into the sample cell, with the truncated veins and arteries being tied shut so as to prevent leakage of blood. The cell itself was filled with a salt solution, to reduce the large phase shifts which would otherwise be incurred by thickness variations in the liver itself. Five interferograms were obtained with the X-ray image sensor, these subsequently being processed using the phase-shifting method described earlier. A correction was then made to the resulting unwrapped phase map, to remove the background phase variations due to thickness variations in the liver, thereby highlighting the phase-shifting effects of the blood vessels within the liver. The resulting phase map is shown in Fig. 4.24 (Momose 2002). This clearly shows the blood-filled vessels in the mouse liver, with diameters as small as 50 μm.

4.6.2 *Coherence measurement using the Young interferometer*

In the latter parts of the first chapter, we studied the mutual coherence function (Section 1.8) associated with a statistically stationary partially coherent scalar electromagnetic field. We saw that the modulus of the associated complex degree

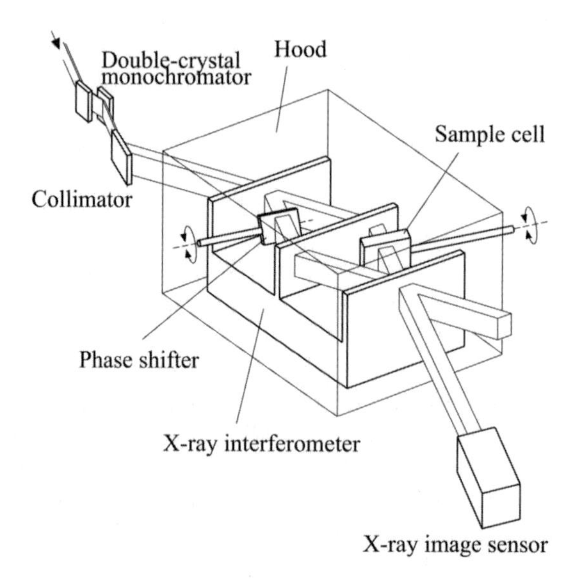

Fig. 4.23. Bonse–Hart interferometer for phase-stepped interferometric imaging of mouse liver. Image taken from Momose (2002). Used with permission.

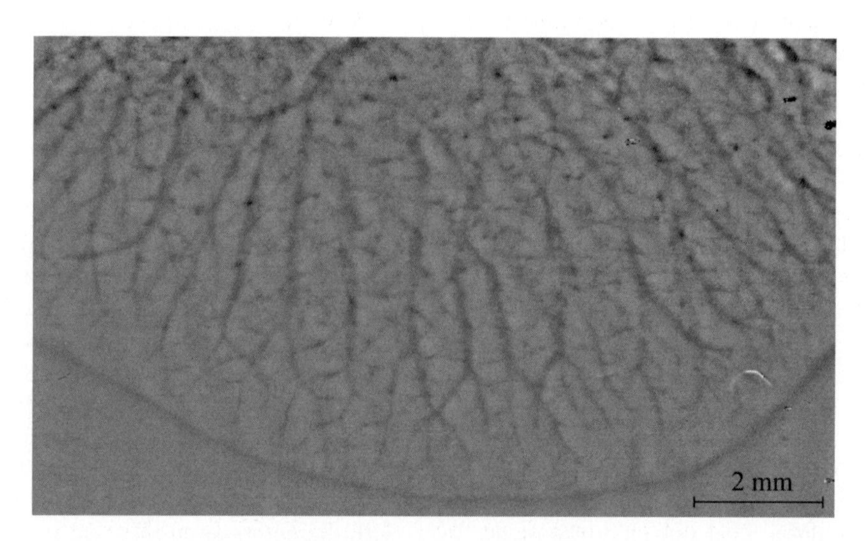

Fig. 4.24. Phase map of exit-surface wave-field of fresh blood-filled mouse liver, obtained using 17.7 keV X-ray radiation. Map obtained using phase-shifting X-ray interferometry, as described in the text. Image taken from Momose (2002). Used with permission.

of coherence, which is a normalized form of the mutual coherence function, is directly related to the visibility of the interference fringes which may be obtained using a Young two-pinhole interferometer (see eqns (1.99) and (1.100)). This observation is related to the possibility of measuring the modulus of the complex degree of coherence, using an X-ray version of Young's experiment. It is to this subject that we now turn.

In the realm of visible-light optics, the use of a Young interferometer to determine the modulus of the complex degree of coherence is well known (Thompson and Wolf 1957; see also Born and Wolf 1999). In the soft X-ray regime key papers on Young interferometry include Ditmire *et al.* (1996), Takayama *et al.* (1998, 2000), Burge *et al.* (1999), Liu *et al.* (2001), and Paterson *et al.* (2001). In the hard X-ray regime, see Leitenberger *et al.* (2001, 2003, 2004).

Here we describe the experiment of Leitenberger *et al.* (2004) which was performed using white bending-magnet radiation from the synchrotron source BESSY-II in Berlin, Germany. Their Young pinholes were drilled into a 30 μm thick tantalum foil, using a focused ion beam. This yielded a pair of near-circular holes in the foil, with approximately equal diameters of 2.6 μm, and a separation of approximately 12.7 μm. This was then normally illuminated with the white synchrotron beam, in a square patch of size 500 μm which was defined using slits upstream of the interferometer. The path between the interferometer and the detector was filled with helium, so as to reduce the absorptive and scattering effects of air. The radiation detector was a silicon drift chamber with energy resolution of 200 eV in the range from 2 to 40 keV. A 5 μm pinhole was placed in front of this detector, this being scanned to produce the images obtained in the experiment. The effective height of this scanning pinhole was approximately halved using a horizontal straight edge in front of the pinhole, implying the horizontal scans to have a resolution of 5 μm and the vertical scans to have a resolution of 2.5 μm. Since the detector was energy sensitive, Leitenberger and colleagues were able to measure Young interferograms at a variety of different X-ray energies. Such images were registered with the source-to-slit distance at 30.6 m, and with the slit-to-pinholes and pinholes-to-detector distances being 30 cm and 1.4 m, respectively. This experiment was performed with the pinholes above one another, and once again for the pinholes horizontally aligned. The visibility of the resulting Young fringes was thereby determined as a function of energy, for both vertical and horizontal orientations, in the range from 5 to 15 keV. The resulting data, of visibility versus X-ray energy for vertical and horizontal orientations of the pinholes, are respectively shown as the square points in Figs. 4.25(a) and (b) (Leitenberger *et al.* 2004). After a correction was imposed for the finite size of the detector pinhole in the vertical and horizontal directions, the circular points in the figure were obtained. The visibility in either direction was seen to vary with increasing energy, according to eqn (1.76) in Chapter 1. This formula was then used to determine the effective size of the X-ray source ('s' in the figure) in the vertical and horizontal directions respectively. This calculation yielded the respective values of 116 and 145 μm.

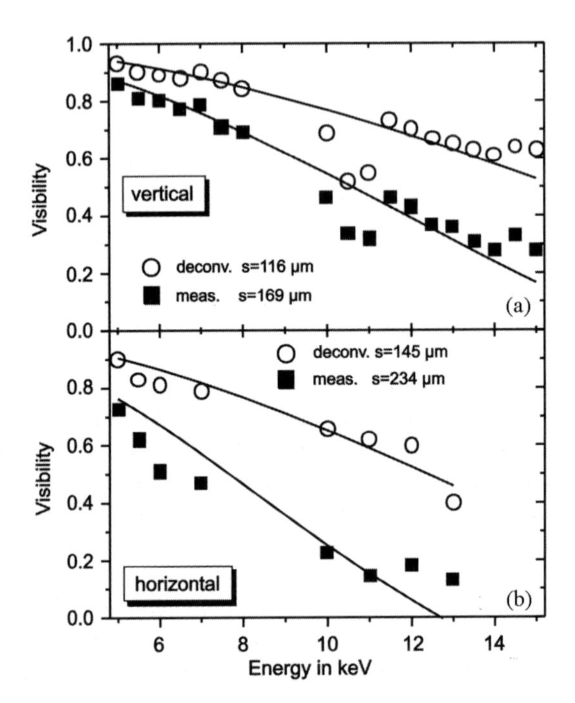

Fig. 4.25. Fringe visibility versus energy for the hard-X-ray Young interference experiment described in the main text. (a) Young pinholes in vertical orientation. (b) Young pinholes in horizontal orientation. Square dots represent raw data, with corresponding open circles obtained by correcting for the finite resolution of the detector pinhole. Solid curves fitted using eqn (1.76) to yield indicated measurements of the effective source size s, in both the vertical and horizontal directions. Image taken from Leitenberger *et al.* (2004). Used with permission.

4.6.3 *Coherence measurement using the intensity interferometer*

Thus far all our discussions on interferometry have been focused on interference which occurs at the level of fields, namely interference patterns which are formed by measuring the time-averaged intensity which results when two scalar fields are superposed. As seen in the latter parts of Chapter 1, the resulting interference phenomena are closely related to the second-order correlation function known as the mutual coherence function. Higher-order correlation functions may also be formed, as was briefly discussed in Section 1.10. A certain fourth-order correlation function is closely related to the notion of intensity interferometry, the subject of the present sub-section.

Consider two points $\mathbf{x}_1$ and $\mathbf{x}_2$ in a statistically stationary partially coherent scalar X-ray field. Suppose further that a non-zero mutual coherence function $\Gamma(\mathbf{x}_1, \mathbf{x}_2, \tau)$ exists between the two points, where τ is the usual time lag (see Section 1.8). Such a non-zero mutual coherence function is indicative of a non-zero degree of correlation between the time-dependent disturbances at each spatial point. On account of this correlation between the field quantities at the two points, one would also in general expect a non-zero degree of correlation between the instantaneous intensities at each of the two points. Such an intensity correlation will be of fourth order in the field quantities.

A twin significance is enjoyed by correlation functions of order higher than second, in the context of the present discussion. (i) As mentioned in Section 1.10, an arbitrary statistically stationary partially coherent field is not completely characterized by its mutual coherence function, requiring instead an infinite hierarchy of correlation functions of ever-increasing order for a full statistical description of the field. If the field is not statistically stationary, then the same statement still holds, with the mutual coherence function being evaluated through an ensemble average rather than through a time average, as a function of two time arguments. Thus, in general, higher-order correlation functions such as those of fourth order furnish additional information regarding a partially coherent field. Such higher-order correlation functions are not measured in the Young interferometer, and must therefore be measured using a different form of interferometer. The intensity interferometer, which shall be described in due course, is one such device. (ii) If one can assume the statistics of a given partially-coherent field to be Gaussian, then the mutual coherence function is sufficient to completely characterize the statistical aspects of the field (see Section 1.10). In this case, higher-order correlations functions either vanish, or may be expressed in terms of the mutual coherence function.[135] This implies a reduced significance for the higher-order correlation functions, since they carry no further information regarding the statistical aspects of the field, beyond that already provided by the mutual coherence function. Having said this, we shall see that under certain circumstances it may be more convenient to measure the mutual coherence function via measurement of a higher-order correlation function such as that recorded in intensity interferometry.

Of central importance, in intensity interferometry, is the measurement of the instantaneous intensity of a partially coherent field. If the field has a very small bandwidth, as will be the case for quasi-monochromatic X-rays, then this instantaneous intensity will appreciably fluctuate over timescales on the order of the coherence time of the field. This timescale being proportional to the reciprocal of the frequency bandwidth of the beam, we conclude that the instantaneous intensity fluctuates appreciably over timescales much longer than the field quantities themselves.[136] Thus, while existing detectors are too slow to directly register temporal variation of field quantities at optical and higher frequencies, this is not the case for temporal variations in the intensity of quasi-monochromatic

[135]This relation is formalized through the 'Gaussian moment theorem'—see, for example, Goodman (1985) or Mandel and Wolf (1995).

[136]This can be seen for example, by comparing eqn (1.78) to its squared modulus.

fields.

Such temporal intensity fluctuations may be registered using a detector with sufficiently high temporal resolution. To study the associated signal, one can employ a semi-classical treatment based on the first-order time-dependent perturbation theory of the photon-detector interaction. This analysis shows that the signal, registered by an otherwise ideal detector with finite time resolution, is proportional to the instantaneous intensity of the incident optical field, temporally smeared out in accord with the finite response time of the device (see, for example, Mandel and Wolf (1995) and references therein).

In the following discussion we shall ignore, for simplicity, the effects of the finite response time of the detector. Suppose that one has two time-sensitive detectors, whose small entrance apertures are respectively located at the points $\mathbf{x}_1$ and $\mathbf{x}_2$ in a quasi-monochromatic statistically stationary partially coherent scalar X-ray wave-field. Suppose that the time-dependent signal, recorded by each of these detectors, is proportional to the instantaneous intensity $I(\mathbf{x}_1, t)$ and $I(\mathbf{x}_2, t)$ of the optical field over the respective apertures. The recorded intensity signals may then be processed through a 'correlator', either online or offline, so as to produce the time average of the product of the difference of each signal from its mean value. The measured signal $S(\mathbf{x}_1, \mathbf{x}_2, \tau)$ of the resulting 'intensity interferometer' is therefore:

$$S(\mathbf{x}_1, \mathbf{x}_2, \tau) = \langle [I(\mathbf{x}_1, t+\tau) - \langle I(\mathbf{x}_1, t+\tau) \rangle] \, [I(\mathbf{x}_2, t) - \langle I(\mathbf{x}_2, t) \rangle] \rangle, \qquad (4.197)$$

where angular brackets denote time averaging. The time lag τ indicates the relative time delay of the two signals prior to being fed into the correlator. Expanding out the right side of this expression, and making use of the fact that the field is statistically stationary, we see that:

$$S(\mathbf{x}_1, \mathbf{x}_2, \tau) \equiv \langle I(\mathbf{x}_1, t+\tau) I(\mathbf{x}_2, t) \rangle - \overline{I}(\mathbf{x}_1) \overline{I}(\mathbf{x}_2), \qquad (4.198)$$

where:

$$\overline{I}(\mathbf{x}_1) \equiv \langle I(\mathbf{x}_1, t) \rangle, \quad \overline{I}(\mathbf{x}_2) \equiv \langle I(\mathbf{x}_2, t) \rangle. \qquad (4.199)$$

The first quantity, appearing on the right side of eqn (4.198), is the promised correlation function which is of fourth order in the fields. Explicitly, we may write this quantity in terms of the field $\Psi(\mathbf{x}, t)$ as:

$$\langle I(\mathbf{x}_1, t+\tau) I(\mathbf{x}_2, t) \rangle = \langle \Psi(\mathbf{x}_1, t+\tau) \Psi^*(\mathbf{x}_1, t+\tau) \Psi(\mathbf{x}_2, t) \Psi^*(\mathbf{x}_2, t) \rangle. \qquad (4.200)$$

This may be compared to eqn (1.88) for the mutual coherence function, which is second order at the level of fields.

As mentioned earlier, for Gaussian light all non-vanishing higher-order correlation functions may be determined from the mutual coherence function. This assumption of Gaussian statistics is typically an excellent approximation for laboratory X-ray sources, together with the three generations of synchrotron light source (see, for example, Yabashi *et al.* (2004)). For chaotic polarized radiation, for which the statistics are Gaussian, it can be shown that our fourth-order correlation function is related to the mutual coherence function as follows:

$$\langle I(\mathbf{x}_1, t+\tau) I(\mathbf{x}_2, t)\rangle = \overline{I}(\mathbf{x}_1)\overline{I}(\mathbf{x}_2) + |\Gamma(\mathbf{x}_1, \mathbf{x}_2, \tau)|^2 . \tag{4.201}$$

If we insert the above expression into eqn (4.198), we see that:

$$S(\mathbf{x}_1, \mathbf{x}_2, \tau) = |\Gamma(\mathbf{x}_1, \mathbf{x}_2, \tau)|^2 . \tag{4.202}$$

Thus, under the assumption of Gaussian statistics, the signal $S(\mathbf{x}_1, \mathbf{x}_2, \tau)$ from the intensity correlator gives a measure of the modulus of the mutual coherence function.

Rather than using the Young interferometer to measure the modulus of the mutual coherence function associated with a Gaussian field, the intensity interferometer may be employed (Hanbury Brown and Twiss 1954, 1956a,b, 1957a,b). This is known as the 'Hanbury Brown–Twiss effect', after its inventors. As they pointed out in the context of both radio-wave and visible-light optics, the intensity interferometer has some advantages over the Young interferometer. These advantages include a relative insensitivity of the device to fluctuations in the phase of each of the beams (indeed, such phase information is discarded), together with the fact that the two beams need not be physically superposed in order to register an intensity interferogram. These two advantages were seen to be of particular utility in the context of determining the angular diameter of radio stars in both the visible and radio regions of the electromagnetic spectrum (Hanbury Brown and Twiss 1956b).

These advantages are also of utility in the context of the X-ray incarnation of the Hanbury Brown–Twiss effect, which has had to wait for the advent of third-generation synchrotron sources in order to become feasible. Proposals for X-ray intensity interferometry include Shuryak (1975), Howells (1989), Ikonen (1992), and Gluskin *et al.* (1992, 1994). For experimental realizations see Kunimune *et al.* (1997), Gluskin *et al.* (1999), Tai *et al.* (1999), and Yabashi *et al.* (2001, 2004).

4.6.4 *Other means for interferometric coherence measurement*

In the previous pair of sub-sections we have considered two different means for measuring the coherence properties of partially coherent X-ray radiation, namely Young interferometry and intensity interferometry. Other means for coherence measurement exist. All are necessarily interferometric, or may be construed as such. Here, we give a listing of four such methods, together with recent papers to provide an entry point into the literature. (i) X-ray prisms (Section 3.4.1) may be used as interferometers for measuring X-ray coherence. Here, an interference pattern is obtained by overlapping an undiffracted plane beam with one which has been refracted using an X-ray prism. Mirror optics, or a pair of prisms, may be used to the same effect. Analysis of the resulting interferogram yields information regarding the coherence properties of the X-ray beam. See, for example, Suzuki (2004) and references therein. (ii) As an extension of the Young double-slit, the so-called 'uniformly redundant array' (Fenimore and Cannon 1978) may be used for X-ray coherence measurement (Nugent and Trebes 1992). These devices are random apertures which contain a series of pinholes located over a discrete lattice of points, which are such that all possible pinhole separations occur the same number of times. A one-dimensional variant of the uniformly redundant array, with points replaced by slits, is also possible. These arrays have the utility that, in a sense, one can perform many Young-type interference experiments in parallel. The Fresnel diffraction patterns of such masks are convenient for X-ray coherence measurement—see, for example, Lin *et al.* (2003), together with references therein. (iii) Fresnel diffraction patterns of an illuminated sample may be viewed in interferometric terms, as the interference pattern formed by the superposition of the scattered and unscattered waves. Since they may be interpreted as interferograms, the fringes in Fresnel diffraction patterns (in-line holograms) of simple objects (e.g. slits and fibres) may be used to infer certain information regarding the state of coherence of the illuminating radiation. See, for example, Snigireva *et al.* (2001) and Kohn *et al.* (2001). (iv) Regarding the use of the Talbot effect for the measurement of coherence, see, for example, Guigay *et al.* (2004) and references therein.

4.7 Virtual optics for coherent X-ray imaging

Below we augment the discussions on virtual X-ray optics which were given in Section 3.5. Section 4.7.1 revisits and extends these comments in light of the material in the present chapter, while Section 4.7.2 gives an experimental demonstration in the context of coherent X-ray imaging.

4.7.1 *General remarks on virtual optics for coherent X-ray imaging*

The relatively recent advent of powerful computers has enabled them to become an intrinsic part of optical imaging systems, as virtual optics which augment the hardware optical elements of a given system.

At the simplest level, such computing power can be used for the purposes of data analysis. One such example is computational interferogram analysis using the Fourier-transform method or the phase-shifting method (Section 4.6.1). Other examples include the use of computers in X-ray crystallography, the digital reconstruction of in-line holograms (Sections 4.3.1) and Fourier holograms (Section 4.3.3), coherent diffractive imaging based on generalized forms of the Gerchberg–Saxton algorithm (Section 4.5.1), and phase retrieval using the transport-of-intensity equation (Section 4.5.2). These examples constitute a trivially generalized form of Gabor's monumental idea of imaging via a two step process, namely recording followed by reconstruction (Gabor 1948). The salient point of difference is that, rather than using an analogue optical computer (e.g. a Fresnel or Fraunhofer diffraction pattern) in order to optically decode a hologram, one uses a digital computer to the same end in analysing one or more phase-contrast images of a sample.[137]

At a deeper level, and focusing for the moment on coherent imaging systems, one can seek total knowledge (up to a given resolution) regarding the two-dimensional coherent field which is input into a two-dimensional imaging system. As we have seen, phase-retrieval methods of coherent X-ray imaging exist which are able to reconstruct both the amplitude and phase of an input two-dimensional disturbance. Irrespective of the means used to reconstruct the complex input disturbance, this total knowledge of the input allows one to use a computational embodiment of the requisite operator theory of imaging in order to emulate in software the subsequent action of an arbitrary coherent imaging system. In such software optics, the computer is playing a deeper role than merely data analysis. Rather, it is performing the same optical-information processing as a hardware optic, albeit in digital form (Lichte *et al.* 1992, 1993; Yaroslavsky and Eden 1996; Bajt *et al.* 2000; Barone-Nugent *et al.* 2002; Paganin *et al.* 2004b).

At the deepest level to be considered here, the question of virtual optics may be considered for the case of partially-coherent fields. If such fields are Gaussian, then total knowledge regarding the statistical properties of a scalar input field is contained within the mutual coherence function, as a function of all pairs of points in the input disturbance, for all time lags. For statistically stationary fields, this mutual coherence function is defined in terms of a suitable time average, with ensemble averages being required for fields which do not possess this property. One can go further, in considering the various electromagnetic correlation tensors for partially coherent fields for which the vectorial nature of the radiation is important. Lastly, if the statistics are not Gaussian—as will be the case, for example, for non-classical states of the X-ray wave-field—then a hierarchy of higher-order correlation functions is required for a complete description of the input field (see, for example, Mandel and Wolf (1995), and references therein).

Return to the case of virtual optics for a partially coherent scalar field. We have given a very brief discussion (Sections 4.6.2, 4.6.3, and 4.6.4) of some means for determining partial knowledge regarding the mutual coherence function of an X-ray wave-field. This only served to scratch the surface of the subject of coherence measurement, but the salient point is nevertheless clear: means exist for determining the mutual coherence function of an X-ray wave-field, and if such wave-fields are Gaussian then this function carries complete information regarding the statistical properties of the field. Such methods have been known for some time—see, for example, the discussion on means of using the Young interferometer to measure the mutual coherence function (not just its modulus)

[137]Here, 'phase contrast' is to be understood in a suitably broad sense of the term, as any output image which has some dependence on the phase of the input field, in addition to its amplitude. This broad use of the term encompasses holograms, interferograms, phase-contrast images, far-field and Fresnel diffraction patterns, and so forth.

of a quasi-monochromatic field which is given in Wolf (1954). Recent progress in the area of measuring mutual coherence functions for X-ray fields thereby open the way for virtual optics using partially coherent radiation. Again, once one has total knowledge of the input field—in this case, the mutual coherence function for all pairs of points in the input plane to a given two-dimensional imaging system, as a function of all time lags—the operator theory of partially coherent imaging (see, for example, Goodman (1985)) can be used to emulate in software the subsequent action of an arbitrary partially coherent imaging system.

4.7.2 *Example of virtual optics in coherent X-ray imaging*

Here we outline a simple case study of virtual optics for the imaging of coherent X-ray wave-fields (Paganin *et al.* 2004b). Our presentation is broken into three parts. Section 4.7.2.1 considers the forward problem for determining the propagation-based phase-contrast image of a single-material object, under the paraxial approximation. In Section 4.7.2.2 we treat the associated inverse problem of determining the exit-surface wave-field from a single propagation-based phase-contrast image of the same. In Section 4.7.2.3 we show how this reconstructed field may be used for virtual X-ray optics.

4.7.2.1 *Forward problem: calculation of propagation-based phase contrast image*

Consider a thin single-material object, with projected thickness $T(x, y)$ in the direction of a nominal optic axis z. Here, (x, y) are the usual Cartesian coordinates in the plane perpendicular to z. Assume the sample, which is illuminated by unit-strength z-directed monochromatic plane waves of angular frequency ω, to be sufficiently gently varying for the projection approximation to hold (Section 2.2). Under these assumptions, the wave-field $\psi_\omega(x, y, z = 0)$ at the exit-surface $z = 0$ of the sample is seen from eqns (2.41) and (2.43) to be given by:

$$\begin{aligned}\psi_\omega(x, y, z = 0) &= \sqrt{\exp\left[-\mu_\omega T(x, y)\right]} \exp\left[-ik\delta_\omega T(x, y)\right] \\ &= \exp\left[-\left(\frac{\mu_\omega}{2} + ik\delta_\omega\right) T(x, y)\right]. \end{aligned} \tag{4.203}$$

Here, μ_ω is the linear attenuation coefficient of the material in the sample, $k = 2\pi/\lambda$ is the wave-number of the radiation corresponding to a wavelength of λ, and δ_ω is related to the real part of the material's refractive index n_ω via $\mathrm{Re}(n_\omega) = 1 - \delta_\omega$.

Assume this exit-surface wave-field to be paraxial, so that it may be inserted into the operator form (1.28) of the Fresnel diffraction integral so as to propagate the field through a specified distance $\Delta \geq 0$. The resulting propagated field is:

$$\psi_\omega(x,y,z=\Delta) = \exp(ik\Delta) \\ \times \mathcal{F}^{-1} \exp\left[\frac{-i\Delta(k_x^2+k_y^2)}{2k}\right] \mathcal{F} \exp\left[-\left(\frac{\mu_\omega}{2}+ik\delta_\omega\right)T(x,y)\right]. \tag{4.204}$$

Here, $\mathcal{F}$ denotes Fourier transformation with respect to x and y under the convention given in Appendix A, $\mathcal{F}^{-1}$ denotes the corresponding inverse transformation and (k_x, k_y) are the Fourier-space coordinates reciprocal to (x, y).

Let a be the smallest non-negligible transverse length scale present in the exit-surface wave-field of the object. More precisely, a may be defined as the smallest distance such that the two-dimensional Fourier transform of the exit-wave-field is negligible for all points (k_x, k_y) in Fourier space which obey the following inequality:

$$\sqrt{k_x^2+k_y^2} \geq \frac{2\pi}{a}. \tag{4.205}$$

Suppose a to be sufficiently large that the following condition holds:

$$\frac{\Delta}{2k} \times \left(\frac{2\pi}{a}\right)^2 \ll 1. \tag{4.206}$$

If this is so then we may replace the Fresnel propagator $\exp[-i\Delta(k_x^2+k_y^2)/(2k)]$, in eqn (4.204), by the following first-order Taylor approximation:

$$\exp\left[\frac{-i\Delta(k_x^2+k_y^2)}{2k}\right] \approx 1 - \frac{i\Delta(k_x^2+k_y^2)}{2k}. \tag{4.207}$$

Note that, upon introduction of the Fresnel number N_{F} via (cf. eqn (1.47)):

$$N_{\mathrm{F}} \equiv \frac{a^2}{\lambda\Delta}, \tag{4.208}$$

condition (4.206) assumes the simple form (cf. eqn (1.48)):

$$N_{\mathrm{F}} \gg \pi. \tag{4.209}$$

Assuming the propagation distance Δ to be sufficiently small for the Fresnel number to obey inequality (4.209), we may insert the Taylor approximation (4.207) for the Fresnel propagator into eqn (4.204). Invoking the Fourier derivative theorem we obtain the following expression for the propagated field:

$$\psi_\omega(x,y,z=\Delta) = \exp(ik\Delta)\left(1+\frac{i\Delta}{2k}\nabla_\perp^2\right)\exp\left[-\left(\frac{\mu_\omega}{2}+ik\delta_\omega\right)T(x,y)\right], \tag{4.210}$$

where $\nabla_\perp^2 \equiv \partial^2/\partial x^2 + \partial^2/\partial y^2$ is the Laplacian in the xy plane.

The intensity $I_\omega(x, y, z = \Delta)$ of the propagated field is obtained by taking the squared modulus of the above expression. To first order in Δ, this yields[138]:

$$I_\omega(x, y, z = \Delta) = \exp[-\mu_\omega T(x, y)] + \frac{\Delta}{k} \mathrm{Re}\left\{ i \exp\left[-\left(\frac{\mu_\omega}{2} - ik\delta_\omega \right) T(x, y) \right] \nabla_\perp^2 \exp\left[-\left(\frac{\mu_\omega}{2} + ik\delta_\omega \right) T(x, y) \right] \right\}. \tag{4.211}$$

After a good deal of vector algebra, but with no further approximations, the above equation yields the following expression for the propagation-based phase-contrast image of our thin single-material sample[139]:

$$I_\omega(x, y, z = \Delta) = \left(1 - \frac{\delta_\omega \Delta}{\mu_\omega} \nabla_\perp^2 \right) \exp[-\mu_\omega T(x, y)]. \tag{4.212}$$

4.7.2.2 *Inverse problem: determination of the exit-surface wave-field* Take the Fourier transform of eqn (4.212) with respect to x and y. Making use of the Fourier derivative theorem, we obtain:

$$\mathcal{F}\left[I_\omega(x, y, z = \Delta) \right] = \left[1 + \frac{\delta_\omega \Delta}{\mu_\omega} (k_x^2 + k_y^2) \right] \mathcal{F}\left\{ \exp[-\mu_\omega T(x, y)] \right\}. \tag{4.213}$$

The differential equation has thereby been transformed into an algebraic equation, which is readily solved for $\mathcal{F}\{\exp[-\mu_\omega T(x, y)]\}$. One thus obtains the following expression for the projected thickness of the single-material object (Paganin *et al.* 2002):

$$T(x, y) = -\frac{1}{\mu_\omega} \log_e \left(\mathcal{F}^{-1} \left\{ \frac{\mathcal{F}\left[I_\omega(x, y, z = \Delta) \right]}{1 + (\delta_\omega \Delta / \mu_\omega)(k_x^2 + k_y^2)} \right\} \right). \tag{4.214}$$

Inserting this result into eqn (4.203), we arrive at an equation for the exit-surface wave-field of the object as a function of the single propagation-based phase contrast image $I_\omega(x, y, z = \Delta)$ (Paganin *et al.* 2004b):

$$\psi_\omega(x, y, z = 0) = \left(\mathcal{F}^{-1} \left\{ \frac{\mathcal{F}\left[I_\omega(x, y, z = \Delta) \right]}{1 + (\delta_\omega \Delta / \mu_\omega)(k_x^2 + k_y^2)} \right\} \right)^{(1/2) + (ik\delta_\omega / \mu_\omega)}. \tag{4.215}$$

4.7.2.3 *Virtual optics for coherent X-ray imaging* At the ID22 beamline of the European Synchrotron Radiation Facility (ESRF) in Grenoble, France, a 100 μm thick slice of dried human femur bone was normally illuminated with

[138] Note that, when $\Delta = 0$, eqn (4.211) reduces to the projection-approximation form of Beer's law of absorption given in eqn (2.43).

[139] For an alternative derivation see Paganin *et al.* (2002).

near-plane monochromated 20 keV X-rays (wavelength 0.62 Å). Allowing the exit-surface wave-field to propagate through a distance of 20 cm, the propagation-based phase-contrast image in Fig. 4.26(a) was registered using a CCD detector. This detector had an effective pixel size of 0.33 μm, with the phase contrast image being 2048×2048 pixels $= 676 \times 676$ μm in size (referred to the plane of the object). The image clearly displays haversian canals in cross section, together with osteocytes in the bone.

For the purposes of analysis the bone sample was assumed to be composed solely of apatite. At 20 keV this material has a refractive index decrement $\delta_\omega = 1.66 \times 10^{-6}$ and a linear attenuation coefficient $\mu_\omega = 1.95 \times 10^3$ m^{-1}. Equation (4.214) was then used to recover the projected thickness $T(x, y)$ of apatite in the sample, given Fig. 4.26(a) as input, with the result in Fig. 4.26(b). The maximum value of the retrieved thickness was 90 μm, which may be compared to the fact that the thickness of the sample was 100 μm. Using eqn (4.215), the complex wave-field at the exit-surface of the sample was then reconstructed. This was subsequently used to emulate phase-contrast images for a number of modalities, as shown in Figs. 4.26(c) though (i). In Fig. 4.26(c) we see an emulated Zernike phase contrast image, for phase bias of the Zernike 'spot' equal to π radians, and Fourier-space radius of the spot at 10^{-3} μm^{-1} (see Section 4.4.1). In Fig. 4.26(d) we see an emulated Schlieren image, corresponding to a vertical 'knife edge' blocking out the left half of the two-dimensional Fourier spectrum of the retrieved exit-surface wave-field. Fig. 4.26(e) shows an emulated propagation-based phase contrast image (in-line hologram) corresponding to a propagation distance 1.00 m, which was computed using the angular-spectrum representation of the free-space diffraction operator in eqn (1.25) (see Section 4.4.4). Figure 4.26(f) gives an emulated differential interference contrast image, corresponding to a phase bias of π radians and a horizontal transverse shift of the displaced copy of the field at 0.33 μm (see Section 4.4.2). Figures 4.26(g) and (h) give emulated analyser-based phase-contrast images, corresponding to respective detunings of $1.5''$ and $-1.5''$ from a symmetrical silicon 111 reflection for the unperturbed incident beam, with the detunings being in the yz plane (see Section 4.4.3). Lastly Fig. 4.26(i) shows an emulated interferogram corresponding to interfering the retrieved exit-surface wave-field with the tilted plane wave $\exp[i(k_x^{(0)} x + k_y^{(0)} y + B)]$, where $k_x^{(0)} = 1.75$ μm^{-1}, $k_y^{(0)} = 0$ μm^{-1}, and $B = \pi$ radians (see Section 4.6.1).

4.8 Summary

We began our explorations, on the fascinating subject of coherent X-ray imaging, with a treatment of the operator theory of imaging in Section 4.1. Here, a two-dimensional X-ray imaging system was viewed in operational terms as mapping a given input field to an associated output. The operator, which maps input to output, was seen to completely characterize a given imaging system. Our main emphasis was on the operator theory of coherent imaging systems, with particular attention being paid to shift-invariant coherent linear systems, although

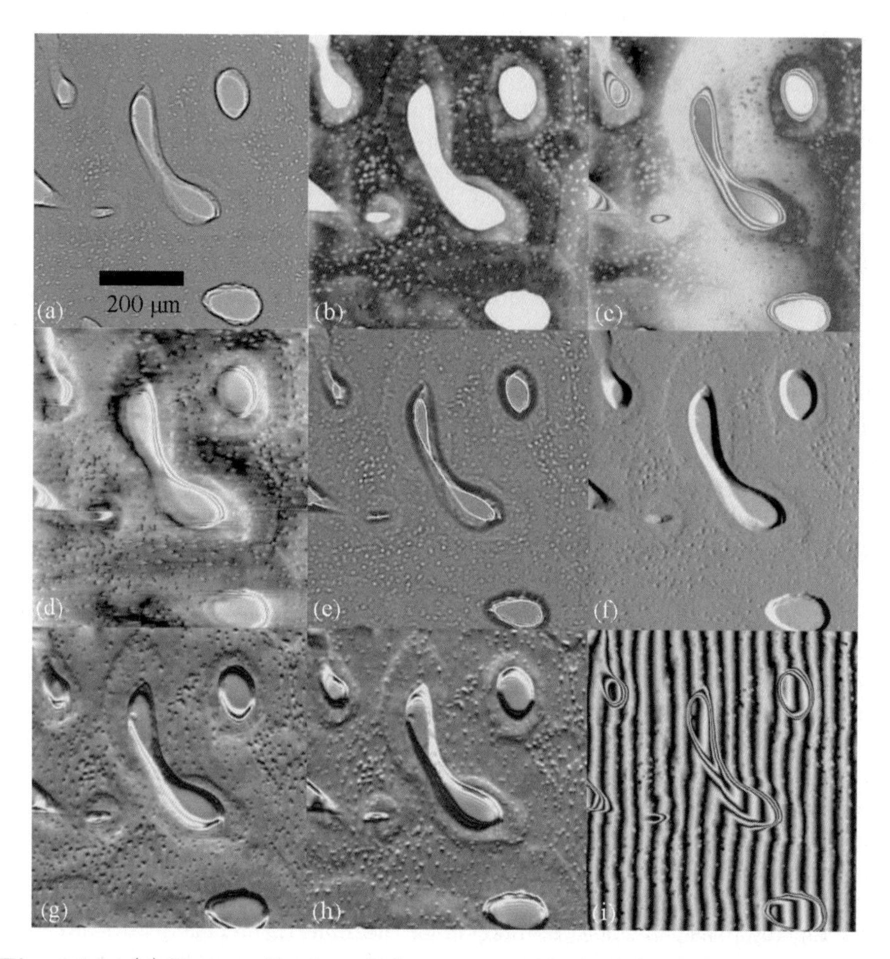

Fig. 4.26. (a) Propagation-based phase contrast image of a 100 μm thick slice of dried human femur, obtained using 20 keV X-rays. Scale bar is 200 μm, with all subsequent images at same magnification. (b) Reconstruction of projected thickness $T(x, y)$ of apatite in bone sample, obtained by applying eqn (4.214) to image in (a). Using eqn (4.215), the complex wave-field at the exit surface of the object was then reconstructed. This wave-field was then used as input into virtual-optics algorithms to emulate in software a variety of phase contrast imaging systems: (c) Zernike phase contrast, (d) Schlieren phase contrast, (e) propagation-based phase contrast, (f) differential interference contrast, (g, h) analyser-based phase contrast for two different detunings of crystal from Bragg angle, and (i) interferometric phase contrast. Further details in main text. Image taken from Paganin *et al.* (2004b). Used with permission.

polychromatic and partially coherent imaging systems were also examined. This formalism laid much of the groundwork employed in later sections of the chapter.

In Section 4.2 we treated self imaging, whereby the act of free-space propagation from plane to parallel plane may serve to form an image of the time-averaged disturbance of the unpropagated intensity. Two such self-imaging systems were considered. In the Talbot effect, we saw that a two-dimensional transversely periodic paraxial coherent field may exhibit self-imaging for propagation distances equal to integer multiples of the so-called Talbot distance. In the stronger form of the Montgomery effect we saw that if the two-dimensional Fourier spectrum of an input forward-propagating coherent input disturbance is non-zero only over a certain circle in Fourier space, then one has a 'diffraction free beam' with the remarkable property that it is self-imaging for all in-vacuo propagation distances. In the weaker form of the Montgomery effect the non-vanishing of the Fourier spectrum over two circles was seen to yield a free-space propagated field with longitudinally periodic intensity. Polychromatic generalizations of both the Talbot and Montgomery effects were also considered.

We then moved onto the subject of holography, in Section 4.3. Three different forms of holography were considered: in-line holography, off-axis holography, and Fourier holography. All of these were seen to have the common feature of seeking to reconstruct both the amplitude and phase of a given incident two-dimensional coherent disturbance. This was achieved via a two-step process: recording followed by reconstruction. This notion has strong parallels with both phase retrieval and virtual optics, subjects discussed later in the chapter.

Phase contrast imaging systems, which are engineered such their output intensity maps are sensitive to transverse phase shifts in the input field, were examined in Section 4.4. Such systems are of particular utility for the X-ray imaging of weakly absorbing or non-absorbing specimens, whose primary effect on an incident beam is to introduce transverse phase shifts but minimal transverse intensity variations. We separately considered several common methods for achieving phase contrast: Zernike phase contrast, differential phase contrast, analyser-based phase contrast, and propagation-based phase contrast. In-line propagation-based phase contrast was seen to be very closely related to in-line holography, with the qualitative difference that the former typically works with propagation distances that are sufficiently small to avoid multiple Fresnel fringes in the output image. We also offered some remarks on hybrid phase-contrast imaging systems, in which two or more different forms of phase contrast may be united into a single system which compensates for the individual shortcomings of each method considered separately. All of these phase-contrast imaging systems were seen to have associated transfer functions which were deliberately 'aberrated' from unity. This conforms to the slightly paradoxical statement that a perfect imaging system, which by definition has a transfer function of unity, yields no information regarding the phase of an X-ray disturbance incident upon it.

Phase retrieval, treated in Section 4.5, was seen to represent a natural follow-on from the physics of phase-contrast imaging. Rather than merely seeking to qualitatively interpret phase-contrast images, the problem of phase retrieval seeks to reconstruct both the phase and the amplitude of the incident disturbance given one or more phase contrast images of the same. Three methods for phase retrieval were examined: the Gerchberg–Saxton algorithm and its subsequent generalizations, deterministic phase retrieval using the transport-of-intensity equation, and the use of complex analysis for the phase retrieval of one-dimensional fields. Generalized forms of the Gerchberg–Saxton algorithm are of current topical interest, in the context of the ongoing quest to solve the non-crystallographic phase problem of reconstructing aperiodic samples given one or more far-field diffraction patterns of the same. The transport-of-intensity equation was seen to yield a deterministic solution to the problem of phase retrieval using two slightly defocused propagation-based phase contrast images of a paraxial field with continuous wave fronts. Lastly, the problem of phase retrieval given the diffraction pattern of a one-dimensional coherent field was seen to be very closely related to the theory of analytic functions.

Interferometric means of both phase and coherence measurement were treated in Section 4.6. Regarding phase measurement, interferometric phase contrast was seen to be effected by superposing a given two-dimensional coherent wave-field with a tilted plane wave, the intensity of which has a series of linear fringes distorted according to the transverse phase variations in the input field. Two methods of phase determination from such interferograms were treated: Takeda's Fourier-transform method and the phase-shifting method. One may note the evident parallels between the former method and Fourier holography. Two methods of interferometric coherence measurement—where one 'images' the coherence properties of X-ray radiation, as opposed to imaging a sample—were treated. The first of these uses the Young interferometer to determine the modulus of the complex degree of coherence associated with a quasi-monochromatic partially coherent X-ray field. The second of these is based on intensity interferometry, where interference occurs at the level of intensity rather than at the level of fields. Many other means of coherence measurement exist, including methods based on Fresnel diffraction, uniformly redundant arrays, prism interferometers, and the Talbot effect. These were not examined in the text, beyond providing some entry points into the recent literature.

We closed with the topic of virtual optics, in Section 4.7. Here, we argued that the computer is now firmly entrenched as an intrinsic part of optical imaging systems. Examples were given, drawn from holography, crystallography, tomography, and propagation-based phase retrieval. At a weaker level, it was argued that computers are convenient tools for optical data analysis. At a deeper level, the techniques of phase retrieval—for both coherent and partially coherent fields—allow the computer to process optical information at the field level, albeit in digital form. An example was given, of the utility of virtual optics in the context of X-ray micro-radiography.

Further reading

The theory of coherent shift-invariant linear optical imaging systems is covered in many optics texts, such as Papoulis (1968). For aberrated linear systems, see Cowley (1995). Regarding the generalization of this approach to partially coherent systems, which was only touched upon in the text, see Beran and Parrent (1964) and Goodman (1985). The Talbot effect is described in Cowley (1995), with a well-referenced discussion in the context of X-ray optics in Erko *et al.* (1996). See also Cloetens *et al.* (1999) and Guigay *et al.* (2004), together with references therein. The Montgomery effect is outlined in the discussion on diffraction-free beams in the text by Nieto-Vesperinas (1991). Key original papers include Montgomery (1967, 1968), Durnin (1987), and Durnin *et al.* (1987), although we note that none of these works treat the possible realization of diffraction-free beams in the X-ray regime. For the early papers on X-ray holography see Erko *et al.* (1996), together with references to the primary literature which are contained therein. An important variant on Fourier holography, known as internal source holography, was not discussed in the text. For more on this topic, see, for example, the reviews by Faigel and Tegze (1999, 2003). For more information on the various methods of phase contrast (Zernike phase contrast, differential interference contrast, analyser-based phase contrast, propagation-based phase contrast, and hybrid means for phase contrast), together with the Gerchberg–Saxton algorithm and its later variants, see the journal articles cited in the main text. For a review of phase retrieval using the transport-of-intensity equation, see Paganin and Nugent (2001). Methods of one-dimensional phase retrieval are reviewed in Klibanov *et al.* (1995). Entry points, to the literature which applies and extends such methods in the context of X-ray diffractometry of one-dimensional layered scatterers under the first Born approximation, include Petrashen' and Chukhovskii (1989), Reiss and Lipperheide (1996), Nikulin (1998), Nikulin and Zaumseil (1999), Siu *et al.* (2001), and Nikulin and Zaumseil (2004). Hariharan (2003) is a good general reference on interferometry. For X-ray interferometry, see the early papers by Bonse and Hart (1965) and Hart and Bonse (1970), together with the reviews by Hart (1980), Bonse (1988), and Momose (2002), and the well-referenced textbook account by Authier (2001). Regarding both Young interferometry and intensity interferometry, we refer the reader to the references cited in the main text. For more on virtual optics see the book on the subject by Yaroslavsky and Eden (1996) and the paper by Paganin *et al.* (2004b), together with references therein.

References

L.J. Allen and M.P. Oxley, *Phase retrieval from series of images obtained by defocus variation*, Opt. Commun. **199**, 65–75 (2001).

G. Anger, *Some remarks on the relationship between mathematics and physics and on inverse problems*, in P.C. Sabatier (ed.), *Inverse methods in action (Proceedings of the multicentennials meeting on inverse problems, Montpellier, November 27th—December 1st, 1989)*, Springer-Verlag, Berlin (1990).

S. Aoki and S. Kikuta, *X-ray holographic microscopy*, Jpn. J. Appl. Phys. **13**, 1385–1392 (1974).

A. Authier, *Dynamical theory of X-ray diffraction*, Oxford University Press, Oxford (2001).

A.V. Baez, *A study in diffraction microscopy with special reference to X-rays*, J. Opt. Soc. Am. **42**, 756–762 (1952).

S. Bajt, A. Barty, K.A. Nugent, M. McCartney, M. Wall, and D. Paganin, *Quantitative phase-sensitive imaging in a transmission electron microscope*, Ultramicroscopy **83**, 67–73 (2000).

H.P. Baltes, *Introduction*, in H.P. Baltes (ed.), *Inverse source problems in optics*, Springer-Verlag, Berlin, p. 2, (1978).

E.D. Barone-Nugent, A. Barty, and K.A. Nugent, *Quantitative phase-amplitude microscopy I: optical microscopy*, J. Microsc. **206**, 194–203 (2002).

R.H.T. Bates, *Fourier phase problems are uniquely soluble in more than one dimension. I.*, Optik **61**, 247–262 (1982).

M.J. Beran and G.B. Parrent, *Theory of partial coherence*, Prentice-Hall, Englewood Cliffs (1964).

M. Bertero, C. De Mol, and E.R. Pike, *Application of singular systems to some data reduction problems in modern optics*, in P.C. Sabatier (ed.), *Inverse methods in action (Proceedings of the multicentennials meeting on inverse problems, Montpellier, November 27th—December 1st, 1989)*, Springer-Verlag, Berlin, pp. 248–261 (1990).

R.P. Boas, *Entire functions*, Academic Press, New York (1954).

U. Bonse, *Recent advances in X-ray and neutron interferometry*, Physica B **151**, 7–21 (1988).

U. Bonse and M. Hart, *An X-ray interferometer*, Appl. Phys. Lett. **6**, 155–156 (1965).

M. Born and E. Wolf, *Principles of optics*, seventh edition, Cambridge University Press, Cambridge (1999).

J. Boyes-Watson, K. Davidson, and M.F. Perutz, *An X-ray study of horse methaemoglobin. I.*, Proc. R. Soc. London Ser. A **191**, 83–137 (1947).

H. Bremmer, *On the asymptotic evaluation of diffraction integrals with a special view to the theory of defocusing and optical contrast*, Physica **18**, 469–485 (1952).

R.E. Burge, X.-C. Yuan, G.E. Slark, M.T. Browne, P. Charalambous, C.L.S. Lewis, *et al.*, *Optical source model for the 23.2–23.6 nm radiation from the multielement germanium soft X-ray laser*, Opt. Commun. **169**, 123–133 (1999).

D. Chapman, W. Thomlinson, R.E. Johnston, D. Washburn, E. Pisano, N. Gmür, *et al.*, *Diffraction enhanced X-ray imaging*, Phys. Med. Biol. **42**, 2015–2025 (1997).

P. Cloetens, R. Barrett, J. Baruchel, J.-P. Guigay, and M. Schlenker, *Phase objects in synchrotron radiation hard X-ray imaging*, J. Phys. D: Appl. Phys. **29**, 133–146 (1996).

P. Cloetens, W. Ludwig, J. Baruchel, J.-P. Guigay, P. Pernot-Rejmánková, M.

Salomé-Pateyron, *et al.*, *Hard X-ray phase imaging using simple propagation of a coherent synchrotron radiation beam*, J. Phys. D: Appl. Phys. **32**, A145–A151 (1999).

P. Coan, E. Pagot, S. Fiedler, P. Cloetens, J. Baruchel, and A. Bravin, *Phase-contrast X-ray imaging combining free space propagation and Bragg diffraction*, J. Synchrotron Rad. **12**, 241–245 (2005).

J.M. Cowley, *Diffraction physics*, third revised edition, North-Holland, Amsterdam (1995).

J.M. Cowley and A.F. Moodie, *Fourier images: I—the point source*, Proc. Phys. Soc. B **70**, 486–496 (1957a).

J.M. Cowley and A.F. Moodie, *Fourier images: II—the out-of-focus patterns*, Proc. Phys. Soc. B **70**, 497–504 (1957b).

J.M. Cowley and A.F. Moodie, *Fourier images: III—finite sources*, Proc. Phys. Soc. B **70**, 505–513 (1957c).

T.J. Davis, D. Gao, T.E. Gureyev, A.W. Stevenson, and S.W. Wilkins, *Phase contrast imaging of weakly absorbing materials using hard X-rays*, Nature **373**, 595–598 (1995a).

T.J. Davis, T.E. Gureyev, D. Gao, A.W.Stevenson, and S.W.Wilkins, *X-ray image contrast from a simple phase object*, Phys. Rev. Lett. **74**, 3173–3176 (1995b).

T.J. Davis, *X-ray diffraction imaging using perfect crystals*, J. X-Ray Sci. Technol. **6**, 317–342 (1996).

T.J. Davis and A.W. Stevenson, *Direct measure of the phase shift of an X-ray beam*, J. Opt. Soc. Am. A **13**, 1193–1198 (1996).

T. Ditmire, E.T. Gumbrell, R.A. Smith, J.W.G. Tisch, D.D. Meyerhofer, and M.H.R. Hutchinson, *Spatial coherence measurement of soft X-ray radiation produced by high order harmonic generation*, Phys. Rev. Lett. **77**, 4756–4759 (1996).

J. Durnin, *Exact solutions for nondiffracting beams. I. The scalar theory*, J. Opt. Soc. Am. A **4**, 651–654 (1987).

J. Durnin, J.J. Miceli Jr., and J.H. Eberly, *Diffraction-free beams*, Phys. Rev. Lett. **58**, 1499–1501 (1987).

S. Eisebitt, J. Lüning, W.F. Schlotter, M. Lörgen, O. Hellwig, W. Eberhardt, *et al.*, *Lensless imaging of magnetic nanostructures by X-ray spectro-holography*, Nature **432**, 885–888 (2004).

A.I. Erko, V.V. Aristov, and B. Vidal, *Diffraction X-ray optics*, Institute of Physics Publishing, Bristol (1996).

G. Faigel and M. Tegze, *X-ray holography*, Rep. Prog. Phys. **62**, 355–393 (1999).

G. Faigel and M. Tegze, *X-ray holography*, Struct. Chem. **14**(1), 15–21 (2003).

E.E. Fenimore and T.M. Cannon, *Coded aperture imaging with uniformly redundant arrays*, Appl. Opt. **17**, 337–347 (1978).

J.R. Fienup, *Phase retrieval algorithms: a comparison*, Appl. Opt. **21**, 2758–2769 (1982).

E. Förster, K. Goetz, and P. Zaumseil, *Double crystal diffractometry for the characterization of targets for laser fusion experiments*, Kristall und Technik **15**, 937–945 (1980).

D. Gabor, *A new microscopic principle*, Nature **161**, 777–778 (1948).

R.W. Gerchberg and W.O. Saxton, *A practical algorithm for the determination of phase from image and diffraction plane pictures*, Optik **35**, 237–246 (1972).

E. Gluskin, I. McNulty, P.J. Viccaro, and M.R. Howells, *X-ray intensity interferometer for undulator radiation*, Nucl. Instr. Meth. Phys. Res. A **319**, 213–218 (1992).

E. Gluskin, I. McNulty, L. Yang, K.J. Randall, Z. Xu, and E.D. Johnson, *Intensity interferometry at the X13A undulator beamline*, Nucl. Instr. Meth. Phys. Res. A **347**, 177–181 (1994).

E. Gluskin, E.E. Alp, I. McNulty, W. Sturhahn, and J. Sutter, *A classical Hanbury Brown–Twiss experiment with hard X-rays*, J. Synchrotron Rad. **6**, 1065–1066 (1999).

J.W. Goodman, *Statistical optics*, John Wiley & Sons, New York (1985).

J.-P. Guigay, S. Zabler, P. Cloetens, C. David, R. Mokso, and M. Schlenker, *The partial Talbot effect and its use in measuring the coherence of synchrotron X-rays*, J. Synchrotron Rad. **11**, 476–482 (2004).

T.E. Gureyev, *Transport of intensity equation for beams in an arbitrary state of temporal and spatial coherence*, Optik **110**, 263–266 (1999).

T.E. Gureyev, *Composite techniques for phase retrieval in the Fresnel region*, Opt. Commun. **220**, 49–58 (2003).

T.E. Gureyev, A. Roberts, and K.A. Nugent, *Partially coherent fields, the transport of intensity equation, and phase uniqueness*, J. Opt. Soc. Am. A **12**, 1942–1946 (1995).

T.E. Gureyev and K.A. Nugent, *Phase retrieval with the transport-of-intensity equation. II. Orthogonal series solution for nonuniform illumination*, J. Opt. Soc. Am. A **13**, 1670–1682 (1996).

T.E. Gureyev and K.A. Nugent, *Rapid quantitative phase imaging using the transport of intensity equation*, Opt. Commun. **133**, 339–346 (1997).

T.E. Gureyev and S.W. Wilkins, *Regimes of X-ray phase-contrast imaging with perfect crystals*, Nuovo Cimento **19D**, 545–552 (1997).

T.E. Gureyev, C. Raven, A. Snigirev, I. Snigireva, and S.W. Wilkins, *Hard X-ray quantitative non-interferometric phase-contrast microscopy*, J. Phys. D: Appl. Phys. **32**, 563–567 (1999).

T.E. Gureyev, D.M. Paganin, A.W. Stevenson, S.C. Mayo, and S.W. Wilkins, *Generalized eikonal of partially coherent beams and its use in quantitative imaging*, Phys. Rev. Lett. **93**, 068103 (2004).

J. Hadamard, *Lectures on Cauchy's problem in linear partial differential equations*, Yale University Press, New Haven (1923).

R. Hanbury Brown and R.Q. Twiss, *A new type of interferometer for use in radio astronomy*, Phil. Mag. **45**, 663–682 (1954).

R. Hanbury Brown and R.Q. Twiss, *Correlation between photons in two coherent beams of light*, Nature **177**, 27–29 (1956a).

R. Hanbury Brown and R.Q. Twiss, *A test of a new type of stellar interferometer on Sirius*, Nature **178**, 1046–1048 (1956b).

R. Hanbury Brown and R.Q. Twiss, *Interferometry of the intensity fluctuations in light I. Basic theory: the correlation between photons in coherent beams of radiation*, Proc. Roy. Soc. **242**, 300–324 (1957a).

R. Hanbury Brown and R.Q. Twiss, *Interferometry of the intensity fluctuations in light II. An experimental test of the theory for partially coherent light*, Proc. Roy. Soc. **243**, 291–319 (1957b).

P. Hariharan, *Optical interferometry*, second edition, Academic Press, San Diego, CA (2003).

M. Hart, *The application of synchrotron radiation to X-ray interferometry*, Nucl. Instr. Meth. Phys. Res. **172**, 209–214 (1980).

M. Hart and U. Bonse, *Interferometry with X rays*, Phys. Today, 26–31 (August 1970).

M.R. Howells, *The X-ray Hanbury Brown and Twiss intensity interferometer: a new physics experiment and a diagnostic for both X-ray and electron beams at light sources*, Advanced Light Source Technical Report, LSBL-27 (1989).

E. Ikonen, *Interference effects between independent gamma rays*, Phys. Rev. Lett. **68**, 2759–2761 (1992).

V.N. Ingal and E.A. Beliaevskaya, *X-ray plane-wave topography observation of the phase contrast from a non-crystalline object*, J. Phys. D: Appl. Phys. **28**, 2314–2317 (1995).

E.T. Jaynes, *Papers on probability, statistics and statistical physics*, D. Reidel Publishing Company, Dordrecht (1983).

E.T. Jaynes, *Probability theory: the logic of science*, Cambridge University Press, Cambridge (2003).

B. Kaulich, T. Wilhein, E. Di Fabrizio, F. Romanato, M. Altissimo, S. Cabrini, *et al.*, *Differential interference contrast X-ray microscopy with twin zone plates*, J. Opt. Soc. Am. A **19**, 797–806 (2002a).

B. Kaulich, F. Polack, U. Neuhaeusler, J. Susini, E. Di Fabrizio, and T. Wilhein, *Diffracting aperture based differential phase contrast for scanning X-ray microscopy*, Opt. Expr. **10**, 1111–1117 (2002b).

M.V. Klibanov, P.E. Sacks, and A.V. Tikhonravov, *The phase retrieval problem*, Inverse Problems **11**, 1–28 (1995).

V. Kohn, I. Snigireva, and A. Snigirev, *Interferometric characterization of spatial coherence of high energy synchrotron X-rays*, Opt. Commun. **198**, 293–309 (2001).

R. Kress, *Linear integral equations*, Springer-Verlag, Berlin, p. 221 (1989).

Y. Kunimune, Y. Yoda, K. Izumi, M. Yabashi, X.-W. Zhang, T. Harami, *et al.*, *Two-photon correlations in X-rays from a synchrotron radiation source*, J. Synchrotron Rad. **4**, 199–203 (1997).

R. Landauer, *The physical nature of information*, Phys. Lett. A **217**, 188–193 (1996).

W. Leitenberger, S.M. Kuznetsov, and A. Snigirev, *Interferometric measurements with hard X-rays using a double slit*, Opt. Commun. **191**, 91–96 (2001).

W. Leitenberger, H. Wendrock, L. Bischoff, T. Panzner, U. Pietsch, J. Grenzer, *et al.*, *Double pinhole diffraction of white synchrotron radiation*, Physica B **336**, 63–67 (2003).

W. Leitenberger, H. Wendrock, L. Bischoff, and T. Weitkamp, *Pinhole interferometry with coherent hard X-rays*, J. Synchrotron Rad. **11**, 190–197 (2004).

E.N. Leith and J. Upatneiks, *Reconstructed wavefronts and communication theory*, J. Opt. Soc. Am. **52**, 1123–1130 (1962).

E.N. Leith and J. Upatneiks, *Wavefront reconstruction with continuous-tone objects*, J. Opt. Soc. Am. **53**, 1377–1381 (1963).

E.N. Leith and J. Upatneiks, *Wavefront reconstruction with diffused illumination and three-dimensional objects*, J. Opt. Soc. Am. **54**, 1295–1301 (1964).

R.A. Lewis, K.D. Rogers, C.J. Hall, A.P. Hufton, S. Evans, R.H. Menk, *et al.*, *Diffraction enhanced imaging: improved contrast, lower dose X-ray imaging*, in L.E. Antonuk and M.J. Yaffe (eds), *Medical imaging 2002: physics of medical imaging*, Proc. SPIE **4682**, 286–297 (2002).

H. Lichte, E. Völkl, and K. Scheerschmidt, *First steps of high resolution electron holography into materials science*, Ultramicroscopy **47**, 231–240 (1992).

H. Lichte, P. Kessler, D. Lenz, and W.-D. Rau, *0.1 nm information limit with the CM30FEG–Special Tübingen*, Ultramicroscopy **52**, 575–580 (1993).

J.J.A. Lin, D. Paterson, A.G. Peele, P.J. McMahon, C.T. Chantler, K.A. Nugent, *et al.*, *Measurement of the spatial coherence function of undulator radiation using a phase mask*, Phys. Rev. Lett. **90**, 074801 (2003).

Y. Liu, M. Seminario, F.G. Tomasel, C. Chang, J.J. Rocca, and D.T. Attwood, *Achievement of essentially full spatial coherence in a high-average-power soft-X-ray laser*, Phys. Rev. A **63**, 033802 (2001).

L. Mandel and E. Wolf, *Optical coherence and quantum optics*, Cambridge University Press, Cambridge (1995).

A.I. Markusevich, *The Theory of analytic functions: a brief course*, Mir Publishers, Moscow (1983).

P.J. McMahon, A.G. Peele, D. Paterson, J.J.A. Lin, T.H.K. Irving, I. McNulty, *et al.*, *Quantitative X-ray phase tomography with sub-micron resolution*, Opt. Commun. **217**, 53–58 (2003).

J. Miao, D. Sayre, and H.N. Chapman, *Phase retrieval from the magnitude of the Fourier transforms of nonperiodic objects*, J. Opt. Soc. Am. A **15**, 1662–1669 (1998).

J. Miao, P. Charalambous, J. Kirz, and D. Sayre, *Extending the methodology of X-ray crystallography to allow imaging of micrometre-sized non-crystalline specimens*, Nature **400**, 342–344 (1999).

A. Momose, *Phase-contrast X-ray imaging based on interferometry*, J. Synchrotron Rad. **9**, 136–142 (2002).

W.D. Montgomery, *Self-imaging objects of infinite aperture*, J. Opt. Soc. Am. **57**, 772–778 (1967).

W.D. Montgomery, *Algebraic formulation of diffraction applied to self-imaging*, J. Opt. Soc. Am. **58**, 1112–1124 (1968).

P.M. Morse and H. Feshbach, *Methods of theoretical physics, volume 1*, McGraw-Hill, New York, chapter 4 (1953).

U. Neuhäusler, G. Schneider, W.Ludwig, M.A. Meyer, E. Zschech, and D. Hambach, *X-ray microscopy in Zernike phase contrast mode at 4 keV photon energy with 60 nm resolution*, J. Phys. D: Appl Phys. **36**, A79–A82 (2003).

M. Nieto-Vesperinas, *Scattering and diffraction in physical optics*, John Wiley & Sons, New York (1991).

A.Yu. Nikulin, *Uniqueness of the complex diffraction amplitude in X-ray Bragg diffraction*, Phys. Rev. B. **57**, 11178–11183 (1998).

A.Y. Nikulin and P. Zaumseil, *Phase retrieval X-ray diffractometry in the case of high- or low-flux radiation sources*, Phys. Stat. Sol. (a) **172**, 291–301 (1999).

A.Y. Nikulin and P. Zaumseil, *Fast nondestructive technique to determine the content of components in a strain-compensated crystalline ternary alloy*, J. Appl. Phys. **95**, 5249–5251 (2004).

K.A. Nugent and J.E. Trebes, *Coherence measurement technique for short-wavelength light sources*, Rev. Sci. Instrum. **63**, 2146–2151 (1992).

K.A. Nugent, T.E. Gureyev, D. Cookson, D. Paganin, and Z. Barnea, *Quantitative phase imaging using hard X-rays*, Phys. Rev. Lett. **77**, 2961–2964 (1996).

K.A. Nugent and D. Paganin, *Matter-wave phase measurement: a noninterferometric approach*, Phys. Rev. A **61**, 063614 (2000).

D. Paganin and K.A. Nugent, *Noninterferometric phase imaging with partially coherent light*, Phys. Rev. Lett. **80**, 2586–2589 (1998).

D. Paganin and K.A. Nugent, *Non-interferometric phase determination*, in P. Hawkes (ed.), *Advances in imaging and electron physics* **118**, Harcourt Publishers, Kent, pp. 85-127 (2001).

D. Paganin, S.C. Mayo, T.E. Gureyev, P.R. Miller, and S.W. Wilkins, *Simultaneous phase and amplitude extraction from a single defocused image of a homogeneous object*, J. Microsc. **206**, 33–40 (2002).

D. Paganin. T.E. Gureyev, K.M. Pavlov, R.A. Lewis, and M. Kitchen, *Phase retrieval using coherent imaging systems with linear transfer functions*, Opt. Commun. **234**, 87–105 (2004a).

D. Paganin, T.E. Gureyev, S.C. Mayo, A.W. Stevenson, Ya.I. Nesterets, and S.W. Wilkins, *X-ray omni microscopy*, J. Microsc. **214**, 315–327 (2004b).

A. Papoulis, *Systems and transforms with applications in optics*, Robert E. Krieger Publishing Company, Florida (1981).

D. Paterson, B.E. Allman, P.J. McMahon, J. Lin, N. Moldovan, K.A. Nugent, *et al.*, *Spatial coherence measurement of X-ray undulator radiation*, Opt. Commun. **195**, 79–84 (2001).

W. Pauli, *Die allgemeinen Prinzipien der Wellenmechanik*, in H. Geiger and K. Scheel (eds), *Handbuch der Physik*, Springer, Berlin (1933).

K.M Pavlov, T.E Gureyev, D. Paganin, Ya.I. Nesterets, M.J. Morgan, and R.A. Lewis, *Linear systems with slowly varying transfer functions and their application to X-ray phase-contrast imaging*, J. Phys. D: Appl. Phys. **37**, 2746–2750 (2004).

P.V. Petrashen' and F.N. Chukhovskii, *Restoration of the phase of an X-ray wave diffracted in a single-crystal layered structure*, Sov. Phys. Dokl. **34**, 957–959 (1989).

A. Pogany, D. Gao, and S.W. Wilkins, *Contrast and resolution in imaging with a microfocus X-ray source*, Rev. Sci. Instrum. **68**, 2774–2782 (1997).

Lord Rayleigh, *On copying diffraction gratings and some phenomena connected therewith*, Phil. Mag. **11**, 196–205 (1881).

G. Reiss and R. Lipperheide, *Inversion and the phase problem in specular reflection*, Phys. Rev. B **53**, 8157–8160 (1996).

W.C. Röntgen, *On a new kind of rays*, Nature **53**, 274–276 (1896).

D. Sayre, *Some implications of a theorem due to Shannon*, Acta Crystallogr. **5**, 843–843 (1952).

G. Schmahl, D. Rudolph, G. Schneider, P. Guttmann, and B. Niemann, *Phase contrast X-ray microscopy studies*, Optik **97**, 181–182 (1994).

E.V. Shuryak, *Two-photon correlations in synchrotron radiation as a method of studying the beam*, Sov. Phys. JETP **40**, 30–31 (1975).

K. Siu, A.Y. Nikulin, K. Tamasaku, and T. Ishikawa, *X-ray phase retrieval in high-resolution refraction data from amorphous materials*, Appl. Phys. Lett. **79**, 2112–2114 (2001).

A. Snigirev, I. Snigireva, V. Kohn, S. Kuznetsov, and I. Schelekov, *On the possibilities of X-ray phase contrast microimaging by coherent high-energy synchrotron radiation*, Rev. Sci. Instrum. **66**, 5486–5492 (1995).

A. Snigirev, I. Snigireva, V.G. Kohn, and S.M. Kuznetsov, *On the requirements to the instrumentation for the new generation of synchrotron radiation sources. Beryllium windows*, Nucl. Instr. Meth. Phys. Res. A **370**, 634–640 (1996).

I. Snigireva, V. Kohn, and A. Snigirev, *Interferometric techniques for characterization of coherence of high-energy synchrotron X-rays*, Nucl. Instr. Meth. Phys. Res. A **467-468**, 925–928 (2001).

V.A. Somenkov, A.K. Tkalich, and S. Shil'stein, *Refraction contrast in X-ray microscopy*, Sov. Phys. Tech. Phys. **3**, 1309–1311 (1991).

Y. Suzuki, *Measurement of X-ray coherence using two-beam interferometer with prism optics*, Rev. Sci. Instrum. **75**, 1026–1029 (2004).

A. Szöke, *X-ray and electron holography using a local reference beam*, in D.T. Attwood and J. Boker (eds), *Short Wavelength Coherent Radiation: Generation and Applications* (AIP conference proceedings number 147), American Institute of Physics, New York, pp. 361–367 (1986).

R.Z. Tai, Y. Takayama, N. Takaya, T. Miyahara, S. Yamamoto, H. Sugiyama, et al., *Chaotic nature of the stored current in an electron storage ring detected by a two-photon correlator for soft-X-ray synchrotron radiation*, Phys. Rev. A **60**, 3262–3266 (1999).

Y. Takayama, R.Z. Tai, T. Hatano, T. Miyahara, W. Okamoto, and Y. Kagoshima, *Measurement of the coherence of synchrotron radiation*, J. Synchrotron Rad. **5**, 456–458 (1998).

Y. Takayama, H. Shiozawa, N. Takaya, T. Miyahara, and R. Tai, *Electron-beam diagnosis with Young's interferometer in soft X-ray region*, Nucl. Instr. Meth. Phys. Res. A **455**, 217–221 (2000).

M. Takeda, H. Ina, and S. Kobayashi, *Fourier-transform method of fringe-pattern analysis for computer-based topography and interferometry*, J. Opt. Soc. Am. **72**, 156–160 (1982).

H.F. Talbot, *Facts relating to optical science. No. IV*, Phil. Mag. **9**, 401–407 (1836).

M.R. Teague, *Deterministic phase retrieval: a Green's function solution*, J. Opt. Soc. Am. **73**, 1434–1441 (1983).

M. Tegze, G. Faigel, S. Marchesini, M. Belakhovsky, and O. Ulrich, *Imaging light atoms by X-ray holography*, Nature **407**, 38–38 (2000).

B.J. Thompson and E. Wolf, *Two-beam interference with partially coherent light*, J. Opt. Soc. Am. **47**, 895–902 (1957).

A. Vasara, J. Turunen, and A. Friberg, *Realization of general nondiffracting beams with computer-generated holograms*, J. Opt. Soc. Am. A **6**, 1748–1754 (1989).

N. Watanabe, K. Sakurai, A. Takeuchi, and S. Aoki, *Soft-X-ray Gabor holography by means of a backilluminated CCD camera*, Appl. Opt. **36**, 7433–7436 (1997).

V. White and D. Cerrina, *Metal-less X-ray phase-shift masks for nanolithography*, J. Vac. Sci. Technol. B **10**, 3141–3144 (1992).

S.W. Wilkins, *Simplified conditions and configurations for phase-contrast imaging with hard X-rays*, Australian Patent Application PN 2112/95 (1995); PCT Patent Application PCT/AU96/00178 (1996).

S.W. Wilkins, T.E. Gureyev, D. Gao, A. Pogany, and A.W. Stevenson, *Phase-contrast imaging using polychromatic hard X-rays*, Nature **384**, 335–338 (1996).

E. Wolf, *A macroscopic theory of the interference and diffraction of light from finite sources. I. Fields with narrow spectral range*, Proc. Roy. Soc. Ser. A **225**, 96–111 (1954).

E. Wolf, *Is a complete determination of the energy spectrum of light possible from measurements of the degree of coherence?*, Proc. Phys. Soc. **80**, 1269–1272 (1962).

N. Wu, *The maximum entropy method*, Springer-Verlag, New York (1997).

M. Yabashi, K. Tamasaku, and T. Ishikawa, *Characterization of the transverse coherence of hard synchrotron radiation by intensity interferometry*, Phys. Rev. Lett. **87**, 140801 (2001).

M. Yabashi, K. Tamasaku, and T. Ishikawa, *Intensity interferometry for the study of X-ray coherence*, Phys. Rev. A **69**, 023813 (2004).

L. Yaroslavsky and M. Eden, *Fundamentals of digital optics*, Birkhauser, Boston, MA (1996).

F. Zernike, *Phase contrast, a new method for the microscopic observation of transparent objects*, Physica **9**, 686–693 (1942).

5

Singular X-ray optics

In this chapter we examine the singularities of various theories describing X-ray radiation. We place particular emphasis on the singularities of the scalar wave theory known as 'nodal lines', these being lines along which the intensity of the wave vanishes. To a lesser extent we examine singularities of the ray theory known as 'caustics', at which the theory predicts an infinite intensity.

A number of familiar phenomena in coherent X-ray optics are viewed in a more sophisticated light as a result of acquaintance with the notions of singular X-ray optics. These phenomena include speckle patterns, Airy rings, the fine structure of caustics, the focal region of imperfect lenses, and the interference of multiple plane waves. It is hoped that the present chapter will serve as an introduction to the theory of singular optics, which has been rather well developed in other fields (e.g. visible-light optics) but which, to date, has received relatively little attention from the X-ray optics community.

Section 5.1 gives an introduction to vortices in complex scalar fields, with some emphasis on the fact that they are nigh-ubiquitous rather than exotic. The notion of nodal lines is examined in Section 5.2, in the general setting of continuous complex scalar wave-fields whose equation of evolution is unspecified. We see in Section 5.3 that these nodal lines may be associated with vortex cores, about which optical energy 'swirls', in a manner somewhat analogous to smoke swirling about a smoke ring or a hurricane whirling about its central axis. The analysis of Sections 5.2 and 5.3 is powerful, insofar as it obtains certain results of a very general character which are independent of the functional form of the differential equation governing the evolution of a continuous complex scalar wave-field. Having said this, details of the dynamics of vortices will depend on the particular form of differential equation obeyed by the wave-field in which they are embedded. Accordingly, in Section 5.4 we consider exact polynomial vortex solutions to the d'Alembert wave equation governing a complex scalar electromagnetic field. While these solutions diverge when one is sufficiently far from the nodal line threading an associated vortex, they may be regarded as good local descriptors of certain vortical electromagnetic scalar fields. Vortex dynamics, which are rather reminiscent of the dynamics of elementary particles, are the subject of Section 5.5. Therein we study vortex creation and annihilation (Section 5.5.1), vortex stability and decay (Section 5.5.2), and the response of vortices to the background scalar field in which they are immersed (Section 5.5.3). We then consider means of creating X-ray wave-field vortices, in Section 5.6. Such methods include vortex creation via the interference of three plane waves (Section 5.6.1), the use of synthetic holograms (Section 5.6.2), and spiral phase masks (Section 5.6.3). Vortices may also be formed spontaneously, as discussed in Section 5.6.4. Certain other classes of wave-theory singularity, besides vortices, are discussed in Section 5.7. Lastly, Section 5.8 introduces the singularities of the ray theory known as 'caustics', drawing several connections between these singularities and those of the wave theory. The notion of a 'singularity hierarchy' is also examined. Here, a singularity is viewed as a breakdown of a given theory of optics, ushering in the need for a more general theory. The passage from ray to wave optics, or from classical wave optics to quantum optics, is discussed in light of this idea.

5.1 Vortices in complex scalar fields

Vortices abound in the natural world—of the many examples which immediately spring to mind let us single out the eddies in a country stream or a plume of smoke, hurricanes, whirlpools, spiral galaxies, the angular-momentum eigenstates of the hydrogen atom, vortices in uncharged superfluids such as liquid helium and charged superfluids such as type-II superconductors in the Meissner state, the red spot of Jupiter, and the vortex state of Bose–Einstein condensates. As these examples are intended to suggest, the vortex is ubiquitous rather than exotic—indeed, we shall argue on rather powerful topological grounds that vortices are present in most non-trivial complex scalar wave-fields.

5.2 Nodal lines

In what follows much use will be made of topological arguments, in which one works with wave-field form rather than with quantity. As shall be seen, these rather powerful arguments allow one to draw a number of conclusions based solely on the continuity and single-valuedness of a given complex wave-function, without needing to make reference to the particular form of partial differential equation that it obeys. Indeed, such topological arguments can be applied to a rich variety of time-dependent and time-independent complex scalar fields, including fields which obey non-linear equations (as will be the case, for example, for intense X-ray wave-fields in a material medium).

Following an argument due to Dirac (1931) let us consider a complex scalar function $\psi(x, y, \tau_1, \tau_2, \cdots)$, which depends on the Cartesian spatial coordinates x, y, together with a number of other continuous real control parameters $\tau_1, \tau_2, \cdots$. Examples of these control parameters include free-space propagation distance z or evolution time t for both time-dependent and time-independent fields, defocus, spherical aberration, thickness of medium (such as a crystal) through which a given wave-field travels, energy of a given monochromatic field which is input into a coherent imaging system, etc. (cf. Section 4.1).

Suppose, in the first instance, that all control parameters have a certain fixed value. Dropping the explicit functional dependence on these fixed control parameters, denote the resulting complex wave-field by $\psi(x, y) = \sqrt{I(x, y)} \exp[i\phi(x, y)]$. Assuming there are no sharp boundaries present, $\psi(x, y)$ must be both continuous and single-valued over the xy plane. Since $\psi(x, y)$ is continuous, the intensity

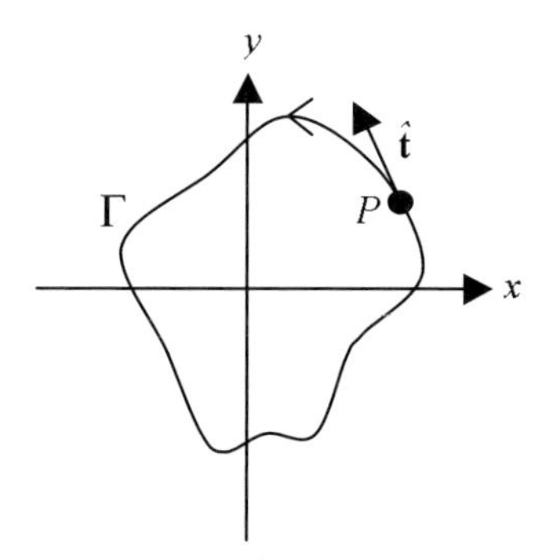

Fig. 5.1. Simple closed curve Γ in the xy plane, which is traversed in an anti-clockwise sense, and along every point of which $|\psi(x, y, \tau_1, \tau_2, \cdots)|^2$ is strictly positive. The unit tangent vector at a point P is denoted by $\hat{\mathbf{t}}$.

$I(x, y) = |\psi(x, y)|^2$ is also continuous over the xy plane. Draw a simple closed curve Γ in this plane (i.e. a continuous differentiable closed curve with no self-intersections), over every point of which the intensity is non-zero: see Fig. 5.1. Since a complex number is unchanged by the addition of any integer multiple of 2π to its phase, one is left with the possibility that the single-valued complex function $\psi(x, y)$ may nevertheless have a multi-valued phase $\phi(x, y)$. Therefore, as one traverses the path Γ once (in an anti-clockwise manner, by convention), the phase of the wave-function may change by an integer multiple m of 2π:

$$\oint_\Gamma d\phi = \oint_\Gamma \nabla\phi \cdot \hat{\mathbf{t}}\; ds = 2\pi m. \tag{5.1}$$

Here $\hat{\mathbf{t}}$ is a unit tangent vector to the path Γ in the xy plane, and ds is a differential element of arc length along this path. As shall be seen in due course, the non-vanishing of m heralds the presence of a wave-field vortex.

Now let us introduce an extra dimension into the space occupied by $\psi(x, y)$, by keeping fixed all but one of the continuous real control parameters $\tau_1, \tau_2, \cdots$. The resulting wave-function will be written as $\psi(x, y, \tau)$, with τ denoting the control parameter which is allowed to vary.[140] If τ is allowed to vary between the limits $\tau_{\rm MIN} \leq \tau \leq \tau_{\rm MAX}$, and if this interval is such that there are no non-differentiable terms in the partial differential equation governing the evolution of $\psi(x, y, \tau)$ with τ (Allen *et al.* 2001a), then $\psi(x, y, \tau)$ will be a continuous function of x, y, and τ. In addition, $\psi(x, y, \tau)$ will be taken to be a single-valued complex function of these variables.

Draw a simple closed curve Γ in the $xy\tau$ space, over every point of which the intensity $|\psi(x, y, \tau)|^2$ is strictly positive. With this appropriately generalized path

[140]More generally one could consider a non-self-intersecting smooth curve in the control space coordinatized by $\tau_1, \tau_2, \cdots$, with the curve being parameterized by the real number τ. If this is the case then the conclusions of the main text remain unchanged.

Γ, eqn (5.1) still holds. Adopting a suitably 'hydrodynamic' language, namely a terminology evocative of fluid mechanics, we will speak of the left side of eqn (5.1) as the 'circulation' of the wave-field about the path Γ. Evidently circulation is quantized,[141] as it is constrained to be an integer multiple of 2π.

Suppose that we infinitesimally and smoothly perturb the contour Γ, to give a deformed smooth contour $\Gamma + \delta\Gamma$ which is such that $|\psi(x, y, \tau)|^2$ is strictly positive at every point on the deformed contour. Continuity of the wave-function, together with strict positivity of the intensity at each point on both deformed and undeformed contours, implies the continuity of the phase of the wave-function at each point on both the deformed and undeformed contours. Since the phase is continuous at every point of both the deformed and undeformed contours, the circulation over the two infinitesimally different contours Γ and $\Gamma + \delta\Gamma$ can at most differ by an infinitesimally small amount. However, since the circulation is quantized, this implies that the circulation has not changed at all.

To proceed further, consider a continuous and smooth finite deformation of the contour Γ, from the initial simple closed curve Γ to a final simple closed curve Γ'. This finite deformation can be viewed as an infinite sequence of infinitesimally small perturbations. Suppose that the intensity of the wave-function is strictly positive, not only along every point of the initial and final curves, but also along every point of every intermediate smooth curve which exists as Γ is continuously deformed to Γ'. Evidently, the circulation is unchanged as one passes between any neighbouring members of our sequence of curves, implying that the circulation is the same for both Γ and Γ'. We therefore see that $\psi(x, y, \tau)$ possesses the same circulation for all allowed closed curves Γ which can be continuously deformed into one another, with the class of 'allowed' curves being those simple closed curves in $xy\tau$ space which are such that $|\psi(x, y, \tau)|^2$ is non-vanishing over every point of the curve.

Consider once more a given allowed curve Γ in $xy\tau$ space, over which the circulation is non-vanishing, and over every point of which $|\psi(x, y, \tau)|^2$ is strictly positive. We now show that this curve cannot be continuously contracted to a point without encountering a zero of the wave-field $\psi(x, y, \tau)$. To arrive at this conclusion, we use a proof by contradiction. Let us (incorrectly) assume, therefore, that Γ *can* be continuously contracted to point, without ever encountering a zero of the field. Having contracted the curve Γ to a point, the circulation about the contracted curve must vanish. This is a contradiction, as it runs counter to the previous paragraph's conclusion that the circulation must be the same as the non-zero circulation obtained for the initial curve Γ. Therefore, the initial assumption is incorrect, from which we conclude that Γ cannot be contracted to a point without encountering a zero of the field.

[141]There are some evident parallels here, with the Bohr–Sommerfeld quantization rule of the old quantum theory. Similarly, there are parallels with the standard argument for obtaining hydrogenic wave-function solutions to the Schrödinger equation, which leads to the quantization of orbital angular momentum.

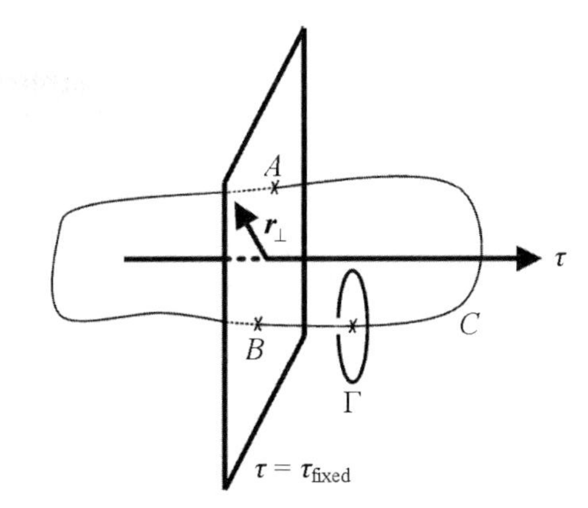

Fig. 5.2. Nodal line ABC in $xy\tau$ space. The wave-function $\psi(x, y, \tau)$ vanishes at each point on the nodal line. Fixing τ to a particular value, $\tau = \tau_{\text{fixed}}$, defines a plane in $xy\tau$ space, with $\mathbf{r}_\perp \equiv (x, y)$ being a coordinate vector in this plane. The nodal line punctures this plane at points A and B, which are identified with a pair of counter-propagating wave-field vortices. Note that the simple closed curve Γ can never be unthreaded from the nodal line ABC, without a point on Γ crossing a zero of the field. Image taken from Allen *et al.* (2001b). Used with permission.

This conclusion leads us to the concept of 'nodal lines', which have been rather evocatively dubbed 'lines of nothingness' (Nye 1999). To introduce this notion, consider Fig. 5.2. Here we see an allowed curve Γ in $xy\tau$ space, which possesses non-zero circulation. We have just shown that this cannot be continuously deformed to a point without encountering a zero of the field. Since the space occupied by the curve is three-dimensional, this implies that the wave-function must vanish over a closed curve of points, which form as it were the thread of a 'necklace' from which the 'bead' Γ can never be removed without encountering a zero of the field. This 'necklace', which forms a connected closed curve of points where the wave-field vanishes, is known as a nodal line.

Before continuing, we make three remarks: (i) Nodal lines may terminate at the sharp edges of scatterers, if they are present. (ii) In the figure, we have shown an elementary example of a nodal line, which constitutes a simple closed curve. In addition to forming such simple closed curves, more complex topologies are possible, such as branch points and knots (Berry and Dennis 2000a; Freund 2000; Berry 2001; Berry and Dennis 2001). As an example of the former, our preceding analysis evidently allows an $m = 2$ nodal line to branch into a pair of $m = 1$ nodal lines. (iii) A nodal line which extends to infinity need not be closed (although, if spatial infinity is treated as a single point, such nodal lines may also be viewed as closed).

In view of the above comments, we are led to the possibility of a complicated nodal-line network permeating a given complex field. These nodal-line networks may possess non-trivial topologies such as branches or knots. Individual nodal lines may only terminate on sharp scattering edges, or at spatial infinity, forming closed loops otherwise. Whether such a network exists depends on the particular form of differential equation obeyed by the field, together with the initial conditions that are specified. Having said this, many familiar fields lacking particular symmetries do possess such a tangled web of dislocation lines. Examples include: (i) turbulent states of trapped Bose–Einstein condensates; (ii) the Meissner state of a type-II superconductor; (iii) the speckled pattern which is observed when a laser beam is reflected from a rough wall or when a coherent X-ray wave is passed through a rough transparent screen; (iv) partially coherent scalar fields possessing no particular spatial symmetries. Note that, in the last-mentioned example, the nodal-line network will evolve through time in a complex manner. A number of further examples, in the field of coherent X-ray optics, will be given later in this chapter.

5.3 Nodal lines are vortex cores

At this point the reader may be asking what the preceding analyses have to do with wave-field vortices. The connection is rather simple: nodal lines are vortex cores. To see this let us consider Fig. 5.2 once again. Imagine that the path Γ, over every point of which the modulus of the wave-function is strictly positive and around which the circulation has the non-zero integer value m, is contracted to a very small path enclosing the nodal line. Suppose the path is such that it lies entirely within a small simply-connected region Σ of a certain plane, which is pierced by the nodal line exactly once, at position $O \in \Sigma$. Introduce plane polar coordinates (r, θ) in Σ, with r denoting radial distance from O, and θ denoting azimuthal angle. Since the circulation is non-zero, this implies a singular character for the phase $\phi(r, \theta)$ of the wave-field over the region Σ, with the phase expressible in the form:

$$\phi(r, \theta) = m\theta + g(r, \theta). \tag{5.2}$$

Here, $g(r, \theta)$ is a continuous and differentiable function that makes no contribution to the circulation evaluated in eqn (5.1).

Since the circulation (around $\Gamma \in \Sigma$) is unchanged by a continuous deformation of the phase effected by the form of $g(r, \theta)$, it is evidently the 'topological' part $m\theta$ of the phase which contributes to the non-zero circulation. A sketch of this helicoidal 'vortex' phase is shown in Fig. 5.3. The top surface of the rectangular prism in (a) contains the region Σ, in which is embedded the contour Γ. The height of the multi-valued helicoidal surface represents the phase, as a function of position in Σ. One branch of this surface, corresponding to phase values between 0 and 2π, is given in the contour plot shown in panel (b). The arrows represent a plan view of the direction of flow of optical energy, which is normal to the surfaces of constant phase. Evidently, these energy-flow vectors constitute a vortex, in the hydrodynamic sense of the term, with the core located on the nodal line. Indeed, small particles which are placed on a nodal line

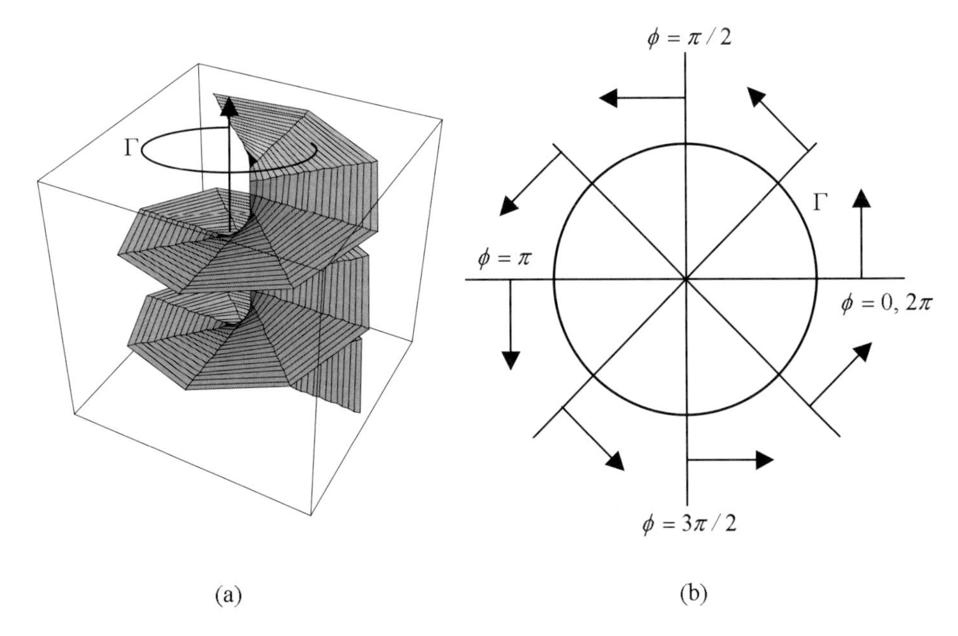

Fig. 5.3. (a) Multi-valued helicoidal surface of constant phase, which represents a wave-field vortex threaded by a nodal line, together with a contour Γ, which may be used to calculate the circulation about the nodal line. After Mair *et al.* (2001). (b) Plan view of single branch of helicoidal phase surface, showing the contour Γ, which is threaded by the vortex core. Lines of constant phase emanate radially outward from the vortex core. Arrows denote local energy flow vectors in plan view, which are perpendicular to lines of constant phase.

of a sufficiently intense electromagnetic field, can be made to rotate under the torque imparted by the optical vortex on the particle (see, for example, He *et al.* (1995)).

5.4 Polynomial vortex solutions to the d'Alembert equation

Thus far we have relied on topological arguments in our discussion of nodal lines, making use of the continuity and single-valuedness of complex scalar wave-functions without reference to the particular partial differential equations that govern their evolution. While such arguments allow us to proceed up to a certain point in a very general setting, one must invoke particular equations of evolution in order to study such questions as the structure of the wave-function in the vicinity of vortex cores, together with vortex dynamics.

In this section, we restrict ourselves to complex wave-functions $\Psi(x, y, z, t)$ which are solutions to the d'Alembert wave equation for time-dependent complex scalar electromagnetic fields (see eqn (1.13)):

$$\left(\frac{1}{c^2}\frac{\partial^2}{\partial t^2} - \nabla^2\right)\Psi(x,y,z,t) = 0. \tag{5.3}$$

Here, (x, y, z) are the usual triad of Cartesian coordinates, t is time, c is the speed of light in vacuum, and $\nabla^2 \equiv \partial^2/\partial x^2 + \partial^2/\partial y^2 + \partial^2/\partial z^2$ is the three-dimensional Laplacian.

The main aim of this section is to construct exact polynomial solutions to the above d'Alembert equation, corresponding to vortical fields. To this end, consider the following polynomial $q_\pm(x, y)$ in $x + iy$:

$$q_\pm(x,y) = \sum_{n=0}^{\infty}(x \pm iy)^n q_n, \tag{5.4}$$

where q_n are arbitrary complex coefficients indexed by the integers $n = 0, 1, 2, \cdots$. It is easy to show that:

$$\left(\frac{\partial^2}{\partial x^2} + \frac{\partial^2}{\partial y^2}\right) q_\pm(x,y) = 0, \tag{5.5}$$

namely that $q_\pm(x, y)$ is a solution to the two-dimensional Laplace equation (see, for example, Freund (1999)). One can also readily see that the d'Alembert equation (5.3) possesses the following elementary z-directed plane wave solutions $\Psi_z(x, y, z, t)$:

$$\Psi_z(x,y,z,t) \equiv \Psi_z(z,t) = \sqrt{I_0}\exp[i(kz - \omega t)], \tag{5.6}$$

where I_0 is a real constant indicative of the uniform intensity of the plane wave with wave-number $k = \omega/c$.

The product of the wave-fields $\Psi_z(z, t)$ and $q_\pm(x, y)$ is a solution to the d'Alembert equation, as shown below:

$$\begin{aligned}
&\left(\frac{1}{c^2}\frac{\partial^2}{\partial t^2} - \nabla^2\right)[\Psi_z(z,t)q_\pm(x,y)] \\
&= \frac{1}{c^2}\frac{\partial^2}{\partial t^2}[\Psi_z(z,t)q_\pm(x,y)] - \nabla^2[\Psi_z(z,t)q_\pm(x,y)] \\
&= \frac{1}{c^2}\left[\frac{\partial^2}{\partial t^2}\Psi_z(z,t)\right]q_\pm(x,y) - \Psi_z(z,t)\frac{\partial^2}{\partial x^2}q_\pm(x,y) \\
&\quad - \Psi_z(z,t)\frac{\partial^2}{\partial y^2}q_\pm(x,y) - \left[\frac{\partial^2}{\partial z^2}\Psi_z(z,t)\right]q_\pm(x,y) \\
&= q_\pm(x,y)\left(\frac{1}{c^2}\frac{\partial^2}{\partial t^2} - \frac{\partial^2}{\partial z^2}\right)\Psi_z(z,t) - \Psi_z(z,t)\left(\frac{\partial^2}{\partial x^2} + \frac{\partial^2}{\partial y^2}\right)q_\pm(x,y) \\
&= 0.
\end{aligned} \tag{5.7}$$

Consider the following class of wave-functions $\Psi_\pm(x, y, z, t)$, which obey eqn (5.3):

$$\Psi_{\pm}(x,y,z,t) \equiv \Psi_z(z,t)q_{\pm}(x,y)$$
$$= \sqrt{I_0}\exp[i(kz-\omega t)]\sum_{n=0}^{\infty}(x\pm iy)^n q_n. \quad (5.8)$$

Write the complex number $x \pm iy$ in the polar form $x \pm iy = r\exp(\pm i\theta)$, where $r = \sqrt{x^2+y^2}$ and θ denotes the angle that the complex number $x+iy$ makes with the positive real axis in the complex plane. At the risk of stating the obvious, note that: (i) r is equal to both the modulus of the complex number $x \pm iy$ and the radial distance of the point (x,y) from the origin of two-dimensional Cartesian coordinates; (ii) θ is equal to both the phase of the complex number $x+iy$ (modulo 2π) and the polar angle which (x,y) makes with the positive x-axis of two-dimensional Cartesian coordinates. Bearing this in mind, and setting q_n equal to unity for a specified integer m and equal to zero otherwise, the cylindrical-polar-coordinates form of eqn (5.8) becomes (Berry 1980):

$$\Psi_{\pm}^{m}(r,\theta,z,t) = r^m\sqrt{I_0}\exp[i(kz-\omega t\pm m\theta)]. \quad (5.9)$$

This solution to the d'Alembert equation possesses a nodal line coinciding with the z-axis, about which the field has a circulation of $\pm m$. The phase of this solution is evidently a special case of eqn (5.2), with the intensity of the wave-field proportional to the mth square of the radial distance from the core. While one may object that these solutions are unphysical because the intensity grows without bound as one increases the radial distance from the core, this is of no concern since we are interested in eqn (5.9) as a certain local model for a vortex core (Berry 1980; Berry and Dennis 2001).

We close this section by obtaining an expression for the current vector associated with the polynomial vortex in eqn (5.9). Multiply the d'Alembert equation by the complex conjugate of $\Psi(x,y,z,t)$. Then, multiply the complex conjugate of the d'Alembert equation by $\Psi(x,y,z,t)$. Subtract the resulting pair of equations to give:

$$\frac{\Psi^*}{c^2}\frac{\partial^2\Psi}{\partial t^2} - \Psi^*\nabla^2\Psi - \frac{\Psi}{c^2}\frac{\partial^2\Psi^*}{\partial t^2} + \Psi\nabla^2\Psi^* = 0, \quad (5.10)$$

where $\Psi \equiv \Psi(x,y,z,t)$, and an asterisk superscript denotes complex conjugation. Next, make use of the following readily proved identities:

$$\frac{\partial}{\partial t}\left(\Psi^*\frac{\partial\Psi}{\partial t} - \Psi\frac{\partial\Psi^*}{\partial t}\right) = \Psi^*\frac{\partial^2\Psi}{\partial t^2} - \Psi\frac{\partial^2\Psi^*}{\partial t^2}, \quad (5.11)$$
$$\nabla\cdot(\Psi\nabla\Psi^* - \Psi^*\nabla\Psi) = \Psi\nabla^2\Psi^* - \Psi^*\nabla^2\Psi, \quad (5.12)$$

in order to transform eqn (5.10) into:

$$\nabla \cdot (\Psi \nabla \Psi^* - \Psi^* \nabla \Psi) + \frac{\partial}{\partial t} \left(\frac{1}{c^2} \Psi^* \frac{\partial \Psi}{\partial t} - \frac{1}{c^2} \Psi \frac{\partial \Psi^*}{\partial t} \right) = 0. \tag{5.13}$$

This has the form of a continuity equation, implying that the first term in brackets is proportional to the current-flow vector for the wave-field, while the second term in brackets is proportional to the energy density. Denoting the associated proportionality constant by the complex number Θ, we conclude that the current vector $\mathbf{j}$ is:

$$\mathbf{j} = \Theta (\Psi \nabla \Psi^* - \Psi^* \nabla \Psi) . \tag{5.14}$$

Next, we state that Θ is a pure imaginary number, a fact which can be demonstrated by first substituting the elementary z-directed plane wave (5.6) into the expression

$$\frac{\Theta}{c^2} \left(\Psi^* \frac{\partial \Psi}{\partial t} - \Psi \frac{\partial \Psi^*}{\partial t} \right) \tag{5.15}$$

for the energy density, and then demanding that the result be real. Hence we may write $\Theta = i\eta$, where η is a real number. If we now write Ψ in terms of its intensity I and phase ϕ as

$$\Psi = \sqrt{I} \exp(i\phi), \tag{5.16}$$

then the current vector in eqn (5.14) becomes the following :

$$\mathbf{j} = i\eta \left[\sqrt{I} e^{i\phi} \nabla \left(\sqrt{I} e^{-i\phi} \right) - \sqrt{I} e^{-i\phi} \nabla \left(\sqrt{I} e^{i\phi} \right) \right] = 2\eta I \nabla \phi. \tag{5.17}$$

Evidently the current vector is orthogonal to the surfaces of constant phase, as was assumed in the construction shown in Fig. 5.3(b). Further, we can now write down an expression for the energy-flow vector $\mathbf{j}_{\pm}^{m}(r, \theta, z, t)$ associated with the polynomial vortex in eqn (5.9):

$$\mathbf{j}_{\pm}^{m}(r, \theta, z, t) = 2\eta I_0 r^{2m} \nabla (kz - \omega t \pm m\theta) = 2\eta I_0 r^{2m} \left(k\hat{\mathbf{z}} \pm \frac{m\hat{\theta}}{r} \right). \tag{5.18}$$

Here $\hat{\mathbf{z}}$ denotes a unit vector in the direction of the positive z-axis, $\hat{\theta}$ is a unit vector in the direction of increasing θ, and we have used the fact that $\nabla \theta = \hat{\theta}/r$. The vortical nature of $\mathbf{j}_{\pm}^{m}(r, \theta, z, t)$ is evident from the term proportional to $\hat{\theta}$ that appears in the equation above.

5.5 Vortex dynamics

Here we explore three separate aspects of vortex dynamics, all of which are tantalisingly suggestive of the dynamics of elementary particles: (i) vortex nucleation and annihilation; (ii) decay of unstable high-order vortices into stable low-order vortices; (iii) interaction of vortices with the background wave-field in which they are embedded.

5.5.1 *Vortex nucleation and annihilation*

Consider Fig. 5.2 once again, this time in the context of vortex nucleation and annihilation for complex wave-functions $\psi(x, y, \tau_1, \tau_2, \cdots)$ which depend continuously on two spatial coordinates (x, y), together with a number of real control parameters $\tau_1, \tau_2, \cdots$. With the exception of the single control parameter denoted by τ in this diagram, all other control parameters are considered to be kept fixed.[142] The diagram shows a closed nodal line puncturing the plane $\tau = \tau_{\text{fixed}}$ at the points A and B. Let Γ_B be the small simple closed curve which results when Γ is slid along the nodal line towards the point B, such that Γ_B lies in the plane $\tau = \tau_{\text{fixed}}$ and encloses the point B. Similarly, let Γ_A be the small simple closed curve which results when Γ is slid along the nodal line towards the point A, such that Γ_A lies in the plane $\tau = \tau_{\text{fixed}}$ and encloses the point A. The same circulation $2\pi m$ is possessed by all closed curves that can be continuously deformed into one another without encountering a zero of the field, so:

$$\oint_{\Gamma_A} d\phi = \oint_{\Gamma_B} d\phi = 2\pi m. \tag{5.19}$$

However, when viewed from the right side (i.e. from the half-space $\tau > \tau_{\text{fixed}}$), the curve Γ_B is traversed in an anti-clockwise fashion, while Γ_A is traversed in a clockwise sense. We conclude that the 'point vortices', in the plane $\tau = \tau_{\text{fixed}}$, form a vortex–antivortex pair.

Let us study the process of vortex nucleation in more detail, in order to deduce that there is no meaning to the notion of vortex–antivortex nucleation at a point (Berry 1998; Freund 2000). For concreteness, rather than considering a wave-function which depends on two spatial variables (x, y) and a number of other control parameters $\tau_1, \tau_2, \cdots$, let us for the purposes of illustration consider a complex scalar wave-function $\psi(x, y, z)$ which depends on the Cartesian coordinates (x, y, z) in three-dimensional space.

Consider the scenario shown in Fig. 5.4(a). Here we see a smooth closed nodal line of the complex wave-function $\psi(x, y, z)$, with the z direction of the Cartesian coordinate system (x, y, z) as shown. Several parallel two-dimensional planes of constant z are as indicated. Over plane A there will be no discontinuities in the phase of the wave-function ψ. Over plane B however, a pair of counter-propagating point vortices will be 'nucleated', at the point C. These vortices

[142] See the first footnote to this chapter, regarding parameterized one-dimensional curves within the control space, which may also be applied in the present context.

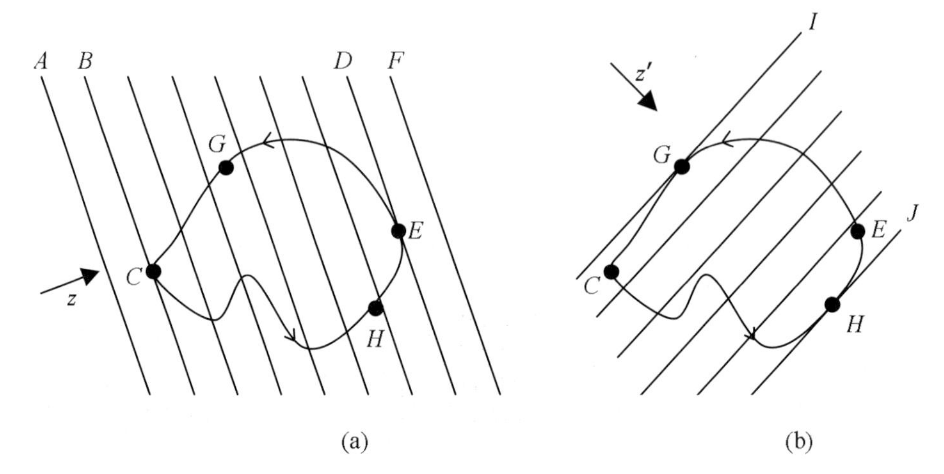

Fig. 5.4. Both (a) and (b) show the same closed nodal line, for a complex scalar wave-function ψ in three-dimensional space. In (a), the space is coordinatized by (x, y, z), while in (b) the space is coordinatized by (x', y', z'). Several planes of constant z (z') are shown, together with a vector pointing in the z (z') direction.

move apart and then coalesce, 'annihilating' once we reach plane D, at the point E. Finally, over plane F there will be no point vortices.

Now, one could certainly choose to coordinatize the space with different Cartesian coordinates (x', y', z'), as shown in Fig. 5.4(b). Several parallel two-dimensional planes of constant z' are shown, these being pierced by the same nodal line as was shown in Fig. 5.4(a). Now, according to an observer who registers the phase over the illustrated sequence of parallel planes from I to J, a pair of counter-propagating point vortices will be 'nucleated' at the point G in plane I, with the pair subsequently 'annihilating' at the point H in plane J.

According to the description accompanying Fig. 5.4(a), a counter-propagating vortex pair was 'created' at C and 'annihilated' at E, while the description accompanying Fig. 5.4(b) concluded that counter-propagating point vortices were 'nucleated' at G and 'annihilated' at H. By appropriately choosing the coordinate system, one could choose any point along the nodal line to be a point where counter-propagating point vortices are 'created' or 'annihilated'. Thus the concept of two-dimensional vortices 'nucleating' or 'annihilating' at a point does not have a precise meaning, as it depends on the set of parallel planes ('foliation') with which one chooses to 'slice up' the space. Rather than point vortices, represented by screw dislocations over two-dimensional planes lying within the three-dimensional space, one should instead consider the complete nodal line as the fundamental entity (Berry 1998; Freund 2000). Our earlier statements, regarding the 'nucleation' of counter-propagating point vortex pairs, should be tempered in light of the above discussion.

5.5.2 *Stability with respect to perturbations: decay of higher-order vortices*

Consider a coherent X-ray scalar optical wave-field that contains vortices. Suppose that one perturbs either or both of (i) the optical system used to generate such a vortical wave-field; (ii) the 'initial' two-dimensional wavefront, which was input into an optical system (cf. Section 4.1) used to generate the vortical wave-field. Are wave-field vortices stable with respect to such perturbations? Focusing on the polynomial solutions outlined in Section 5.4, we shall see that first order polynomial vortices (i.e. those with $m = \pm 1$) are stable with respect to perturbations, whereas those with higher m are unstable with respect to such perturbations. Note that we will henceforth speak of m, which appears in eqn (5.1), as the 'topological charge' of a nodal line.

Consider the first-order polynomial vortex, obtained by setting $I_0 = 1, t = 0$, and $q_n = \delta_{1n}$ (Kronecker delta) in eqn (5.8). In the plane $z = 0$, the resulting wave-function $\Psi_{\pm}(x, y, z = 0, t = 0) \equiv \psi(x, y)$ is:

$$\psi(x, y) = x \pm iy. \tag{5.20}$$

Following the explanation in Freund (1999) it is evident that the real part of $\psi(x, y)$ vanishes along the line $x = 0$ in the xy plane, as indicated by the solid line in Fig. 5.5(a), while the imaginary part of $\psi(x, y)$ vanishes along the line $y = 0$ in the xy plane, as indicated by the dashed line in Fig. 5.5(a). These two lines ('zero lines') cross at $(x, y) = (0, 0)$ ('zero crossing'), at which point both the real and imaginary parts of the wave-function vanish. This vanishing, of both the real and imaginary parts of a wave-function at a given point, is a necessary but not sufficient condition for the existence of a wave-field vortex at that point. In fact there is a vortex located at the origin of coordinates $(x, y) = (0, 0)$, as shown earlier in the chapter, where we saw that the above polynomial wave-function corresponds to a point vortex with topological charge $m = \pm 1$, which is centred at $(x, y) = (0, 0)$.

With reference to Fig. 5.5(b) let us now suppose that our wave-function is perturbed in some way, such that the two zero lines are continuously deformed in a manner consistent with the equation of motion governing the wave-function. Assuming the deformation to be such that the two zero lines still cross, the resulting zero crossing will correspond to a first-order vortex with the same topological charge as the undeformed vortex. The location of the vortex core will have shifted, in general, but this does not alter the fact that the vortex is stable with respect to such perturbations.

Stability with respect to perturbations is not enjoyed by polynomial vortices with topological charges of modulus greater than unity. For example, consider the following second-order polynomial vortex, which is a special case of eqn (5.8) with $z = t = 0$, $I_0 = 1$, and $q_n = \delta_{2n}$ (Freund 1999):

$$\psi(x, y) = (x \pm iy)^2 = x^2 - y^2 \pm 2ixy. \tag{5.21}$$

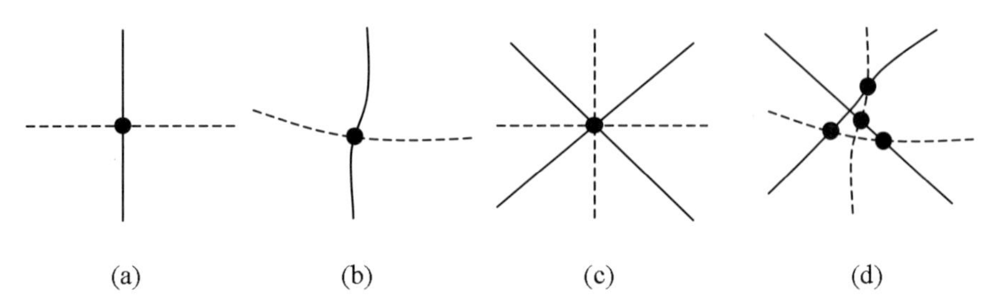

Fig. 5.5. Solid and dashed lines, respectively, denote lines along which the real or imaginary parts of certain two-dimensional wave-functions $\psi(x, y)$ vanish. Associated vortex and antivortex cores are denoted by black dots. (a) First-order polynomial vortex; (b) perturbed first-order polynomial vortex; (c) second-order polynomial vortex; (d) perturbed second-order polynomial vortex has decayed into three first-order vortices and one first-order anti-vortex.

In this case, the real part of the wave-function vanishes when $x^2 - y^2 = 0$ (i.e. along the curves $y = \pm x$), while the imaginary part vanishes when $xy = 0$ (i.e. along the curves $x = 0$ and $y = 0$). These zero lines are respectively indicated by the solid and dashed lines in Fig. 5.5(c). As is evident from Fig. 5.5(d), continuous perturbation of the real and imaginary parts of the wave-function, in a manner consistent with the equation governing the evolution of that wave-function, leads to the decay of the second-order vortex into three first-order vortices and one first-order antivortex. This phenomenon, indicative of the instability of higher-order polynomial vortices leading to their decay into stable first-order vortices and anti-vortices, has been termed a 'critical-point explosion' by Freund (1999). Such instability of higher-order vortices is not restricted to polynomial vortices—see Freund (1999), and references therein, for a discussion. Note, further, that critical-point explosions are a source of the branched nodal lines mentioned earlier in this chapter. Lastly, we note that topological charge is conserved during the process of vortex decay.

5.5.3 *Vortex interaction with a background field*

We complete our discussions on vortex dynamics, which have already considered phenomena such as vortex nucleation/annihilation and higher-order vortex decays, with a treatment of how a single first-order vortex interacts with the background field in which it is embedded. In this context, recall our earlier discussions on the stability of first-order vortices with respect to perturbations (Section 5.5.2). In the present sub-section we go a step further and ask: how does a first-order polynomial vortex move in response to amplitude or phase variations in the immediate vicinity of its vortex core?

Following the methodology outlined in Rozas *et al.* (1997), begin by assuming that we have paraxial monochromatic scalar electromagnetic radiation whose nominal direction of propagation coincides with the z-axis of a Cartesian coordinate system. As discussed in Chapter 2, such radiation obeys the paraxial (parabolic) equation, which it will prove convenient to write in the form (see eqn (2.34)):

$$\frac{\partial}{\partial z}\psi(x,y,z) = \frac{i}{2k}\nabla_{\perp}^{2}\psi(x,y,z). \tag{5.22}$$

Written in this way, we see that the transverse Laplacian of the complex disturbance, across any plane $z =$ constant, is proportional to the z rate of change of that complex wave-function. This allows us to take any 'initial' disturbance in a given plane of constant z and then study how it evolves as it propagates through a small distance δz, to first order in that quantity:

$$\begin{aligned}\psi(x,y,z=\delta z) &= \psi(x,y,z=0) + \delta z\frac{\partial}{\partial z}\psi(x,y,z)|_{z=0} + \mathrm{O}(\delta z^2)\\ &= \psi(x,y,z=0) + \frac{i\delta z}{2k}\nabla_{\perp}^{2}\psi(x,y,z=0) + \mathrm{O}(\delta z^2).\end{aligned} \tag{5.23}$$

This formula will allow us to study the effects of certain perturbations on the polynomial first-order point vortex given in eqn (5.20). For example, if we multiply this point vortex by the perturbation $1+\alpha x$, where α is a small real number, we have the following amplitude-perturbed vortex in the plane $z=0$:

$$\psi(x,y,z=0) = (1+\alpha x)(x \pm iy). \tag{5.24}$$

Substituting this boundary value into eqn (5.23), we can compute the wave-function that results when the field is propagated through an infinitesimally small distance δz, to first order in that quantity:

$$\psi(x,y,z=\delta z) = x + \alpha x^2 + i\left(\pm y \pm \alpha x y + \frac{\alpha\delta z}{k}\right) + \mathrm{O}(\delta z^2). \tag{5.25}$$

Separating out real and imaginary parts, we see that the real part of the wave-function vanishes along the line $x=0$, while the imaginary part vanishes along the curve:

$$y = \mp\frac{\alpha\delta z}{k(1+\alpha x)}. \tag{5.26}$$

The vortex core, for a given δz, is located at the intersection of these two curves, which is at:

$$(x,y) = \left(0, \mp\frac{\alpha\delta z}{k}\right). \tag{5.27}$$

We see that the vortex moves in a direction perpendicular to the background amplitude gradient, that is, the vortex moves in the direction of the y-axis in response to a local variation in amplitude along the x direction. Vortices and antivortices move in opposite directions. When α vanishes, the core runs along the z-axis.

Having studied the response of first-order polynomial vortices to a linear background amplitude gradient, let us consider the response of such vortices to a linear background phase gradient. To this end consider the following unpropagated value for a phase-perturbed first-order polynomial vortex in the plane $z = 0$:

$$\psi(x, y, z = 0) = \exp(i\beta x)(x \pm iy). \tag{5.28}$$

Here, β is a small real number equal to the phase gradient, at the vortex core, in the x direction. Substituting eqn (5.28) into eqn (5.23), we see that the infinitesimally propagated wave-function is:

$$\psi(x, y, z = \delta z) = \exp(i\beta x)\left[x \pm iy - \frac{\beta\delta z}{k}\left(1 + \frac{i\beta x}{2} \pm \frac{\beta y}{2}\right)\right] + \mathrm{O}(\delta z^2). \tag{5.29}$$

Setting the real part of $\psi(x, y, z = \delta z)$ to zero we see that, for a propagation distance δz, the real part of the wave-function vanishes along the line:

$$x = \frac{\beta\delta z}{k}\left(1 \pm \frac{\beta y}{2}\right). \tag{5.30}$$

Similarly, setting the imaginary part of the wave-function to zero, for a given propagation distance δz, we see that the imaginary part of the wave-function vanishes along the line:

$$y = \pm\frac{\beta^2 x\delta z}{2k}. \tag{5.31}$$

The vortex core is located at the intersection of these two lines. To order δz, and for a given propagation distance δz, the vortex core is located at

$$(x, y) = \left(\frac{\beta\delta z}{k}, 0\right). \tag{5.32}$$

We conclude that both polynomial vortices and antivortices, of first order, travel in the direction of the linear background phase gradient. Note that this background phase gradient may be produced by another vortex, in which case vortices may orbit one another.

The above methodology is rather general, and may also be used to study: (i) transverse phase-amplitude perturbations of linear and higher orders; together with (ii) critical-point explosions; and (iii) response of vortices, to phase and/or amplitude background perturbations, for wave-fields that obey partial differential equations other than the paraxial equation.

5.6 Means of generating wave-field vortices

Having outlined some of the key theoretical ideas necessary for an appreciation of the role played by vortices in coherent X-ray optics, we are ready to outline four different means of generating wave-field vortices in coherent X-ray optical systems. The first of these methods is remarkable simple, indicating that a lattice of vortices can be formed from the coherent superposition of three plane waves, which are inclined with respect to one another. The second method will outline a form of 'forked' diffraction grating that encodes vortices in its various diffracted orders. The third method passes from a diffractive to a refractive optical element, this being the 'spiral-phase plate' used to imprint a vortical phase on a coherent plane wave passing through it. The last method to be considered uses free-space propagation in order to spontaneously nucleate phase vortices in a coherent X-ray optical field.

5.6.1 *X-ray vortex generation by interference of three coherent plane waves*

For our first means of generating coherent X-ray vortices, we consider the remarkably simple method of interfering three coherent plane waves. As we shall see, this method is able to yield a lattice of infinitely many vortices. Our treatment is adapted from Masajada and Dubik (2001).

As discussed in Chapter 1 the spatial wave-function $\psi(x, y, z)$ for a monochromatic scalar electromagnetic wave-field, not necessarily paraxial, is governed by the Helmholtz equation given in eqn (1.16):

$$(\nabla^2 + k^2)\psi(x, y, z) = 0. \tag{5.33}$$

Here, $k = 2\pi/\lambda$, where λ is the radiation wavelength, and (x, y, z) is the usual triad of Cartesian coordinates. One can readily see that this equation possesses the elementary plane-wave solutions:

$$\exp[i(k_x x + k_y y + k_z z)], \tag{5.34}$$

where

$$k_x^2 + k_y^2 + k_z^2 = k^2. \tag{5.35}$$

This last equation can be solved for k_z, to give:

$$k_z = \sqrt{k^2 - k_x^2 - k_y^2}. \tag{5.36}$$

Note that the positive square root has been chosen, as we shall only be considering plane waves that are forward propagating with respect to the z-axis. Hence our elementary plane waves may be written in the form (cf. eqn (1.20)):

$$\exp\left[i\left(k_x x + k_y y + z\sqrt{k^2 - k_x^2 - k_y^2}\right)\right]. \tag{5.37}$$

We shall henceforth assume that $k^2 > k_x^2 + k_y^2$. This ensures that: (i) k_z is real and (ii) we only consider plane waves whose modulus is constant for all (x, y, z) in a space free of sources (i.e. evanescent plane waves are excluded from the analysis).

For the moment, restrict our consideration to the plane $z = 0$. Consider the z-directed plane wave $A\exp(ikz)$, which has a value of A at every point in the plane $z = 0$, where A is a complex constant. Add a second, tilted plane wave to the first. Now, we can always choose the x-axis such that the second plane wave is tilted only along the x-axis. Hence we write the wave-field, due to the second plane-wave in the plane $z = 0$, as $B\exp(ik_x^B x)$, where B is a complex constant and $k_x^B < k$ is a real constant (this may be positive or negative, but not zero). We then add a third plane wave, which produces a disturbance $C\exp[i(k_x^C x + k_y^C y)]$ in the plane $z = 0$, where k_x^C and k_y^C are real non-zero constants with $(k_x^C)^2 + (k_y^C)^2 < k^2$, and C is a complex constant. Bearing in mind the results of the previous paragraph, we can now write down our three-dimensional disturbance $\psi(x, y, z)$, which consists of a sum of three elementary plane-wave solutions to the Helmholtz equation, as:

$$\begin{aligned}\psi(x, y, z) =& A\exp(ikz) + B\exp\left[i\left(k_x^B x + z\sqrt{k^2 - (k_x^B)^2}\right)\right] \\ &+ C\exp\left[i\left(k_x^C x + k_y^C y + z\sqrt{k^2 - (k_x^C)^2 - (k_y^C)^2}\right)\right].\end{aligned} \tag{5.38}$$

In the ensuing paragraphs we shall show the range of conditions under which the above superposition, of three elementary plane-wave solutions to the Helmholtz equation, is able to yield a lattice of wave-field vortices ('screw dislocations' in the phase). Choose any plane of constant z, denoted by $z = z_0$. Let us absorb all z-dependent phase factors into appropriately redefined versions of A, B, and C, which are respectively denoted D, E, and F:

$$\begin{aligned}D \equiv A\exp(ikz_0), \qquad E \equiv B\exp\left[iz_0\sqrt{k^2 - (k_x^B)^2}\right], \\ F \equiv C\exp\left[iz_0\sqrt{k^2 - (k_x^C)^2 - (k_y^C)^2}\right].\end{aligned} \tag{5.39}$$

Write the complex numbers D, E, and F in terms of their modulus and phase as:

$$D = |D|\exp(i\phi_D), \qquad E = |E|\exp(i\phi_E), \qquad F = |F|\exp(i\phi_F). \tag{5.40}$$

Therefore, in the plane $z = z_0$, eqn (5.38) becomes:

$$\psi(x,y,z=z_0) = |D|\exp(i\phi_D) + |E|\exp\left[i(k_x^B x + \phi_E)\right] + |F|\exp\left[i\left(k_x^C x + k_y^C y + \phi_F\right)\right]. \quad (5.41)$$

Factoring out $\exp(i\phi_D)$ from the right side of this equation, and then introducing the real quantities:

$$\alpha \equiv \phi_E - \phi_D, \qquad \beta \equiv \phi_F - \phi_D, \quad (5.42)$$

we obtain

$$\psi(x,y,z=z_0) = \exp(i\phi_D)\{|D| + |E|\exp\left[i(k_x^B x + \alpha)\right] + |F|\exp\left[i\left(k_x^C x + k_y^C y + \beta\right)\right]\}. \quad (5.43)$$

A necessary condition for a point vortex to be present, at (x, y, z_0), is that the quantity in braces vanish. Thus we seek ordered pairs $(\tilde{x}, \tilde{y})$, if they exist, which solve the following equation:

$$|D| + |E|\exp\left[i(k_x^B \tilde{x} + \alpha)\right] + |F|\exp\left[i\left(k_x^C \tilde{x} + k_y^C \tilde{y} + \beta\right)\right] = 0. \quad (5.44)$$

Assume that $|D| > |E| > |F|$, which implies no loss of generality if the intensity of each of the three interfering plane waves is different.[143] With reference to the geometric construction shown in Fig. 5.6, we separately consider the following three cases: Case #1: $|D| < |E| + |F|$; Case #2: $|D| = |E| + |F|$; Case #3: $|D| > |E| + |F|$.

Case #1: This case, where $|D| > |E| > |F|$ and $|D| < |E| + |F|$, is shown in Fig. 5.6(a). The first term of eqn (5.44) is represented by the complex-plane phasor OA. Next, noting that the second term of eqn (5.44) is a complex number of modulus $|E|$, we draw a circle of radius $|E|$ which is centred at A. The phasor, representing the second term of the equation, can be placed so as to reach from A to some point on the perimeter of the circle that has just been described. Next, we draw a circle of radius $|F|$ which is centred at O. Since $|D| < |E|+|F|$, our two circles will intersect at two points denoted by B and C. Evidently, if eqn (5.44) is to be obeyed, the corresponding phasors of the three terms on the left side must form either the closed triangle OAB, or the closed triangle OAC. Since we are here interested in demonstrating the existence of vortices in the plane $z = z_0$, we shall concentrate on the upper triangle OAB, leaving the case of the lower

[143] When $|D| \neq |E| \neq |F|$, this ordering of plane-wave moduli can always be achieved, by an appropriate re-ordering of the plane waves used in constructing our three-plane-wave disturbance, together with an appropriate rotation of Cartesian coordinate system. Note also that consideration of the following cases is left as an exercise for the reader: (a) $|D| = |E| > |F|$; (b) $|D| > |E| = |F|$; (c) $|D| = |E| = |F|$.

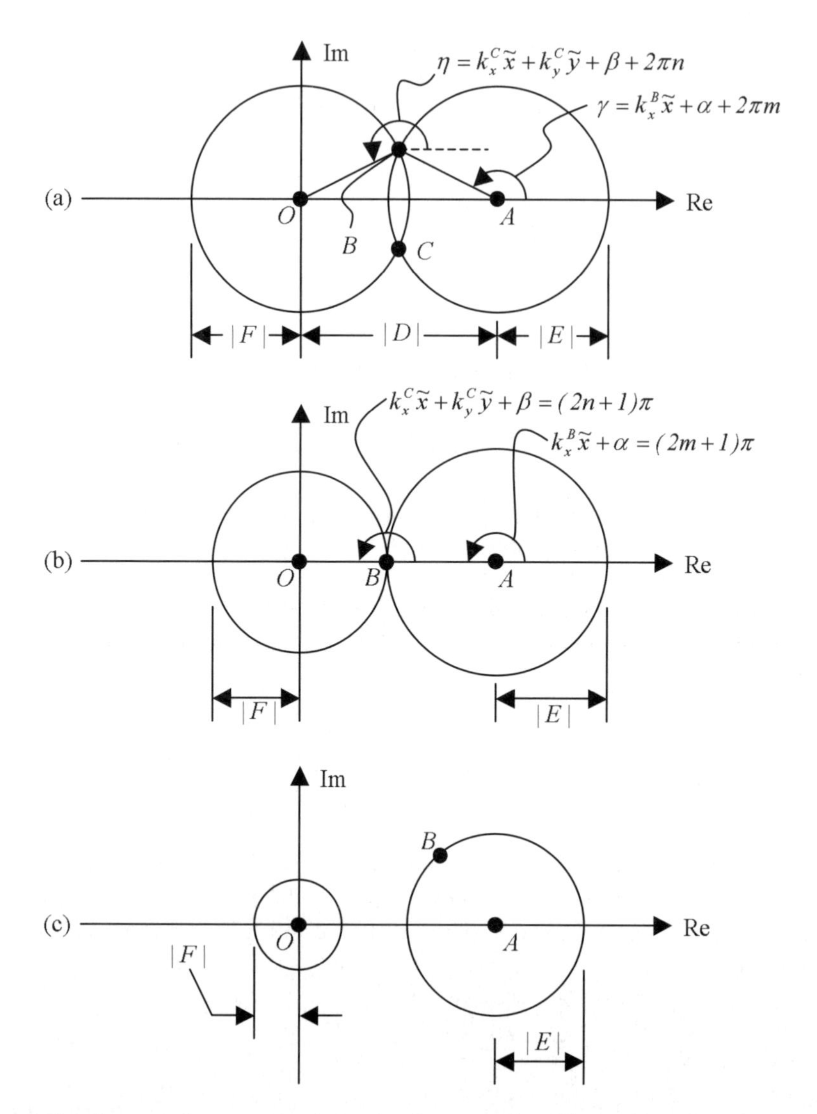

Fig. 5.6. Geometric constructions, in the complex plane, to aid in finding ordered pairs $(x, y) = (\tilde{x}, \tilde{y})$ which solve eqn (5.44), where $|D| > |E| > |F|$. The three terms on the left side of this equation are respectively denoted by the complex-plane phasors OA, AB, and BO. Three separate cases are shown: (a) $|D| < |E| + |F|$; (b) $|D| = |E| + |F|$; (c) $|D| > |E| + |F|$. Note that the phasor BO cannot be constructed for the last of these cases, as the point B must lie on both circles in the diagram if eqn (5.44) is to be fulfilled.

triangle as an exercise for the reader. Specifically, we shall now determine the points $(\tilde{x}, \tilde{y})$ in the xy plane, corresponding to the upper triangle in Fig. 5.6(a), which are such that eqn (5.44) is obeyed. With reference to this figure, note that the anti-clockwise angle, made by the phasor AB with respect to the positive real axis, must be equal to $k_x^B \tilde{x} + \alpha$, modulo 2π. This angle, denoted here by γ, can evidently be calculated by application of the cosine rule to the triangle OAB, whose side lengths are known (by taking the modulus of each member of eqns (5.39)). Hence one can determine γ, which obeys:

$$\gamma = k_x^B \tilde{x} + \alpha + 2\pi m, \tag{5.45}$$

where m is any integer. Using a similar argument we can write down:

$$\eta = k_x^C \tilde{x} + k_y^C \tilde{y} + \beta + 2\pi n, \tag{5.46}$$

where η is the known angle at B which is indicated in Fig. 5.6(a), and n is an integer. This may also be computed using the cosine rule. Solving the last two simultaneous equations for $\tilde{x}$ and $\tilde{y}$, we see that the wave-function $\psi(x, y, z = z_0)$, formed by the coherent superposition of three plane waves which is given in eqn (5.38), vanishes at each member of a certain lattice of points $(\tilde{x}, \tilde{y})$. These points are given by:

$$\tilde{x} = \frac{\gamma - \alpha - 2\pi m}{k_x^B}, \qquad \tilde{y} = \frac{k_x^B(\eta - \beta - 2\pi n) + k_x^C(\alpha + 2\pi m - \gamma)}{k_x^B k_y^C}, \tag{5.47}$$

where m and n are integers. Having determined certain points $(\tilde{x}, \tilde{y})$ at which the wave-function vanishes, we need to test for phase vortices at each of these points.[144] Accordingly, denote a point in the vicinity of a wave-function zero by $(\tilde{x} + \delta x, \tilde{y} + \delta y)$, where δx and δy are small real numbers. Evaluate $\psi(\tilde{x} + \delta x, \tilde{y} + \delta y, z = z_0)$ with the use of eqn (5.43), then isolate complex exponential terms with exponents proportional to linear combinations of δx and/or δy, before Taylor expanding these exponentials to first order in δx and δy. Then, make use of eqn (5.44) in order to discard three terms in the resulting equation. Up to an irrelevant multiplicative complex constant, which may be factored out of the resulting expression, the approximate equation for $\psi(\tilde{x} + \delta x, \tilde{y} + \delta y, z = z_0)$ will equal a linear combination of δx and δy, with the coefficient of δy being real and the coefficient of δx in general containing both real and imaginary components:

[144] We have shown that the wave-function, in the plane $z = z_0$, vanishes at certain points $(\tilde{x}, \tilde{y})$. This is a necessary but not sufficient condition for the existence of vortices at these points. To show that there are indeed phase vortices at these points, we need to show that the phase winding about such points is equal to an integer multiple of 2π.

$$\begin{aligned}
&\psi(\tilde{x} + \delta x, \tilde{y} + \delta y, z = z_0) \\
&\quad \approx i \exp(i\phi_D)|F| \exp[i(k_x^C \tilde{x} + k_y^C \tilde{y} + \beta)] \\
&\quad \times \left(\left\{ k_x^C + \frac{|E| k_x^B \exp[i(k_x^B \tilde{x} + \alpha)]}{|F| \exp[i(k_x^C \tilde{x} + k_y^C \tilde{y} + \beta)]} \right\} \delta x + k_y^C \delta y \right). \qquad (5.48)
\end{aligned}$$

Provided the term in braces is not real, the lowest line of the above expression indicates that there exists a phase vortex about the points $(\tilde{x}, \tilde{y})$ where $\delta x = \delta y = 0$. This completes the proof that there are 'upper triangle' point vortices located at the lattice of points $(\tilde{x}, \tilde{y})$ given in eqn (5.47). We leave it as an exercise to both deduce the handedness of these point vortices, and to study the lattice of wave-field zeros associated with the 'lower triangle' OAC in Fig. 5.6(a).

Case #2: This case, where $|D| > |E| > |F|$ and $|D| = |E| + |F|$, is shown in Fig. 5.6(b). Once again, the three terms on the left side of eqn (5.44) are respectively denoted by the phasors OA, AB, and BO. Since $|D| = |E| + |F|$, it is evident that there is a single point of intersection, denoted by B, between (i) the circle of radius $|E|$ centred at A and (ii) the circle of radius $|F|$ centred at O. Evidently the second term of eqn (5.44) must be purely real and negative, so that:

$$k_x^B \tilde{x} + \alpha = (2m + 1)\pi, \qquad (5.49)$$

where m is an integer. Applying a similar argument to the phasor BO, we see that:

$$k_x^C \tilde{x} + k_y^C \tilde{y} + \beta = (2n + 1)\pi, \qquad (5.50)$$

where n is an integer. Solving these last two simultaneous equations for $\tilde{x}$ and $\tilde{y}$, we see that:

$$\tilde{x} = \frac{(2m + 1)\pi - \alpha}{k_x^B}, \qquad \tilde{y} = \frac{(2n + 1)\pi - \beta}{k_y^C} - \frac{k_x^C[(2m + 1)\pi - \alpha]}{k_x^B k_y^C}. \quad (5.51)$$

Since m and n can take on any integer values, it is evident from eqn (5.51) that this defines an infinite lattice of points $(\tilde{x}, \tilde{y})$ which obey eqn (5.44). In order to show that the points in eqn (5.51) correspond to vortex cores, it is necessary for us to compute the phase winding about any one of these cores. Accordingly, denote a point in the vicinity of a wave-function zero by $(\tilde{x} + \delta x, \tilde{y} + \delta y)$, where δx and δy are small real numbers. Using similar logic to that presented for the previous case, one concludes that there are point vortices coinciding with the lattice of points $(\tilde{x}, \tilde{y})$ given in eqn (5.51), provided that the braced term in eqn (5.48) is not real. Once again, we leave it as an exercise to the reader to deduce the handedness of these point vortices.

Case #3: This case, where $|D| > |E| > |F|$ and $|D| > |E| + |F|$, is shown in Fig. 5.6(c). The first term on the left side of eqn (5.44) is denoted by the phasor

OA; the second term, being of modulus $|E|$, stretches from A to the edge of the circle centred at A; the third term, being of modulus F, stretches from the edge of the circle centred at O, to O itself. For this case, where one of the three plane waves has an intensity that is sufficiently greater than that of the other two, no solution exists for eqn (5.44). Hence no vortices are generated. From a physical point of view, this is to be expected: if the second and third plane waves both have an intensity that is sufficiently smaller than that of the first plane wave, they will never be able to completely cancel the complex amplitude (at any point in space) of the first plane wave. Since such a complete cancellation (wave-function zero) is a necessary condition for the existence of a wave-field vortex, no vortices are produced in this case.

To conclude, in the present sub-section we have explored the remarkable result that the coherent superposition of three monochromatic scalar plane waves can yield a lattice of phase vortices in the resulting wave-field (Masajada and Dubik 2001). Of the many X-ray-optics implementations of this idea that immediately spring to mind, one of the simplest is the following. Take a Fresnel zone plane, and block off all but three simply-connected regions (e.g. by holes in a suitable aperture), each of which is sufficiently small for the frequency of the zones to not change appreciably over that region, while being sufficiently large to contain enough zones for the region to comprise a diffraction grating. Upon illumination with paraxial plane monochromatic X-ray radiation, the three illuminated regions of the zone plate will each contribute a roughly plane wave to the focal volume of the zone plate. In this focal volume, one will therefore have three approximately plane waves, the relative intensities and phases of which can be tuned using thin slabs of material placed over two of the three apertures in the zone plate. One will thereby have created a vortex lattice in this focal volume. Rather than zone plates, multi-faceted X-ray prisms with an irregular pyramidal shape could also be used, to similar effect.

5.6.2 *X-ray vortex generation using synthetic holograms*

Pushing the previous paragraph's simple zone-plate example a little further, one can seek to actively design two-dimensional masks which, when illuminated with coherent X-ray radiation, lead to vortical structures in the X-ray wave-field downstream of the mask. Means for doing so are outlined below, using an extension of the fundamentals of off-axis holography which were presented in Section 4.3.2. Note that there are also some overlaps with ideas employed in our earlier discussions on linear and circular diffraction gratings, in Sections 3.2.1 and 3.2.2, respectively.

Consider the two-dimensional first-order monochromatic point vortex:

$$\psi(r, \theta) = f(r) \exp(i\theta), \tag{5.52}$$

where (r, θ) are plane polar coordinates and $f(r)$ is any real non-negative function that vanishes when $r = 0$. For simplicity we assume $f(r)$ to be real. We wish to holographically create this vortical two-dimensional wave-field downstream of a suitable mask, which is assumed to be located in the plane $z = 0$ of a Cartesian coordinate system (x, y, z). Accordingly, let us mathematically interfere this desired field with the $z = 0$ disturbance created by a tilted monochromatic plane wave solution $\exp[i(-k_x x + z\sqrt{k^2 - k_x^2})]$ to the Helmholtz equation (5.33), corresponding to the same wavelength λ as the desired first-order vortex, where $k \equiv 2\pi/\lambda$ and the real number k_x quantifies the tilt of the plane wave. Note that we speak of 'mathematically' interfering the two complex amplitudes, since this procedure evidently corresponds to the off-axis holographic procedure of physically interfering $\psi(r, \theta) = f(r)\exp(i\theta)$ with a tilted plane wave. The intensity $I_{\mathrm{H}}(x, y, z = 0) \equiv I_{\mathrm{H}}(x, y)$ of the resulting wave-field over the plane $z = 0$ is:

$$\begin{aligned} I_{\mathrm{H}}(x, y) &= |f(r)\exp(i\theta) + \exp(-ik_x x)|^2 \\ &= 1 + |f(r)|^2 + f(r)\exp[i(k_x x + \theta)] + f(r)\exp[-i(k_x x + \theta)] \\ &= 1 + |f(r)|^2 + 2f(r)\sin\left(k_x x + \theta + \frac{\pi}{2}\right). \end{aligned} \tag{5.53}$$

As should be clear from the second line of the previous equation, if one should make a purely amplitude-shifting mask with a real transmission function equal to $I_{\mathrm{H}}(x, y)$, then illumination of the mask (which may be viewed as a holographic diffraction grating) with normally incident plane monochromatic X-rays $\exp(ikz)$ will result in a real exit-surface wave-function that contains a superposition of three terms (two of which are complex).[145] The first of these terms, namely $1+|f(r)|^2$, corresponds to the zeroth-order mode of the diffraction grating, which will travel in the z direction provided that $f(r)$ is not too strongly varying. The second term, namely $f(r)\exp(i\theta)\exp(ik_x x)$, is equal to a product of the vortical wave-field $f(r)\exp(i\theta)$ which we wished to synthesize, and a tilted plane wave $\exp(ik_x x)$; evidently, this corresponds to a tilted version of the desired first-order vortex beam, which will become separated from the other diffraction orders upon propagation through a sufficient distance. Finally, the third term is equal to the complex conjugate of the second term, which is a tilted version of the complex conjugate of the desired vortical wave-field, corresponding to an first-order anti-vortex tilted in the opposite direction to the reconstructed vortex.

Regarding the preceding analysis, one may object that it is extremely difficult to fabricate a two-dimensional grey-tone mask that affects the intensity but not the phase of the illuminating radiation passing through it. In addition, there is the technical difficulty of constructing a mask with a continuously varying transmission function. It is often more practical to fabricate binary masks, of which the simplest instance is a single-material mask whose projected thickness (projected along the z-axis) is either A or B, with A and B being different real numbers. With this in mind, let us binarize the function $\sin[k_x x + \theta + (\pi/2)]$

[145] The projection approximation has been assumed in making this statement (see Section 2.2).

which appears in the expression for $I_{\mathrm{H}}(x,y)$, to yield a two-level holographic diffraction grating that is composed of a single material and whose projected thickness $T_{\mathrm{H}}^{\mathrm{binary}}(x,y)$ is given by:

$$T_{\mathrm{H}}^{\mathrm{binary}}(x,y) = \begin{cases} A, & \sin\left(k_x x + \theta + \frac{\pi}{2}\right) \geq 0, \\ B, & \text{otherwise.} \end{cases} \tag{5.54}$$

Assuming the projection approximation to be valid, one may ask what wave-field results when the binarized mask in the above equation is illuminated with normally incident monochromatic plane waves. To answer this question, we make an apparent digression by considering the Fourier-series decomposition of a periodic square wave with period 2π. The square wave may be written as $\tilde{\Upsilon}(t)$, where t is a real parameter and[146]:

$$\tilde{\Upsilon}(t) = \begin{cases} +1, & 0 \leq t < \pi, \\ -1, & \pi \leq t < 2\pi, \end{cases} \qquad \tilde{\Upsilon}(t + 2\pi) = \tilde{\Upsilon}(t). \tag{5.55}$$

The Fourier-series decomposition of this square wave is (cf. eqn (3.12)):

$$\frac{\pi}{4}\tilde{\Upsilon}(t) = \sin t + \frac{1}{3}\sin 3t + \frac{1}{5}\sin 5t + \cdots = \sum_{m=1}^{\infty} \frac{\sin[(2m-1)t]}{2m-1}. \tag{5.56}$$

Now, let us view this series in a slightly unorthodox manner. If one keeps only the $m = 0$ term in the Fourier series above, one has a purely sinusoidal function with period 2π. Inclusion of all remaining terms in the Fourier series serves to change this sinusoidal function of period 2π, into a square wave with period 2π and amplitude $\pi/4$. Stated differently, one can binarize the function $\sin(t)$ by the mapping:

$$\sin t \to \sin t + \sum_{m=2}^{\infty} \frac{\sin[(2m-1)t]}{2m-1} = \begin{cases} +\frac{\pi}{4}, & 0 \leq t(\mathrm{mod} 2\pi) < \pi, \\ -\frac{\pi}{4}, & \text{otherwise.} \end{cases} \tag{5.57}$$

The binary function, given above, takes on the two values $\{\pi/4, -\pi/4\}$. Suppose, instead, that we wanted the binary function to take on the two real values A and B, where $A \neq B$. With this in mind, note that the pair of numbers $\{\pi/4, -\pi/4\}$ can be transformed into the pair of numbers $\{A, B\}$ by first multiplying each number by $2(A - B)/\pi$ and then adding $(A + B)/2$. Hence, the mapping required to binarize the function $\sin t$ such that the resulting binary function takes on either the values A or B, is given by:

[146]Note that this is the $L = 2\pi$ case of eqn (3.11).

$$\sin t \to \frac{A+B}{2} + \frac{2(A-B)}{\pi}\left[\sin t + \sum_{m=2}^{\infty} \frac{\sin[(2m-1)t]}{2m-1}\right] = \begin{cases} A, & 0 \leq t(\text{mod} 2\pi) < \pi, \\ B, & \text{otherwise.} \end{cases} \quad (5.58)$$

Now, while we have hitherto viewed t as an independent real variable, we could instead view t as a dependent variable. For example, if we let:

$$t = t(x, y) = k_x x + \theta + \frac{\pi}{2}, \quad (5.59)$$

in the equation above, then we will have written down a representation for the projected thickness $T_{\text{H}}^{\text{binary}}(x, y)$ of the binarized holographic grating, appearing in eqn (5.54). Therefore:

$$T_{\text{H}}^{\text{binary}}(x, y) = \frac{A+B}{2} + \frac{2(A-B)}{\pi}\left\{\sin\left(k_x x + \theta + \frac{\pi}{2}\right) + \sum_{m=2}^{\infty} \frac{\sin\left[(2m-1)\left(k_x x + \theta + \pi/2\right)\right]}{2m-1}\right\}. \quad (5.60)$$

Now let us adopt the projection approximation and assume that this grating is illuminated with normally incident monochromatic plane waves $\exp(ikz)$, with the grating itself lying in the plane $z = 0$. If the binary mask is made of a single material with complex refractive index n and linear attenuation coefficient μ, then the exit-surface wave-function will take on one of two complex values:

$$\xi = \exp\{ik[\text{Re}(n) - 1]A\}\sqrt{\exp(-\mu A)}, \quad (5.61)$$

$$\zeta = \exp\{ik[\text{Re}(n) - 1]B\}\sqrt{\exp(-\mu B)}. \quad (5.62)$$

Therefore, the exit-surface wave-function $\psi_{\text{H}}(x, y)$, which results when the grating specified by eqn (5.54) is illuminated under the previously-stated conditions, is:

$$\psi_{\text{H}}(x, y) = \frac{\xi+\zeta}{2} + \frac{2(\xi-\zeta)}{\pi}\left\{\sin\left(k_x x + \theta + \frac{\pi}{2}\right) + \sum_{m=2}^{\infty} \frac{\sin[(2m-1)\left(k_x x + \theta + \pi/2\right)]}{2m-1}\right\}. \quad (5.63)$$

If we now make use of the fact that:

$$\sin x = \frac{\exp(ix) - \exp(-ix)}{2i}, \tag{5.64}$$

we arrive at the central result for this sub-section:

$$\begin{aligned}\psi_{\mathrm{H}}(x,y) =& \frac{\xi+\zeta}{2} + \frac{\xi-\zeta}{\pi} e^{ik_x x} e^{i\theta} + \frac{\xi-\zeta}{\pi} e^{-ik_x x} e^{-i\theta} \\ &+ \frac{\xi-\zeta}{\pi i} \sum_{m=2}^{\infty} \frac{e^{i(2m-1)(k_x x+\theta+\pi/2)} - e^{-i(2m-1)(k_x x+\theta+\pi/2)}}{2m-1}.\end{aligned} \tag{5.65}$$

This equation gives the exit-surface wave-function which results when our binarized hologram is illuminated with normally incident plane waves. We now give a brief description of the physical meaning of each term appearing in this expression. (i) The first term, on the right-hand side of eqn (5.65), corresponds to the zeroth diffracted order of the grating—that is, to the undiffracted beam. Note that this term can be made to vanish by having $\xi = -\zeta$, which will be the case if one has a non-absorbing grating ('phase grating'), so that the exit-surface wave-function is everywhere constant in modulus. (ii) The second term, on the right side of eqn (5.65), corresponds to the first positive diffracted order of the grating. This order carries a first-order vortex $\exp(i\theta)$, tilted at an angle of:

$$\tan^{-1}\left(\frac{k_x}{\sqrt{k^2 - k_x^2}}\right) \tag{5.66}$$

with respect to the xy plane. (iii) The third term corresponds to the first negative diffracted order of the grating. This order carries a first-order anti-vortex $\exp(-i\theta)$, tilted at an angle of:

$$-\tan^{-1}\left(\frac{k_x}{\sqrt{k^2 - k_x^2}}\right) \tag{5.67}$$

with respect to the xy plane. (d) The final term, on the right side of eqn (5.65), corresponds to diffracted orders higher than the first. Evidently, these orders contain vortices and anti-vortices with odd topological charge, contained within a diffraction order making an angle to the xy plane, whose modulus:

$$\tan^{-1}\left[\frac{(2|m|-1)k_x}{\sqrt{k^2 - k_x^2(2|m|-1)^2}}\right] \tag{5.68}$$

increases with the magnitude of the topological charge of the vortex contained within it. Note that when the angle becomes complex, which will evidently be the case when m is sufficiently large in magnitude, no propagating diffracted beam is created for the corresponding order.

We close this sub-section with some computer modelling, to demonstrate the use of diffractive optical elements to form X-ray vortices. Figure 5.7(a) shows a 500×500 pixel image of a continuous diffraction grating which is such that its transmission is purely real, and given by $\cos(k_x x + \theta)$.[147] The salient feature of this figure is the fork in what would otherwise be considered a sinusoidal diffraction grating. A binarized form of the same mask is given in Fig. 5.7(b). For the purposes of numerical modelling, we have assumed that the two masks, mentioned above, are (i) illuminated with a Gaussian distribution of intensity, such that the phase at the exit surface of the mask is approximately uniform; (ii) such that the complex transmission function of the masks are respectively given by $\cos(k_x x + \theta)$, and the binarized form of $\cos(k_x x + \theta)$. The intensity distribution in the far field of the continuous mask was obtained by zero-padding the unpropagated wave-function in a 1000×1000 pixel array, taking the squared modulus of the Fourier transform of the wave-field at the exit surface of the mask, and then cropping the resulting image to a central stripe which was 1000 pixels wide and 200 pixels high. The result is given in Fig. 5.7(c). The intensity of this diffraction pattern is plotted on a logarithmic scale, with the corresponding phase map[148] being given in Fig. 5.7(d). The central diffracted order, together with the +1 and -1 diffracted orders, are clearly visible. Further, the +1 and -1 orders evidently carry oppositely charged first-order vortices, as can be seen from the phase map in Fig. 5.7(d). Passing over to the binarized form of the diffraction grating, the simulated far-field intensity and phase are respectively given in Figs 5.7(e) and (f). We see that the effect of binarization has introduced additional odd diffracted orders: in addition to the central order and the ± 1 orders, we see the ± 3 and ± 5 orders that were predicted by eqn (5.65). The +1, +3, and +5 diffracted orders are marked with white arrows in Fig. 5.7(e)—note that each of these orders has a dark core. As was the case with the continuous grating, the ± 1 orders are associated with a first-order vortex and a first-order anti-vortex. Third order phase vortices and anti-vortices are associated with the ± 3 diffracted orders, with fifth-order vortices and anti-vortices being associated with the ± 5 diffracted orders. The charge-one, charge-three, and charge-five nature of the three positive diffracted orders, are indicated by white arrows in Fig. 5.7(f). In addition to the vortex nature of the various diffracted orders, note that there is a low-intensity speckled background in the far-field diffraction pattern of the binary grating, containing a rich structure of propagation-induced phase vortices (cf. Section 5.6.4).

We close this sub-section with some remarks on the feasibility of the experimental situation which has just been modelled. In order to be in the far-field, one

[147]This assumption is, of course, rather unrealistic. We are unconcerned with this fact, however, in light of the fact that we will later binarize the mask.

[148]By invoking the two-dimensional form of the Fourier shift theorem, it is evident that a transverse shift in the unpropagated wave-field will lead to a linear ramp in the phase of the far-field diffraction pattern. It is undesirable to have such a linear phase ramp in our plots of the phase of the far-field diffraction pattern. Therefore, in the simulations presented here, this linear phase ramp was removed by forcing the 'centre of mass' of the modulus of the undiffracted wave-function to coincide with the central pixel in the image.

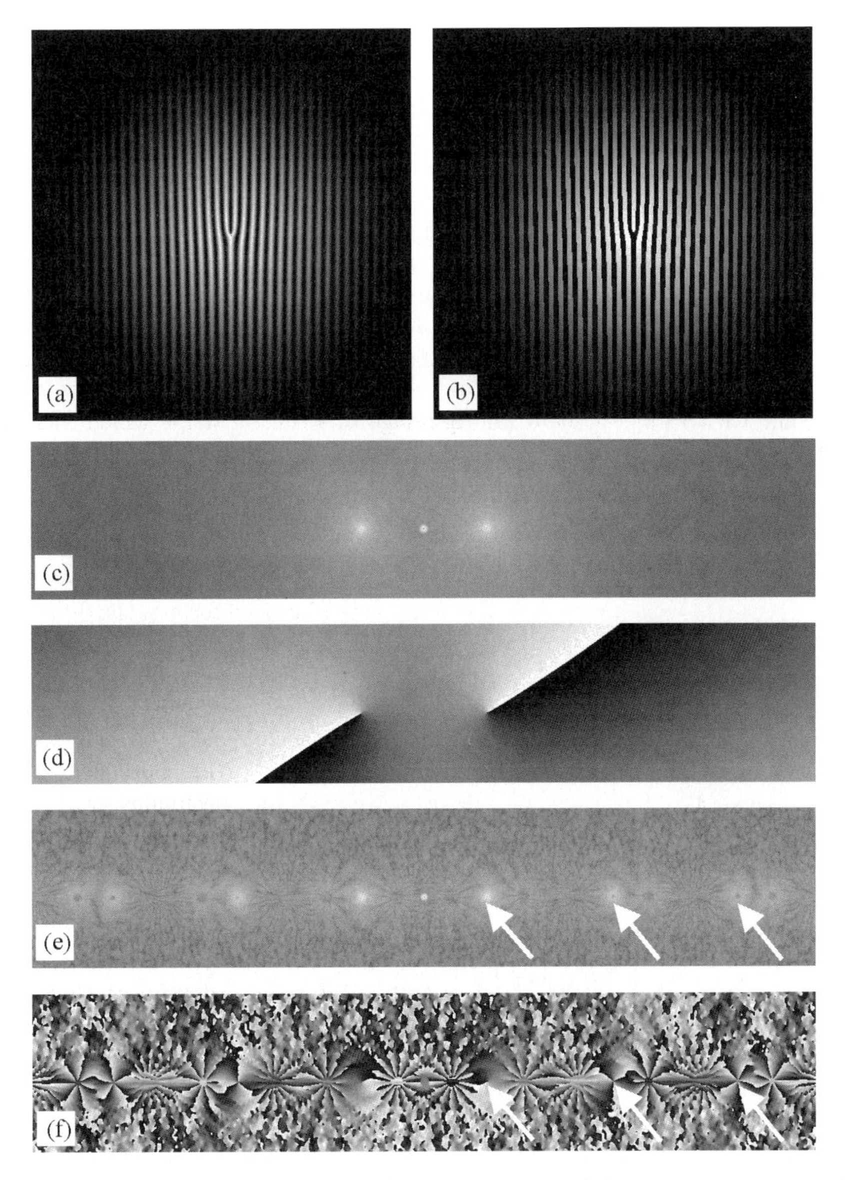

Fig. 5.7. Exit-surface intensity of (a) continuous and (b) binary vortex mask for Gaussian illumination; (c) far-field diffraction pattern corresponding to (a), on logarithmic scale, showing undiffracted beam and ± 1 orders; (d) phase corresponding to (a), showing first-order vortex and first-order anti-vortex associated with ± 1 orders; (e) far-field diffraction pattern corresponding to (b), on logarithmic scale, showing zeroth and odd diffracted orders (positive odd diffracted orders marked with white arrows); (f) phase corresponding to (b) (charge $+1$, $+3$, and $+5$ singularities, associated with first three odd diffracted orders, marked with white arrows). All phase maps are modulo 2π.

requires the Fresnel number N_{F} to be much smaller than unity (see eqn (1.48)). Therefore,

$$N_{\mathrm{F}} \equiv \frac{a^2}{\lambda z} \ll 1. \tag{5.69}$$

Here, λ is the wavelength of the X-rays, z is the propagation distance between the screen and the detector and a is the characteristic length scale of fluctuations in the mask. Next, we note that the energy E of a photon is related to the magnitude of its momentum p via the relation

$$E = pc, \tag{5.70}$$

where c is the speed of light in vacuum. Now,

$$E = Ve, \tag{5.71}$$

where V is the energy of the photon in electron volts and e is the magnitude of the charge on the electron. Also, the de Broglie relation states that:

$$p = \frac{h}{\lambda}, \tag{5.72}$$

where h is Planck's constant. Utilizing these last two equations, eqn (5.70) becomes:

$$Ve = \frac{hc}{\lambda}, \tag{5.73}$$

which may be solved for λ. If this is then used to eliminate λ from eqn (5.69), we arrive at:

$$z \gg \frac{a^2 Ve}{hc}. \tag{5.74}$$

Now, $a = 50 \times 10^{-9}$ m is readily achievable; indeed, this dimension is typical of the width of outermost zones in contemporary Fresnel zone plates. Inserting numerical values for e, h, and c into the above equation, we obtain the requirement that $z \gg 0.002 V_{\mathrm{keV}}$, where V_{keV} is the photon energy in keV. For photon energies in the range 10–50 keV, this requirement is always satisfied for propagation distances z that are greater than 2 m. Note also that the efficiency of the forked grating may be studied using methods analogous to those given in Section 3.2.1, in the context of linear diffraction gratings.

5.6.3 *X-ray vortex generation using spiral phase masks*

Here we discuss the use of spiral phase masks as a means for generating phase vortices in coherent X-ray wave-fields. To this end, briefly return attention to Fig. 5.3(b), which shows constant-phase contours for one Riemann sheet of a simple vortex. Adopting plane polar coordinates (r, θ) for this figure, we may write the simple expression:

$$\phi(r, \theta) = \theta, \tag{5.75}$$

for the phase of a first-order vortex. Next, recall the projection approximation, which was introduced in Section 2.2. Under this approximation, the phase shift induced on a monochromatic plane wave of wavelength λ which is normally incident on a thin single-material mask of projected thickness $T(x, y)$, is given by (see eqn (2.41)):

$$\phi(r, \theta) = -\frac{2\pi\delta}{\lambda} T(r, \theta). \tag{5.76}$$

Here,

$$\mathrm{Re}(n) \equiv 1 - \delta \tag{5.77}$$

is the real part of the refractive index n of the material from which the mask is composed (see eqn (2.37)).

We seek a mask—a so-called 'spiral phase mask', so termed for reasons that shall become clear in a moment—which will produce an optical vortex when illuminated with normally incident monochromatic plane waves. The projected thickness $T(r, \theta)$ of such a mask is readily obtained by equating the right-hand sides of eqns (5.75) and (5.76), and then solving for $T(r, \theta)$. This leads to:

$$T(r, \theta) = \frac{\lambda\theta}{2\pi\delta}, \quad 0 \leq \theta < 2\pi, \tag{5.78}$$

where we have discarded a minus sign as it influences only the sense of the vortex, and the restriction on θ is imposed because of the single-valuedness of the projected thickness of the mask.

The thickness of the spiral phase mask is proportional to the angle θ about a given origin, with a discontinuity along the line $\theta = 0 = 2\pi$ (see, for example, Kristensen *et al.* (1994)). In practice, workers often use a stepped approximation to such a mask, such as that sketched in Fig. 5.8. Here we see a spiral phase plate with 16 steps, the structure of which bears a close resemblance to a spiral staircase. Note that the height difference h between the first and last steps is:

$$h = T(r, \theta = 2\pi) - T(r, \theta = 0) = \frac{\lambda}{\delta}. \tag{5.79}$$

The first experimental demonstration, of the use of spiral phase plates to produce an X-ray vortex, was given by Peele and co-workers (Peele *et al.* 2002; see also Peele and Nugent 2003 and Peele *et al.* 2004). Figure 5.9(a) shows an

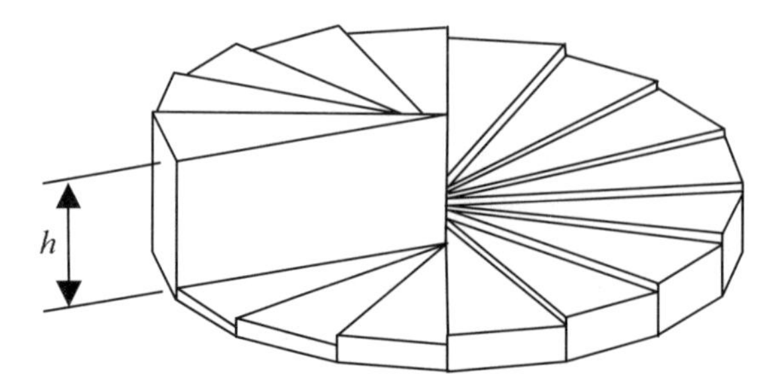

Fig. 5.8. Sketch of a stepped spiral phase mask, which will produce a vortex when normally illuminated with monochromatic plane waves of a suitable wavelength. After Beijersbergen *et al.* (1994).

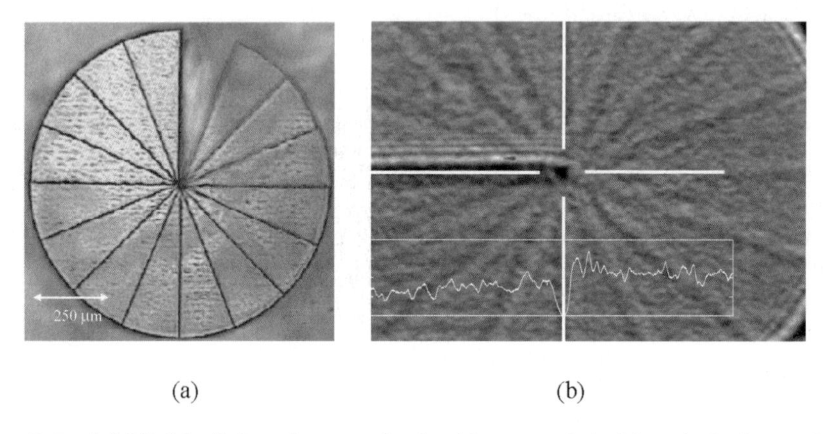

(a) (b)

Fig. 5.9. (a) Visible-light micrograph of a 16-step polyimide spiral phase plate, designed to produce a first-order vortex when normally illuminated with near-plane 9 keV X-rays. The difference in projected thickness, between thinnest and thickest steps, is $h = 34.2 \pm 0.5\mu\text{m}$. (b) Fresnel diffraction pattern located 5.8 m from the spiral phase plate, when illuminated with 9 keV plane X-rays. Inset shows an average of horizontal and vertical profiles recorded through the centre of the vortex core, which is evident as a black spot between the white cross-hairs. Image taken from Peele *et al.* (2002). Used with permission.

optical micrograph of their 16-level polyimide spiral phase plate. Viewed from above, this image is focussed on the step furthest from the viewer, so that the top-most step is most blurred, with the thickness of the mask increasing as one moves anti-clockwise about its axis. The thickness difference between the highest and lowest steps was $h = 34.2 \pm 0.5$ μm, which equates to a total phase winding of $(1.90 \pm 0.03)\pi$ at a photon energy of 9 keV.

When illuminated with double-crystal monochromated X-ray photons with a mean energy of 9 keV and an energy spread of 1 part in 7000, the mask resulted in the image of Fig. 5.9(b) being recorded 5.8 m downstream of the spiral phase plate. Note that, in obtaining this image, a flat-field correction was performed (i.e. to compensate for non-uniformity in the intensity of the illuminating beam, the image in the presence of the spiral phase plate was divided by an image obtained in the absence of the spiral phase plate).

The top-right and bottom-right corners show the effects of Fresnel diffraction from the edges of the mask. Fifteen weaker and one stronger black-white fringes can be seen to emanate radially from the central dark spot in the image. The weaker black-white fringes correspond to diffraction from the interface between steps whose projected thickness varies by $h/15$, with the stronger fringe corresponding to diffraction from the interface between steps whose projected thickness varies by h. Note that multiple Fresnel diffraction fringes are evident in this feature. The central black spot in the image, which is located at the centre of the white cross hairs, corresponds to the vortex core created by the spiral phase plate. The inset graph gives an average of two graphs: a horizontal profile through the core of the vortex, and a vertical profile through the core of the vortex. Note that the vortex character of the beam was ascertained using an interferometer, constructed by interfering the X-ray vortex with the conical waves scattered by a tungsten wire (diameter = 7.5 μm) placed 3 mm behind the spiral phase plate. The resulting interferogram displayed a forked appearance characteristic of a vortex phase. Recall, in this context, the previous section's result that a forked interferogram results when a vortex beam is interfered with an off-axis plane wave (see Fig. 5.7(a)).

5.6.4 *Spontaneous vortex formation in coherent X-ray imaging systems*

In the preceding sub-sections we described three different methods to engineer vortex states in a coherent X-ray wave-field. In this final sub-section, we examine the notion of spontaneous vortex formation, whereby wave-field X-ray phase vortices are 'naturally' or 'accidentally' formed in the absence of any specific optical elements being designed to create them (see, for example, Nye (1999)). This notion of spontaneous vortex formation is in harmony with our earlier discussions emphasizing the ubiquity of the optical vortex.

Before considering some specific instances of spontaneous vortex creation in X-ray wave-fields, let us recall the operator theory of two-dimensional imaging systems described in Section 4.1. In the first equation of that section, we wrote an expression characterizing the action of an X-ray imaging system, which takes as input a given coherent two-dimensional complex scalar electromagnetic field $\psi_{\rm IN}(x,y)$, in order to produce a field $\psi_{\rm OUT}(x,y,\tau_1,\tau_2,\cdots)$ as output. The action of this imaging system was represented in operational terms as:

$$\psi_{\rm OUT}(x,y,\tau_1,\tau_2,\cdots)=\mathcal{W}(\tau_1,\tau_2,\cdots)\psi_{\rm IN}(x,y), \tag{5.80}$$

with $\mathcal{W}(\tau_1,\tau_2,\cdots)$ being an operator completely characterizing the imaging system, in a state specified by the set $\tau_1,\tau_2,\cdots$ of real control parameters. By allowing only one control parameter (or a given combination of control parameters) τ to vary, while keeping all others fixed, wave-field phase vortices may be induced in the wave-function $\psi(x, y, \tau_1, \tau_2, \cdots)$ that emerges from the exit surface of the coherent imaging system. In the following paragraphs, we shall consider two such examples of optical systems, which spontaneously create such wave-field vortices: those that yield far field diffraction patterns, and those involving scattering from random media.

As a first example of spontaneous vortex formation in coherent X-ray imaging systems, we consider the deceptively simple case of a far-field diffraction pattern. In this context consider the far-field diffraction pattern of a finite-sized but otherwise perfect crystal. As discussed in Section 2.5.1, the far-field diffraction pattern of a perfect crystalline slab will consist of series of spots (see Fig. 2.7). This series of spots will be smeared out by a 'shape function' (see, for example, Cowley (1995)), which will be considered negligibly small for our present purposes. Evidently, such a far-field diffraction pattern is inconsistent with the existence of vortices, since a necessary condition for the existence of a point vortex is that there be an intensity zero at the centre of this vortex. Since the Fraunhofer diffraction pattern of a perfect crystal is non-zero only at a given set of spatially separated points, there will be no vortices present in the far zone. Note that there may be vortex–antivortex pairs present at the exit-surface of the crystalline slab; however, the preceding analysis implies that all vortices will have annihilated once the wave-field has propagated to the far zone.

The situation changes when one considers the far-field diffraction pattern of a non-crystalline object. In this case, vortices may be present in the far field. As a first example of this fact, consider the case of a coherent exit-surface wave-function $\psi(x, y)$ which consists of three bright points, each with arbitrary amplitudes and phases. Such a disturbance can be written in the form:

$$\begin{aligned}\psi(x,y) =& A_1\delta\left(x - x_0^{(1)}, y - y_0^{(1)}\right) + A_2\delta\left(x - x_0^{(2)}, y - y_0^{(2)}\right)\\ &+ A_3\delta\left(x - x_0^{(3)}, y - y_0^{(3)}\right),\end{aligned} \tag{5.81}$$

where $x_0^{(1)}, y_0^{(1)}, x_0^{(2)}, y_0^{(2)}, x_0^{(3)}, y_0^{(3)}$ are real numbers specifying the Cartesian coordinates of the three distinct bright points in the exit-surface wave-function, and A_1, A_2, A_3, are non-zero complex numbers specifying the amplitude and phase of the disturbance at each of these points. Taking the Fourier transform of this exit wave-function, and then making use of the Fourier shift theorem, it is evident that the resulting far-field diffraction pattern consists of three plane waves, each of which is inclined with respect to the other two, and each of which has an arbitrary amplitude and phase fixed by A_1, A_2, and A_3. Now, we have already seen in Section 5.6.1 that it is possible for vortices to form as a result of the interference between three mutually inclined plane waves, thereby proving by example the assertion that the far-field diffraction pattern of a given two-dimensional wave-function (which may itself be free of vortices) may contain propagation-induced vortices.

The above example may appear rather artificial, as the possibility of propagation-induced vortices in the far field does not imply that these are a generic feature of far-field diffraction patterns for non-crystalline objects. Notwithstanding this caveat, a stronger statement is in order: propagation-induced vortices appear to be a generic feature of far-field diffraction patterns of non-crystalline objects that possess no particular symmetries. The author is unaware of any proof for this conjecture.

As a simple computational example of propagation-induced phase vortices in the far field, which constitutes our second example of propagation-induced phase vortices in coherent X-ray optics, consider the simulations shown in Fig. 5.10. In Fig. 5.10(a), we have plotted the intensity distribution of a simple two-dimensional wave-function, the associated phase for which is taken to be constant. The image is 1000×1000 pixels in size, with black corresponding to zero intensity and white corresponding to an intensity of unity. At the centre of this image is an 84×84 pixel white square, inside which is the letter ‘a’. Taking the given intensity and trivial phase to correspond to the exit-surface wave-function of a given coherent X-ray wave-field, one obtains the Fraunhofer diffraction pattern by taking the squared modulus of the Fourier transform of this disturbance. The resulting simulated far-field diffraction pattern is given in Fig. 5.10(b), on a logarithmic scale.[149] Salient features of this image include a strong cross-shaped intensity distribution, which is due to the square shape present in the unpropagated intensity. Outside of this cross-shaped region the far-field diffraction pattern has a speckled appearance, which may be compared—on a qualitative level, of course—with the speckled appearance of the far-field diffraction pattern given in Fig. 4.18(b). The phase of the far-field diffraction pattern is shown in Fig. 5.10(c), with a 300×300 pixel central sub-image of the same given in Fig. 5.10(d).[150] The phase of this simulated far-field diffraction pattern contains a complex network of propagation-induced phase vortices, including the pair of counter-propagating vortices, which are marked with white arrows in Fig. 5.10(d).

As a third example of spontaneous vortex formation, we briefly treat the case of scattering from a random phase screen. In both the near and far fields of a random phase screen, phase vortices will typically be present. For example, Berry (1978) has calculated the density of phase vortices in the far-field of a random phase screen, upon which a scalar plane wave is normally incident, under the

[149] Using an identical argument to that which was presented in Section 5.6.2 on X-ray vortex generation using synthetic holograms, a far-field diffraction pattern can be achieved using propagation distances of no more than 2 m if the characteristic length scale of the diffracting structure is on the order of 50 nm, and the radiation energy is between 10 and 50 keV.

[150] The previously footnoted method, of eliminating a linear ramp in the phase of the far-field diffraction pattern by forcing the ‘centre of mass’ of the undiffracted field to coincide with the centre of the pixellated image array, was employed in the present analysis (see Section 5.6.2).

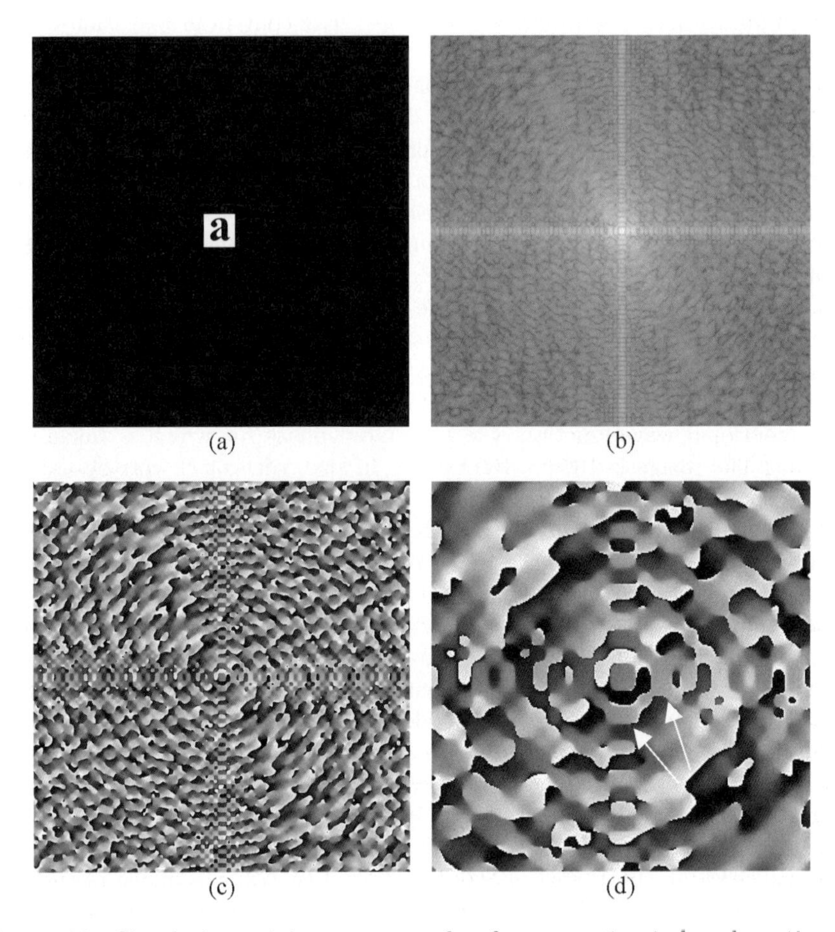

Fig. 5.10. Simulations giving an example of propagation-induced vortices in the far-field diffraction pattern corresponding to a vortex-free unpropagated wave-function. Intensity of unpropagated two-dimensional coherent scalar wave-function is given in (a), which consists of the black letter 'a' embedded in a white 84×84 pixel sub-image of a 1000×1000 pixel array. The phase of the unpropagated wave-function was chosen to be constant. (b) Intensity of far-field diffraction pattern, corresponding to previously described unpropagated field, plotted on a logarithmic greyscale. (c) Phase of far-field diffraction pattern, modulo 2π. (d) Central 300×300 pixel sub-image of (c), with counter-propagating vortex pair indicated by white arrows. Note that, in all intensity images, black corresponds to the minimum intensity (or logarithm of intensity, for the diffraction pattern) with white corresponding to the maximum. For the phase maps, which are all modulo 2π, black corresponds to a phase of 0 and white corresponds to a phase of 2π.

assumptions that: (i) the wave-field in the half-space downstream of the screen is both quasi-monochromatic and paraxial; (ii) the wave-field downstream of the screen is a Gaussian random function of both position and time; and (iii) the phase screen is sufficiently strong such that its root-mean-square value is at least several radians. In the same paper, Berry remarks that propagation-induced vortices are to be expected in the Fresnel zone, although the calculation of such quantities as vortex densities is then made considerably more difficult by the non-Gaussian nature of the wave-field statistics in this zone. For more detail on phase singularities in random wave-fields, see Freund (1996), Freund (1998), and Berry and Dennis (2000b), together with references therein.

As an experimental example of the use of random phase screens to induce vortices in a coherent X-ray wave-field, we treat the case of lung imaging using propagation-based phase contrast. We can treat this form of imaging as arising when a normally-incident monochromatic X-ray plane wave passes through a random phase screen (in this case, the lung of an animal such as a rabbit or a mouse), with the resulting exit-surface wave function being allowed to propagate through some distance before the resulting intensity is registered over the surface of a two-dimensional imaging device. An example of such an image is given in Fig. 5.11 (Suzuki *et al.* (2002); see also Yagi *et al.* (1999)). Here we see both a contact image[151] and a propagation-based phase contrast image (Section 4.4.4) of a hairless rat's thorax. Both images were obtained using monochromated 35 keV synchrotron X-rays, normally incident upon the rat, with both contact and phase-contrast images being recorded using a two-dimensional position-sensitive imaging detector placed downstream of the sample. The salient feature of these images is the strong speckled appearance of the lungs in the phase contrast image, together with the absence of speckle in the corresponding contact image.

Kitchen and co-workers have investigated the origin of the speckled appearance of such phase-contrast images of lung tissue, within the contexts of both the ray and wave theories of X-ray radiation (Kitchen *et al.* 2004). For both ray and wave treatments, they chose—as one of the simplest non-trivial models intended to capture the essence of the speckle-formation phenomenon—to model the lung as a uniform-thickness slab of a single material, inside which was embedded a random distribution of spherical cavities which serve as an approximation to the air-filled alveolar sacs in the lung. Having modelled the lung in this manner, the projection approximation was made, allowing them to relate the projected thickness of lung material to the phase of the exit-surface wave-function, which would result if the lung were illuminated with normally incident monochromatic scalar X-ray radiation.

Specifically, the simplified model for lung tissue comprised a slab of material with complex refractive index $n = 1-\delta+i\beta$, with $\delta = 2.21\times10^{-7}$ and $\mu = 2k\beta =$

[151]A 'contact image' is an X-ray image of a specimen, obtained by causing X-rays to pass through the specimen prior to being registered by a position-sensitive detector, with the distance between the exit-surface of the object and the surface of the detector being sufficiently small for the effects of free-space diffraction to be negligible.

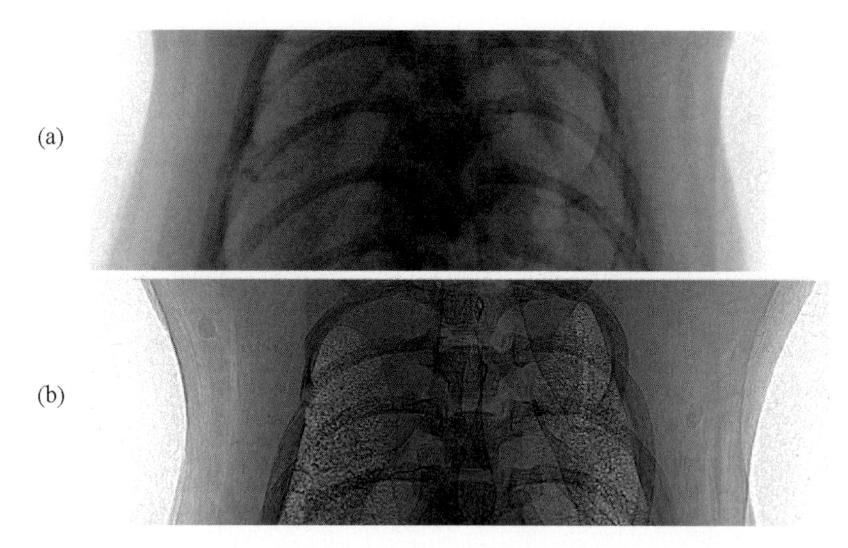

Fig. 5.11. (a) Contact image and (b) propagation-based phase contrast image of a hairless rat's thorax. Both images were obtained using 35 keV monochromated synchrotron radiation, with the propagation-based phase contrast image being recorded 5.5 m downstream of the exit surface of the rat. In the phase contrast image, the lungs have a strongly speckled appearance, due to the diffraction of the X-rays by the near-random phase screen provided by the network of alveoli in the lung. Image taken from Suzuki *et al.* (2002). Used with permission.

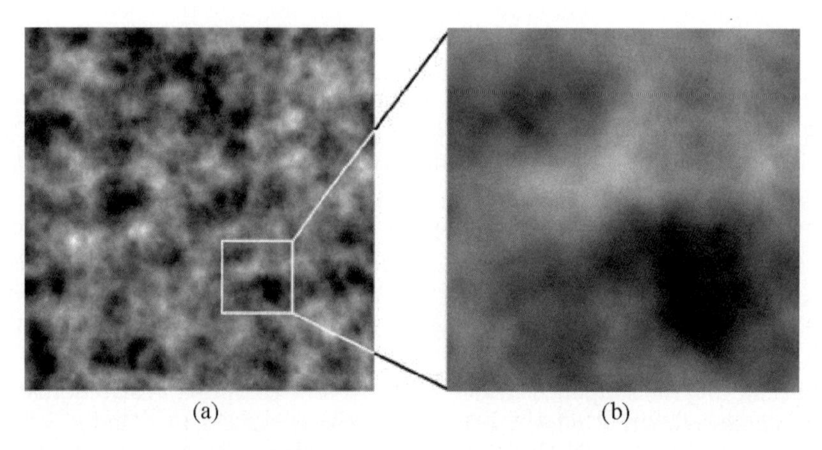

Fig. 5.12. (a) Projected thickness and (b) magnification of inset, for a simple model of lung tissue. Details in main text. Image taken from Kitchen *et al.* (2004). Used with permission.

12.6 m^{-1} (see eqns (2.37) and (2.44)). This complex refractive index matches that of lung tissue illuminated with 33 keV X-rays. The slab of tissue was taken to be 11.6 mm thick, with the slab containing randomly-positioned spherical cavities. These spherical cavities had a range of radii, following a Gaussian distribution with a mean radius of 60 μm and a standard deviation of 10 μm. A packing fraction of 74% was chosen, that is, 74% of the model lung volume was considered to be filled with air. As one realization of a random object generated in accord with the above specifications, there is the projected thickness of 'lung' tissue shown in Fig. 5.12(a). This image, which is 0.6 mm $\times$ 0.6 mm in size, shows the projected thickness of lung tissue associated with the model described above. The minimum and maximum values of the projected thickness are respectively denoted by black and white pixels, with a linear interpolation of greyscale for projected thickness values that lie between these two extremes. A magnified version of the area in the white square is given in Fig. 5.12(b).

Proceeding further, one can take the projected thickness $T(x, y)$ shown in the inset to Fig. 5.12(b) and then use the projection approximation to estimate the phase $\phi(x, y)$ of the exit-surface wave-function using eqn (2.41):

$$\phi(x, y) = -k\delta T(x, y). \tag{5.82}$$

Here, $k = 2\pi/\lambda$, with λ being the wavelength of the 33 keV plane X-rays which are assumed to be normally incident on the sample. Similarly, the projection approximation allows us to estimate the intensity of the exit-surface wave-function using eqn (2.43):

$$I(x, y) = \exp[-\mu T(x, y)]. \tag{5.83}$$

Uniform unit intensity, of a normally incident monochromatic scalar plane wave, has been assumed. The intensity and phase, so modelled, are given in the left column of Fig. 5.13. Minimum and maximum values of the intensity correspond to black and white, respectively, with all phase maps being modulo 2π. Note how the maxima of the exit-surface intensity correspond to the minima in the projected thickness, and *vice versa*.

Using the angular spectrum formalism outlined in Section 1.3, to computationally propagate the exit-surface wave-function through distances of 0.67 m and 2.22 m, one obtains the images for intensity and phase shown in the second and third columns of Fig. 5.13. Propagation-induced phase vortices are evident in all propagated phase maps, with one counter-propagating pair of vortices being indicated by white arrows in the figure. However, the detectors used in the experimental component of the present investigation were of too coarse a resolution to resolve the fine speckles associated with these propagation-induced vortices. Notwithstanding this fact, the rebinning of a speckled pattern was itself seen to produce a speckled pattern. Specifically, when these vortex-riddled X-ray wave-functions are binned to coarser-resolution pixels which are 12 μm in size, there is broad agreement between the previously mentioned computer

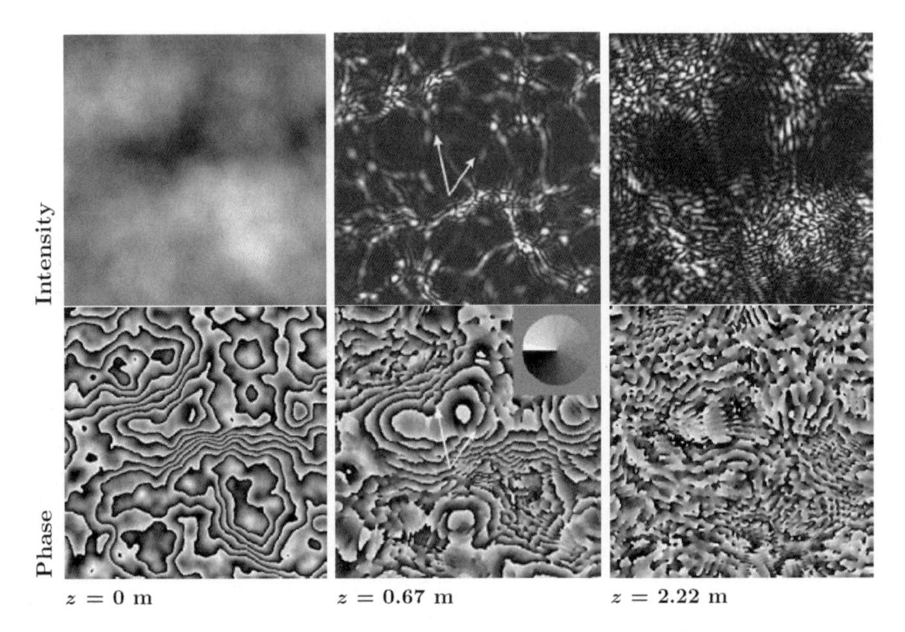

Fig. 5.13. Using the projected thickness in Fig. 5.12(b), and the projection approximation, results in the intensity and phase for the exit-surface wave-function given in the left column. Pixel size is 0.3 μm × 0.3 μm, with radiation energy at 33 keV. Upon using the angular-spectrum formalism to computationally propagate this wave-field through distances of z=0.67 m and z=2.22 m, one obtains the simulated maps of intensity and phase given in the second and third columns, respectively. Inset shows a greyscale map of the phase associated with a first-order vortex. Location of two phase vortices indicated by white arrows. Image taken from Kitchen *et al.* (2004). Used with permission.

modelling, and the speckled appearance of lung tissue observed in experimental data (Kitchen *et al.* 2004).

5.7 Domain walls and other topological phase defects

Thus far the vortex has been our only example of a topological defect in the phase of a coherent wave-field. A simpler form of defect, known as the domain wall, will be briefly considered here. For an example of a domain wall, consider the wave-field $\psi(x, y, z)$ associated with two counter-propagating monochromatic plane waves:

$$\psi(x, y, z) = \frac{\exp(ikz) + \exp(-ikz)}{2} = \cos(kz). \tag{5.84}$$

The cusped but continuous amplitude of this wave-function is given by:

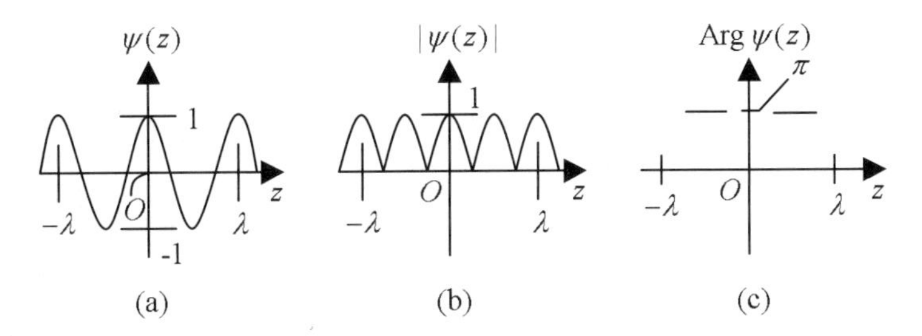

Fig. 5.14. (a) Graph of the real wave-function $\psi(x, y, z) = \psi(z) = \cos(kz)$ as a function of z, which corresponds to two counter-propagating plane waves; (b) Modulus of $\psi(z)$, as a function of z; (c) Phase of $\psi(z)$, as a function of z.

$$|\psi(x, y, z)| = |\cos(kz)|, \tag{5.85}$$

with the discontinuous phase of the wave-function equal to 0 when z is such that:

$$\cos(kz) > 0, \tag{5.86}$$

and equal to π when z is such that:

$$\cos(kz) < 0. \tag{5.87}$$

Plots of this wave-function, together with its amplitude and phase, are given in Fig. 5.14.

The phase of this wave-field has a discontinuity along the planes $z = (2m + 1)\lambda/4$, where m is any integer and $\lambda = 2\pi/k$ is the wavelength of the field. These planes are examples of domain walls, these being two-dimensional surfaces across which the phase of the wave-field changes discontinuously. To generate domain walls which are not planar, one might superpose a given coherent scalar wave-field with its complex conjugate—the resulting superposition will yield a real wave-function, whose phase is equal to zero over volumes where the wave-function is positive, and equal to π over volumes where it is negative. Irrespective of the form of the domain wall, there is an evident analogy with domain walls in hard three-dimensional ferromagnetic materials, these being two-dimensional surfaces which are such that the magnetization changes very rapidly (from one value, corresponding to the bulk of a given magnetic domain on one side of the wall, to another value, corresponding to the bulk of a given magnetic domain on the other side of the wall) as one passes from one side to the other side of the wall.

By appropriately adapting our earlier discussions of the stability of phase vortices with respect to continuous perturbations in the wave-fields carrying such defects, the reader can readily show that domain walls (in the phase of scalar X-ray fields) are unstable with respect to perturbation. They, therefore, do not represent generic features of the phase of an optical wave-field, unlike the case of first-order optical vortices. Note, further, that in this context the analogy with ferromagnetic materials breaks down, for in this case the domain walls are in general stable with respect to perturbation.

Despite their instability with respect to perturbation, domain walls are worth mentioning for at least two reasons. First, if we restrict ourselves to two-dimensional sections through a coherent scalar X-ray wave-field, then domain walls (which, in this case, are one-dimensional rather than two-dimensional) will be present whenever a portion of a nodal line lies in the planar section. For example, when looking at a planar section in the focal region of a perfect rotationally symmetric lens of finite size, with the planar section being normal to the optic axis, each of the Airy rings will be separated by a zero-intensity circle that coincides with a domain wall.

As a second reason for introducing the notion of a domain wall, we note that the domain wall and the vortex are the first two members in a hierarchy of topological defects, which can be categorized and studied with the use of homotopy theory. This classification of defects is applicable to a vast range of wave-functions and order-parameter fields, ranging from cosmology and particle physics through to condensed-matter physics and optics. Richly referenced introductions to the use of this theory for defect classification are given in the books by Vilenkin and Shellard (1994) and Pismen (1999). Regarding the application of these ideas to X-ray optics, note that defects other than vortices and domain walls are possible when one takes into account the vector nature of the electromagnetic field: for overviews and references, see Nye (1999).

5.8 Caustics and the singularity hierarchy

Thus far we have made little mention of the ray formalism for modelling X-ray disturbances. This is appropriate since, with the exception of semi-classical theories of diffraction,[152] geometric optics ignores the fact that light beams are able to coherently interfere with one another. Notwithstanding this, for reasons that will soon become clearer, let us turn to the ray theory of X-rays. We are particularly concerned with the singularities of the ray theory, where this theory predicts regions of infinite intensity known as 'caustics'.

Consider the sketches in Fig. 5.15. Figure 5.15(a) shows the elementary geometric-optics example, which will be revisited several times in the course of this section, of a source of rays A being focused by a thin perfect lens BC in order to produce a point focus D. According to the ray theory, the intensity of the rays will be infinite at this point focus, constituting a singularity of the ray theory. This furnishes our first example of a caustic. Specifically, we have here a 'point caustic', since the region of infinite intensity occupies a single point in space. Note that this is a singularity of the theory rather than one of Nature, since the ray theory's prediction of infinite intensity is inconsistent with experimental observations.

[152] See, for example, the papers on the 'geometric theory of diffraction' in the collection edited by Oughstun (1992).

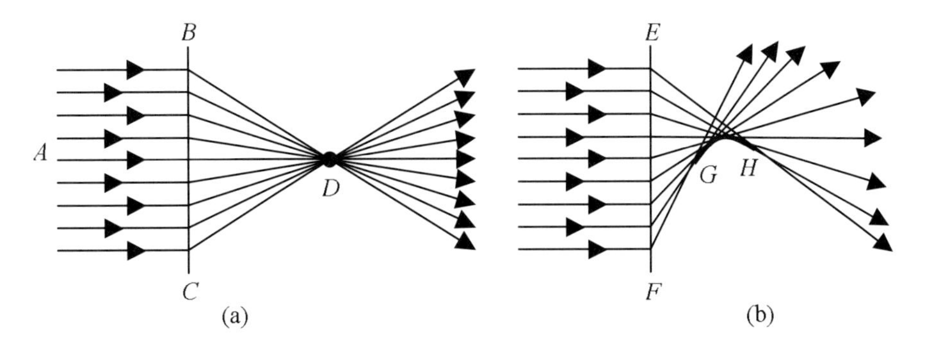

Fig. 5.15. (a) A source of parallel rays A is focussed by a perfect thin lens BC to a point D. This focal point is an example of a caustic. (b) As a result of making use of an aberrated lens EF, the point caustic D is transformed into the caustic surface GH.

Returning to the main thread of the argument, one might enquire into the stability of this point caustic with respect to perturbations in the lens used to create the caustic (Berry 1980).[153] If the lens is aberrated, as denoted by EF in Fig. 5.15(b), then the point caustic will be destroyed. Rather than having a point at which the intensity is predicted to be infinite, one will instead have a surface (known as a 'caustic surface') over which geometric optics predicts an infinite intensity. Rather than having infinitely many rays passing through the point caustic D, one now has the caustic surface forming an envelope for a 'family' of rays. A cross-section through this two-dimensional caustic surface is indicated by the bold line GH in Fig. 5.15(b). To visualize this line in an actual experiment, one might place a piece of white card in the vertical plane containing the optic axis, and then directly observe the light that is scattered from the card.

Unlike the point caustic, which is unstable with respect to perturbations in the ray-field which created it, caustic surfaces may be stable with respect to perturbations. The stable caustic surfaces may be classified with the help of catastrophe theory, introductions to which are given in the texts of Thom (1975), Arnol'd (1986, 1990), and Nye (1999). In three dimensions,[154] the theory

[153]Compare the discussions, earlier in this chapter, on the stability of wave-field vortices with respect to perturbations in the coherent X-ray wave-field in which they are embedded (Section 5.5.2).

[154]Note that the dimensionality of a caustic surface can be arbitrarily large. To appreciate this point, consider once more the point caustic indicated by the point D in Fig. 5.15(a). This point caustic occupies a point (i.e. a zero-dimensional region) of the three-dimensional half-space downstream of the lens. Rather than viewing this caustic as being embedded in three dimensions let us erect a plane Π, passing through the point D and parallel to the plane BC occupied by the thin lens. Notwithstanding the fact that we are here dealing with geometric optics rather than wave optics, one may view this optical setup in the operational terms introduced in Section 4.1, so that the 'input' ray-field at the entrance surface of the lens is mapped to the 'output' ray-field (which may be multi-valued, since more than one ray can pass through a given point in the output plane) over the plane Π. We can consider this imaging

predicts five types of stable caustic surfaces, the simplest two of which are the 'fold' (whose essence is one-dimensional, in the sense described in the preceding footnote) and the 'cusp' (whose essence is two-dimensional). The remaining three caustic surfaces, which are in essence three-dimensional, are the 'swallowtail', the 'hyperbolic umbilic', and the 'elliptic umbilic'. Note that these five equivalence classes of caustic surface are the first in an infinite hierarchy of catastrophes, the remainder of which are higher than three-dimensional. This sequence of caustic surfaces, which are stable with respect to perturbation, may be thought of as exhibiting 'natural focusing' (a term due to Hannay—see Nye (1999)), as it is a form of focussing, which occurs without the need to engineer a particular optical element or system for the purpose of effecting a focus.

To illustrate the above notions, consider the 'coffee-cup cusp caustic' shown in Fig. 5.16. As a result of a suitably distant source of light being reflected from the inside edge of a coffee cup, a two-dimensional caustic surface (this is a cusp, which has already been mentioned as being one of the five possible forms for stable caustic surfaces in three or fewer dimensions) is formed in the reflected light. This caustic intersects the surface of the coffee along a cusp-shaped line, which is seen as a bright line along the surface of the liquid. The stability of this structure with respect to perturbations can be experimentally demonstrated if the coffee cup is sufficiently flexible to be deformed by hand in real time. Provided that the deformations of the coffee cup are not too extreme, one will notice that the location of the caustic is moved about but remains otherwise stable as the cup is deformed. For another everyday example of the stability of caustics with respect to perturbations, one may consider the pattern of propagation-induced caustics (a 'caustic network') which dances about the bottom of a swimming pool floor on a sunny day—small perturbations of the water surface, which arise as water waves evolve through a small time interval, serve only to deform most of the caustic surfaces that intersect the plane formed by the floor of the pool.

At this point, the reader may be enquiring into the relevance of the preceding discussion on singularities of the ray theory, given that this book is concerned with coherent X-ray optics. With this in mind let us again return to the example of a point caustic, with which we opened the present section. As mentioned

system to be characterized by a single control parameter τ, this being equal to the focal length f of the lens BC. The point caustic is, in essence, zero-dimensional because one can 'see' the entire caustic surface over a single point in a single output plane Π, when—and only when—the previously mentioned control parameter has a specific value. Moving up two dimensions, we have the cusp caustic that is illustrated in Fig. 5.16. This caustic is, in essence, two-dimensional because it can be completely 'seen' in a single two-dimensional output plane Π, for a given set of control parameters in the imaging system which leads to that caustic, but cannot be completely 'seen' across a one-dimensional line that forms a subset of the output plane. Moving up a dimension, one can consider a caustic surface which is intrinsically three-dimensional, so that one would need to sweep through a given range of values of a suitable control parameter in order to entirely 'see' such a caustic in the output two-dimensional image plane, as a sequence of two-dimensional slices through the three-dimensional caustic surface. A four-dimensional caustic would require one to sweep through two appropriate control parameters, in order to 'see' the entire caustic over a two-dimensional output plane, and so on *ad infinitum*.

Fig. 5.16. Visible-light cusp caustic over the surface of a cup of coffee. After Nye (1999).

earlier, the ray theory incorrectly predicts a point of infinite intensity at the focal point of a finite but otherwise perfect lens. This singularity of the ray theory is an infinity which heralds the breakdown of geometric optics, ushering in the need for a more general formalism—in this case, the wave theory. The singularity of the ray theory (the point caustic) is 'tamed' by the wave theory, which instead predicts a sharply peaked but nevertheless finite intensity in the focal volume of the lens. The caustic therefore appears softened when a wave-optics calculation is taken into account.

Pushing this notion further, we have already made mention of the Airy rings which surround the point focus of a finite but otherwise perfect rotationally symmetric converging lens, according to the scalar wave theory of electromagnetic radiation. In the focal plane of the lens, each of these Airy disks is separated by a domain wall, across which the wave-function intensity vanishes, coinciding with a discontinuous jump of π radians in the phase of the wave-field over the nominal focal plane of the lens.[155] This is a singularity of the wave theory, for it is at the points of vanishingly small intensity where the photon nature of the radiation must be taken into account, with scalar wave optics needing to be re-

[155]Note that the wave theory can never predict propagation-induced vortices to form in a coherent field that exhibits rotational symmetry.

placed by quantum optics. Intermediate between the scalar wave theory and the quantum theory of light, there is the vector theory of light, which possesses its own singularities (disclinations and C-lines).

What all of this points to is the notion of a hierarchy of singularities, associated with a hierarchy of theories—in order of increasing generality, these are the ray theory of optics, the scalar wave theory of optics based on the inhomogeneous d'Alembert equation, the vector wave theory of optics based on Maxwell's equations, and the equations for the quantized electromagnetic field (Berry 1998). Caustics are the singularities at which the ray theory breaks down, ushering in the need for the scalar wave theory of optics, which softens or tames the singularities of the ray theory. When the scalar wave theory breaks down, near the phase singularities, the vector theory based on Maxwell's equations will need to be invoked. When classical electromagnetism breaks down, in regions of zero energy density as predicted by that theory, the photon nature of light will become important and one will need to invoke a photon theory based on the quantized electromagnetic field (Berry 1998).

One can view the caustic surfaces of the ray theory as a singular 'skeleton', which becomes 'decorated' with vortices as one makes a transition from the ray to the wave theory (see, for example, Angelsky *et al.* (2004), together with references therein). In this context, we point out the well-known fact that the focal volume of an aberrated lens is typically littered with propagation-induced phase vortices whose location depends sensitively on both the nature and magnitude of the lens aberrations, when the optical system is not rotationally symmetric (see Born and Wolf (1999) and references therein). As another example, one can view the propagation-induced vortices in X-ray speckle patterns, such as those arising in phase contrast images of lung tissue, or those found in the far field diffraction patterns of aperiodic objects, as being nucleated in the vicinity of aberrated foci created in regions downstream of the sample where rays cross (Kitchen *et al.* 2004).

Associated with the notion of a hierarchy of theories, is the notion of a hierarchy of length scales (Berry and Upstill 1980). At the largest length scales, and for X-ray wave-fields that possess no particular symmetry, the effects of the finite wavelength of the radiation may be ignored. Accordingly, the ray theory is appropriate for describing the field, with the associated singularities of the ray theory—namely the caustics—being the dominant feature of the field at this length scale. If one zooms in, as it were, to a length scale which is sufficiently small for the effects of the finite wavelength of the radiation to be significant, the singularities of the ray theory—the caustics—will appear softened by diffraction. Thus, the singularities of the ray theory will no longer dominate; rather, the singularities of the wave theory—namely the nodal lines—will now dominate the field, which will be speckled in appearance. One can zoom in still further, passing onto the theory of X-ray waves based on classical electromagnetism and then quantum electrodynamics, although we will not examine this here. Rather, we remark that the wave-field is neither a ray, nor a scalar wave, nor an electric and magnetic field, nor a quantized electromagnetic field. Each of these constructs associated with four different theories, together with their associated singularities, will be the dominant feature of a given nontrivial partially coherent electromagnetic field, depending on the level of detail at which one chooses to examine the field.

As an example of this notion of multiple length scales, consider once again the computer simulations shown in Fig. 5.13. At a propagation distance of 0.67 m, the resulting simulated intensity has the form of a network of caustics, which is viewed at a sufficiently fine scale that the softening effects of diffraction are evident, replacing the very sharp caustic skeleton of the ray theory with strongly peaked lines decorated with fine diffraction structure. The resolution at which these images are sampled is just fine enough to resolve propagation-induced vortices. Thus at the coarser length scale we see a caustic-dominated pattern, while at sufficiently fine length scales the caustics dissolve, being replaced with a vortex-dominated speckle field. At this fine scale, the vortex density is seen to be particularly strong in the focal volumes, which may be associated with the aberrated compound refractive lenses (see Section 3.4.2) that can be thought to form as the result of many spherical cavities overlapping in projection (Kitchen *et al.* 2004).

As a corollary to the idea of multiple length scales, at each of which a particular type of singularity is seen to be dominant, one has the concept of caustics and vortices being complementary singularities (Berry 1980; Berry and Upstill 1980; Berry 1998). This complementarity exists on at least two levels. First, there is the obvious counterpoint of the caustic's infinite intensity versus the zero intensity of the phase singularity. Second, there is the fact (which Berry has conjectured to bear a deep relationship to the wave-particle duality of quantum mechanics (Berry 1980)) that these singularities are not simultaneously observable, since caustics dominate at length scales sufficiently coarse for the finite wavelength of the radiation to be neglected, while vortices dominate at finer length scales for which the finite radiation wavelength gives rise to diffraction effects that serve to soften the caustics.

5.9 Summary

The near-ubiquity of vortices, cutting across a very wide range of fields of study in physics, was emphasized in Section 5.1. In Section 5.2, we followed Dirac's analysis for the study of nodal lines in complex scalar fields, these being 'lines of nothingness' along which the wave-function vanishes. Section 5.3 connected these nodal lines with the notion of vortex cores, which thread wave-field phase vortices. The topological analysis of the just-listed pair of sections was seen to be rather general, as its results were not predicated on the particular functional form of the differential equation governing the evolution of the complex scalar field under study. Rather, these results were obtained solely from the demand that the wave-function should be both single-valued and continuous.

Notwithstanding the powerful generalities mentioned above, a detailed analysis of vortex dynamics and core structure requires a statement of the differential equation obeyed by a given complex scalar field. Accordingly, in Section 5.4 we saw how to construct exact polynomial vortex solutions to the d'Alembert wave equation governing complex scalar electromagnetic waves. These allowed us to gain some deeper analytical insights into the nature of vortical X-ray wave-fields. While these exact solutions were seen to diverge as one moves sufficiently far from the vortex core, it was argued that this was of no consequence on account of the fact that the polynomial solutions were considered to be local models for vortical behaviour in coherent X-ray optics.

The vortical polynomial solutions to the d'Alembert wave equation were also seen to provide some insight into the question of vortex dynamics, as discussed in Section 5.5. We saw that the notion of 'vortex nucleation at a point', together with that of 'vortex annihilation at a point', is devoid of meaning. Rather, we saw that the nodal lines are the fundamental entities. Higher-order polynomial vortices were seen to be unstable with respect to perturbations, decaying into first-order vortices in so-called 'critical-point explosions'. We also examined the question of how vortices interact with the background field in which they are embedded.

We then moved onto a discussion of different means for generating wave-field vortices, in Section 5.6. Remarkably, we saw that a vortex lattice may be generated by interfering three mutually inclined monochromatic scalar plane waves (Section 5.6.1). Synthetic holograms (Section 5.6.2) and spiral phase plates (Section 5.6.3) were seen to be two additional means for generating X-ray vortices. Lastly, propagation-based means for spontaneous vortex generation were treated in Section 5.6.4.

Wave-field phase vortices are not the only class of phase defect. Domain walls, a second such class, were introduced in Section 5.7. Mention was also made of other classes of defect, such as C-lines and disclinations, associated with a vector theory of electromagnetic fields.

The singularities of the ray theory, known as caustics, were examined in Section 5.8. We also discussed the notion of a singularity hierarchy, with the singularities of a less general theory of optics being the gateway, as it were, to a more general optical theory in which such singularities are softened. For example, we saw that a scalar wave theory of optics was able to tame the caustic singularities of the ray theory. The wave theory itself was seen to have singularities, associated with nodal lines and nodal surfaces, where the wave-function vanishes. In the vicinity of such nodal lines and surfaces, the weak field intensities call for a quantum description of the electromagnetic field.

Further reading

Regarding general references for a deeper treatment of many of the topics in this chapter, see Dirac (1931), Riess (1970), Nye and Berry (1974), Hirschfelder *et al.* (1974), Berry (1980), Berry and Upstill (1980), and Nye (1999). This last-mentioned text covers most of the material given in the first four sub-sections of the chapter. Regarding both vortex dynamics and exact polynomial solutions to the d'Alembert equation, see Berry (1980, 1998) and Freund (1999, 2000), together with references therein. For vortex generation via the interference of three plane waves, see Masajada and Dubik (2001). For more on the use of holograms to generate phase vortices, see Vasara *et al.* (1989), Heckenberg *et al.* (1992), and Brand (1999). Regarding X-ray vortex generation using spiral phase masks, see Peele *et al.* (2002), Peele and Nugent (2003), and Peele *et al.* (2004). Regarding spontaneous vortex creation, see Nye (1999) and references therein, together with the X-ray demonstration by Kitchen *et al.* (2004). Vilenkin and Shellard (1994) and Pismen (1999) give good general introductions to the use of homotopy theory to study domain walls, vortices, and more exotic topological defects in a general setting. For more on caustics and the singularity hierarchy, see Berry (1980, 1998).

References

L.J. Allen, M.P. Oxley, and D. Paganin, *Computational aberration correction for an arbitrary linear imaging system*, Phys. Rev. Lett. **87**, 123902 (2001a).

L.J. Allen, H.M.L. Faulkner, M.P. Oxley, and D. Paganin, *Phase retrieval and aberration correction in the presence of vortices in high-resolution transmission electron microscopy*, Ultramicroscopy **88**, 85–97 (2001b).

O.V. Angelsky, P.P. Maksimyak, A.P. Maksimyak, S.G. Hanson, and Y.A. Ushenko, *Role of caustics in the formation of networks of amplitude zeros for partially developed speckle fields*, Appl. Opt. **43**, 5744–5753 (2004).

V.I. Arnol'd, *Catastrophe theory*, second edition, Springer, Berlin (1986).

V.I. Arnol'd, *Singularities of caustics and wavefronts*, Kluwer Academic, Dordrecht (1990).

M.W. Beijersbergen, R.P.C. Coerwinkel, M. Kristensen, and J.P. Woerdman, *Helical-wavefront laser beams produced with a spiral phaseplate*, Opt. Commun. **112**, 321–327 (1994).

M.V. Berry, *Disruption of wavefronts: statistics of dislocations in incoherent Gaussian random waves*, J. Phys. A: Math. Gen. **11**, 27–37 (1978).

M.V. Berry, *Singularities in waves and rays*, in R. Balian *et al.* (eds), *Les Houches, session XXXV, 1980—physics of defects*, North Holland, Amsterdam, pp. 453–543 (1980).

M.V. Berry, *Much ado about nothing: optical dislocations lines (phase singularities, zeros, vortices...)*, in M.S. Soskin (ed.), *Singular optics*, SPIE Proceedings **3487**, SPIE International Society for Optical Engineering, Bellingham, pp. 1–5 (1998).

M.V. Berry, *Knotted zeros in the quantum states of hydrogen*, Found. Phys. **31**, 659–667 (2001).

M.V. Berry and C. Upstill, *Catastrophe optics: morphologies of caustics and their diffraction patterns*, in E. Wolf (ed.), *Progress in optics XVIII*, North-Holland, Amsterdam, Chapter 4 (1980).

M.V. Berry and M.R. Dennis, *Knotted and linked phase singularities in monochromatic waves*, Proc. R. Soc. A **457**, 2251–2263 (2000a).

M.V. Berry and M.R. Dennis, *Phase singularities in isotropic random waves*, Proc. R. Soc. Lond. A **456**, 2059–2079 (2000b).

M.V. Berry and M.R. Dennis, *Knotting and unknotting of phase singularities: Helmholtz waves, paraxial waves and waves in 2+1 spacetime*, J. Phys. A: Math. Gen. **34**, 8877–8888 (2001).

M. Born and E. Wolf, *Principles of optics*, seventh edition, Cambridge University Press, Cambridge (1999).

G.F. Brand, *Phase singularities in beams*, Am. J. Phys. **67**, 55–60 (1999).

J.M. Cowley, *Diffraction physics*, third revised edition, North-Holland, Amsterdam (1995).

P.A.M. Dirac, *Quantised singularities in the electromagnetic field*, Proc. Roy. Soc. Lond. A **133**, 60–72 (1931).

I. Freund, *Phase autocorrelation of random wave fields*, Opt. Commun. **124**, 321–332 (1996).

I. Freund, *'1001' correlations in random wave fields*, Waves in Random Media **8**, 119–158 (1998).

I. Freund, *Critical point explosions in two-dimensional wave fields*, Opt. Commun. **159**, 99–117 (1999).

I. Freund, *Optical vortex trajectories*, Opt. Commun. **181**, 19–33 (2000).

H. He, N.R. Heckenberg, and H. Rubinsztein-Dunlop, *Optical particle trapping with higher-order doughnut beams produced using high efficiency computer generated holograms*, J. Mod. Opt. **42**, 217–223 (1995).

N.R. Heckenberg, R. McDuff, C.P. Smith, H. Rubinsztein-Dunlop, and M.J. Wegener, *Laser beams with phase singularities*, Opt. Quant. Electr. **24**, S951–S962 (1992).

J.O. Hirschfelder, C.J. Goebel, and L.W. Bruch, *Quantized vortices around wavefunction nodes. II*, J. Chem. Phys. **61**, 5456–5459 (1974).

M.J. Kitchen, D. Paganin, R.A. Lewis, N. Yagi, K. Uesugi, and S.T. Mudie, *On the origin of speckle in X-ray phase contrast images of lung tissue*, Phys. Med. Biol. **49**, 4335–4348 (2004).

M. Kristensen, M.W. Beijersbergen, and J.P. Woerdman, *Angular momentum and spin-orbit coupling for microwave photons*, Opt. Commun. **104**, 229–233 (1994).

A. Mair, A. Vaziri, G. Weihs, and A. Zeilinger, *Entanglement of the orbital angular momentum states of photons*, Nature **412**, 313–316 (2001).

J. Masajada and B. Dubik, *Optical vortex generation by three plane wave interference*, Opt. Commun. **198**, 21–27 (2001).

J.F. Nye, *Natural focusing and fine structure of light: caustics and wave dislocations*, Institute of Physics Publishing, Bristol (1999).

J.F. Nye and M.V. Berry, *Dislocations in wave trains*, Proc. Roy. Soc. A **336**, 165–190 (1974).

K.E. Oughstun (ed.), *Selected papers on scalar wave diffraction*, SPIE Milestone Series volume M51, SPIE Optical Engineering Press, Bellingham (1992).

A.G. Peele, P.J. McMahon, D. Paterson, C.Q. Tran, A.P. Mancuso, K.A. Nugent, *et al.*, *Observation of an X-ray vortex*, Opt. Lett. **27**, 1752–1754 (2002).

A.G. Peele and K.A. Nugent, *X-ray vortex beams: a theoretical analysis*, Opt. Express **11**, 2315–2322 (2003).

A.G. Peele, K.A. Nugent, A.P. Mancuso, D. Paterson, I. McNulty, and J.P. Hayes, *X-ray phase vortices: theory and experiment*, J. Opt. Soc. Am. A **21**, 1575–1584 (2004).

L.M. Pismen, *Vortices in nonlinear fields: from liquid crystal to superfluids, from non-equilibrium patterns to cosmic strings*, Oxford University Press, Oxford (1999).

J. Riess, *Nodal structure, nodal flux fields, and flux quantization in stationary quantum states*, Phys. Rev. D **2**, 647–653 (1970).

D. Rozas, C.T. Law, and G.A. Swartzlander Jr., *Propagation dynamics of optical vortices*, J. Opt. Soc. Am. B **14**, 3054–3065 (1997).

Y. Suzuki, N. Yagi, and K. Uesugi, *X-ray refraction-enhanced imaging and a method for phase retrieval for a simple object*, J. Synchrotron Rad. **9**, 160–165 (2002).

R. Thom, *Structural stability and morphogenesis*, Benjamin, Reading, MA (1975).

A. Vasara, J. Turunen, and A.T. Friberg, *Realization of general nondiffracting beams with computer-generated holograms*, J. Opt. Soc. Am. A **6**, 1748–1754 (1989).

A. Vilenkin and E.P.S. Shellard, *Cosmic strings and other topological defects*, Cambridge University Press, Cambridge (1994).

N. Yagi, Y. Suzuki, and K. Umetani, *Refraction-enhanced X-ray imaging of mouse lung using synchrotron radiation source*, Med. Phys. **26**, 2190–2193 (1999).

Appendix A

Review of Fourier analysis

Here we review the fundamentals of Fourier analysis which are employed throughout the main text. We begin by defining the one-dimensional and two-dimensional forms of the Fourier transform, together with the corresponding inverse transformations. We then outline the convolution theorem, the shift theorem, and the derivative theorem. Lastly, we mention the sifting property of the Dirac delta.

A.1 Fourier transforms in one and two dimensions

Consider a function $f(x)$ of one real variable x, which is assumed to be sufficiently well behaved to admit representation as a Fourier integral:

$$f(x) = \frac{1}{\sqrt{2\pi}} \int_{-\infty}^{\infty} \breve{f}(k_x) \exp(ik_x x) dk_x, \quad \breve{f}(k_x) \equiv \mathcal{F}[f(x)]. \tag{A.1}$$

The function $\breve{f}(k_x)$ is said to be the 'Fourier transform' of $f(x)$ with respect to x, with k_x being the Fourier variable conjugate to (dual to, or reciprocal to) x. The Fourier-transform operator $\mathcal{F}$ maps $f(x)$ to its Fourier transform $\breve{f}(k_x)$. To derive the corresponding inverse transform, multiply eqn (A.1) by $\exp(-ik'_x x)/\sqrt{2\pi}$, integrate over all x and then make use of the Fourier integral representation of the Dirac delta $\delta(x)$:

$$\delta(x) = \frac{1}{2\pi} \int_{-\infty}^{\infty} \exp(ik_x x) dk_x. \tag{A.2}$$

One thereby arrives at:

$$\breve{f}(k_x) = \frac{1}{\sqrt{2\pi}} \int_{-\infty}^{\infty} f(x) \exp(-ik_x x) dx, \quad f(x) \equiv \mathcal{F}^{-1}[\breve{f}(k_x)]. \tag{A.3}$$

Equations (A.1) and (A.3) form a Fourier transform pair. The second member of the pair is an integral transform which maps the function $f(x)$ to its Fourier transform $\breve{f}(k_x)$, with the first member of the pair effecting the corresponding inverse transformation.

The above can be generalized to two-dimensional Fourier transformations of a suitably well-behaved function $f(x, y)$ of two real variables x and y:

$$f(x,y) = \frac{1}{2\pi}\iint_{-\infty}^{\infty} \breve{f}(k_x,k_y)\exp[i(k_x x + k_y y)]dk_x dk_y,$$
$$\breve{f}(k_x,k_y) \equiv \mathcal{F}[f(x,y)], \tag{A.4}$$
$$\breve{f}(k_x,k_y) = \frac{1}{2\pi}\iint_{-\infty}^{\infty} f(x,y)\exp[-i(k_x x + k_y y)]dx dy,$$
$$f(x,y) \equiv \mathcal{F}^{-1}[\breve{f}(k_x,k_y)]. \tag{A.5}$$

The function $\breve{f}(k_x,k_y)$ is said to be the Fourier transform of $f(x,y)$ with respect to x and y, with k_x and k_y being the Fourier variables which are respectively conjugate to x and y. The Fourier-transform operator $\mathcal{F}$ maps $f(x,y)$ to its Fourier transform $\breve{f}(k_x,k_y)$.

Since it will always be clear from the context of a particular calculation, no confusion should arise from the use of $\mathcal{F}$ to denote both one- and two-dimensional Fourier transformation.

A.2 Convolution theorem

In one dimension the convolution $f(x)\star g(x)$ of two suitably well-behaved functions $f(x)$ and $g(x)$ is defined to be:

$$f(x)\star g(x) \equiv \int_{-\infty}^{\infty} f(x')g(x-x')dx'. \tag{A.6}$$

One can readily establish the one-dimensional form of the 'convolution theorem', which states that the Fourier transform of $f(x)\star g(x)$ is equal to the product of the Fourier transform of $f(x)$ with the Fourier transform of $g(x)$, multiplied by $\sqrt{2\pi}$:

$$\mathcal{F}\left[f(x)\star g(x)\right] = \sqrt{2\pi}\left\{\mathcal{F}\left[f(x)\right]\right\}\times\left\{\mathcal{F}\left[g(x)\right]\right\}. \tag{A.7}$$

The above formulae are readily generalized to functions of two variables. Accordingly, the two-dimensional convolution $f(x,y)\star g(x,y)$, of two suitably well-behaved functions $f(x,y)$ and $g(x,y)$, is given by:

$$f(x,y)\star g(x,y) \equiv \iint_{-\infty}^{\infty} f(x',y')g(x-x',y-y')dx'dy'. \tag{A.8}$$

The corresponding two-dimensional form of the convolution theorem is then:

$$\mathcal{F}\left[f(x,y)\star g(x,y)\right] = 2\pi\left\{\mathcal{F}\left[f(x,y)\right]\right\}\times\left\{\mathcal{F}\left[g(x,y)\right]\right\}. \tag{A.9}$$

It will sometimes prove convenient to write down an alternative form of eqn (A.9), obtained by taking the inverse Fourier transform of both sides:

$$f(x,y)\star g(x,y) = 2\pi\mathcal{F}^{-1}\left(\left\{\mathcal{F}\left[f(x,y)\right]\right\}\times\left\{\mathcal{F}\left[g(x,y)\right]\right\}\right). \tag{A.10}$$

A.3 Fourier shift theorem

Take eqn (A.3), writing it in the form:

$$\mathcal{F}[f(x)] = \frac{1}{\sqrt{2\pi}} \int_{-\infty}^{\infty} f(x) \exp(-ik_x x) dx. \tag{A.11}$$

Replace $f(x)$ with $f(x - x_0)$ in the above expression, where x_0 is a real constant. Make the following change of variables:

$$X = x - x_0, \tag{A.12}$$

so that eqn (A.11) becomes:

$$\mathcal{F}[f(x - x_0)] = \frac{\exp(-ik_x x_0)}{\sqrt{2\pi}} \int_{-\infty}^{\infty} f(X) \exp(-ik_x X) dX. \tag{A.13}$$

Substituting eqn (A.11) into eqn (A.13), we arrive at the one-dimensional form of the Fourier shift theorem:

$$\mathcal{F}[f(x - x_0)] = \exp(-ik_x x_0) \mathcal{F}[f(x)]. \tag{A.14}$$

The two-dimensional form of the Fourier shift theorem is similarly derived:

$$\mathcal{F}[f(x - x_0, y - y_0)] = \exp[-i(k_x x_0 + k_y y_0)] \mathcal{F}[f(x, y)], \tag{A.15}$$

where both x_0 and y_0 are real constants.

A.4 Fourier derivative theorem

Take eqn (A.1), writing it in the form:

$$f(x) = \frac{1}{\sqrt{2\pi}} \int_{-\infty}^{\infty} \mathcal{F}[f(x)] \exp(ik_x x) dk_x. \tag{A.16}$$

Differentiate both sides with respect to x a total of m times, where m is an integer. Taking the derivative operator inside the integral of the right side of the resulting expression, and noting that it acts only on the term $\exp(ik_x x)$ within this integral, we see that:

$$\frac{d^m}{dx^m} f(x) = \frac{1}{\sqrt{2\pi}} \int_{-\infty}^{\infty} \{\mathcal{F}[f(x)](ik_x)^m\} \exp(ik_x x) dk_x. \tag{A.17}$$

The term in braces is equal to the Fourier transform of the left side of the equation, leading us to the one-dimensional form of the Fourier derivative theorem:

$$\mathcal{F}\left[\frac{d^m}{dx^m} f(x)\right] = (ik_x)^m \mathcal{F}[f(x)]. \tag{A.18}$$

By taking the inverse Fourier transform of this equation, we obtain a useful alternative form of the one-dimensional Fourier derivative theorem:

$$\frac{d^m}{dx^m} f(x) = \mathcal{F}^{-1}(ik_x)^m \mathcal{F} f(x). \tag{A.19}$$

Note that cascaded operators are assumed to act from right to left, a convention assumed throughout the text. Thus the right side of the above equation should be read as follows: 'The function $f(x)$ is first acted upon by the Fourier transform operator $\mathcal{F}$, with the result being multiplied by $(ik_x)^m$. The resulting expression is then acted upon by the inverse Fourier transform operator $\mathcal{F}^{-1}$'.

The above argument can be generalized to give the two dimensional Fourier derivative theorem. The two-dimensional generalization of eqn (A.18) is:

$$\mathcal{F}\left[\frac{\partial^m}{\partial x^m}\frac{\partial^n}{\partial y^n} f(x,y)\right] = (ik_x)^m (ik_y)^n \mathcal{F}[f(x,y)], \tag{A.20}$$

where m and n are integers. The corresponding generalization of eqn (A.19) is:

$$\frac{\partial^m}{\partial x^m}\frac{\partial^n}{\partial y^n} f(x,y) = \mathcal{F}^{-1}(ik_x)^m (ik_y)^n \mathcal{F} f(x,y). \tag{A.21}$$

A.5 Sifting property of Dirac delta

The 'sifting property' of the Dirac delta is expressed as follows:

$$f(x_0) = \int_a^b \delta(x - x_0) f(x) dx. \tag{A.22}$$

Here x_0 is any real number lying between the real numbers a and b.

Appendix B

Fresnel scaling theorem

As shown in Fig. B.1(a), let us consider an in-vacuo point source A of monochromatic scalar X-ray waves, which lies on a nominal optic axis z at a distance R upstream of a sample B. A diffraction pattern of the object is to be recorded over the plane $z = \Delta$, this being parallel to the nominal exit surface $z = 0$ of the object. The source is considered to be sufficiently distant from the sample for the radiation to be paraxial over its entrance, with the object being sufficiently weakly scattering for the projection approximation to hold (see eqns (2.39), (2.40), and (2.42)). Further, we assume that Fresnel diffraction theory is adequate for calculation of the diffraction pattern over the plane $z = \Delta$, given the wave-field over the exit surface $z = 0$ of the object (see Section 1.4). The Fresnel diffraction pattern ('point-projection image') will have a geometric magnification M, which depends on the source-to-sample distance R, together with the sample-to-detector distance Δ. A limiting case of this scenario is obtained by taking the source-to-sample distance R to infinity, as sketched in Fig. B.1(b). This unit-magnification limiting case corresponds to a Fresnel diffraction pattern of the sample which is obtained for the case of plane-wave illumination.

Subject to the approximations listed above, the Fresnel scaling theorem yields a simple mapping between the Fresnel diffraction pattern of an object which is obtained for the case of illumination by a point source, and a Fresnel diffraction pattern obtained for the limiting case of plane-wave illumination. This theorem states that the point-source-illuminated Fresnel diffraction pattern, obtained over a plane at a propagation distance $z = \Delta$ from the exit surface of the object with a source-to-sample distance of R corresponding to geometric magnification M, is related to a certain plane-wave-illuminated diffraction pattern by the following sequence of steps: (i) take the plane-wave-illuminated Fresnel diffraction pattern over a plane at a propagation distance of $z = \Delta/M$ downstream of the exit surface of the sample, before (ii) transversely magnifying it by a factor of M and then (iii) dividing the intensity at each point in the resulting image by M^2. This scaling theorem, which will be derived using Fresnel diffraction theory in the succeeding paragraphs, is of great utility in the computation and analysis of point-projection X-ray images.

With a view to deriving the Fresnel scaling theorem, recall that one may expand the free-space propagator in the convolution formulation (1.45) of the Fresnel diffraction integral, thereby recasting it in the manner given by eqn (1.46):

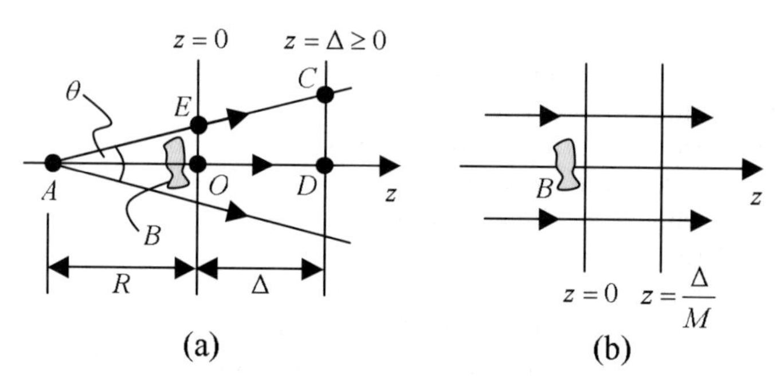

Fig. B.1. (a) A monochromatic point source A emits X-rays into vacuum, which subsequently pass through an object B. A Fresnel diffraction pattern is measured over a plane at distance Δ from the exit surface of the sample. The geometric magnification M of this image is $(R+\Delta)/R$. (b) The same object is illuminated by plane waves, with a diffraction pattern being recorded at a distance of Δ/M from the exit surface of the object. According to the Fresnel scaling theorem, the diffraction patterns of (a) and (b) are identical to one another, up to transverse and multiplicative scale factors, if: (i) θ is sufficiently small for the paraxial approximation to be valid for all rays/streamlines crossing the detection plane; (ii) the object is sufficiently weakly scattering to obey the projection approximation.

$$\begin{aligned}\psi_\omega^{(R)}(x, y, z = \Delta \geq 0) = & -\frac{ik \exp(ik\Delta)}{2\pi\Delta} \exp\left[\frac{ik}{2\Delta}(x^2+y^2)\right] \\ & \times \iint_{-\infty}^{\infty} \psi_\omega^{(R)}(x', y', z = 0) \exp\left[\frac{ik}{2\Delta}(x'^2+y'^2)\right] \\ & \times \exp\left[\frac{-ik}{\Delta}(xx'+yy')\right] dx'dy'. \qquad \text{(B.1)}\end{aligned}$$

Here, $\psi_\omega^{(R)}(x, y, z = 0)$ denotes the exit-surface monochromatic scalar X-ray wave-field over the plane $z = 0$ in Fig. B.1(a), $\psi_\omega^{(R)}(x, y, z = \Delta \geq 0)$ denotes the corresponding propagated disturbance over the plane $z = \Delta \geq 0$, $k = 2\pi/\lambda$ is the usual wave-number corresponding to a wavelength λ and angular frequency ω of the illuminating radiation, (x, y) denotes a Cartesian coordinate system in the plane perpendicular to the optic axis z, and the R superscript on exit-surface and propagated wave-fields indicates the distance between the illuminating on-axis point source and the exit surface of the object.

Under the paraxial and projection approximations, the exit-surface wave-field $\psi_\omega^{(\infty)}(x, y, z = 0)$ for the case of plane-wave illumination is related to the exit-surface wave-field $\psi_\omega^{(R)}(x, y, z = 0)$ for point-source illumination via:

$$\psi_\omega^{(R)}(x, y, z = 0) = \psi_\omega^{(\infty)}(x, y, z = 0) \exp\left[\frac{ik}{2R}(x^2+y^2)\right]. \qquad \text{(B.2)}$$

Note that the phase factor, on the right side of the above equation, is the second-order approximation to the phase variation over the exit surface $z = 0$ due to an on-axis point source at distance R upstream of this plane. Note, also, that in writing the above approximation we have taken R to be sufficiently large that one may neglect the intensity variations introduced into the exit-surface wave-field in moving from plane-wave to point-source illumination.

Having written down the relation (B.2) between the unpropagated exit-surface fields for the case of plane-wave and point-source illumination, we are ready to seek a corresponding relation between the intensities of the propagated fields, under the Fresnel approximation. To this end, insert eqn (B.2) into eqn (B.1) to give:

$$\begin{aligned}\psi_\omega^{(R)}(x, y, z = \Delta \geq 0) = &-\frac{ik\exp(ik\Delta)}{2\pi\Delta}\exp\left[\frac{ik}{2\Delta}(x^2+y^2)\right] \\ &\times \iint_{-\infty}^{\infty} \psi_\omega^{(\infty)}(x', y', z = 0) \\ &\qquad \times \exp\left[\frac{ik}{2}(x'^2+y'^2)\left(\frac{1}{\Delta}+\frac{1}{R}\right)\right] \\ &\qquad \times \exp\left[\frac{-ik}{\Delta}(xx'+yy')\right] dx'dy'. \end{aligned} \tag{B.3}$$

To proceed further, we note from Fig. B.1(a) that the geometric magnification M for point-source illumination is equal to the ratio of the length CD to the length EO. By similar triangles, this is equal to the ratio of the length of AD to that of AO. Thus:

$$M = \frac{R+\Delta}{R}, \tag{B.4}$$

so that:

$$\frac{1}{\Delta}+\frac{1}{R} = \frac{1}{\Delta}\left(\frac{R+\Delta}{R}\right) = \frac{M}{\Delta}. \tag{B.5}$$

Substituting the above expression into the squared modulus of eqn (B.3), and letting $I_\omega^{(R)}(x, y, z = \Delta) \equiv |\psi_\omega^{(R)}(x, y, z = \Delta)|^2$ denote the intensity of the propagated coherent wave-field, we arrive at:

$$\begin{aligned} I_\omega^{(R)}(x, y, z = \Delta \geq 0) = &\frac{k^2}{4\pi^2\Delta^2} \\ \times &\left|\iint_{-\infty}^{\infty} \psi_\omega^{(\infty)}(x', y', z = 0)\exp\left[\frac{ikM}{2\Delta}(x'^2+y'^2) - \frac{ik}{\Delta}(xx'+yy')\right] dx'dy'\right|^2. \end{aligned} \tag{B.6}$$

If we take the limit $R \to \infty$ in the above formula, so that $M \to 1$, we see that:

$$I_\omega^{(\infty)}(x,y,z=\Delta\geq 0)=\frac{k^2}{4\pi^2\Delta^2}$$
$$\times\left|\iint_{-\infty}^{\infty}\psi_\omega^{(\infty)}(x',y',z=0)\exp\left[\frac{ik}{2\Delta}(x'^2+y'^2)-\frac{ik}{\Delta}(xx'+yy')\right]dx'dy'\right|^2. \tag{B.7}$$

As may be readily checked by direct substitution, the left sides of the above pair of equations are related to one another through the Fresnel scaling theorem:

$$I_\omega^{(R)}(x,y,z=\Delta\geq 0)=M^{-2}I_\omega^{(\infty)}\left(\frac{x}{M},\frac{y}{M},z=\frac{\Delta}{M}\geq 0\right). \tag{B.8}$$

To supplement the verbal description of this result given in the second paragraph of this appendix, we note that: (i) The factor of M^{-2} embodies energy conservation; (ii) R may be negative, corresponding to the object being illuminated by a collapsing spherical wave.

Appendix C

Reciprocity theorem for monochromatic scalar fields

Here, we outline a certain reciprocity theorem governing forward-propagating monochromatic scalar electromagnetic fields, a result due to Lukosz (1968) and Shewell and Wolf (1968). Use is made of this theorem in the discussion on in-line holography given in Section 4.3.1.

With reference to Fig. C.1 let us consider a set of sources A lying in the half-space $z < -\Delta$, which radiate monochromatic scalar electromagnetic waves. With specified optic axis z, the volume $z \geq -\Delta$ is assumed to be vacuum. Denote by $\psi_\omega^{(1)}(x, y, z)$ the spatial part of the monochromatic field radiated by the sources. Here, ω denotes the angular frequency of the radiation corresponding to wave-number k, and (x, y) are Cartesian coordinates in planes perpendicular to the optic axis z. The radiated field obeys the Helmholtz equation, given in eqn (1.16):

$$(\nabla^2 + k^2)\psi_\omega^{(1)}(x, y, z) = 0. \tag{C.1}$$

Assuming the field to be forward propagating, we saw in Section 1.3 that the problem of diffracting the field from $z = 0$ to $z = \Delta \geq 0$ is given by the following angular-spectrum representation which exactly solves the Helmholtz equation (see eqn (1.23)):

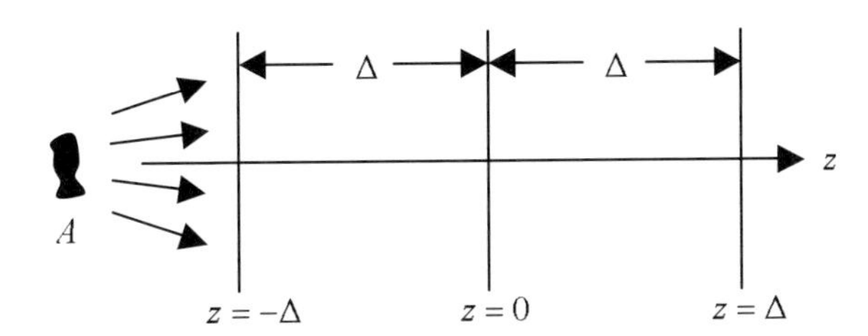

Fig. C.1. Sources A radiate a monochromatic scalar electromagnetic field into a vacuum-filled space. The field is assumed to be forward propagating over the plane $z = 0$, which is perpendicular to a specified optic axis z. Further, the field is assumed to be free from evanescent waves. According to the reciprocity theorem, if one propagates the field from the plane $z = 0$ to the plane $z = \Delta$, the intensity of the resulting field will be equal to the intensity obtained if one propagates the complex conjugate of the field, from the plane $z = 0$ to the plane $z = -\Delta$.

$$\psi_\omega^{(1)}(x,y,z=\Delta) = \frac{1}{2\pi}\iint_{-\infty}^{\infty} \breve{\psi}_\omega^{(1)}(k_x,k_y,z=0)\exp\left[i\Delta\sqrt{k^2-k_x^2-k_y^2}\right] \times \exp\left[i(k_x x + k_y y)\right] dk_x dk_y. \tag{C.2}$$

Here $\breve{\psi}_\omega^{(1)}(k_x,k_y,z=0)$ represents the Fourier transform of the unpropagated field $\psi_\omega^{(1)}(x,y,z=0)$ with respect to x and y, using the convention for two-dimensional Fourier transforms outlined in Appendix A. As specified there, the Fourier variables conjugate to x and y are denoted by k_x and k_y, respectively.

Next we use eqn (A.5) to write $\breve{\psi}_\omega^{(1)}(k_x,k_y,z=0)$ as:

$$\breve{\psi}_\omega^{(1)}(k_x,k_y,z=0) = \frac{1}{2\pi}\iint_{-\infty}^{\infty} \psi_\omega^{(1)}(x',y',z=0)\exp[-i(k_x x' + k_y y')]dx'dy', \tag{C.3}$$

which may be substituted into eqn (C.2) to give:

$$\psi_\omega^{(1)}(x,y,z=\Delta) = \frac{1}{(2\pi)^2}\iiiint_{-\infty}^{\infty} \psi_\omega^{(1)}(x',y',z=0)\exp\left[i\Delta\sqrt{k^2-k_x^2-k_y^2}\right] \times \exp\left\{i[k_x(x-x') + k_y(y-y')]\right\} dk_x dk_y dx'dy'. \tag{C.4}$$

Next, consider a second forward-propagating monochromatic scalar electromagnetic field $\psi_\omega^{(2)}(x,y,z)$, which also obeys the Helmholtz equation. By definition we assume that the boundary value of this field, over the plane $z=0$, is equal to the complex conjugate of the boundary value taken by $\psi_\omega^{(1)}(x,y,z)$ over the same plane. Thus:

$$\psi_\omega^{(1)}(x,y,z=0) = \left[\psi_\omega^{(2)}(x,y,z=0)\right]^*. \tag{C.5}$$

Substitute this into eqn (C.4) and then take the complex conjugate of the resulting expression, leaving:

$$\left[\psi_\omega^{(1)}(x,y,z=\Delta)\right]^* = \frac{1}{(2\pi)^2}\iiiint_{-\infty}^{\infty} \psi_\omega^{(2)}(x',y',z=0) \times \exp\left[-i\Delta\left(\sqrt{k^2-k_x^2-k_y^2}\right)^*\right] \times \exp\left\{-i[k_x(x-x') + k_y(y-y')]\right\} \times dk_x dk_y dx'dy'. \tag{C.6}$$

In the above expression let us consider the integrals over x' and y' to be performed before those over k_x and k_y. These first two integrals will serve to map $\psi_\omega^{(2)}(x',y',z=0)$ into its two-dimensional Fourier transform $\breve{\psi}_\omega^{(2)}(k_x,k_y,z=0)$. Now suppose that evanescent waves (see Section 1.3) are absent, which amounts to the requirement that $\breve{\psi}_\omega^{(2)}(k_x, k_y, z = 0)$ vanishes at every point (k_x, k_y) for which $k_x^2 + k_y^2 > k^2$. There will therefore be a zero contribution, to the integral over k_x and k_y, from points (k_x, k_y) for which $k_x^2 + k_y^2 > k^2$. The only non-zero contribution to the integral over k_x and k_y will be due to points (k_x, k_y) for which $k_x^2 + k_y^2 \leq k^2$. At all such points, the square root in the second line of eqn (C.6) is real. We may therefore drop the star from this square root, leaving:

$$\begin{aligned}\left[\psi_\omega^{(1)}(x, y, z = \Delta)\right]^* = \frac{1}{(2\pi)^2} \iiiint_{-\infty}^{\infty} & \psi_\omega^{(2)}(x', y', z = 0) \\ & \times \exp\left(-i\Delta\sqrt{k^2 - k_x^2 - k_y^2}\right) \\ & \times \exp\{-i[k_x(x - x') + k_y(y - y')]\} \\ & \times dk_x dk_y dx' dy'. \end{aligned} \tag{C.7}$$

To proceed further make the change of variables:

$$k_x \longrightarrow -k_x, \quad k_y \longrightarrow -k_y, \tag{C.8}$$

in eqn (C.7), to see that it reduces to:

$$\begin{aligned}\left[\psi_\omega^{(1)}(x, y, z = \Delta)\right]^* = \frac{1}{(2\pi)^2} \iiiint_{-\infty}^{\infty} & \psi_\omega^{(2)}(x', y', z = 0) \\ & \times \exp\left(-i\Delta\sqrt{k^2 - k_x^2 - k_y^2}\right) \\ & \times \exp\{i[k_x(x - x') + k_y(y - y')]\} \\ & \times dk_x dk_y dx' dy'. \end{aligned} \tag{C.9}$$

Compare the right side of the above equation with the right side of the equation which results when $\psi_\omega^{(1)}$ is replaced by $\psi_\omega^{(2)}$ in eqn (C.4). This shows us that the right side of eqn (C.9) is equal to the field that results when $\psi_\omega^{(2)}(x, y, z = 0)$ is inverse propagated from the plane $z = 0$, to the plane $z = -\Delta$ (see Fig. C.1). Thus we arrive at the desired reciprocity theorem:

$$\left[\psi_\omega^{(1)}(x, y, z = \Delta)\right]^* = \psi_\omega^{(2)}(x, y, z = -\Delta). \tag{C.10}$$

This theorem states that if one propagates through a certain positive distance Δ a given forward-propagating monochromatic scalar electromagnetic field which is free of evanescent waves, the propagation being from plane to parallel plane, then the propagated field is equal to the complex conjugate of the field one would have obtained if the complex conjugate of the unpropagated field had been (inverse) diffracted through a distance equal to the negative of Δ.

Taking the squared modulus of eqn (C.10), and denoting the intensities of $\psi_\omega^{(1)}(x, y, z)$ and $\psi_\omega^{(2)}(x, y, z)$ by $I_\omega^{(1)}(x, y, z) = |\psi_\omega^{(1)}(x, y, z)|^2$ and $I_\omega^{(2)}(x, y, z) = |\psi_\omega^{(2)}(x, y, z)|^2$, we see that:

$$I_\omega^{(1)}(x, y, z = \Delta) = I_\omega^{(2)}(x', y', z = -\Delta). \tag{C.11}$$

Thus if one propagates the field from the plane $z = 0$ to the plane $z = \Delta$, the resulting intensity will be equal to the intensity obtained if one were to propagate the complex conjugate of the field, from the plane $z = 0$ to the plane $z = -\Delta$ (see Fig. C.1).

References

W. Lukosz, *Equivalent-lens theory of holographic imaging*, J. Opt. Soc. Am. **58**, 1084–1091 (1968).

J.R. Shewell and E. Wolf, *Inverse diffraction and a new reciprocity theorem*, J. Opt. Soc. Am. **58**, 1596–1603 (1968).

Index